Space Sciences Series of ISSI

Volume 71

More information about this series at http://www.springer.com/series/6592

Klaus Scipal · Anny Cazenave · Teodolina Lopez
Editors

Forest Properties and Carbon Cycle Studies from Earth Observations

Previously published in *Surveys in Geophysics*, Volume 40, Issue 4, 2019

Springer

Editors
Klaus Scipal
European Space Agency (ESA-ESTEC)
Noordwijk, The Netherlands

Teodolina Lopez
International Space Science Institute
Bern, Switzerland

Anny Cazenave
International Space Science Institute
Bern, Switzerland

LEGOS-CNES
Toulouse, France

ISSN 1385-7525
Space Sciences Series of ISSI
ISBN 978-3-030-32838-2

Cover illustration: Radar view of a mangrove forests in the Pongara National Park, Gabon.
Image credit information: © DLR

This Springer imprint is published by the registered company Springer Nature Switzerland AG
The registered company address is: Gewerbestrasse 11, 6330 Cham, Switzerland

Contents

Surveys in Geophysics (2019) 40:689–691
https://doi.org/10.1007/s10712-019-09552-w

GUEST EDITORIAL

Guest Editorial: International Space Science Institute (ISSI) Workshop on Space-Based Measurement of Forest Properties for Carbon Cycle Research

Klaus Scipal[1] · Anny Cazenave[2,3] · Teodolina Lopez[3]

Received: 29 May 2019 / Accepted: 16 June 2019 / Published online: 26 June 2019

Through their active roles in ecological processes and resources, forests are crucial elements of the Earth system. They sustain a myriad of ecosystems, especially tropical forests that provide habitats for many flora and fauna. Thus, forests play a central role in the local and global carbon cycles. Indeed, forests are an important sink for atmospheric carbon (Bonan 2008; Grace et al. 2014; Pan et al. 2011), and changes in terms of forest biomass (as forest carbon stocks) and forest productivity (forest carbon fluxes) impact the climate change (Foley et al. 2005). Consequently, aboveground forest biomass (AGB) has been recognised as a Global Climate Observing System (GCOS) Essential Climate Variable (ECV), a critical input to the United Nations' Reducing Emissions from Deforestation and Degradation-plus (REDD+) program, and an important input to Earth system models.

In response to the need to monitor forest ecosystems, the number of space-borne missions with a capability to map ecosystem structure steadily increases (e.g., NASA's GEDI and ICESat, NASA/ISRO's NISAR, ESA's BIOMASS, JAXA's ALOS-4 and MOLI, CONAE's SAOCOM and DLR's Tandem-L); the approaches to AGB retrieval using these data diversify. These missions use innovative SAR and LiDAR technologies to measure forest structure parameters (such as forest height and AGB) and their change with time on a global scale with much more precision than is possible today, particularly in areas where insufficient ground or airborne lidar data are available such as in the tropics or the vast Siberian forests. Some of these missions are already in orbit (e.g., GEDI, ALOS, SAOCOM) or are expected to be in orbit within the next 3 years (BIOMASS, NISAR), providing a unique and powerful combination of techniques to monitor Earth's forests.

These missions meet a pressing need for information on forests globally. Analysis of the global carbon cycle shows that the annual emissions of carbon from fossil fuels and land-use change are larger than the annual accumulations of carbon in the atmosphere and oceans. This suggests a largely unknown terrestrial sink for carbon, which has never been measured. Measurements provided by these missions will give a unique opportunity to reduce the uncertainties in both the global net emissions of carbon from forest

✉ Klaus Scipal
klaus.scipal@esa.int

1 European Space Agency, Keplerlaan 1, 2201 AZ Noordwijk, The Netherlands

2 LEGOS, 14 Avenue Edouard Belin, 31400 Toulouse, France

3 International Space Science Institute, Hallerstrasse 6, 3012 Bern, Switzerland

cover change and global changes in aboveground forest biomass. This will significantly improve our understanding of the global carbon cycle, which will be of essential value for climate modelling and policy adaptation actions.

The papers presented in this Special Issue are the outcome of a Workshop on "Space-Based Measurement of Forest Properties for Carbon Cycle Research" held at the International Space Science Institute (ISSI) in Bern, Switzerland, from 6 to 9 November 2017. The main objective of the workshop was to provide an overview of the related science questions, community needs and Earth Observations technologies and capabilities. A specific emphasis was put on the synergies and complementarities of the different observation types and on the potential benefits of a joint exploitation of these data.

The first part of this Special Issue presents the role and benefits of forests and their properties. The first paper, by Reichstein and Carvalhais (2019), provides an overview of the interaction between forest biomass and major ecosystem functions. It also investigates the potential of future Earth Observations missions to better understand the Earth system dynamics. The second article, by Fischer et al. (2019), investigates the role of forest structure heterogeneity for biomass and productivity estimations in temperate forests from lidar remote sensing. The next paper, by Exbrayat et al. (2019), provides an overview of how remote sensing observations of the vegetation have improved our understanding of the terrestrial carbon cycle. The last article in this part on Science and Policy, by Herold et al. (2019), deals with policy needs for environmental management and biomass data provided by space-based biomass monitoring.

The second part contains reviews of the potential of space-based observation techniques to provide information of forest structure. This part paves the way to the synergistic use of the different space-borne techniques, with articles such on SAR tomography, by Tebaldini et al. (2019), or combining Lidar and Multi-baseline SAR data by Pardini et al. (2019), and the benefits of very high resolution visual imagery as Schepashenko et al. (2019).

The third part highlights the challenges and new opportunities of collecting AGB measurements, which are critically needed for the interpretation and validation of the space-based observations. First, Chave et al. (2019) highlight the importance of ground-data for this purpose. Réjou-Méchain et al. (2019) next review the error structure of up-scaled ground-based observations. Phillips et al. (2019) discuss the criticality of tree species' knowledge in the precise estimation of ground-based AGB. The next two articles explore new opportunities in using Terrestrial Laser Scanning (TLS) (Disney et al. 2019) and Ultra-High-Density Drone Lidar (Kellner et al. 2019). The last article of this part provides a review of biomass validation practices from the Committee of Earth Observing Satellites (CEOS) perspective (Duncanson et al. 2019).

This Special Issue concludes with two papers. McRoberts et al. (2019) discuss the statistics of regional, map-based estimates of AGB merging ground and remotely sensed observations; a paper by Albinet et al. (2019) describes the strategy of ESA and NASA to make data available to the scientific community via their Multi-Mission Algorithm and Analysis Platform (MAAP).

We are most grateful to the ISSI for hosting this Workshop, to Thuy Le Toan, Shaun Queqan and Ralph Dubayah for co-organising this Workshop, to those scientists who organised the reviewing of the submitted papers, to all the referees, to the Editor in Chief, and to the staff of Springer Nature for publishing this record of the event.

References

Albinet C, Whitehurst AS, Jewell LA et al (2019) A joint ESA-NASA multi-mission algorithm and analysis platform (MAAP) for biomass, NISAR, and GEDI. Surv Geophys. https://doi.org/10.1007/s10712-019-09541-z

Bonan GB (2008) Forests and climate change: forcings, feedbacks, and the climate benefits of forests. Science 320(5882):1444–1449. https://doi.org/10.1126/science.1155121

Chave J, Davies SJ, Phillips OL, Lewis SL, Sist P, Schepaschenko D, Armston J, Baker TR, Coomes D, Disney M, Duncanson L, Hérault B, Labrière N, Meyer V, Réjou-Méchain M, Scipal K, Saatchi S (2019) Ground data are essential for biomass remote sensing missions. Surv Geophys. https://doi.org/10.1007/s10712-019-09528-w

Disney M, Burt A, Calders K et al (2019) Innovations in ground and airborne technologies as reference and for training and validation: terrestrial laser scanning (TLS). Surv Geophys. https://doi.org/10.1007/s10712-019-09527-x

Duncanson L, Armston J, Disney M et al (2019) The importance of consistent global forest aboveground biomass product validation. Surv Geophys. https://doi.org/10.1007/s10712-019-09538-8

Exbrayat JF, Bloom AA, Carvalhais N et al (2019) Understanding the land carbon cycle with space data: current status and prospects. Surv Geophys. https://doi.org/10.1007/s10712-019-09506-2

Fischer R, Knapp N, Bohn F, Shugart HH, Huth A (2019) The relevance of forest structure for biomass and productivity in temperate forests: new perspectives for remote sensing. Surv Geophys. https://doi.org/10.1007/s10712-019-09519-x

Foley JA, DeFries R, Asner GP, Barford C, Bonan G, Carpenter SR, Chapin FS, Coe MT, Daily GC, Gibbs HK, Helkowski JH, Holloway T, Howard EA, Kucharik CJ, Monfreda C, Patz JA, Prentice IC, Ramankutty N, Snyder PK (2005) Global consequences of land use. Science 309(5734):570–574. https://doi.org/10.1126/science.1111772

Grace J, Mitchard E, Gloor E (2014) Perturbations in the carbon budget of the tropics. Glob Change Biol 20(10):3238–3255. https://doi.org/10.1111/gcb.12600

Herold M, Carter S, Avitabile V et al (2019) The role and need for space-based forest biomass-related measurements in environmental management and policy. Surv Geophys. https://doi.org/10.1007/s10712-019-09510-6

Kellner JR, Armston J, Birrer M et al (2019) New opportunities for forest remote sensing through ultra-high-density drone lidar. Surv Geophys. https://doi.org/10.1007/s10712-019-09529-9

McRoberts RE, Næsset E, Liknes GC et al (2019) Using a finer resolution biomass map to assess the accuracy of a regional, map-based estimate of forest biomass. Surv Geophys. https://doi.org/10.1007/s10712-019-09507-1

Pan Y, Birdsey RA, Fang J, Houghton R, Kauppi PE, Kurz WA, Phillips OL, Shvidenko A, Lewis SL, Canadell JG, Ciais P, Jackson RB, Pacala S, McGuire AD, Piao S, Rautiainen A, Sitch S, Hayes D (2011) A large and persistent carbon sink in the world's forests. Science 333(6045):988–993. https://doi.org/10.1126/science.1201609

Pardini M, Armston J, Qi W et al (2019) Early lessons on combining lidar and multi-baseline SARmeasurements for forest structure characterization. Surv Geophys. https://doi.org/10.1007/s10712-019-09553-9

Phillips OL, Sullivan MJP, Baker TR, Mendoza AM, Vargas PN, Vásquez R (2019) Species matter: wood density influences tropical forestbiomass at multiple scales. Surv Geophys. https://doi.org/10.1007/s10712-019-09540-0

Reichstein M, Carvalhais N (2019) Aspects of forest biomass in the Earth system: its roleand major unknowns. Surv Geophys. https://doi.org/10.1007/s10712-019-09551-x

Réjou-Méchain M, Barbier N, Couteron P, Ploton P, Vincent G, Herold M, Mermoz S, Saatchi S, Chave J, de Boissieu F, Féret J-B, Takoudjou S, Pélissier R (2019) Upscaling Forest biomass from field to satellite measurements: sources of errors and ways to reduce them. Surv Geophys. https://doi.org/10.1007/s10712-019-09532-0

Schepaschenko D, See L, Lesiv M, Bastin J-F, Mollicone D, Tsendbazar N-E, Bastin L, McCallum I, Bayas JCL, Baklanov A, Perger C, Dürauer M, Fritz S (2019) Recent advances in forest observation with visual interpretation of very high-resolution imagery. Surv Geophys. https://doi.org/10.1007/s10712-019-09533-z

Tebaldini S, Minh DHT, d'Alessandro MM, Villard L, Toan TL, Chave J (2019) The status of technologies to measure forest biomass and structural properties: state of the art in SAR tomography of tropical forests. Surv Geophys. https://doi.org/10.1007/s10712-019-09539-7

Surveys in Geophysics (2019) 40:693–707
https://doi.org/10.1007/s10712-019-09551-x

Aspects of Forest Biomass in the Earth System: Its Role and Major Unknowns

Markus Reichstein[1] · Nuno Carvalhais[1,2]

Received: 16 October 2018 / Accepted: 17 June 2019 / Published online: 9 July 2019

Abstract

Forests are a major and diverse land cover occupying a third of the terrestrial vegetated surface; they store 50 to 65% of terrestrial organic carbon (including the soil) and contribute half to terrestrial productivity. Forest biomass stores close to 80% of all the biomass on Earth. As noted earlier, forests play an important role in the Earth system as carbon stocks, carbon sinks, mediator of the water cycle and as modifier of land surface roughness and albedo. Moreover, forests play a role as habitat for many species, are an economic source of timber and firewood and have recreational value for local populations and touristic visitors. Here, we appraise how ecosystem functions are influenced in particular by biomass and its vertical and horizontal distribution and hypothesize that almost all functions are directly or indirectly related to biomass, in addition to other factors. At landscape or regional scale, heterogeneity of biomass presumably has an important influence on a variety of processes, but there are gaps both in quantifying the heterogeneity of forests globally and in quantifying the effect of this heterogeneity. Similarly, while the role of forests for the global carbon cycle is important, large uncertainties exist regarding stocks, turnover times and the carbon sink function in forest, as an analysis of state-of-the-art carbon cycle and vegetation models shows. Upcoming global satellite missions such as GEDI, NISAR and BIOMASS will be able to address the above uncertainties and lack of understanding in combination with modeling approaches, in particular by exploiting information on vertical and horizontal heterogeneity.

Keywords Forest biomass · Earth System Models · Turnover times of carbon

✉ Nuno Carvalhais
ncarval@bgc-jena.mpg.de

Markus Reichstein
mreichstein@bgc-jena.mpg.de

1 Max-Planck-Institut für Biogeochemie, Hans-Knöll-Str. 10, 07745 Jena, Germany

2 Departamento de Ciências e Engenharia do Ambiente, DCEA, Faculdade de Ciências e Tecnologia, FCT, Universidade Nova de Lisboa, 2829-516 Caparica, Portugal

1 Introduction

Forests play important biophysical, biogeochemical, hydrological, economic and cultural roles in the Earth system (Fig. 1). Particularly, forests have a central role in the local and global carbon and water cycles with feedbacks into the climate system. Typically, an increase in forest cover is most often associated with a carbon sink from the atmosphere to forest biomass and soils (Guo and Gifford 2002; Houghton et al. 1999; Scurlock and Hall 1998). The impact of forests extends to the hydrological cycle, e.g., via the increased interception of water, deeper rooting and thus more extended sustenance of transpiration than shallow-rooted vegetation (e.g., grasslands and elsewhere) (Bonan 2008; Ellison et al. 2017). An extension in forest cover can further impact belowground water resources, as well as hydrological flows, via a reduction in groundwater recharge, and less runoff is also observed in forested catchments (Zhang et al. 2017). In particular, it has been hypothesized that floods are mitigated by forests through a buffering effect (e.g., Dixon et al. 2016 and elsewhere). Forests also impact the radiation balance and surface roughness, implying climate feedbacks which are variable, and can differ between regional and global scales (Betts et al. 1997; Bonan 2008). The effects on biodiversity can be both positive and negative, and depend on the region and the aspect of biodiversity (e.g., structural, functional, species richness and elsewhere), considered (Schulze and Mooney 2012). For example, structural biodiversity is higher in unmanaged, e.g., old growth, forests compared to, for instance,

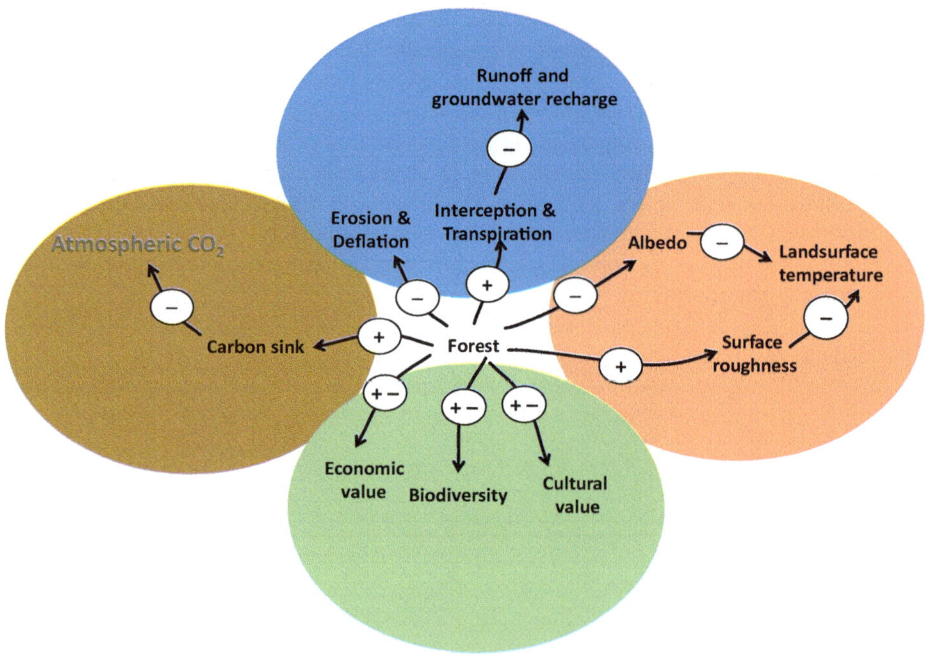

Fig. 1 How a hypothetical increase in forest cover in a landscape will influence various aspects of the Earth system regionally and globally. A ⟶ + B means, when A increases, B also increases. A ⟶ − B means when A increases B decreases. +− indicates variable responses are possible. The colors refer to water cycle and geomorphological processes (blue), energy balance and transfers (orange), biological and societal aspects (green), and carbon cycle (brown)

heavily managed even-aged plantations, implying a richer habitat for animal wildlife and elsewhere. On the other hand, grazed forests and shrubland systems are often more biodiverse than old growth (e.g., beech forests and elsewhere). Hence, it is expected that rather the heterogeneity of biomass than the mean biomass itself relates to many ecosystem properties and functions (e.g., biodiversity and elsewhere). Also, biomass certainly is neither the single factor influencing ecosystem properties and functions, but rather plant traits and their spatial arrangement and temporal dynamics in the ecosystem (e.g., Reichstein et al. 2014a, b; Yguel et al. 2019).

Ecosystem services are defined as the benefits people obtain from ecosystems (Millennium Ecosystem Assessment 2005)—via their functioning. Despite the clear difficulty in quantifying the global benefits of such an ubiquitous good, the economic value of forest ecosystem services has been estimated to be up to more than 10,000 Int$/ha/year (International dollars per hectare per year, De Groot et al. 2012) only for currently known and discovered services. Yet, certainly by far not all of these values are accounted for in regular markets and some values, such as medicinal plants and animals in tropical forest, are even yet undiscovered (Balunas and Kinghorn 2005). And also, the perceived cultural value of forests is not uniform and depends on region, culture and historical epoch. For further insight, the reader is referred to Bengston (1994).

In this short paper, we provide a concise overview of the links between forest biomass and major ecosystem functions with emphasis on carbon cycle aspects. We further illustrate how major processual uncertainties are emerging in global modeling approaches, with emphasis on diagnostics in the global carbon cycle. We finish by highlighting the potential of new Earth observation (EO) data streams to constrain such uncertainties, which will gain from synergistic approaches with complementary data streams for a better understanding of Earth system dynamics.

2 Forest Biomass and Ecosystem Services

Many of the mentioned effects and ecosystem services of forests are not directly observable spatially at the global scale or even at regional scales. For instance, photosynthesis or transpiration cannot be directly observed, but inferred via modeling biophysical relations (Running et al. 2004). Yet, with satellite missions, in particular upcoming LiDAR and RADAR missions, aspects of forests' structure and functioning can be inferred and correlated with the effects briefly discussed above. The relation between land carbon storage and biomass is trivial, since biomass simply is part of total ecosystem storage. Globally, more than 75% of biomass is stored in forests (Bar-On et al. 2018). Current estimates show that the proportion of carbon in biomass and in soils varies between biomes and is highest in tropical forests ($C_{veg}/C_{soil} \approx 0.45$), while in high-latitude boreal forests the soil carbon storage can overtake vegetation carbon stocks ($C_{veg}/C_{soil} \approx 1/12$) (Carvalhais et al. 2014). There is no unique relationship between the current carbon sink and the biomass stock. While classical theory predicts that forests reach an equilibrium after a certain time (Odum 1969), empirical evidence points to carbon sinks in even old-growth forests (Luyssaert et al. 2008). Also, it is easy to build plausible theories and models which predict an absence of a dynamic equilibrium state (steady state) in the soils (Reichstein et al. 2009; Wutzler and Reichstein 2008). These models simply have to relax the classical assumption of first-order decay kinetics. Interestingly, forest age and species diversity has been shown to dampen the interannual variability of gross carbon uptake, which is likely also related to the vertical structure and horizontal heterogeneity of the forest, but also to belowground heterogeneity, although the mechanisms have to be further investigated (Musavi et al. 2017).

Despite the sociocultural role of forests, the concrete relation of the biomass variable to cultural value has to our knowledge not been explicitly empirically studied. Yet it can be hypothesized that the cultural value in industrialized countries increases both with increasing biomass and increasing structure, today, because these forests are relatively rare there. One hypothesis to test is that protected forest areas have larger biomass and larger heterogeneity than non-protected areas (already at the start of protection). As mentioned above, the cultural value is certainly very much society dependent—for instance the tropical forests sometimes have connotations of being dangerous, and in the past "the dark forest" was also associated with danger, e.g., in fairy tales (Lüthi 1986 and elsewhere), negating a high value of forest in former times. However, in society it has been clear the imperative character of quantifying the present and predicting the future of forests. Observation and simulation research has been developed in different directions, e.g., biodiversity, productivity and most recently on the biophysical–chemical feedbacks in the Earth system. In this aspect, the description of the global carbon cycle has been having a central role, especially due to the relevance in prognostics and to the apparent uncertainties in the current knowledge (Friedlingstein et al. 2006, 2013).

3 Uncertainties Related to Carbon Cycle Aspects

Generally, in dynamic global vegetation models (DGVMs) and global circulation models (GCMs), biomass emerges as the long-term net integral of carbon gains via gross primary productivity (GPP) and losses through autotrophic respiration, litter fall and plant mortality. Changes in vegetation biomass stocks are tightly linked to the interannual variability and decadal trends in net ecosystem C fluxes, given the source–sink strength controls during forest growth trajectories and the substantial emissions of C to the atmosphere from forest disturbances (for a global synthesis see Bonan 2008). At the global scale, the net carbon sink modeled by DGVMs is still very uncertain but is now consistent with the atmospheric growth rate (Le Quéré et al. 2017 and elsewhere). Yet, spatial distribution is much more homogeneous than, for example, observed at flux sites (Fig. 2). There may be several possible reasons for the difference, including spatial aggregation to larger grid cells, which caused the higher variability in ecosystem in local (~ 1 km^2) development stages across sites to be dampened (averaged out) for modeled grid cells (~ 2500 km^2) that encompass different development stages. Likely, that moderate-resolution biomass change estimates from upcoming missions will also provide an independent benchmark to this mismatch in variability in net carbon balance. However, even at larger scales, substantial uncertainties are observed in different regions (Table 1).

Globally, depending on the modeling ensemble, the estimated global vegetation carbon stocks can deviate from observations on average from 14 to 27%, i.e., for the CMIP5 (Taylor et al. 2012) and TRENDY[1] ensembles, respectively. Noteworthy is the spread within the modeling ensemble. The ensembles can range from 62 to 70% (range over mean) for the CMIP5 and TRENDY, respectively. In general, the lower the bias in modeled biomass, the better the models capture the global spatial variability that widely spans across biomes

[1] http://dgvm.ceh.ac.uk/node/9.

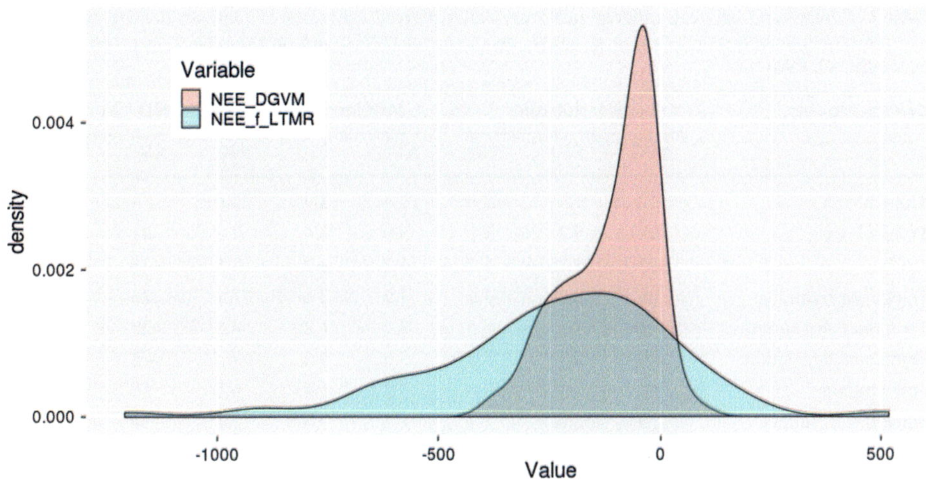

Fig. 2 Distribution of net ecosystem exchange (NEE) as modeled by a state-of-the-art dynamic global vegetation model (DGVM) and as observed at eddy covariance sites

Table 1 Ecosystem services and how they are related to overall biomass and its vertical and horizontal variability

Ecosystem service	Relation to mean biomass	Relation to vertical structure	Relation to horizontal heterogeneity
Carbon storage	Direct relationship, $C_{storage} \sim 0.5 \times$ biomass; soil C and coarse woody debris not	Vulnerability?	Vulnerability?
Carbon sink	Biomass *changes* directly related, not directly related to stock itself	Variability decreasing?	Variability decreasing?
Biodiversity	Unclear	Positive effect	Positive effect
Cultural value (industrialized countries)	Often positively related	Often positively related	Often positively related

(Table 2). But for some regions, even in large geographic regions, the biomass stocks can range more than three times across the TRENDY models, as for example in Central and South America (Li et al. 2017). Particularly in the tropical forests, Negron-Juarez et al. (2015) highlight that the limitations in modeling the C allocation of net primary productions propagate to errors in biomass estimates from CMIP5 models. Further, this analysis reveals that biomass steadily increases with NPP, while observations show a saturation effect, emerging from shorter turnover times of highly productive forests, or lower wood density of fast growing species. This strong relationship between biomass and productivity in models compared to observations is also apparent outside of the tropical regions (Fig. 3), even if the patterns may be confounded by varying forest fraction in the grid cells. This needs to be further investigated. The relationships strongly vary across models, but throughout the majority of them temperature appears to impose the sensitivity on the C_{veg} to GPP relationship, which in nature could be related to controls on respiratory processes

Table 2 Comparison between models and observation-derived estimates of biomass density (C_{veg}, kg C/m^2) per bioclimatic class (Kottek et al. 2006) on the mean (AVG) and on the ranges (percentile 10, P10, percentile 90, P90)

Bioclimatic class	Observation based			CMIP5 models			RD [%]	
	AVG	P10	P90	AVG	P10	P90	AVG	RANGE
Tropical wet	12.2	9.4	15.0	16.4	8.9	26.6	34	218
Tropical dry	4.2	3.2	5.1	5.9	4.4	7.3	41	51
Arid	1.5	1.2	1.7	1.2	0.2	2.5	− 18	302
Temperate humid	3.8	2.9	4.6	4.6	2.6	7.1	23	171
Temperate dry summers	2.1	1.7	2.5	2.7	1.2	4.3	29	264
Temperate dry winters	6.0	4.7	7.3	5.6	3.6	7.4	− 8	44
Snow humid	2.1	1.5	2.6	3.2	1.4	6.5	56	376
Snow dry summers	2.3	2.0	2.7	1.5	0.6	2.7	− 34	219
Snow dry winters	2.1	1.6	2.6	3.4	1.5	6.1	63	360
Polar	0.6	0.5	0.7	0.5	0.1	1.5	− 9	658
Global	3.4	2.6	4.2	3.8	2.6	5.3	12	78

The relative differences (RD, shown in percentage) are estimated on the mean (AVG) and on the ranges (RANGE)

(Amthor 2000), to gradients in tree cover imposed by precipitation and groundwater, or to increasing mortality patterns associated with drought or heat (Thurner et al. 2016).

The contrast between modeled and observed biomass, once fluxes are known, highlights discrepancies in simulating loss processes in C cycle models. Regionally, the spatial patterns of background mortality in temperate forests can be associated with climatic drought or with potential frost damage in boreal forests (Thurner et al. 2016), but these patterns do not emerge in state-of-the-art DGVMs (Thurner et al. 2017). Model-based approaches have also shown the 50% reduction in turnover times of carbon in biomass via land use (Erb et al. 2016). Hence, the mismatch between observed and modeled carbon cycle dynamics could explain the disagreement in estimates of forest biomass, although it can also emerge from the challenges in simulating human-driven land use land cover change dynamics, which impose ~ 50% reductions in forest carbon stocks at the global scale (Erb et al. 2018). These estimates imply that, using a bookkeeping approach and assuming no climate effect, land use change emissions have been thrown into the atmosphere circa 375 to 525 PgC. A model ensemble-based estimate, using contemporaneous biomass estimates to constrain modeled trajectories regionally, estimates that one-third of these emissions occurred in the last century (155 ± 50 PgC; Li et al. 2017).

Processes controlling the turnover times of carbon have been shown to be the highest structural discrepancy between models in projections of forest carbon cycle (Friend et al. 2014). Also in CMIP5 models, despite the moderate to substantial agreement between GPP and biomass patterns, the modeled turnover times of carbon in vegetation (T_{veg}) compared poorly with observations ($T_{veg} = C_{veg}$/GPP; Carvalhais et al. 2014). Globally, the correlation between T_{veg} and tree cover fraction is significant across the CMIP5 models, though it can range between 0.50, i.e., inmcm4, and 0.88 in the Max Planck Institute for Meteorology Earth System Model at Low Resolution (MPI-ESM-LR). However, between different tree cover ranges the relationship between tree cover and T_{veg} can change substantially (Table 3). These can emerge from different climate

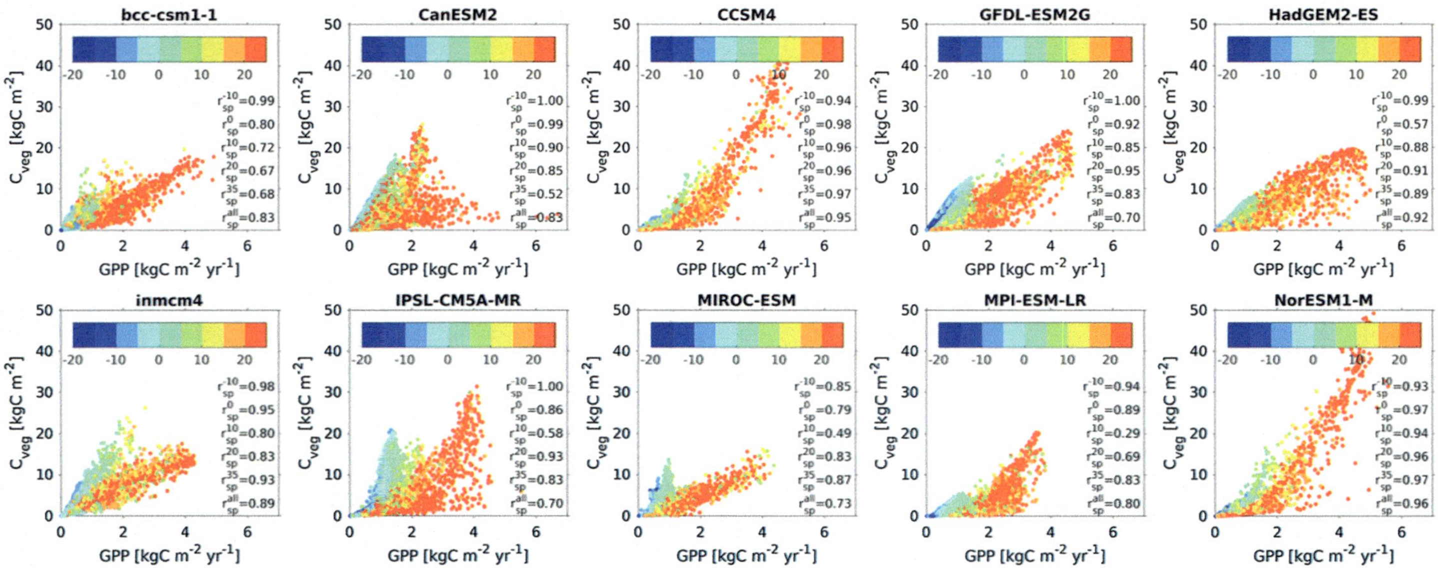

Fig. 3 Global relationship between vegetation carbon (C_{veg}) and gross primary productivity (GPP) across the different CMIP5 models. Colors represent different mean annual surface temperatures used by the models. Reported in the lower right corner is the Spearman correlation coefficient (r_{sp}) globally (r_{sp}^{all}), and for different temperature ranges: 10°C and 0°C (r_{sp}),^{0}0°C and 10°C (r_{sp}^{10}), 10°C and 20°C (r_{sp}^{20}) and above 20°C (r_{sp}^{35}). Significant correlations are observed across temperature intervals, although the association strength between GPP and C_{veg} per temperature interval can diverge substantially between models

Table 3 Spearman rank partial correlation (r) between T_{veg} and CUE, and between T_{veg} and tree cover fraction (f_tree) across CMIP5 models globally ([0 100]) and for different ranges of tree cover

Model	X	Tree cover [%] interval				
		[0 100]	[0 25]	[25 50]	[50 75]	[75 100]
HadGEM2-ES	CUE	−0.23	0.03 (p=0.248)	−0.47	−0.04 (p=0.349)	0.06 (p=0.283)
	f_tree	0.84	0.54	0.39	0.42	0.29
inmcm4	CUE	0.18	0.35	0.38	0.31	0.14 (p=0.010)
	f_tree	0.5	0.44	0.27	0.16	−0.10 (p=0.077)
IPSL-CM5A-MR	CUE	−0.29	−0.26	−0.26	−0.08 (p=0.104)	−0.16 (p=0.006)
	f_tree	0.68	0.65	0.19	0.31	0.27
MIROC-ESM	CUE	−0.14	−0.38	−0.34	−0.02 (p=0.698)	−0.04 (p=0.337)
	f_tree	0.58	0.61	0.17	0.12	0.05 (p=0.225)
MPI-ESM-LR	CUE	0.27	0.33	0.27	0.23	0.56
	f_tree	0.88	0.7	0.5	0.68	0.6

Temperature and precipitation were used as control variables, and when calculating the correlation between T_{veg} and CUE, f_tree is considered also as a control variable, and vice versa. Only p values higher than 0.01 are reported

effects on modeled dynamics affecting mortality and, consequently, tree density. For instance, the pattern of the relationship between vegetation carbon turnover times and temperature across the CMIP5 models differs substantially (Fig. 4a), even showing contrasting patterns: see, for example, the Geophysical Fluid Dynamics Laboratory Earth System Model, with generalized ocean layer dynamics (GFDL-ESM2G) compared with the Hadley Centre Global Environmental Model, version 2, Earth System (GEM2-ES) below 0 °C; or the strong control of precipitation on T_{veg} beyond 20 °C for only half of the models (i.e., CanESM2, CCSM4, IPSL-CM5A-MR, MPI-ESM-LR and NorESM1-M). At global scales, the controls on T_{veg} do not emerge exclusively from mortality dynamics, but also from carbon allocations strategies and the controls of autotrophic respiration patterns. It would be expected that carbon use efficiency (CUE = NPP/GPP) correlates positively with T_{veg}, though if the spatial patterns of CUE mostly emerge from changes in tree-to-grass continuum, the emergent pattern should be opposite, since CUE of herbaceous vegetation is higher though its turnover is lower when compared to woody vegetation. But this relationship seems to be far from agreement across models as well (Fig. 4b): In some cases across temperature ranges the CUE seems to be positively related to T_{veg} (i.e., CanESM2 and MPI-ESM-LR), in other cases it is negatively associated (i.e., bcc-csm1-1, HadGEM2-ES, IPSL-CM5A-MR, MIROC-ESM), while in other cases the patterns are inconclusive (i.e., CCSM4, NorESM1-M). It is worth noting that CUE, despite the lesser association T_{veg} when compared to tree cover fraction (Table 3), strongly differs between modeling approaches, even showing opposite signs at the global scales. Both MPI-ESM-LR and HadGEM2-ES show a strong rank correlation between tree fraction and T_{veg}, but show opposite relationships in CUE-to-T_{veg}, likely underpinning the contrasting patterns of T_{veg} with different covariates (Fig. 4). These patterns emphasize that a primary focus on vegetation mortality dynamics, from which tree density and other ecosystem properties emerge, should not cast shadow on the importance of understanding the controls carbon allocation and respiratory costs have in predicting long-term changes in turnover times of carbon.

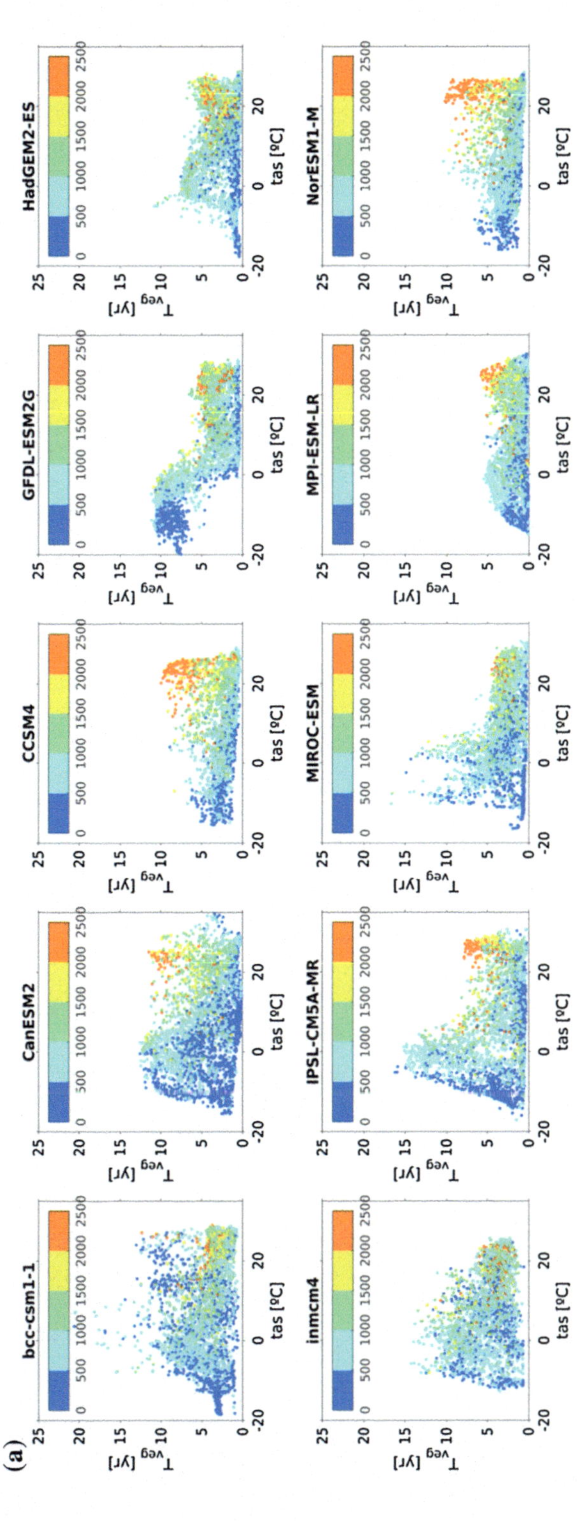

Fig. 4 Contrasting the turnover time of vegetation carbon (T_{veg}, year, estimated as the ratio between C_{veg} and GPP [*Carvalhais* et al. 2014]) against temperature (tas, °C), colored by precipitation (mm/year) in CMIP5 models **(a)**. In **(b)**, comparing T_{veg} against carbon use efficiency (CUE, dimensionless). For the models where tree cover fraction was available, the regions for high tree coverage (higher than 75%) are highlighted in green; while in gray, tree cover fraction is lower than 75% **(c)**

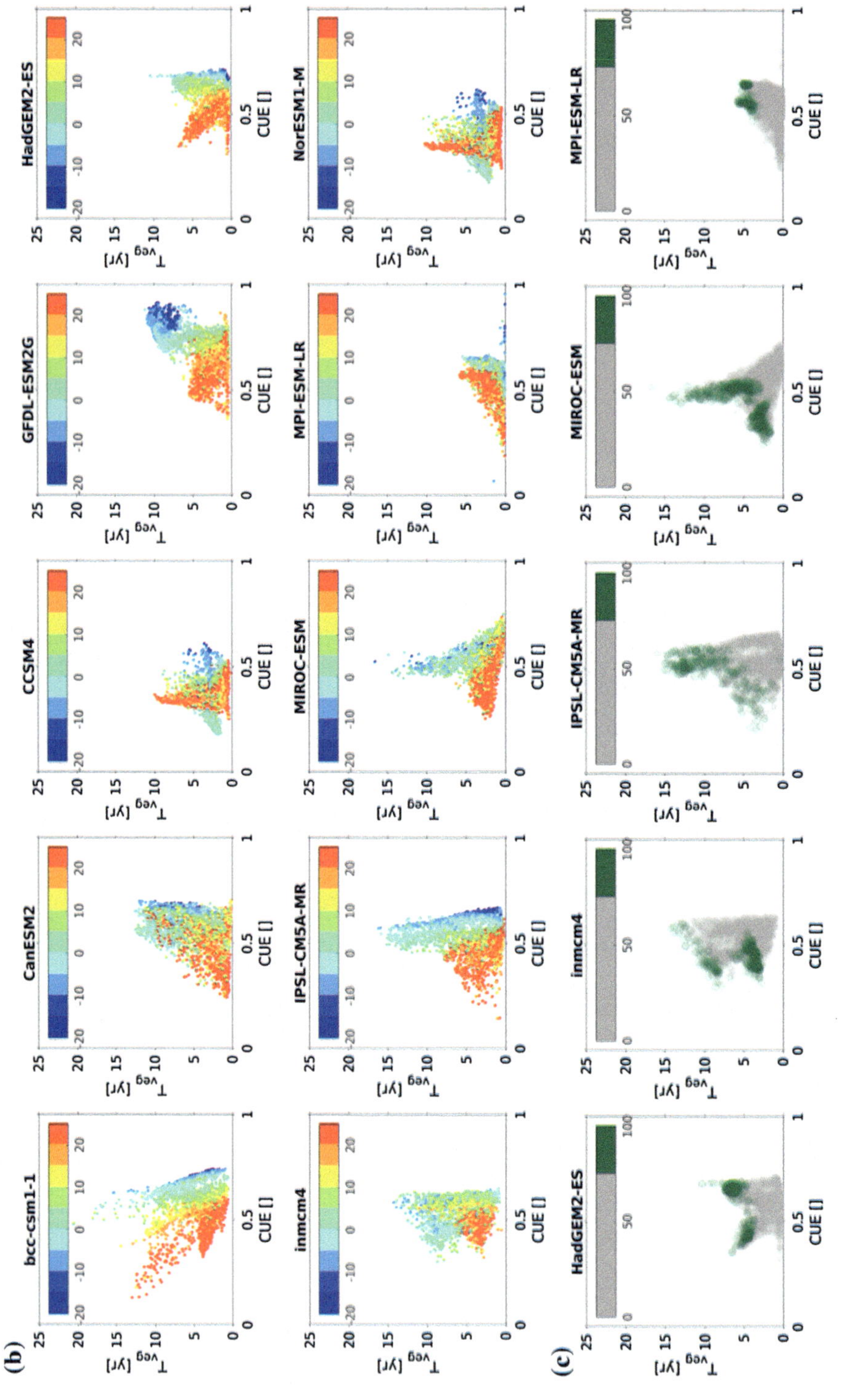

Fig. 4 (continued)

Each of the CMIP5 models originates from a different Earth system modeling group, team, or even country, and can differ in terms of spatial representativeness as well as in process integration and representation. For instance, models may differ in terms of spatial resolution and number of layers and depths used to simulate water and energy fluxes in the soil. In terms of ecosystem dynamics, three models do not include dynamic vegetation components (i.e., do not explicitly simulate the processes of vegetation mortality and succession; bcc-csm1-1, CCSM4 and NorESM1-M, the latter two force it with land use land cover change datasets); with the exception of Inmcm4, all models embed a representation for crop dynamics; only four models do not include pastures (bcc-csm1-1, CanESM2, Inmcm4 and IPSL-CM5A-MR); and only four include wood harvest (CCSM4, GFDL-ESM2G, MPI-ESM-LR and NorESM1-M); and three include deforestation (GFDL-ESM2G, Inmcm4 and MPI-ESM-LR). In addition, the description of processes of photosynthesis generally relies on Farquhar and Sharkey (1982), or similar representations, but other processes that control the stocking of carbon in vegetation, like autotrophic respiration, litterfall and other metabolic processes, lack a unifying or generally acceptable framework. Apart from the similarities between CCSM4 and NorESM1-M, which may stem from the fact that Nor ESM1-M is partly based on the CCSM4 (Tjiputra et al. 2013), the description of qualitative differences between models falls short in understanding the reasons behind the wide diversity in the relationships between stocks and fluxes, or on the responses of T_{veg} to climate or CUE across models.

Overall, repeated biomass measurements are key observational constraints on models for both contemporaneous and prognostic simulations of the global carbon cycle. The ability to converge on future projections of the coupled carbon–climate cycle requires a better understanding of phenomena and mechanisms controlling the mortality dynamics of vegetation, as well as processes leading to changes in allocation of assimilates to organ growth and maintenance processes. Here, probably, simulation approaches that omit land use land cover change dynamics, or the target of processes outside that realm, require that DGVMs and GCMs should better compare with patterns of potential biomass (Erb et al. 2018; Exbrayat et al. 2017). Furthermore, continuous monitoring approaches have unique information content at scales that are unprecedented and embed the potential to shed light on controls of growth at larger scales. Ultimately, combining time series at high resolution of remotely sensed biomass fields with models can help disentangle both plant physiology and demography processes at larger scales, ultimately leading to a better understanding of coupled biosphere–atmosphere dynamics and assessment of global forest function.

Acknowledgements Open access funding provided by Max Planck Society.

Appendix: Data and Methods

The analysis presented here focuses on the outputs from the historical simulations of ten Earth System Models included in the 5th Climate Model Intercomparison Project (CMIP5) (Taylor et al. 2012). From these historical scenario simulations were processed climate variables (temperature and precipitation), but also carbon fluxes (gross and net primary productivity) and vegetation carbon pools. For each variable from each model were computed mean annual values per grid cell between 1982 and 2005. These model outputs were always processed at the native spatial resolution of each model, but, in order to harmonize the spatial resolution between models, the results were aggregated to a common grid of intermediate resolution across models which corresponded to the native resolution of the NorESM1-M model (~ 1.89° × 2.5°, latitude × longitude). The aggregation was based on an area weighted mean per grid cell. Five of those ten models reported tree, grass and bare soil coverages per grid cell. These were also processed to determine the tree fraction per grid cell, defined as the fraction of trees over the fraction of vegetated area.

The observation-based estimates of total vegetation carbon were obtained by putting together spatially explicit estimates for the pantropical regions by Saatchi et al. (2011) and estimated for the northern and temperate forests as estimated by Thurner et al. (2014) as described in Carvalhais et al. (2014). Gross primary production (GPP) is based on the data-driven estimates of Jung et al. (2011). Desert areas were excluded from the analysis, for the comparison between models and data, but also within models. Deserts were defined according to the Köppen–Geiger classification described in Kottek et al. (2006) and to a maximum GPP of 10 $gC/m^2/year$. The original Köppen–Geiger classification was simplified to represent larger bioclimatic gradients, but still translating the seasonality covariation between temperature and precipitation (Table 2). For the CUE, grid cells with GPP values lower than 100 $gC\ m^{-2}\ year^{-1}$ were filtered out (Fig. 4).

The relative differences (RD) are estimated as the difference between models and observations normalized by observations (shown in percentage). For the RD on the ranges (RANGE), ranges are estimated based on the differences between P90 and P10 (Table 2). The correlation analysis is based on the nonparametric Spearman's rank correlation coefficient for robustness in associating variables that may not have a linear relationship. The same is assumed in the partial correlation analysis. Partial correlation was used to measure the degree of association between two variables, controlling for the covariation between the dependent variable and a second independent covariate (Table 3).

Brief Glossary

Carbon pool is the representation of a reservoir of carbon that can match plant organs (e.g., leafs, stems or roots) or soil reservoirs (e.g., litter or humus). Carbon pools are usually divided according to their physical or chemical (or both) properties, e.g., according to the function that the pools have, or the average time span of particles in that pool, or to the vertical locations in the soil. Different models can have different sets of carbon pools, e.g., some can represent branches, others the division between fine and coarse roots) and the representation of the processes that control gains and losses in those pools can also change between models (Rodhe 1992). A carbon pool operates as a carbon sink (or source) when it accumulates (or loses) carbon in time.

CMIP5-the Coupled Model Intercomparison Project, phase 5—is a worldwide collective activity that involves modeling teams that investigate the dynamics of the Earth System via global coupled ocean–atmosphere general circulation models (GCMs). The project designs harmonized modeling experiments to investigate potential impacts of climate change scenarios on the land, oceans and atmosphere of the Earth (Taylor et al. 2012).

Ecosystem service is a direct or indirect benefit that originates from the presence and functioning of an ecosystem (Chapin et al. 2006; Schulze 2006).

Earth System Model-ESM-is a numerical simulation environment that represents the changes in space and time on the three larger domains in the Earth system—land, ocean and atmosphere.

GPP—gross primary productivity—is the flux of carbon from the atmosphere to the ecosystem mediated by photosynthesis (Chapin et al. 2006; Schulze 2006).

Model ensemble is a dataset consisting of outputs from several different models, or of model outputs with different parameterization or initialization settings. Here, the model ensembles refer to datasets originating from different models ran with the same forcing conditions (either ESMs or just global vegetation models).

NEP—net ecosystem productivity—is the net flux of carbon between the ecosystem and the atmosphere that results from the balance between GPP and RECO. Oppositely to NEE (net ecosystem exchange), NEP is defined as a difference between GPP and RECO, meaning that the ecosystem gains carbon when NEP is positive (Chapin et al. 2006; Schulze 2006).

NPP—net primary productivity—is the net flux of carbon that results from the balance between GPP and the autotrophic respiration flux (Chapin et al. 2006; Schulze 2006).

RECO—ecosystem respiration—is the flux of carbon between an ecosystem and the atmosphere that integrates all the respiratory fluxes, i.e., decomposition fluxes (heterotrophic respiration) and plant (autotrophic) respiration (Chapin et al. 2006; Schulze 2006).

Carbon turnover time can be roughly defined as the average time that one particle spends in an ecosystem pool since it is assimilated until it is released (Rodhe 1992).

References

Amthor JS (2000) The McCree–de Wit-Penning de Vries–Thornley respiration paradigms: 30 years later. Ann Bot 86(1):1–20. https://doi.org/10.1006/anbo.2000.1175

Balunas MJ, Kinghorn AD (2005) Drug discovery from medicinal plants. Life Sci 78(5):431–441

Bar-On YM, Phillips R, Milo R (2018) The biomass distribution on Earth. Proc Natl Acad Sci USA 115(25):6506–6511. https://doi.org/10.1073/pnas.1711842115

Bengston DN (1994) Changing forest values and ecosystem management. Soc Nat Resources 7(6):515–533

Betts RA, Cox PM, Lee SE, Woodward FI (1997) Contrasting physiological and structural vegetation feedbacks in climate change simulations. Nature 387:796–799

Bonan GB (2008) Forests and climate change: forcings, feedbacks, and the climate benefits of forests. Science 320(5882):1444–1449

Carvalhais N et al (2014) Global covariation of carbon turnover times with climate in terrestrial ecosystems. Nature 514:213–217. https://doi.org/10.1038/nature13731

Chapin FS et al (2006) Reconciling carbon-cycle concepts, terminology, and methods. Ecosystems 9(7):1041–1050. https://doi.org/10.1007/s10021-005-0105-7

De Groot R, Brander L, Van Der Ploeg S, Costanza R, Bernard F, Braat L, Christie M, Crossman N, Ghermandi A, Hein L (2012) Global estimates of the value of ecosystems and their services in monetary units. Ecosyst Serv 1(1):50–61

Dixon SJ, Sear DA, Odoni NA, Sykes T, Lane SN (2016) The effects of river restoration on catchment scale flood risk and flood hydrology. Earth Surf Process Landf 41(7):997–1008

Ellison D, Morris CE, Locatelli B, Sheil D, Cohen J, Murdiyarso D, Gutierrez V, Van Noordwijk M, Creed IF, Pokorny J (2017) Trees, forests and water: cool insights for a hot world. Glob Environ Change 43:51–61

Erb KH, Fetzel T, Plutzar C, Kastner T, Lauk C, Mayer A, Niedertscheider M, Korner C, Haberl H (2016) Biomass turnover time in terrestrial ecosystems halved by land use. Nat Geosci 9(9):674. https://doi.org/10.1038/ngeo2782

Erb KH et al (2018) Unexpectedly large impact of forest management and grazing on global vegetation biomass. Nature 553(7686):73. https://doi.org/10.1038/nature25138

Exbrayat JF, Liu YY, Williams M (2017) Impact of deforestation and climate on the Amazon Basin's above-ground biomass during 1993–2012. Sci Rep UK. https://doi.org/10.1038/s41598-017-15788-6
Farquhar GD, Sharkey TD (1982) Stomatal conductance and photosynthesis. Annu Rev Plant Physiol Plant Mol Biol 33:317–345. https://doi.org/10.1146/annurev.pp.33.060182.001533
Friedlingstein P et al (2006) Climate-carbon cycle feedback analysis: results from the (CMIP)-M-4 model intercomparison. J Clim 19(14):3337–3353
Friedlingstein P, Meinshausen M, Arora VK, Jones CD, Anav A, Liddicoat SK, Knutti R (2013) Uncertainties in CMIP5 climate projections due to carbon cycle feedbacks. J Clim 27(2):511–526. https://doi.org/10.1175/JCLI-D-12-00579.1
Friend AD et al (2014) Carbon residence time dominates uncertainty in terrestrial vegetation responses to future climate and atmospheric CO2. Proc Natl Acad Sci USA 111(9):3280–3285. https://doi.org/10.1073/pnas.1222477110
Guo LB, Gifford R (2002) Soil carbon stocks and land use change: a meta analysis. Glob Change Biol 8(4):345–360
Houghton RA, Hackler JL, Lawrence KT (1999) The US carbon budget: contributions from land-use change. Science 285(5427):574–578
Jung M et al (2011) Global patterns of land-atmosphere fluxes of carbon dioxide, latent heat, and sensible heat derived from eddy covariance, satellite, and meteorological observations. J Geophys Res Biogeol 116:1–16
Kottek M, Grieser J, Beck C, Rudolf B, Rubel F (2006) World map of the Köppen-Geiger climate classification updated. Meteorol Z 15(3):259–263
Le Quéré C et al (2017) Global carbon budget 2017. Earth Syst Sci Data Discuss 10(1):1–79
Li W et al (2017) Land-use and land-cover change carbon emissions between 1901 and 2012 constrained by biomass observations. Biogeosciences 14(22):5053–5067. https://doi.org/10.5194/bg-14-5053-2017
Lüthi M (1986) The European folktale: form and nature. Indiana University Press, Bloomington
Luyssaert S, Schulze ED, Börner A, Knohl A, Hessenmöller D, Law BE, Ciais P, Grace J (2008) Old-growth forests as global carbon sinks. Nature 455(7210):213–215. https://doi.org/10.1038/nature07276
Millennium Ecosystem Assessment (2005) Ecosystems and human well-being. Island Press, Washington, DC
Musavi T, Migliavacca M, Reichstein M, Kattge J, Wirth C, Black TA, Janssens I, Knohl A, Loustau D, Roupsard O, Varlagin A (2017) Stand age and species richness dampen interannual variation of ecosystem-level photosynthetic capacity. Nat Ecol Evol 1(2):48
Negron-Juarez RI, Koven CD, Riley WJ, Knox RG, Chambers JQ (2015) Observed allocations of productivity and biomass, and turnover times in tropical forests are not accurately represented in CMIP5 Earth system models. Environ Res Lett. https://doi.org/10.1088/1748-9326/10/6/064017
Odum EP (1969) Strategy of ecosystem development. Science 164(3877):262–270
Reichstein M, Ågren GI, Fontaine S (2009) Is there a theoretical limit to soil carbon storage in old-growth forests? A model analysis with contrasting approaches. In: Wirth C, Gleixner G, Heimann M (eds) Old-growth forests: function, fate and value. Springer, Berlin, pp 267–281
Reichstein M, Bahn M, Mahecha MD, Kattge J, Baldocchi DD (2014a) Linking plant and ecosystem functional biogeography. Proc Natl Acad Sci USA 111(38):13697–13702. https://doi.org/10.1073/pnas.1216065111
Reichstein M, Richardson AD, Migliavacca M, Carvalhais N (2014b) Plant–environment interactions across multiple scales. In: Monson RK (ed) Ecology and the environment. Springer, New York, pp 1–27. https://doi.org/10.1007/978-1-4614-7612-2_22-1
Rodhe H (1992) Modeling biogeochemical cycles. In: Samuel SS, Charlson RJ, Orians GH, Wolfe GV (eds) Global biogeochemical cycles. Academic Press Limited, London
Running SW, Nemani RR, Heinsch FA, Zhao MS, Reeves M, Hashimoto H (2004) A continuous satellite-derived measure of global terrestrial primary production. Bioscience 54(6):547–560. https://doi.org/10.1641/0006-3568(2004)054%5b0547:Acsmog%5d2.0.Co;2
Saatchi SS et al (2011) Benchmark map of forest carbon stocks in tropical regions across three continents. Proc Natl Acad Sci USA 108(24):9899–9904. https://doi.org/10.1073/pnas.1019576108
Schulze ED (2006) Biological control of the terrestrial carbon sink. Biogeosciences 3(2):147–166
Schulze E-D, Mooney HA (2012) Biodiversity and ecosystem function. Springer, Berlin
Scurlock J, Hall D (1998) The global carbon sink: a grassland perspective. Glob Change Biol 4(2):229–233
Taylor KE, Stouffer RJ, Meehl GA (2012) An overview of CMIP5 and the experiment design. Bull Am Meteorol Soc 93(4):485–498. https://doi.org/10.1175/Bams-D-11-00094.1
Thurner M et al (2014) Carbon stock and density of northern boreal and temperate forests. Glob Ecol Biogeogr 23(3):297–310. https://doi.org/10.1111/geb.12125

Thurner M, Beer C, Carvalhais N, Forkel M, Santoro M, Tum M, Schmullius C (2016) Large-scale variation in boreal and temperate forest carbon turnover rate related to climate. Geophys Res Lett 43(9):4576–4585. https://doi.org/10.1002/2016gl068794

Thurner M et al (2017) Evaluation of climate-related carbon turnover processes in global vegetation models for boreal and temperate forests. Global Change Biol 23(8):3076–3091. https://doi.org/10.1111/gcb.13660

Tjiputra JF, Roelandt C, Bentsen M, Lawrence DM, Lorentzen T, Schwinger J, Seland O, Heinze C (2013) Evaluation of the carbon cycle components in the Norwegian Earth System Model (NorESM). Geosci Model Dev 6(2):301–325

Wutzler T, Reichstein M (2008) Colimitation of decomposition by substrate and decomposers—a comparison of model formulations. Biogeosciences 5(3):749–759. https://doi.org/10.5194/bg-5-749-2008

Yguel B, Piponiot C, Mirabel A, Dourdain A, Hérault B, Gourlet-Fleury S, Forget P-M, Fontaine C (2019) Beyond species richness and biomass: impact of selective logging and silvicultural treatments on the functional composition of a neotropical forest. For Ecol Manag 433:528–534

Zhang M, Liu N, Harper R, Li Q, Liu K, Wei X, Ning D, Hou Y, Liu S (2017) A global review on hydrological responses to forest change across multiple spatial scales: importance of scale, climate, forest type and hydrological regime. J Hydrol 546:44–59

Surveys in Geophysics (2019) 40:709–734
https://doi.org/10.1007/s10712-019-09519-x

The Relevance of Forest Structure for Biomass and Productivity in Temperate Forests: New Perspectives for Remote Sensing

Rico Fischer, et al. ***[full author details at the end of the article]***

Received: 17 August 2018 / Accepted: 14 February 2019 / Published online: 4 March 2019

Abstract
Forests provide important ecosystem services such as carbon sequestration. Forest landscapes are intrinsically heterogeneous—a problem for biomass and productivity assessment using remote sensing. Forest structure constitutes valuable additional information for the improved estimation of these variables. However, survey of forest structure by remote sensing remains a challenge which results mainly from the differences in forest structure metrics derived by using remote sensing compared to classical structural metrics from field data. To understand these differences, remote sensing measurements were linked with an individual-based forest model. Forest structure was analyzed by lidar remote sensing using metrics for the horizontal and vertical structures. To investigate the role of forest structure for biomass and productivity estimations in temperate forests, 25 lidar metrics of 375,000 simulated forest stands were analyzed. For the lidar-based metrics, top-of-canopy height arose as the best predictor for describing horizontal forest structure. The standard deviation of the vertical foliage profile was the best predictor for the vertical heterogeneity of a forest. Forest structure was also an important factor for the determination of forest biomass and aboveground wood productivity. In particular, horizontal structure was essential for forest biomass estimation. Predicting aboveground wood productivity must take into account both horizontal and vertical structures. In a case study based on these findings, forest structure, biomass and aboveground wood productivity are mapped for whole of Germany. The dominant type of forest in Germany is dense but less vertically structured forest stands. The total biomass of all German forests is 2.3 Gt, and the total aboveground woody productivity is 43 Mt/year. Future remote sensing missions will have the capability to provide information on forest structure (e.g., from lidar or radar). This will lead to more accurate assessments of forest biomass and productivity. These estimations can be used to evaluate forest ecosystems related to climate regulation and biodiversity protection.

Keywords Forest structure · Biomass · Aboveground wood productivity · Remote sensing · Lidar · Forest model

Electronic supplementary material The online version of this article (https://doi.org/10.1007/s10712-019-09519-x) contains supplementary material, which is available to authorized users.

1 Introduction

Forests are crucial components of the Earth system. They represent an important pool in the global carbon cycle as they bind huge amounts of atmospheric carbon (Bonan 2008; Grace et al. 2014; Pan et al. 2011). Quantifying forest carbon stocks (e.g., forest biomass) and forest carbon fluxes (e.g., forest productivity) is important for understanding the effects of land use and climate change (Foley et al. 2005). However, forests can be heterogeneously structured which lead to variable biomass stocks and carbon fluxes (Rödig et al. 2018; Saatchi et al. 2011). The structural heterogeneity makes it difficult to reliably estimate forest biomass or forest productivity for larger regions. For example, biomass estimations for the Amazon rainforest—the largest intact tropical forest on Earth—range from 39 to 93 GtC (Houghton et al. 2001; Malhi et al. 2006; Saatchi et al. 2007, 2011). The range of estimates arises from different methodological approaches which take into account the landscape heterogeneity in different ways.

Remote sensing is a promising technique that can be used to capture the state of forests with high spatial resolution (e.g., Exbrayat et al. 2019). However, remote sensing cannot measure directly forest biomass and productivity, but it can partly detect the structure of forests. In general, forest structure is related to the spatial distribution of trees and their variability in size (Schall et al. 2018b). Most of current estimates of forest biomass from remote sensing are based on relationships between forest structure and biomass. Forest height characterizes one aspect of forest structure. For example, forest canopy height, derived from lidar remote sensing, is often used for forest biomass estimations (Asner and Mascaro 2014; Dubayah et al. 2010; Lefsky et al. 1999). However, more complex structural characteristics of forests are often ignored, and considerable uncertainties remain.

Forest structure, however, is an important element in forest ecology as it is linked to many ecological processes (Pretzsch 2009; Shugart et al. 2010; Snyder 2010). It is also used as indicator for biodiversity, as vertically structured forests foster the biodiversity of some taxa (Boncina 2000; Ishii et al. 2004; Schall et al. 2018a). Further, horizontal and vertical structural heterogeneities enhance resistance of forest ecosystems against disturbances (Dobbertin 2002; Pretzsch et al. 2016). Some studies also explored the effects of forest structure on forest productivity (Bohn and Huth 2017; Dănescu et al. 2016; Hardiman et al. 2011; Liang et al. 2016; Schall et al. 2018b) and found that variables characterizing forest structure are the main drivers of forest productivity instead of biodiversity variables.

Although forest structure plays an important role in understanding forest dynamics, there are no global forest structure maps available. There are few coarse-resolution maps, but these only show components of forest structure (e.g., forest height from MODIS and ICESat, resolution 1 km; Lefsky 2010; Simard et al. 2011). Clearly, an efficient analysis of the multilayered forest structure for larger regions is required. New satellite missions are being launched (e.g., GEDI, BIOMASS and Tandem-L) which will use novel technologies to measure forest structure on a global scale including forest height and vertical heterogeneity.

A systematic framework for ecologically meaningful structural classes has been recently proposed, which suggests two main elements: the horizontal and the vertical forest structures (Bohn and Huth 2017; Cazcarra-Bes et al. 2017; Fischer et al. 2019; Tello et al. 2014, 2018). However, there are several ways to describe horizontal and vertical forest structures from field data (del Río et al. 2016; Pommerening 2002; Reineke 1933; Shugart 2003; Zenner and Hibbs 2000). It will be difficult to find a suitable definition of forest structure for a wide range of applications, spatial scales and forest types. Moreover, forest structure metrics are different depending on whether they are based on field or remote sensing data. Field-based descriptors for forest structure are derived from individual tree size measurements, while remote sensing descriptors are often based on the heterogeneous canopy structure for a given area (Cazcarra-Bes et al. 2017). A hybrid approach for forest structure measured by field data and by remote sensing is the terrestrial laser scanning, which allows very detailed measurements of single trees and forest canopy structure (Disney 2018). However, this technique is limited to forest with a limited extent due to its high degree of detail. Airborne or satellite-based remote sensing data are therefore a suitable choice for capturing forest structure on a larger scale.

In this study, we combined forest inventory data, forest modeling and airborne lidar remote sensing to answer the question, "How can we estimate forest structure from remote sensing, and what is the role of forest structure for forest biomass and aboveground wood productivity estimations?" The aim is to classify forests into structural categories, using horizontal and vertical structural descriptors that can be measured by remote sensing (Fig. 1). Based on this structural classification, we explore whether forest biomass and aboveground wood productivity can be estimated more accurately if structural information is included. We are using an individual-based model to generate hypotheses on the relations among structural variables and ecosystem variables (aboveground wood productivity and biomass). Since the output from individual-based forest models (a tabulation of each tree, its species and its size) resembles standard forest inventories from field data, obtained results can be directly applied to forest inventories.

We used the forest model FORMIND (Bohn and Huth 2017; Fischer et al. 2016) to generate thousands of forest stands. With this large set of data, we examined in detail the role of forest structure for biomass and aboveground wood productivity estimations. Furthermore, in a case study for Germany we will explore how well forest biomass and aboveground wood productivity can be derived from remote sensing data including information

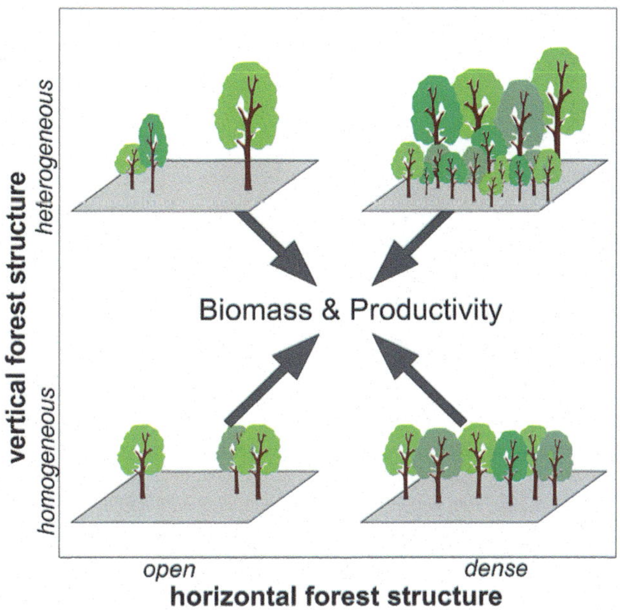

Fig. 1 Biomass and aboveground wood productivity estimations are related to forest structure. In this study, we distinguish between horizontal forest structure, which represents the density of trees in a forest stand, and the vertical forest structure, which quantifies the vertical heterogeneity of tree heights. Forest structure can be estimated by measurements from field data and from remote sensing

on forest structure (Fig. 2). This presents a conceptual approach that can be applied across large areas where lidar data are available.

2 Methods

2.1 Simulating Temperate Forest Stands with a Forest Model

We applied the individual- and process-based forest gap model FORMIND (Fischer et al. 2016) in the "forest factory mode." In contrast to classical long-term simulations, this mode allows the creation of thousands of forest stands based on different stem size distributions and species mixtures. For each forest stand, competitions for light, space and water were calculated depending on the spatial arrangement and the size of the trees (by applying established processes from the FORMIND model). A full forest inventory is then available for each stand, which can include different arrangements of trees leading to different horizontal and vertical structures.

In order to generate such forest stands, trees were randomly selected from the stem size distribution, were given a species identity and were then planted within a stand. Fifteen stem size distributions were applied, which cover a gradient from young to old and disturbed to undisturbed forests (tree height differ between 5 m and 40 m). Species mixtures

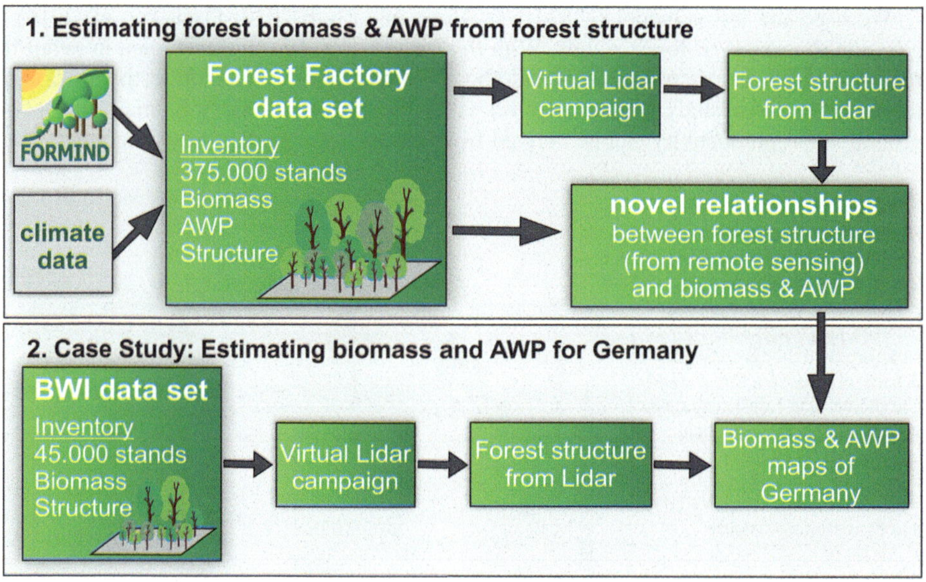

Fig. 2 Concept of this study. (1) Applying the forest model FORMIND, a large data set of virtual forest stands was generated ("forest factory data set"). For each forest stand of this data set, a virtual lidar campaign was simulated. These lidar data were used to estimate forest structure from remote sensing data. Then, relationships were explored to determine aboveground biomass and aboveground woody productivity (AWP) using remote sensing-based forest structure metrics. (2) In a case study for Germany, a lidar campaign was simulated for each forest stand from the national forest inventory data set of Germany (BWI). Forest structure for each stand was estimated from this lidar data. Based on the estimated forest structure and the relationships found in the first part of this study, Germany-wide maps of biomass and AWP were produced

included all combinations of eight common temperate species using the species-specific allometries and model algorithms of the FORMIND model version for temperate forests (Bohn et al. 2014). For each combination of age and species, we assembled 100 simulated forest stands, each stand with a size of 20 m × 20 m. This is done under the conditions that: (1) there is enough space available for the crowns of every tree and (2) every tree in the simulated forest has positive net primary productivity under environmental conditions of the temperate zone (light, temperature, water). The net primary productivity of each forest in this study is calculated using climate date of a site in Germany (climate station of the Hainich, Germany). In total, five different climates over time were applied (years 2000–2004). For more details regarding the forest factory mode of the FORMIND model, see Bohn and Huth (2017).

In this "forest factory mode," we have created a total of 375,000 different virtual forest stands, each characterized by an inventory of all trees. We refer always to this data set if the term "field data" is used within this study. For each stand, the simulated forest structure was described by horizontal and vertical structural characteristics (e.g., basal area, maximal tree height). In addition, aboveground biomass (AGB) and aboveground woody productivity (AWP) of each forest stand were calculated using the FORMIND model. The full data set is available in Electronic supplementary material.

2.2 Remote Sensing Data

For each forest stand, a lidar point cloud was simulated following the approach described by Knapp et al. (2018a). This lidar model uses the positions, heights, crown diameters and crown lengths of all trees to generate a 3D voxel representation of a stand and to simulate lidar measurements. The point clouds obtained for all stands were rasterized to 1 m resolution canopy height models (CHM) by taking the height values of the highest returns in each 1 m pixel, respectively, and setting empty pixels to ground height 0. Data were further aggregated to vertical CHM profiles of 1-m height bins which served as inputs to calculate vertical foliage profiles (VFP, see section "Describing forest structure from field data" of Appendix 1).

2.3 Describing Forest Structure with Field Data and Lidar Remote Sensing

Several structural indices were used to characterize the forest structure. For each 20 m × 20 m forest stand from the forest factory data set, we calculated horizontal and vertical descriptors of the structure from field data and from remote sensing data. Here, we present the most important metrics used in this study. However, other metrics are also possible. In total, 13 metrics were examined to describe forest structure from field data and 25 metrics which estimate forest structure from lidar remote sensing. A full list of all these indices can be found in sections "Describing forest structure from field data" and "Describing forest structure from remote sensing data" of Appendix 1.

2.3.1 Forest Structure Estimation from Field Data

Forest structure can be described by metrics derived from tree-level inventory data—from either real forest inventories or simulated stands. Here, these descriptors of forest structure were inspired by previous studies (Bohn and Huth 2017; Cazcarra-Bes et al.

2017; Fischer et al. 2019). Horizontal structure is described by stand basal area BA (m^2), which is the sum of all tree basal area values of one forest stand:

$$\mathrm{BA} = \sum_{\mathrm{trees}} \frac{\pi}{4} d^2,$$

where d (m) is the stem diameter of a tree. Vertical structure is quantified by the standard deviation of tree heights $\mathrm{SD_{th}}$ (m):

$$\mathrm{SD_{th}} = \sqrt{\frac{1}{n-1} \sum_{\mathrm{trees}} \left(h - \bar{h}\right)^2},$$

where h (m) is the height of a tree and $\bar{h}$ (m) the mean tree height of a stand.

2.3.2 Forest Structure Estimation from Remote Sensing

Horizontal structure for each forest stand from the lidar data set is described by the mean top-of-canopy height (m), which is the mean of the canopy height model (CHM) with 1 m resolution:

$$\mathrm{TCH} = \frac{\sum_{i=1}^{i\mathrm{max}} P_{\mathrm{CHM},i}}{i_{\mathrm{max}}},$$

where $P_{\mathrm{CHM},i}$ is the height of the CHM in pixel i and i_{max} is the total number of pixels. Vertical structure is quantified by the standard deviation of the vertical foliage profile ($\mathrm{SD_{VFP}}$ (m)):

$$\mathrm{SD_{VFP}} = \sqrt{\frac{\sum_{i=1}^{i\mathrm{max}} \left(p_i \cdot \left(h_i - \frac{\sum_{i=1}^{i\mathrm{max}} (p_i \cdot h_i)}{\sum_{i=1}^{i\mathrm{max}} p_i} \right)^2 \right)}{\frac{N-1}{N} \cdot \sum_{i=1}^{i\mathrm{max}} p_i}},$$

where p_i is the foliage profile value (leaf area per m^2) in height bin h_i, i_{max} is the number of height bins and N is the count of all p_i that are not zero. The process of creating the required vertical foliage profile (VFP) from a discrete lidar point cloud is described in section "Calculating the vertical foliage profile from a CHM" of Appendix 1.

2.4 Estimating Forest Biomass and Productivity from Forest Structure

For estimating forest biomass and aboveground wood productivity, we follow a two-step approach. First, horizontal (i.e., TCH) and vertical (i.e., $\mathrm{SD_{VFP}}$) forest structure was estimated based on remote sensing measurements. Second, the derived information on forest structure was used to estimate forest biomass and aboveground wood productivity. We compare three different approaches which differ in the degree of complexity:

(a) H_{struct}: We estimate the aboveground biomass (AGB) and aboveground woody productivity (AWP) only with the information from the horizontal forest structure (here TCH). For this, we used the following power-law approach:

$$\text{AGB}_H = a \cdot \text{TCH}^b, \text{AWP}_H = m \cdot \text{TCH}^n.$$

(b) V_{struct}: We estimate the biomass and productivity only from the vertical forest structure (here SD_{VFP}).

$$\text{AGB}_V = a \cdot \text{SD}_{\text{VFP}}^b, \text{AWP}_H = m \cdot \text{SD}_{\text{VFP}}^p.$$

(c) $H_{\text{struct}} + V_{\text{struct}}$: We estimate the biomass and productivity from the horizontal and vertical forest structures.

$$\text{AGB}_{H+V} = a \cdot \text{TCH}^b \cdot \text{SD}_{\text{VFP}}^p, \text{AWP}_{H+V} = m \cdot \text{TCH}^n \cdot \text{SD}_{\text{VFP}}^p.$$

For all three variants, we applied the artificial forest stand data set described in Sect. 2.1. For each simulated forest stand, a virtual 3D airborne lidar scan was available (Sect. 2.2). Using these lidar data, for each stand all structural metrics (e.g., TCH, SD_{VFP}) were calculated (Sect. 2.3). Beside these metrics, also AGB and AWP are known for all stands. With this information, we fitted the unknown parameters of equations (a)–(c) by linear model fit using the double logarithmic transformation.

2.5 Preparation of Nationwide Biomass and Productivity Maps with the German National Forest Inventory Data Set

The German national forest inventory data (BWI) record forest stands on a sample basis according to a standardized procedure throughout Germany (Thuenen-Institut 2015). It consists of more than 45,000 field plots and is repeated every 10 years. The field plot varies in size (circa 20 m × 20 m) due to the angle count sampling method. For every plot of the BWI data set, the stem diameter (DBH) and species identification of all trees measured were available. Tree height and aboveground biomass were calculated using the same species-specific allometries as applied to the forest factory data set (Sect. 2.1). The BWI plots are distributed on a regular grid across Germany. Plot density varies between regions but there is at least one plot located within each 4 km × 4 km grid cell. For each forest stand of the BWI, all structural and stand features were determined as described above (cf. section "Describing forest structure from field data" of Appendix 1), a virtual lidar campaign was carried out and the remote sensing-based metrics were then calculated (cf. section "Describing forest structure from remote sensing data" of Appendix 1).

Using the information on forest structure from simulated lidar remote sensing for each BWI plot and the equations from Sect. 2.4 (parameters are derived from the forest factory data set), we derived forest biomass and aboveground wood productivity maps for Germany. As the exact positions of the BWI plots are unknown, we used their approximate positions (on the regular grid, i.e., with a precision of at least 4 km) to geolocate the estimated forest attributes for each plot. We used Voronoi tessellation to interpolate between BWI plots. In this procedure, each forested point in Germany got assigned to the closest BWI plot nearby. Gridded maps of the different forest attributes were produced by rasterizing the tessels with 250 m pixel resolution. The rasters were masked with a forest cover mask based on a forest cover map from Hansen et al. (2013; 50% minimum tree cover, resampled to 250 m resolution).

3 Results

3.1 Estimating Forest Structure from Lidar Remote Sensing

Analyzing 375,000 forest stands from the forest factory data set, we investigated whether forest structure can be determined using lidar remote sensing. For this purpose, two field-based metrics were analyzed—basal area (BA) as proxy for the horizontal forest structure (i.e., the density of the forest) and tree height heterogeneity (SD_{th}) as proxy for the vertical heterogeneity of a forest stand. In total, we tested 325 correlations between structural descriptors estimated from field data and remote sensing data. Top-of-canopy height (TCH) as a horizontal descriptor estimated from remote sensing has the highest correlation with the horizontal metric basal area estimated from field data ($r^2=0.86$, see Fig. 3a). The quadratic top-of-canopy height ($r^2=0.76$, see Fig. 4) and the fractional canopy cover with a threshold of 10 m ($r^2=0.81$) show similar results and are also suitable to estimate the horizontal structure.

For the vertical forest structure, the standard deviation of the vertical foliage profile (SD_{VFP}) shows the strongest correlation with the field-based metric SD_{th} ($r^2=0.75$, see Fig. 3b). The Shannon index of the vertical foliage profile also reveals a relevant correlation ($r^2=0.53$, see Fig. 4), but the analysis did not identify any further remote sensing metric that could be used to describe the vertical structure. A full list of all analyzed correlations between field-based structural metrics and remote sensing-based metrics is shown in Fig. 4. For the analyses that follow, the remote sensing-based metrics with the highest correlation were used, i.e., TCH as proxy for the horizontal structure and SD_{VFP} as proxy for the vertical structure.

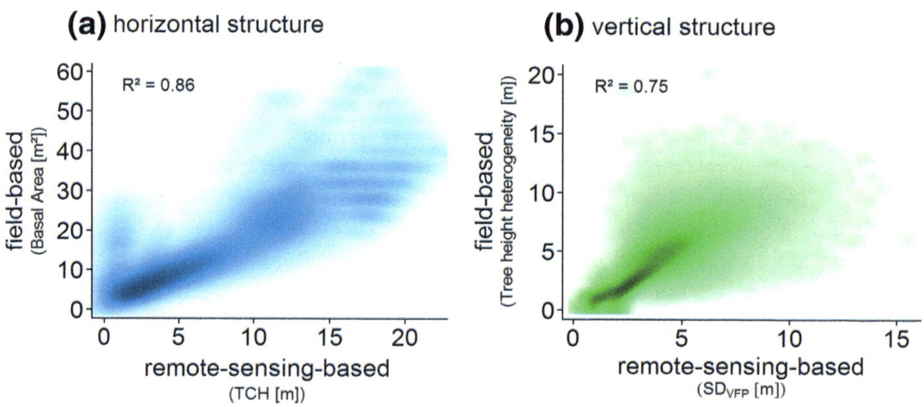

Fig. 3 Remote sensing-based estimation of the forest structure based on the analysis of the forest factory data set. The figure shows the estimate of **a** the horizontal forest structure (basal area) using top-of-canopy height (TCH) derived from lidar and **b** the estimate of the vertical forest structure (tree height heterogeneity) using the standard deviation of the vertical foliage profile (SD_{VFP}) derived also from lidar. For this study, all forest inventories from the forest factory data set (in total 375,000 temperate forest stands with a size of 20 m × 20 m) were analyzed

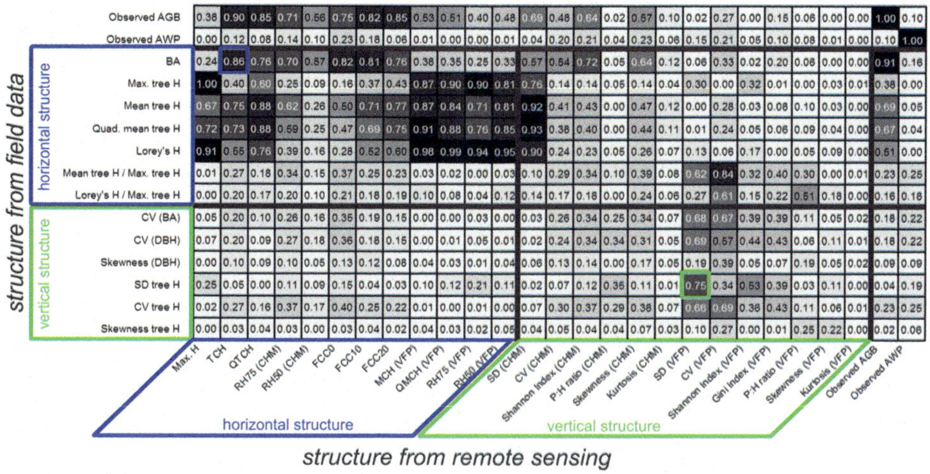

Fig. 4 Correlations (coefficient of determination r^2) between field-based structural metrics and remote sensing-based metrics. Numbers and gray scale indicate the coefficient of determination. For each of the 375 correlations, in total 375,000 forest stands from the forest factory data set (Sect. 2.1) were analyzed. For every virtual forest stand, a forest inventory was available allowing the calculation of the aboveground biomass (AGB) and several structural indices (e.g., maximum tree height, standard deviation of tree height). The productivity AWP for each stand was determined by applying the FORMIND model. All structural indices from remote sensing are derived from the virtual lidar campaign for each forest stand in the forest factory data set. Highlighted are the two correlations from Fig. 3 for the horizontal structure (blue) and vertical structure (green). All structural metrics are explained in section "Describing forest structure from field data" of Appendix 1 (field-based) and "Describing forest structure from remote sensing data" of Appendix 1 (remote sensing-based)

3.2 Analyzing the Role of Remote Sensing-Based Forest Structure for Biomass and Productivity Estimations

All 375,000 temperate forest stands from the forest factory data set were analyzed to investigate the role of forest structure for the determination of forest biomass and aboveground wood productivity. For this purpose, the forest stands were grouped into sixteen forest structure classes with equal spacing, consisting of four horizontal (TCH 0–6 m, 6–12, 12–18, > 18) and four vertical (SD_{VFP} 0–2 m, 2–4, 4–6, > 6) structure classes. It turns out that forest structure is a key factor in estimating forest biomass and aboveground wood productivity. There is an increase in forest biomass with increasing TCH—which is a proxy for the horizontal structure (Fig. 5a). However, biomass in forests with an open horizontal structure (TCH < 12 m) is much more influenced by vertical structure than in closed forests (TCH > 12 m).

The relation between aboveground wood productivity and forest structure is more complex. For stands with a homogeneous vertical structure (SD_{VFP} between 0 and 2 m), the productivity increases with the density of the forest (high horizontal structure; Fig. 5b). The highest productivity is achieved for these single-layer forest stands (SD_{VFP} < 2 m) and a medium dense horizontal structure (TCH around 15 m). However, as the vertical structure increases to more heterogeneous stands (SD_{VFP} > 2 m), the positive effect of the horizontal forest structure on productivity is reduced. For forest stands with heterogeneous vertical structure (SD_{VFP} > 4 m), productivity increases only slightly with increasing density

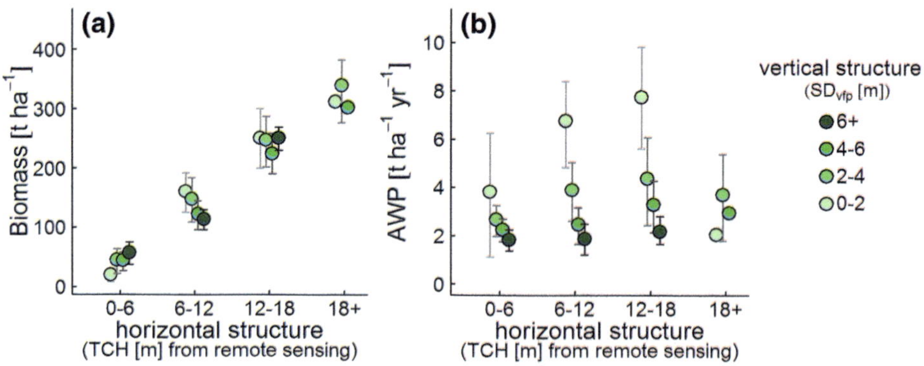

Fig. 5 Role of forest structure for biomass and aboveground wood productivity (AWP) estimations. For all 375,000 forest stands, forest structure was estimated from remote sensing. As horizontal forest structure descriptor, the top-of-canopy height (TCH) was used, and as vertical structure descriptor, the standard deviation of the vertical foliage profile (SD_{VFP}) was used. All forest stands were grouped in 16 structure classes (four horizontal classes and four vertical classes). Shown are **a** the observed mean aboveground biomass and **b** the observed mean aboveground wood productivity (AWP) in relation to the forest structure classes. Biomass and AWP values are taken directly from the forest factory data set. Error bars indicate the standard deviation

of the horizontal structure and can even decline for forests with dense horizontal structure (TCH > 18 m). Forests with a heterogeneous vertical structure represent multilayered forests where large trees shade smaller trees, which then reduce the aboveground wood productivity of smaller trees. To summarize, following the analysis of 375,000 forest stands from the forest factory data set TCH (as a proxy for the horizontal forest structure) plays an important role in forest state monitoring (like forest biomass). However, both horizontal and vertical forest structures are relevant for aboveground wood productivity estimates (Fig. 5b; Table 1).

To quantify the relation between forest structure and biomass/productivity, three statistical approaches between these variables were investigated (cf., methods 2.4). On the one hand, these approaches were created with only one information about forest structure and on the other hand with information on horizontal and vertical forest structures (Table 1).

Table 1 Aboveground biomass (AGB) and aboveground wood productivity (AWP) determined using structural metrics from 3D remote sensing (TCH as horizontal descriptor, SD_{VFP} as vertical descriptor)

Approach	Forest biomass AGB (t/ha)	Forest productivity AWP [t/(ha year)]
H_{struct}	$AGB_H = 9.49 \cdot TCH^{1.22}$ $r^2 = 0.90$, rmse = 28 (30%)	$AWP_H = 1.68 \cdot TCH^{0.31}$ $r^2 = 0.14$, rmse = 2.1 (75%)
V_{struct}	$AGB_V = 34.77 \cdot SD_{VFP}^{0.48}$ $r^2 = 0.01$, rmse = 98 (158%)	$AWP_V = 4.03 \cdot SD_{VFP}^{-0.34}$ $r^2 = 0.09$, rmse = 2.1 (75%)
$H_{struct} + V_{struct}$	$AGB_{H+V} = 7.55 \cdot TCH^{1.20} \cdot SD_{VFP}^{0.23}$ $r^2 = 0.90$, rmse = 29 (31%)	$AWP_{H+V} = 2.55 \cdot TCH^{0.34} \cdot SD_{VFP}^{-0.39}$ $r^2 = 0.31$, rmse = 1.8 (64%)

For the calibration of the approaches, all forest stands (20 m × 20 m) from the forest factory data set were used. R-squared values indicate the correlation between observed value from the forest factory data set and the estimated value using the corresponding approach. For all approaches we got a p value ≤ 0.01. Given are the root-mean-square error (RMSE) and the normalized RMSE (nRMSE; normalized by the mean value of the observation). The detailed scatterplots can be found in Appendix 1 (Figs. 9 and 10)

The best results for biomass estimation are obtained with the H_{struct} and $H_{struct}+V_{struct}$ approach ($r^2=0.90$, nRMSE = 30%; Table 1 and Fig. 9), while only the $H_{struct}+V_{struct}$ approach is somehow suitable for aboveground wood productivity estimation ($r^2=0.31$, nRMSE = 64%; Table 1 and Fig. 10). The error for the productivity estimate for forest stands with a size of 20 m × 20 m can be improved from 75 to 64% by taking both structural dimensions into account (Fig. 6; Table 1).

3.3 Case Study: Estimating Forest Biomass and Productivity at Country Level

We estimate forest structure for whole of Germany based on 45,000 plots from the German national forest inventory data (BWI data set) and relate this information to forest biomass and aboveground wood productivity. For this country-wide analysis, we used lidar data derived from a virtual lidar campaign over all BWI forest stands (for more details see methods 2.5). With this lidar data set, we created two maps for estimated horizontal and vertical structures of forests in Germany (Fig. 7). According to this analysis, 19% of all forest stands in Germany have a dense structure (horizontal structure with TCH > 18 m), which corresponds to high and closed forests. Only 10% of all forests have an open structure (horizontal structure with TCH < 6 m), which means that these stands are either low in height or have a low tree density. The mean value of TCH—the proxy for horizontal structure—for Germany is 12.4 m. The amount of forest area with heterogeneous vertical structure ($SD_{VFP}>6$ m) is with 6% low compared to the amount of forest areas with simple vertical structure (20%, $SD_{VFP}<2$ m). Most stands in Germany represent a medium value of heterogeneity in vertical structure (60%, SD_{VFP} between 2 m and 4 m) which corresponds to forests with a low number of layers. For whole Germany, the mean value of SD_{VFP}—the proxy for the vertical forest structure—is 3.1 m. Overall, according to this study, we have large areas with dense forests in Germany, but these forests are often less vertically structured with only one to two layers.

The densest forests can be found in the northern part and in the southern part of Germany. Within Germany, there is a trend for the horizontal structure to become increasingly dense from north to south (Fig. 16a). Looking at the east–west gradient of the

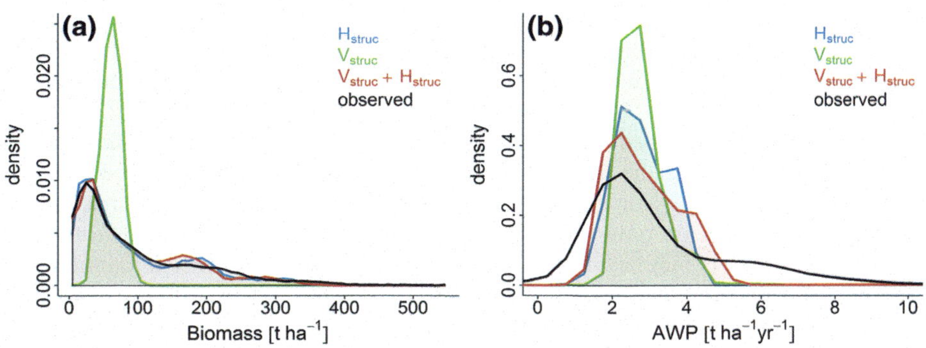

Fig. 6 Histogram for estimated forest biomass (**a**) and aboveground woody productivity (AWP, **b**) for all forest stands from the forest factory data set. Each of this forest attributes was estimated with the three approaches: H_{struc}—horizontal structure, V_{struc}—vertical structure and $H_{struc}+V_{struc}$—horizontal and vertical structures. The histograms of the estimated values were compared with the observed values from all forest stands (black line). The detailed relationships between estimated and observed biomass/productivity are presented in Figs. 9 and 10

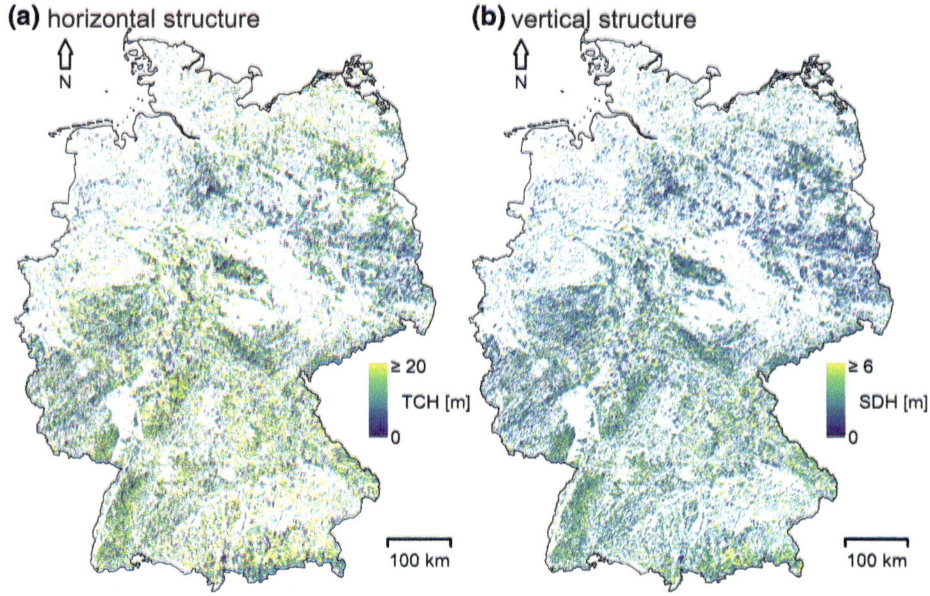

Fig. 7 Horizontal and vertical forest structures for Germany estimated from 3D remote sensing. As horizontal index for forest structure, we used top-of-canopy height (as proxy for the density of a forest stand), and as index for the vertical heterogeneity, the standard deviation of the vertical foliage profile was used. Low values of vertical structure stand for a homogenous structure, higher values for a heterogeneous structure

horizontal structure, however, one finds the densest stands in the middle region; at the outer edges, the horizontal structure is rather open (Fig. 16c). A similar pattern emerges for the vertical structural index with the most heterogeneous stands in the southern part of Germany (Fig. 16b). Regarding the relationship between horizontal forest structure and topographic gradients, no clear trend was observable (Fig. 16e). What can be detected, however, is that the variability of open and dense forests stands increases significantly for altitudes higher than 1000 m (Fig. 16e). The same applies to the vertical structure (Fig. 16f). In addition, a clear trend can be identified with vertically homogeneous stands at lower altitudes and vertically heterogeneous stands at higher altitudes (Fig. 16f).

Three regions in Germany show a high structural heterogeneity: the Black Forest, the Bavarian Forest and the mid-Alpine foothills. Looking at the described structural characteristics especially for forest areas in national parks and for forest areas outside national parks, there are only minor differences (Fig. 17). The horizontal structure of these forest stands is similar (TCH national park: 11.7 m vs. TCH non-national park: 12.4 m) and the vertical structure as well (SD_{VFP} national park: 3.3 m vs. SD_{VFP} non-national park: 3.1 m).

Using the estimated horizontal and vertical forest structures of each stand in Germany (cf., Fig. 7), it is possible now to estimate forest biomass and aboveground wood productivity. Applying the best approach $H_{struct} + V_{struct}$ (as shown in Table 1), we produced a forest biomass and aboveground wood productivity map for Germany (Fig. 8). In southern Germany we found higher biomass values compared to the north. The total

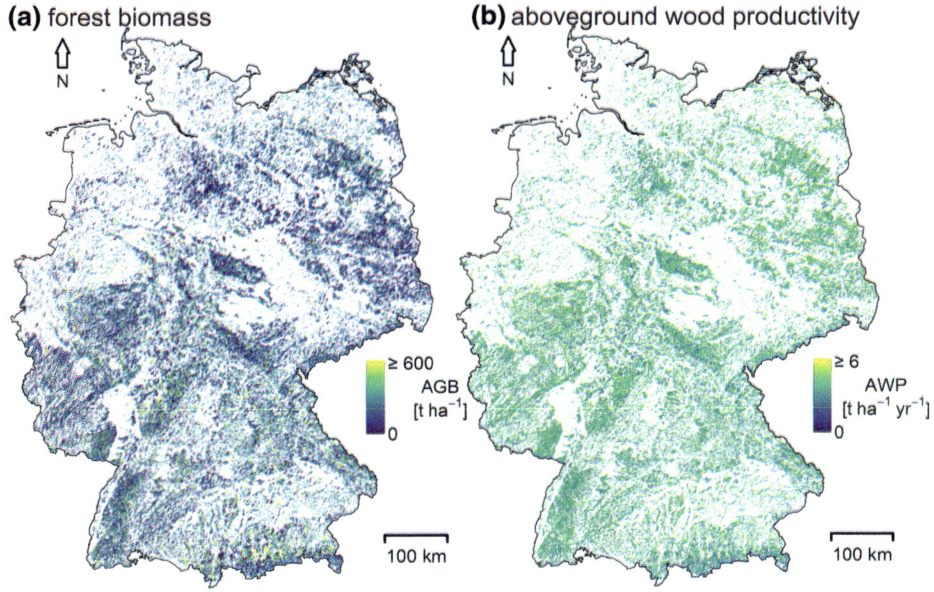

Fig. 8 Estimation of **a** forest biomass and **b** aboveground wood productivity for Germany by applying the $H_{struct}+V_{struct}$ approach. The $H_{struct}+V_{struct}$ approach was calibrated using all forest stands from the forest factory data set (cf., Table 1). Structural information for each forest stand was taken from Fig. 7

biomass of all forested areas in Germany is 2.3 Gt, and the mean biomass of forested areas is 209 t/ha. We estimated a total aboveground wood productivity of 43 Mt/year with a mean value of 4 t/ha/year.

We obtained a good agreement when comparing the estimated biomass ($H_{struct}+V_{struct}$ approach) with the observed biomass from the BWI plots ($r^2=0.76$, RMSE = 65 t/ha; cf. Fig. 14). Using only horizontal structural information produces also good results for the biomass estimation ($r^2=0.76$, RMSE = 66 t/ha, cf. Fig. 14). However, if only vertical structure is used as input, the results are poor ($r^2=0.14$, RMSE = 125 t/ha, cf. Fig. 14). This shows that for the estimation of the biomass, the information about the horizontal structure is sufficient.

4 Discussion

Forest structure is usually quantified using tree sizes from inventory measurements. In this study, we estimated forest structure from remote sensing (lidar) and provide a workflow to generate forest structure maps for Germany.

4.1 Estimating Horizontal and Vertical Forest Structures

For each structural dimension, a metric was found which could be used to determine horizontal and vertical forest structures. It was harder to find a metric for the vertical structure. For the measurement of vertical forest structure by remote sensing, standard deviation of the vertical foliage profile is very well suited to reflect the heterogeneity in the vertical leaf distribution. For the estimation of the horizontal structure, top-of-canopy height (TCH) was selected, which is a height measurement and at first glance not a structural measure. However, TCH is a quite general metric that covers also information on the forest structure. On the one hand, it is affected by the height and shape of the largest trees in the upper canopy and is therefore closely linked to the basal area of these canopy trees. On the other hand, TCH also provides information about the horizontal vegetation density and the openness of a forest stand, if, for example, canopy gaps occur (Lu et al. 2016). For these reasons, TCH is suited to describe the density of a forest and thus its horizontal structure. However, horizontal structure can be also described by other approaches, which more closely include the positions of all trees. For example, the concepts of point pattern analysis could be used for this purpose (Wiegand et al. 2013). This would require identifying all individual trees from remote sensing, which has so far only been tested for small areas (e.g., Ferraz et al. 2016).

4.2 The Role of Forest Structure for Biomass and Productivity

For the estimation of forest biomass and productivity, the concept of forest structure classification in combination with forest modeling and remote sensing has a high potential for applications on larger scales. For 375,000 forest stands, we have investigated the relations between structure, biomass and aboveground wood productivity in forest ecosystems. Forest structure has been shown to be an important factor for estimating biomass and productivity from remote sensing. In particular, horizontal forest structure seems to be a good predictor for forest biomass, while the vertical forest structure showed only weaker relationships with forest biomass. Hence, metrics describing the horizontal structure of forests might be a good choice for forest biomass estimations which is in accordance with other studies (Asner and Mascaro 2014; Knapp et al. 2018a). Metrics describing the vertical structural (e.g., the standard deviation of the vertical foliage profile) are useful for forest productivity estimations. Vertical foliage profile showed a much higher correlation with the ground-based standard deviation of tree heights than the standard deviation of the classical canopy height model profile. With VFP, the weight of larger and smaller canopy trees is the same, while classical lidar profiles are dominated by upper canopy trees. Small canopy trees may only provide a minor contribution to stand biomass, but can play an important role for stand productivity—which emphasizes the role of vertical structure for productivity estimations (Bohn and Huth 2017). In a study by Stark et al. (2012), the Shannon index of the lidar profile was successfully related to forest dynamics like productivity and mortality rates. In our study, this Shannon index performed also very well as vertical metric, as it measures the heterogeneity of leaf densities in different layers.

Our aim was to estimate forest structure from a single remote sensing measurement and use this information for improved biomass and productivity estimations. An alternative approach for productivity estimations would be to analyze the change in forest structure and relate this to biomass change or productivity. For example, with lidar or radar it is possible to detect larger changes in forest height over a certain time period. However, often an exact detection of the height change via remote sensing is not possible, because particularly old-growth forests rarely change their canopy height and the detection error of the sensor is often larger than the actual change in height (e.g., Knapp et al. 2018b). In addition, data are often only available for a single remote sensing campaign (e.g., country-wide airborne lidar campaigns). Therefore, there is a strong need for the estimation of forest productivity from forest structure.

We assume a simplified Central European climate and sites with homogenous soil conditions, without any spatial heterogeneities. This has the advantage of eliminating the influence of changing climate and soil properties on biomass and productivity estimations. It allowed to study the fundamental role of forest structure on biomass and productivity. Future studies will investigate the influence of changing environmental factors on these estimations.

In perspective, information about forest structure can be used to distinguish different forest types, for example natural forests, forests that are disturbed by natural hazards or forests that have been managed by different strategies (Dieler et al. 2017; Müller et al. 2000; Peck et al. 2014; Young et al. 2017). It is also worthwhile to consider forest structure in the context of biodiversity research (Jetz et al. 2016; Pereira et al. 2013; Pettorelli et al. 2016). Forest structure can be a valuable indicator of biodiversity, since habitat structure and habitat heterogeneity can be correlated with animal and plant species diversity (e.g., Boncina 2000; Ishii et al. 2004; Schall et al. 2018a; Tews et al. 2004). All of this emphasizes the need for large-scale forest structure estimations from remote sensing.

4.3 Linking Forest Models with Remote Sensing

In this study, the role of forest structure for biomass and productivity estimations was investigated by analyzing a large data set generated from a forest model. The synergy of forest models with remote sensing is a promising method. Structural realism in a model is a requirement for applying such a concept, which was realized here by the individual-based approach of the FORMIND model (Fischer et al. 2016; Shugart et al. 2015). In particular, forest gap models are helpful tools to understand forest responses to climate change, modified disturbance regimes and structural changes (Shugart et al. 2018). A few studies have tried to link remote sensing and forest modeling, covering model parameterization and initialization (Falkowski et al. 2010; Hurtt et al. 2004; Ranson et al. 2001), exploring several remote sensing metrics for ecosystem service estimations (Knapp et al. 2018a; Köhler and Huth 2010; Palace et al. 2015) and for error quantification (Frazer et al. 2011; Hurtt et al. 2010).

To sum up, forest models can help to find reliable estimates of forest biomass and productivity as presented in this study. Gap models are particularly valuable because they take into account not only large-scale disturbances but also small-scale variations of forest structures (due to gap-building processes). Linking forest models with remote sensing can help to extrapolate local findings to larger scales and better understand ecosystem patterns and processes (Knapp et al. 2018b; Rödig et al. 2017, 2018).

Here, we used information from airborne lidar remote sensing—however, this analysis is not limited to airborne lidar, but can also be applied to other remote sensing techniques such as lidar and radar satellite missions. Therefore, the forest structure classification presented here can be applied also to other regions. Measurements of the upcoming satellite missions will offer a unique opportunity to reduce the uncertainties in the estimation of aboveground carbon emissions from forests (resulting from photosynthesis, respiration, mortality, human disturbances). These missions will also provide the opportunity to identify changes in forest structure which can be relevant for the estimation of forest productivity. This will improve our understanding of the global carbon cycle, which will be relevant for climate modeling and policy adaptation.

5 Conclusion

In this study, we applied a novel approach of linking remote sensing measurements with dynamic forest models to study the role of forest structure for biomass and productivity. While information on horizontal structure seems to be sufficient for estimating biomass, information on horizontal and vertical structures is required for estimating aboveground wood productivity. In a case study for Germany, maps for forest structure, biomass and aboveground wood productivity were provided. The presented workflow is transferable to other regions as remote sensing data are available. Future satellite missions that measure forest structure (like GEDI from 2018, BIOMASS from 2020 and Tandem-L expected from 2022) will allow the derivation of more accurate estimations of forest biomass, productivity and other forest ecosystem services.

Acknowledgements This study originates from the workshop "Space-based Measurement of Forest Properties for Carbon Cycle Research" at the International Space Science Institute in Bern during November 2017. We thank the Thünen Institute for providing the German national forest inventory data. We also want to thank Hans Pretzsch, Peter Biber and Michael Heym (TUM) for their input on forest structure and structure metrics. Kostas Papathanassiou, Victor Cazcarra-Bes, Matteo Pardini and Marivi Tello Alonso (DLR) gave useful insights into linking forest structure and remote sensing. We also thank the anonymous reviewers for their insightful comments and suggestions. This study was part of the HGF-Helmholtz-Alliance "Remote Sensing and Earth System Dynamics" HA-310 under the funding reference RA37012. NK was funded by the German Federal Ministry for Economic Affairs and Energy (BMWi) under the funding reference 50EE1416. FB was funded by the Deutsche Forschungsgemeinschaft (DFG) within the research unit FOR1246 (Kilimanjaro ecosystems under global change: linking biodiversity, biotic interactions and biogeochemical ecosystem processes). HHS was funded by NASA grants 14-TE14-0085 and 16-ESUSPI-16-0015.

Appendix 1: Estimation of Forest Attributes Using Structural Information

Describing Forest Structure from Field Data

The study examined a total of 13 field-based metrics to describe forest structure, which are listed in the following. Forest structure was described, for example, by basal area BA [m^2], which is the sum of all tree basal area values BA_i of a forest stand:

$$\mathrm{BA} = \sum_i \mathrm{BA}_i = \sum_i \frac{\pi}{4} d_i^2,$$

where d_i (m) is the stem diameter of a tree i (in total n trees in a stand). Alternative metrics to describe the horizontal and vertical structures of a forest stand are:

standard deviation of stem diameters: $\mathrm{SD_{DBH}} = \sqrt{\frac{1}{n-1} \sum_i \left(d_i - \bar{d}\right)^2}$

coefficient of variation of all stem diameters: $\mathrm{CV_{DBH}} = \frac{\mathrm{SD_{DBH}}}{\bar{d_i}}$

skewness of the diameter distribution: $\mathrm{Skew_{DBH}} = \frac{\frac{1}{n} \cdot \sum_{i=1}^{n} \left(d_i - \bar{d}\right)^3}{\left(\frac{1}{n} \cdot \sum_{i=1}^{n} \left(d_i - \bar{d}\right)^2\right)^{\frac{3}{2}}}$

Gini coefficient of the diameter distribution: $\mathrm{Gini_{DBH}} = \frac{2 \sum_i i \cdot d_i}{n \sum_i d_i} - \frac{n+1}{n}$, where d_i is the sorted list of stem diameters.

$\bar{d}$ is the mean diameter of all trees within a stand. The same metrics can be calculated also for the tree height distribution (where H_i (m) is the height of a tree) or basal area distribution. Especially for the tree height distribution, we have calculated further metrics.

maximum height: $H_{\max} = \max(H_i)$

mean height: $H_{\text{mean}} = \frac{1}{n}\sum_i H_i$

quadratic mean height: $H_{\text{quad·mean}} = \sqrt{\frac{1}{n}\sum_i H_i^{2_i}}$

Lorey's height: $H_{\text{Lorey's}} = \frac{\sum_i H_i \cdot \text{BA}_i}{\sum_i \text{BA}_i}$.

Describing Forest Structure from Remote Sensing Data

Estimating forest structure from remote sensing is more challenging as remote sensing data are not tree-based as in the field-based case. This study examined a total of 25 remote sensing-based metrics to describe forest structure. The basis for most metrics is the lidar-derived canopy height model (CHM) with a spatial resolution of 1 m × 1 m. In this study, we described horizontal structure for each 20 m × 20 m forest stand mainly by the mean top-of-canopy height TCH (m), which is the mean of the canopy height model (CHM):

$$\text{TCH} = \frac{\sum_{i=1}^{n} P_{\text{CHM},i}}{n},$$

where $P_{\text{CHM},i}$ is the forest height of the CHM in pixel i and n is the number of pixels. Alternative metrics based on the CHM are:

maximum height: $H_{\max} = \max(P_{\text{CHM},i})$

quadratic TCH: $\text{QTCH} = \sqrt{\frac{\sum_{i=1}^{n} P^2_{\text{CHM},i}}{n}}$

relative height of the CHM: $\text{RH}_q = \text{quantile}_q(P_{\text{CHM},i})$

It is also possible to calculate the standard deviation, the coefficient of variation and the skewness of the CHM (functions are described above in the field-based section). In this study, we considered further advanced metrics based on the CHM:

Shannon index of the CHM: $\text{Shannon}_{\text{CHM}} = -\sum_{i=1}^{i_{\max}} \text{CHM}(h_i) \cdot \ln(\text{CHM}(h_i))$,

with CHM (h_i) being the CHM profile value (pixel count) in bin i. CHM (h_i) has to be > 0, and CHM (h_i) = 0 is ignored,

Kurtosis of the CHM: $\text{Kurtosis}_{\text{CHM}} = n \cdot \frac{\sum_{i=1}^{n}\left(P_{\text{CHM},i} - \overline{P_{\text{CHM}}}\right)^4}{\left(\sum_{i=1}^{n}\left(P_{\text{CHM},i} - \overline{P_{\text{CHM}}}\right)^2\right)^2}$,

with n being the total pixel number, $P_{\text{CHM},i}$ the value of pixel i and $\overline{P_{\text{CHM}}}$ the mean value of the CHM (which is the same as TCH),

the p–h ratio of the CHM: $P : H_{\text{CHM}} = \frac{h\left(\max_{i\in[1,i_{\max}]}(\text{CHM}(h_i))\right)}{\max_{i\in[1,i_{\max}]}(h_i)}$, with CHM ($h_i$) being the pixel count in height bin h_i and i_{max} is the highest height bin.

Another class of metrics calculates the fractional canopy cover above a certain threshold x (m) using the CHM: $FCC_x = \frac{\sum_{h_i=x}^{h\max} CHM(h_i)}{\sum_{h_i=0}^{h\max} CHM(h_i)}$, with CHM ($h_i$) the count of CHM pixels in height bin h_i and x the height threshold to distinguish canopy from gap.

Instead of using the CHM as the basic information for calculating all these lidar metrics, we have used the vertical foliage profile (VFP) for a second class of metrics. All the above-described metrics can be calculated using the VFP. For this reason, the VFP was divided into 1-m height classes. This height classes can now be used in the equations described above by replacing the CHM. The generation of a VFP profile from a CHM is described below.

Calculating the Vertical Foliage Profile from a CHM

The vertical foliage profile (VFP) was reconstructed from the CHM profile at 1 m vertical resolution following the approach described by Harding et al. (2001).

$$\mathrm{VFP}(h_i) = \frac{1}{k * \Delta h} * \ln\left(\frac{P(h_i)}{P(h_{i+1})}\right),$$

with k being the light extinction coefficient, Δh the width of one height bin and $P(h_i)$ the value of the cumulative CHM profile in height bin h_i. The method reconstructs the vertical leaf profile by giving more weight to lower parts of the profile. All pixels below 5 m height were regarded as ground and the light extinction coefficient was set to 0.3 which has been shown to result in good LAI estimations (Getzin et al. 2017).

Estimation of Forest Biomass and Productivity Using Forest Structure

See Figs. 9 and 10.

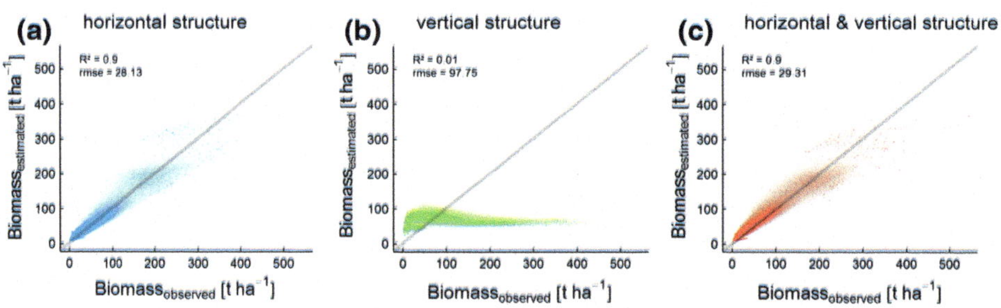

Fig. 9 Relationship between observed biomass and estimated biomass derived by three different approaches (see Table 1). Each point represents one of 375,000 forest stands from the forest factory data set. The observed biomass have been derived by summing up the biomass values of all trees in the 20 m×20 m stand. The estimated biomass was determined using the structural information for each forest stand. **a** Estimation of biomass using only information from the horizontal structural index TCH ($AGB = 9.49 * TCH^{1.22}$, $r^2 = 0.90$), **b** using the vertical structural index SD_{VFP} ($AGB = 34.77 * SD_{VFP}^{0.48}$, $r^2 = 0.01$) and **c** using the vertical and horizontal structural index ($AGB = 7.55 * TCH^{1.20} * SD_{VFP}^{0.23}$, $r^2 = 0.90$). A comparison of the estimated biomass values for the different approaches is shown in Fig. 6a

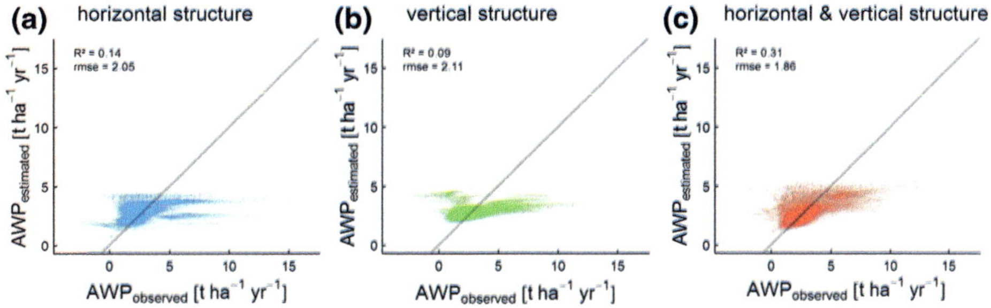

Fig. 10 Relationship between observed and estimated aboveground woody productivity (AWP) for 375,000 forest stands (forest factory data set). Each dot represents one forest stand. **a** Estimation of productivity using only the horizontal structural index TCH ($AWP = 1.68 * TCH^{0.31}$, $r^2 = 0.14$), **b** only the vertical structural index SD_{VFP} ($AWP = 4.03 * SD_{VFP}^{-0.34}$, $r^2 = 0.09$) and **c** using the vertical and horizontal structural index ($AWP = 2.55 * TCH^{0.34} * SD_{VFP}^{-0.39}$, $r^2 = 0.31$). A comparison of the estimated productivity values with the different approaches is shown in Fig. 6b

Appendix 2: Analysis of the German Forest Inventory Data Set

All analyses so far referred to the forest factory data set. This Appendix reproduces all analyses with the empirical BWI data set. For each forest stand of the BWI data set, a virtual lidar campaign was carried out and the remote sensing-based metrics were then calculated.

See Figs. 11, 12, 13, 14, 15, 16 and 17.

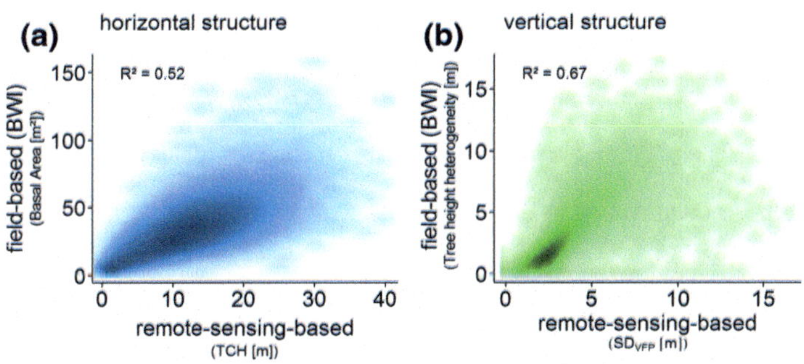

Fig. 11 Remote sensing-based estimation of the forest structure using the BWI data set. Each dot represents one stand of the BWI. The figure shows the estimate of **a** the horizontal forest structure (basal area) from lidar using top-of-canopy height and **b** the vertical forest structure (tree height heterogeneity) from lidar using the standard deviation of the vertical foliage profile

	Max. H	TCH	QTCH	RH75 (CHM)	RH50 (CHM)	FCC0	FCC10	FCC20	MCH (VFP)	QMCH (VFP)	RH75 (VFP)	RH50 (VFP)	SD (CHM)	CV (CHM)	Shannon Index (CHM)	P:H ratio (CHM)	Skewness (CHM)	Kurtosis (CHM)	SD (VFP)	CV (VFP)	Shannon Index (VFP)	Gini Index (VFP)	P:H ratio (VFP)	Skewness (VFP)	Kurtosis (VFP)	Observed AGB
Observed AGB	0.36	0.76	0.66	0.51	0.63	0.60	0.59	0.57	0.33	0.33	0.33	0.33	0.22	0.43	0.36	0.11	0.59	0.00	0.12	0.00	0.18	0.02	0.00	0.06	0.02	1.00
BA	0.14	0.52	0.39	0.30	0.49	0.69	0.60	0.29	0.10	0.10	0.11	0.10	0.06	0.51	0.55	0.13	0.64	0.00	0.07	0.01	0.27	0.13	0.00	0.02	0.01	0.74
Max. tree H	0.97	0.64	0.82	0.72	0.37	0.18	0.33	0.67	0.92	0.94	0.96	0.94	0.87	0.15	0.07	0.08	0.17	0.00	0.33	0.00	0.17	0.03	0.00	0.09	0.03	0.38
Mean tree H	0.69	0.45	0.60	0.52	0.26	0.07	0.20	0.52	0.86	0.84	0.72	0.76	0.70	0.07	0.00	0.01	0.10	0.00	0.01	0.19	0.01	0.20	0.07	0.07	0.02	0.24
Quad. mean tree H	0.74	0.48	0.64	0.56	0.28	0.08	0.22	0.55	0.90	0.88	0.77	0.80	0.74	0.07	0.00	0.00	0.11	0.00	0.02	0.16	0.02	0.19	0.07	0.09	0.02	0.26
Lorey's H	0.90	0.60	0.78	0.68	0.35	0.12	0.28	0.66	0.99	0.99	0.94	0.95	0.87	0.11	0.02	0.03	0.15	0.00	0.15	0.04	0.07	0.11	0.04	0.14	0.04	0.34
Mean tree H / Max. tree H	0.02	0.01	0.01	0.01	0.01	0.07	0.03	0.00	0.01	0.00	0.01	0.00	0.00	0.05	0.18	0.40	0.02	0.01	0.57	0.74	0.33	0.35	0.23	0.00	0.01	0.02
Lorey's H / Max. tree H	0.00	0.00	0.00	0.00	0.00	0.06	0.02	0.01	0.04	0.03	0.00	0.01	0.00	0.06	0.22	0.15	0.02	0.02	0.34	0.63	0.35	0.42	0.42	0.09	0.00	0.01
CV (BA)	0.04	0.04	0.04	0.04	0.03	0.10	0.06	0.02	0.00	0.00	0.03	0.01	0.01	0.08	0.17	0.31	0.05	0.01	0.47	0.51	0.26	0.22	0.13	0.00	0.00	0.05
CV (DBH)	0.05	0.08	0.07	0.07	0.06	0.15	0.10	0.04	0.00	0.01	0.04	0.02	0.02	0.12	0.23	0.43	0.09	0.01	0.48	0.44	0.33	0.27	0.08	0.02	0.01	0.09
Skewness (DBH)	0.00	0.00	0.00	0.00	0.00	0.01	0.00	0.01	0.03	0.02	0.00	0.01	0.01	0.01	0.04	0.01	0.00	0.00	0.12	0.27	0.07	0.09	0.21	0.09	0.00	0.00
SD tree H	0.18	0.17	0.17	0.17	0.10	0.13	0.12	0.13	0.07	0.09	0.16	0.14	0.12	0.10	0.17	0.61	0.08	0.00	0.67	0.37	0.33	0.17	0.02	0.15	0.07	0.13
CV tree H	0.03	0.02	0.02	0.03	0.01	0.05	0.03	0.01	0.00	0.00	0.02	0.01	0.01	0.04	0.12	0.57	0.02	0.00	0.57	0.58	0.24	0.22	0.07	0.09	0.04	0.03
Skewness tree H	0.00	0.02	0.02	0.01	0.03	0.01	0.02	0.02	0.03	0.02	0.00	0.01	0.00	0.01	0.00	0.00	0.02	0.00	0.10	0.23	0.01	0.01	0.20	0.15	0.01	0.02

Fig. 12 Overview of all correlations between field-based structural metrics and remote sensing-based metrics based only on the BWI data set. Numbers and gray scale indicate the coefficient of determination. All structural metrics are explained in Appendix 1

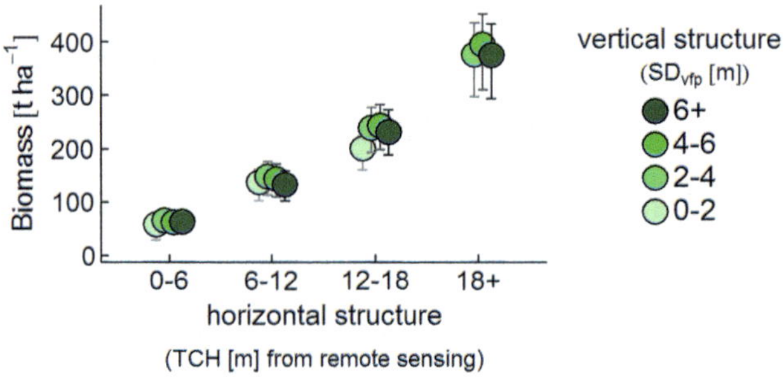

Fig. 13 Role of forest structure for biomass, derived from the BWI data set (more than 45,000 field plots, 20 m × 20 m). Forest structure is estimated from remote sensing. As horizontal forest structure descriptor the top-of-canopy height (TCH) was used and as vertical structure descriptor the standard deviation of the vertical foliage profile (SD_{VFP}). Shown is the mean aboveground biomass in relation to the forest structure classes. Error bars indicate the standard deviation

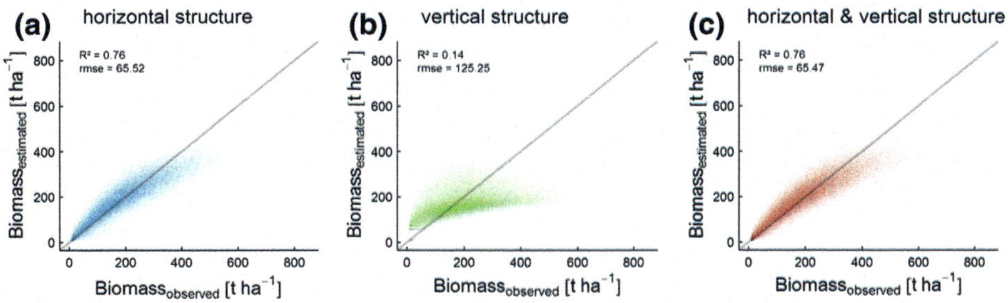

Fig. 14 Relationship between observed biomass and estimated biomass. Each point represents one forest stand from the forest inventory data set BWI. The observed biomass has been taken from the BWI data set. The estimated biomass values were determined using different approaches (cf. Table 1) and information on forest structure. **a** Estimation of biomass using only the horizontal structural index TCH, **b** only the vertical structural index SD_{VFP} and **c** using the vertical and horizontal structural index. A comparison of the estimated values with the different approaches is shown in Fig. 15

Fig. 15 Histogram for forest biomass estimates for Germany based on the BWI data set. The biomass was estimated using three different approaches: *H*—horizontal structure (blue), *V*—vertical structure (green) and *H* + *V*—horizontal and vertical structures (red). The histogram was compared with the measured values from the BWI data set (black line)

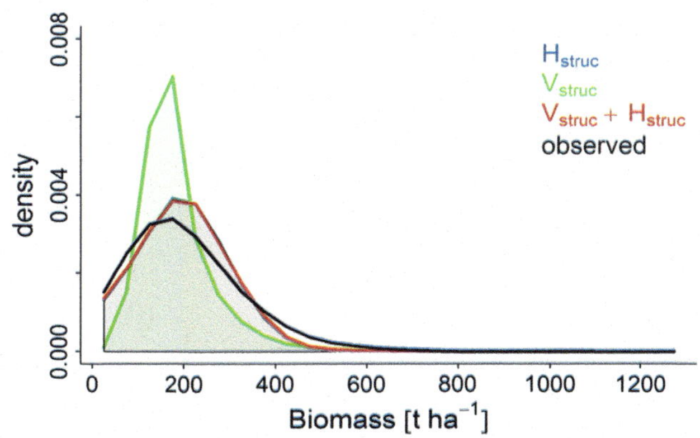

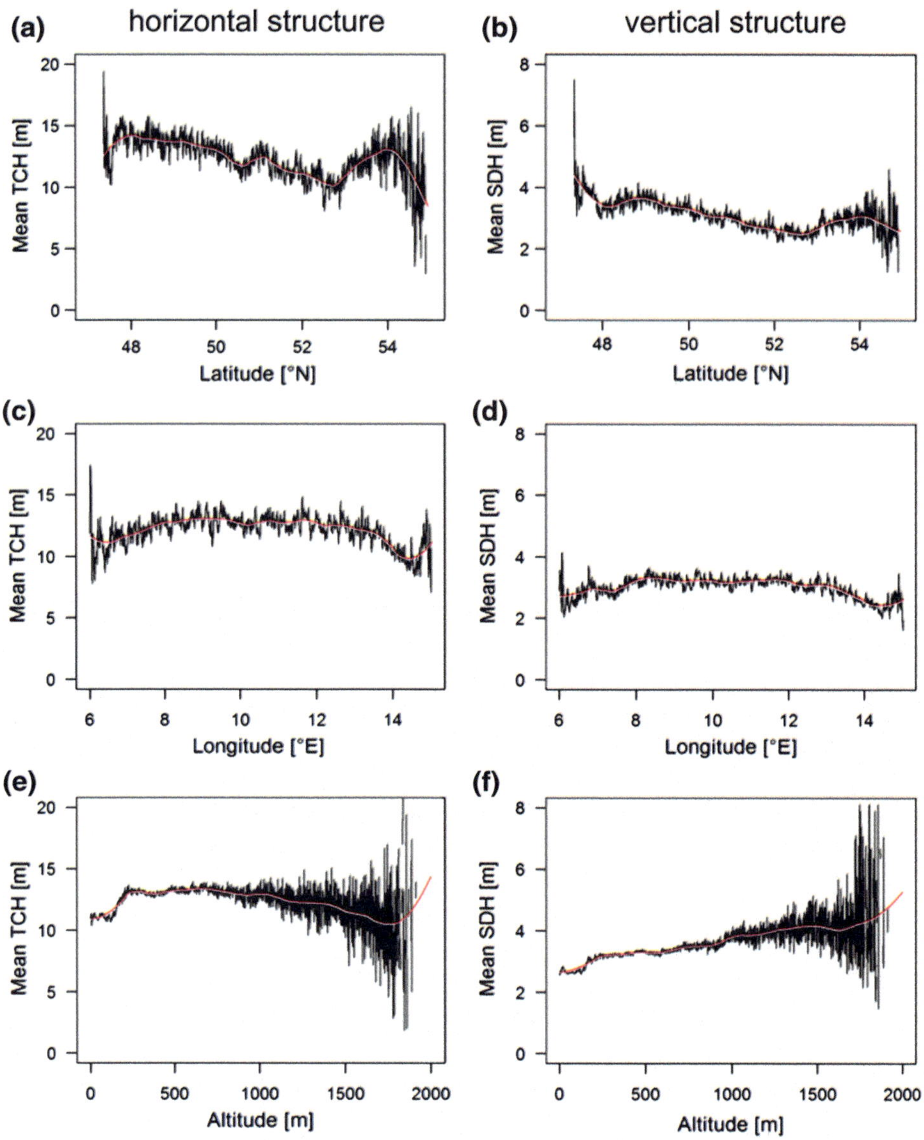

Fig. 16 Forest structure of Germany over different gradients. Mean value of the horizontal structure (TCH) from **a** south to north, **c** west to east in Germany and **e** over the altitudinal gradient. Mean value of the vertical structure (SD_{VFP}) from **a** south to north, **c** west to east in Germany and **e** over the altitudinal gradient. The structure values correspond to Fig. 7 (forest structure maps of Germany)

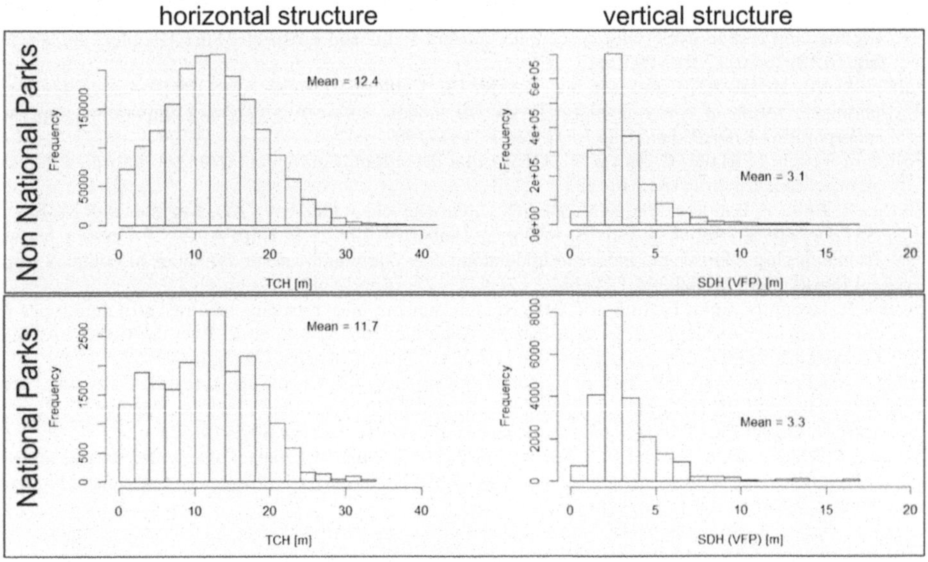

Fig. 17 Histograms of structural metrics for forests outside of national parks and inside national parks. As horizontal forest structure descriptor, the top-of-canopy height (TCH, left) was used, and as vertical structure descriptor, the standard deviation of the vertical foliage profile (SD_{VFP}, right) was used

References

Asner GP, Mascaro J (2014) Mapping tropical forest carbon: calibrating plot estimates to a simple LiDAR metric. Remote Sens Environ 140:614–624

Bohn FJ, Huth A (2017) The importance of forest structure to biodiversity–productivity relationships. R Soc Open Sci 4:160521

Bohn FJ, Frank K, Huth A (2014) Of climate and its resulting tree growth: simulating the productivity of temperate forests. Ecol Model 278:9–17

Bonan GB (2008) Forests and climate change: forcings, feedbacks, and the climate benefits of forests. Science 320:1444–1449

Boncina A (2000) Comparison of structure and biodiversity in the Rajhenav virgin forest remnant and managed forest in the Dinaric region of Slovenia. Glob Ecol Biogeogr 9:201–211

Cazcarra-Bes V, Tello-Alonso M, Fischer R, Heym M, Papathanassiou K (2017) Monitoring of forest structure dynamics by means of L-band SAR tomography. Remote Sens 9:1229

Dănescu A, Albrecht AT, Bauhus J (2016) Structural diversity promotes productivity of mixed, uneven-aged forests in southwestern Germany. Oecologia 182:319–333

del Río M, Pretzsch H, Alberdi I, Bielak K, Bravo F, Brunner A, Condés S, Ducey MJ, Fonseca T, von Lüpke N, Pach M, Peric S, Perot T, Souidi Z, Spathelf P, Sterba H, Tijardovic M, Tomé M, Vallet P, Bravo-Oviedo A (2016) Characterization of the structure, dynamics, and productivity of mixed-species stands: review and perspectives. Eur J For Res 135:23–49

Dieler J, Uhl E, Biber P, Müller J, Rötzer T, Pretzsch H (2017) Effect of forest stand management on species composition, structural diversity, and productivity in the temperate zone of Europe. Eur J For Res 136:739–766

Disney M (2018) Terrestrial Li DAR: a three-dimensional revolution in how we look at trees. New Phytol. https://doi.org/10.1111/nph.15517

Dobbertin M (2002) Influence of stand structure and site factors on wind damage—comparing the storms Vivian and Lothar. For Snow Landsc Res 77:187–205

Dubayah RO, Sheldon SL, Clark DB, Hofton MA, Blair JB, Hurtt GC, Chazdon RL (2010) Estimation of tropical forest height and biomass dynamics using lidar remote sensing at La Selva, Costa Rica. J Geophys Res Biogeosci 115:G00E09

Exbrayat J-F, Bloom AA, Carvalhais N, Fischer R, Huth A, MacBean N, Williams M (2019) Understanding the land carbon cycle with space data: current status and prospects. Surv Geophys. https://doi.org/10.1007/s10712-019-09506-2
Falkowski MJ, Hudak AT, Crookston NL, Gessler PE, Uebler EH, Smith AMS (2010) Landscape-scale parameterization of a tree-level forest growth model: a k-nearest neighbor imputation approach incorporating LiDAR data. Can J For Res 40:184–199
Ferraz A, Saatchi S, Mallet C, Meyer V (2016) Lidar detection of individual tree size in tropical forests. Remote Sens Environ 183:318–333
Fischer R, Bohn F, Dantas de Paula M, Dislich C, Groeneveld J, Gutiérrez AG, Kazmierczak M, Knapp N, Lehmann S, Paulick S, Pütz S, Rödig E, Taubert F, Köhler P, Huth A (2016) Lessons learned from applying a forest gap model to understand ecosystem and carbon dynamics of complex tropical forests. Ecol Model 326:124–133
Fischer R, Knapp N, Bohn F, Huth A (2019) Remote sensing measurements of forest structure types for ecosystem service mapping. In: Schröter M, Bonn A, Klotz S, Seppelt R, Baessler C (eds) Atlas of ecosystem services: drivers, risks, and societal responses. Springer, Cham, pp 63–67
Foley JA, DeFries R, Asner GP, Barford C, Bonan G, Carpenter SR, Chapin FS, Coe MT, Daily GC, Gibbs HK, Helkowski JH, Holloway T, Howard EA, Kucharik CJ, Monfreda C, Patz JA, Prentice IC, Ramankutty N, Snyder PK (2005) Global consequences of land use. Science 309:570–574
Frazer GW, Magnussen S, Wulder MA, Niemann KO (2011) Simulated impact of sample plot size and co-registration error on the accuracy and uncertainty of LiDAR-derived estimates of forest stand biomass. Remote Sens Environ 115:636–649
Getzin S, Fischer R, Knapp N, Huth A (2017) Using airborne LiDAR to assess spatial heterogeneity in forest structure on Mount Kilimanjaro. Landsc Ecol 32:1881–1894
Grace J, Mitchard E, Gloor E (2014) Perturbations in the carbon budget of the tropics. Glob Change Biol 20:3238–3255
Hansen MC, Potapov PV, Moore R, Hancher M, Turubanova SA, Tyukavina A, Thau D, Stehman SV, Goetz SJ, Loveland TR, Kommareddy A, Egorov A, Chini L, Justice CO, Townshend JRG (2013) High-resolution global maps of 21st-century forest cover change. Science 342:850–853
Hardiman BS, Bohrer G, Gough CM, Vogel CS, Curtis PS (2011) The role of canopy structural complexity in wood net primary production of a maturing northern deciduous forest. Ecology 92:1818–1827
Harding DJ, Lefsky MA, Parker GG, Blair JB (2001) Laser altimeter canopy height profiles: methods and validation for closed-canopy, broadleaf forests. Remote Sens Environ 76:283–297
Houghton RA, Lawrence KT, Hackler JL, Brown S (2001) The spatial distribution of forest biomass in the Brazilian Amazon: a comparison of estimates. Glob Change Biol 7:731–746
Hurtt GC, Dubayah R, Drake J, Moorcroft PR, Pacala SW, Blair JB, Fearon MG (2004) Beyond potential vegetation: combining lidar data and a height-structured model for carbon studies. Ecol Appl 14:873–883
Hurtt GC, Fisk J, Thomas RQ, Dubayah R, Moorcroft PR, Shugart HH (2010) Linking models and data on vegetation structure. J Geophys Res Biogeosci. https://doi.org/10.1029/2009JG000937
Ishii HT, Tanabe S, Hiura T (2004) Exploring the relationships among canopy structure, stand productivity, and biodiversity of temperate forest ecosystems. For Sci 50:342–355
Jetz W, Cavender-Bares J, Pavlick R, Schimel D, Davis FW, Asner GP, Guralnick R, Kattge J, Latimer AM, Moorcroft P, Schaepman ME, Schildhauer MP, Schneider FD, Schrodt F, Stahl U, Ustin SL (2016) Monitoring plant functional diversity from space. Nat Plants 2:16024
Knapp N, Fischer R, Huth A (2018a) Linking lidar and forest modeling to assess biomass estimation across scales and disturbance states. Remote Sens Environ 205:199–209
Knapp N, Huth A, Kugler F, Papathanassiou K, Condit R, Hubbell SP, Fischer R (2018b) Model-assisted estimation of tropical forest biomass change: a comparison of approaches. Remote Sens 10:731
Köhler P, Huth A (2010) Towards ground-truthing of spaceborne estimates of above-ground life biomass and leaf area index in tropical rain forests. Biogeosciences 7:2531–2543
Lefsky MA (2010) A global forest canopy height map from the moderate resolution imaging spectroradiometer and the geoscience laser altimeter system. Geophys Res Lett 37:L15401
Lefsky MA, Harding D, Cohen WB, Parker G, Shugart HH (1999) Surface lidar remote sensing of basal area and biomass in deciduous forests of eastern Maryland, USA. Remote Sens Environ 67:83–98
Liang JJ, Crowther TW, Picard N, Wiser S, Zhou M, Alberti G, Schulze ED, McGuire AD, Bozzato F, Pretzsch H, de-Miguel S, Paquette A, Hérault B, Scherer-Lorenzen M, Barrett CB, Glick HB, Hengeveld GM, Nabuurs GJ, Pfautsch S, Viana H, Vibrans AC, Ammer C, Schall P, Verbyla D, Tchebakova N, Fischer M, Watson JV, Chen HYH, Lei XD, Schelhaas MJ, Lu HC, Gianelle D, Parfenova EI, Salas C, Lee E, Lee B, Kim HS, Bruelheide H, Coomes DA, Piotto D, Sunderland T, Schmid B, Gourlet-Fleury S, Sonké B, Tavani R, Zhu J, Brandl S, Vayreda J, Kitahara F, Searle EB, Neldner VJ, Ngugi MR, Baraloto C, Frizzera L, Balazy R, Oleksyn J, Zawiła-Niedźwiecki T, Bouriaud O, Bussotti F, Finér L, Jaroszewicz

B, Jucker T, Valladares F, Jagodzinski AM, Peri PL, Gonmadje C, Marthy W, O'Brien T, Martin EH, Marshall AR, Rovero F, Bitariho R, Niklaus PA, Alvarez-Loayza P, Chamuya N, Valencia R, Mortier F, Wortel V, Engone-Obiang NL, Ferreira LV, Odeke DE, Vasquez RM, Lewis SL, Reich PB (2016) Positive biodiversity–productivity relationship predominant in global forests. Science 354:196
Lu D, Chen Q, Wang G, Liu L, Li G, Moran E (2016) A survey of remote sensing-based aboveground biomass estimation methods in forest ecosystems. Int J Digit Earth 9:63–105
Malhi Y, Wood D, Baker TR, Wright J, Phillips OL, Cochrane T, Meir P, Chave J, Almeida S, Arroyo L, Higuchi N, Killeen TJ, Laurance SG, Laurance WF, Lewis SL, Monteagudo A, Neill DA, Vargas PN, Pitman NCA, Quesada CA, Salomão R, Silva JNM, Lezama AT, Terborgh J, Martinez RV, Vinceti B (2006) The regional variation of aboveground live biomass in old-growth Amazonian forests. Glob Change Biol 12:1107–1138
Müller S, Ammer C, Nüsslein S (2000) Analyses of stand structure as a tool for silvicultural decisions—a case study in a *Quercus petraea—Sorbus torminalis* stand. Forstwiss Cent 119:32–42
Palace MW, Sullivan FB, Ducey MJ, Treuhaft RN, Herrick C, Shimbo JZ, Mota-E-Silva J (2015) Estimating forest structure in a tropical forest using field measurements, a synthetic model and discrete return lidar data. Remote Sens Environ 161:1–11
Pan YD, Birdsey RA, Fang JY, Houghton R, Kauppi PE, Kurz WA, Phillips OL, Shvidenko A, Lewis SL, Canadell JG, Ciais P, Jackson RB, Pacala SW, McGuire AD, Piao SL, Rautiainen A, Sitch S, Hayes D (2011) A large and persistent carbon sink in the world's forests. Science 333:988–993
Peck JE, Zenner EK, Brang P, Zingg A (2014) Tree size distribution and abundance explain structural complexity differentially within stands of even-aged and uneven-aged structure types. Eur J For Res 133:335–346
Pereira HM, Ferrier S, Walters M, Geller GN, Jongman RHG, Scholes RJ, Bruford MW, Brummitt N, Butchart SHM, Cardoso AC, Coops NC, Dulloo E, Faith DP, Freyhof J, Gregory RD, Heip C, Hoft R, Hurtt G, Jetz W, Karp DS, McGeoch MA, Obura D, Onoda Y, Pettorelli N, Reyers B, Sayre R, Scharlemann JPW, Stuart SN, Turak E, Walpole M, Wegmann M (2013) Essential biodiversity variables. Science 339:277–278
Pettorelli N, Wegmann M, Skidmore A, Mücher S, Dawson TP, Fernandez M, Lucas R, Schaepman ME, Wang T, O'Connor B, Jongman RHG, Kempeneers P, Sonnenschein R, Leidner AK, Böhm M, He KS, Nagendra H, Dubois G, Fatoyinbo T, Hansen MC, Paganini M, de Klerk HM, Asner GP, Kerr JT, Estes AB, Schmeller DS, Heiden U, Rocchini D, Pereira HM, Turak E, Fernandez N, Lausch A, Cho MA, Alcaraz-Segura D, McGeoch MA, Turner W, Mueller A, St-Louis V, Penner J, Vihervaara P, Belward A, Reyers B, Geller GN (2016) Framing the concept of satellite remote sensing essential biodiversity variables: challenges and future directions. Remote Sens Ecol Conserv 2:122–131
Pommerening A (2002) Approaches to quantifying forest structures. Forestry 75:305–324
Pretzsch H (2009) Forest dynamics, growth and yield. Springer, Berlin
Pretzsch H, del Río M, Schütze G, Ammer C, Annighöfer P, Avdagic A, Barbeito I, Bielak K, Brazaitis G, Coll L, Drössler L, Fabrika M, Forrester DI, Kurylyak V, Löf M, Lombardi F, Matović B, Mohren F, Motta R, den Ouden J, Pach M, Ponette Q, Skrzyszewski J, Sramek V, Sterba H, Svoboda M, Verheyen K, Zlatanov T, Bravo-Oviedo A (2016) Mixing of Scots pine (*Pinus sylvestris* L.) and European beech (*Fagus sylvatica* L.) enhances structural heterogeneity, and the effect increases with water availability. For Ecol Manag 373:149–166
Ranson KJ, Sun G, Knox RG, Levine ER, Weishampel JF, Fifer ST (2001) Northern forest ecosystem dynamics using coupled models and remote sensing. Remote Sens Environ 75:291–302
Reineke LH (1933) Perfecting a stand-density index for even-aged forests. J Agric Res 46:627–638
Rödig E, Cuntz M, Heinke J, Rammig A, Huth A (2017) Spatial heterogeneity of biomass and forest structure of the Amazon rain forest: linking remote sensing, forest modelling and field inventory. Glob Ecol Biogeogr 26:1292–1302
Rödig E, Cuntz M, Rammig A, Fischer R, Taubert F, Huth A (2018) The importance of forest structure for carbon fluxes of the Amazon rainforest. Environ Res Lett 13:054013
Saatchi SS, Houghton RA, Alvala RCDS, Soares JV, Yu Y (2007) Distribution of aboveground live biomass in the Amazon basin. Glob Change Biol 13:816–837
Saatchi SS, Harris NL, Brown S, Lefsky M, Mitchard ETA, Salas W, Zutta BR, Buermann W, Lewis SL, Hagen S, Petrova S, White L, Silman M, Morel A (2011) Benchmark map of forest carbon stocks in tropical regions across three continents. Proc Natl Acad Sci USA 108:9899–9904
Schall P, Gossner MM, Heinrichs S, Fischer M, Boch S, Prati D, Jung K, Baumgartner V, Blaser S, Böhm S, Buscot F, Daniel R, Goldmann K, Kaiser K, Kahl T, Lange M, Müller J, Overmann J, Renner SC, Schulze ED, Sikorski J, Tschapka M, Türke M, Weisser WW, Wemheuer B, Wubet T, Ammer C (2018a) The impact of even-aged and uneven-aged forest management on regional biodiversity of multiple taxa in European beech forests. J Appl Ecol 55:267–278
Schall P, Schulze E-D, Fischer M, Ayasse M, Ammer C (2018b) Relations between forest management, stand structure and productivity across different types of Central European forests. Basic Appl Ecol 32:39–52

Shugart HH (2003) A theory of forest dynamics. The Blackburn Press, Caldwell
Shugart HH, Saatchi S, Hall FG (2010) Importance of structure and its measurement in quantifying function of forest ecosystems. J Geophys Res Biogeosci. https://doi.org/10.1029/2009JG000993
Shugart HH, Asner GP, Fischer R, Huth A, Knapp N, Le Toan T, Shuman JK (2015) Computer and remote-sensing infrastructure to enhance large-scale testing of individual-based forest models. Front Ecol Environ 13:503–511
Shugart HH, Wang B, Fischer R, Ma J, Fang J, Yan X, Huth A, Armstrong AH (2018) Gap models and their individual-based relatives in the assessment of the consequences of global change. Environ Res Lett 13:033001
Simard M, Pinto N, Fisher JB, Baccini A (2011) Mapping forest canopy height globally with spaceborne lidar. J Geophys Res Biogeosci 116:G04021
Snyder M (2010) What is forest stand structure and how is it measured? North Woodl 64:15
Stark SC, Leitold V, Wu JL, Hunter MO, de Castilho CV, Costa FRC, McMahon SM, Parker GG, Shimabukuro MT, Lefsky MA, Keller M, Alves LF, Schietti J, Shimabukuro YE, Brandao DO, Woodcock TK, Higuchi N, de Camargo PB, de Oliveira RC, Saleska SR (2012) Amazon forest carbon dynamics predicted by profiles of canopy leaf area and light environment. Ecol Lett 15:1406–1414
Tello M, Pardini M, Papathanassiou K, Fischer R (2014) Towards forest structure characteristics retrieval from SAR tomographic profiles. In: Electronic proceedings EUSAR 2014; 10th European conference on synthetic aperture radar, 03–05 June 2014, Berlin, Germany. VDE Verlag, Berlin, pp 1425–1428
Tello M, Cazcarra-Bes V, Fischer R, Papathanassiou K (2018) Multiscale forest structure estimation from SAR tomography. In: Electronic proceedings EUSAR 2018; 12th European conference on synthetic aperture radar, 04–07 June, 2018, Aachen, Germany. VDE Verlag, Berlin, pp 600–603
Tews J, Brose U, Grimm V, Tielbörger K, Wichmann MC, Schwager M, Jeltsch F (2004) Animal species diversity driven by habitat heterogeneity/diversity: the importance of keystone structures. J Biogeogr 31:79–92
Thuenen-Institut (2015) Dritte Bundeswaldinventur - Basisdaten (Stand 20.03.2015)
Wiegand T, He F, Hubbell SP (2013) A systematic comparison of summary characteristics for quantifying point patterns in ecology. Ecography 36:92–103
Young BD, D'Amato AW, Kern CC, Kastendick DN, Palik BJ (2017) Seven decades of change in forest structure and composition in *Pinus resinosa* forests in northern Minnesota, USA: comparing managed and unmanaged conditions. For Ecol Manag 395:92–103
Zenner EK, Hibbs DE (2000) A new method for modeling the heterogeneity of forest structure. For Ecol Manag 129:75–87

Publisher's Note Springer Nature remains neutral with regard to jurisdictional claims in published maps and institutional affiliations.

Affiliations

Rico Fischer[1] · Nikolai Knapp[1] · Friedrich Bohn[1,2] · Herman H. Shugart[3] · Andreas Huth[1,4,5]

✉ Rico Fischer
rico.fischer@ufz.de

1 Department of Ecological Modeling, Helmholtz Centre for Environmental Research GmbH – UFZ, Permoserstasse 15, 04318 Leipzig, Germany

2 Institute for Meteorology and Climate Research, Atmospheric Environmental Research, Karlsruhe Institute of Technology, Kreuzeckbahnstrasse 19, 82467 Garmisch-Partenkirchen, Germany

3 Department of Environmental Sciences, University of Virginia, Charlottesville, VA 22903, USA

4 German Centre for Integrative Biodiversity Research (iDiv) Halle-Jena-Leipzig, Deutscher Platz 5e, 04103 Leipzig, Germany

5 Institute of Environmental Systems Research, University of Osnabrück, Barbarastr. 12, 49076 Osnabrück, Germany

Surveys in Geophysics (2019) 40:735–755
https://doi.org/10.1007/s10712-019-09506-2

Understanding the Land Carbon Cycle with Space Data: Current Status and Prospects

Jean-François Exbrayat[1] · A. Anthony Bloom[2] · Nuno Carvalhais[3,6] · Rico Fischer[4] · Andreas Huth[4,7,8] · Natasha MacBean[5] · Mathew Williams[1]

Received: 20 July 2018 / Accepted: 7 January 2019 / Published online: 6 February 2019

Abstract
Our understanding of the terrestrial carbon cycle has been greatly enhanced since satellite observations of the land surface started. The advantage of remote sensing is that it provides wall-to-wall observations including in regions where in situ monitoring is challenging. This paper reviews how satellite observations of the biosphere have helped improve our understanding of the terrestrial carbon cycle. First, it details how remotely sensed information of the land surface has provided new means to monitor vegetation dynamics and estimate carbon fluxes and stocks. Second, we present examples of studies which have used satellite products to evaluate and improve simulations from global vegetation models. Third, we focus on model data integration approaches ranging from bottom-up extrapolation of single variables to carbon cycle data assimilation system able to ingest multiple types of observations. Finally, we present an overview of upcoming satellite missions which are likely to further improve our understanding of the terrestrial carbon cycle and its response to climate change and extremes.

Keywords Terrestrial carbon cycle · Earth observation · Satellite · Ecosystem modelling · Model data integration

1 Introduction

Terrestrial ecosystems help offset climate change by absorbing 25–30% of anthropogenic emissions of carbon dioxide (CO_2) each year (Canadell et al. 2007; Le Quéré et al. 2018). The global land carbon sink is calculated as the residual between reported fossil-fuel emissions, measurements of the atmospheric CO_2 growth and constrained estimates of ocean carbon uptake (see, for example, Le Quéré et al. 2018 for more details). The spatio-temporal distribution of the land carbon sink is estimated using an ensemble of process-based terrestrial ecosystem models (TEMs; Sitch et al. 2015). However, these largely unconstrained models exhibit significant differences in the location, magnitude and sign of the land carbon balance. This lack of agreement leads to large uncertainties in Earth system models'

✉ Jean-François Exbrayat
j.exbrayat@ed.ac.uk

Extended author information available on the last page of the article

projections of the response of terrestrial ecosystems to future climate change (Friedlingstein et al. 2006; Arora et al. 2013). This is further complicated by the interplay of vegetation CO_2 uptake and emissions from land use and land cover change (Arneth et al. 2017) such as deforestation (van der Werf et al. 2009).

A key issue for TEMs has been the challenge of integrating global observations to calibrate process parameters. Model spread seems to emerge from the lack of understanding of processes that control carbon allocation, turnover and mortality (Friend et al. 2014). Although networks of eddy covariance towers (e.g., FLUXNET, Baldocchi et al. 2001) provide useful information to improve models (Williams et al. 2009; Kuppel et al. 2014), their distribution is highly skewed towards temperate regions of the Northern Hemisphere which challenges their applicability to other ecosystems. However, in recent years the multiplication of continuous Earth Observation (EO) has allowed the production of observational datasets relevant to the biosphere. While satellites do not measure carbon stocks and fluxes directly, they provide covariates for the extrapolation of in situ data to global gridded products related to ecosystem carbon fluxes (e.g., Jung et al. 2009) and biomass stocks (e.g., Saatchi et al. 2011; Baccini et al. 2012; Avitabile et al. 2016) through machine learning. Integrating EO to constrain process-based models has led to breakthrough in our understanding of the terrestrial carbon cycle, allowing a robust attribution of the increasing atmospheric CO_2 amplitude to the response of high latitude productivity (Forkel et al. 2016).

In this paper, we review studies in which satellite-driven datasets have been used to improve our understanding of the terrestrial carbon cycle. We first focus on studies in which EO products have been used to monitor vegetation dynamics, carbon fluxes and stocks. Then, we illustrate how EO products can be used to evaluate TEMs, understand their biases and improve projections of the carbon cycle. Third, we present model data integration approaches in which EO products are used to constrain TEMs using automated model data fusion approaches. Finally, we review the foreseeable improvements future satellite missions are likely to generate.

2 Earth Observation to Monitor Vegetation Dynamics, Carbon Fluxes and Stocks

Satellite observations of the land surface allow a continuous monitoring of vegetation dynamics through the calculation of spectral indices (Myneni et al. 1995). One of the most common metrics is the Normalized Difference Vegetation Index (NDVI). It is calculated such as $NDVI = (NIR + RED)/(NIR - RED)$ where RED and NIR are the spectral reflectance in the visible red and near-infrared region of the photosynthetically active radiation spectrum, respectively. Because chlorophyll strongly absorbs visible light during photosynthesis, active canopies have higher NDVI values.

NDVI has been used in numerous studies to characterize the response of plant phenology and productivity to climate trends and inter-annual variability. Bimonthly NDVI from the Advanced Very High-Resolution Radiometer (AVHRR), supported by the National Oceanic and Atmospheric Administration (NOAA), spans the period since July 1981 until present. This long-term dataset has allowed describing a lengthening of the growing season in temperate regions of the Northern Hemisphere due to an earlier disappearance of snow in warming conditions (Myneni et al. 1997). Using the third generation of the AVHHR-based Global Inventory Modelling and Mapping System (GIMMS) NDVI dataset (Pinzon and Tucker 2014), Buitenwerf et al. (2015) detected significant changes in phenological cycles for more than half of the land surface between 1981 and 2012. This long-term dataset has also been used to identify dominant climate modes driving the inter-annual variability in the start of growing season across North America (Dannenberg et al. 2018).

Higher-resolution NDVI data from the Moderate Resolution Imaging Spectroradiometer (MODIS) sensor (Huete et al. 2002) have been provided by the National Aeronautics and Space Administration (NASA) since the year 2000. These data have been used to attribute the anomalous greening of the Northern Hemisphere land surface in 2015 to a strong state of the Pacific Decadal Oscillation (Bastos et al. 2017). The MODIS archive also includes an Enhanced Vegetation Index (EVI) which performs well better in high biomass regions (Huete et al. 2002). It has been used to describe leaf growth across the Amazon basin during the dry season, which promotes primary productivity in sunnier conditions (Huete et al. 2006). EVI has also been used to describe the phenological response of Australian ecosystems to climate IAV (Broich et al. 2014), in particular the continental greening that followed the extremely wet 2010/11 La Niña episode (Fasullo et al. 2013).

NDVI and EVI are useful proxies for vegetation activity, but they are not biophysical variables directly relatable to TEMs. However, two key state variables for these models can also be derived from EO: the Fraction of Absorbed Photosynthetically Active Radiation (FAPAR) and Leaf Area Index (LAI), the one-sided area of leaves per units of ground. FAPAR and LAI are related to NDVI (Zhu et al. 2013) but can also be retrieved using physically based (Knyazikhin et al. 1998) or machine learning (Baret et al. 2013) methods.

FAPAR is a key driver for light use efficiency models of primary productivity (e.g., Potter et al. 1993; Field et al. 1995; Prince and Goward 1995; Knorr 2000). Global MODIS-based FAPAR (Myneni et al. 2002) is used to produce high-resolution (~ 1 km) estimates of gross primary productivity (GPP) and net primary productivity (NPP) across the global land surface (Running et al. 2004). This dataset has provided insights in a possible reduction in global net primary productivity because of a drying in the Southern Hemisphere (Zhao and Running 2010). It helped characterized the influence of the El Niño—Southern Oscillation on regional and global NPP (Bastos et al. 2013) and the impact of recent European heat waves on productivity (Bastos et al. 2014).

LAI represents the physiologically active part of the vegetation which interacts with the atmosphere and is a key state variable for land surface and ecosystem models (Kala et al. 2014). Figure 1 presents mean annual LAI for the year 2015 derived from the European Space Agency (ESA)'s Proba-V satellite (Baret et al. 2013) as part of the European Union's Copernicus programme and from NASA's MODIS sensor. There is a good agreement in the spatial distribution of LAI between these datasets, but the MODIS products report higher LAI values for tropical which may lead to non-negligible differences in the calculation of energy, water and carbon fluxes (Kala et al. 2014). MODIS LAI data have been helpful to understand the seasonality of the Amazon (Myneni et al. 2007). Zhu et al. (2013) used the relationship between AVHHR NDVI and MODIS FAPAR and LAI products to create the GIMMS FAPAR3 g and LAI3 g dataset extending back to the 1980s. This long-term dataset exhibits a greening trend, i.e. an increase in LAI during the growing season (Zhu et al. 2016).

A recent advance in remote sensing has been the production of solar-induced fluorescence (SIF) retrievals (Frankenberg et al. 2011). SIF is directly related to plant photosynthetic activity; therefore, SIF data provide a more direct measure of gross C uptake than reflectance-based indicators like NDVI or FAPAR (Porcar-Castell et al. 2014). There has been an increase in the availability of global SIF products derived from space-borne instruments like GOSAT, developed by the Japan Aerospace Exploration Agency (JAXA),

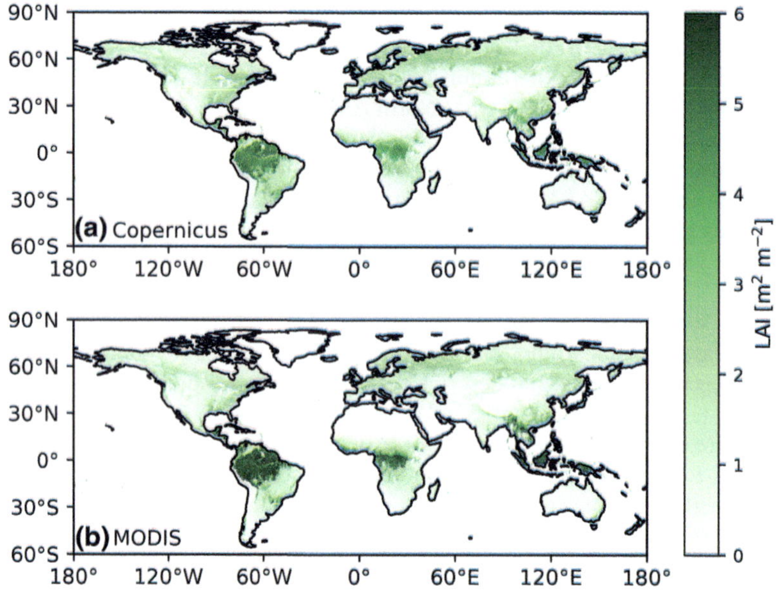

Fig. 1 Mean Leaf Area Index (LAI) in 2015 according to **a** Copernicus and **b** NASA's MODIS. Datasets were resampled at 0.25° spatial resolution for plotting purposes

(Frankenberg et al. 2011; Guanter et al. 2012), GOME-2, designed by the German Aerospace Centre (DLR) (Joiner et al. 2013; Köhler et al. 2015), SCIAMACHY from ESA (Joiner et al. 2016) and OCO-2 from NASA (Sun et al. 2018). More details about SIF are provided in Sect. 3.2.3 of the review by Scholze et al. (2017).

Additionally to plant productivity, satellite datasets have been used for over three decades to monitor fire, a fundamental component of the terrestrial carbon cycle which accounts for a large degree of the inter-annual variability of the terrestrial land sink (van der Werf et al. 2010; Le Quéré et al. 2018). Observations of decadal trends in burned area (Flannigan and Haar 1986; Giglio et al. 2013; Andela et al. 2017) have been used to establish the role of fires as a key component of the long-term C balance evolution (Le Quéré et al. 2018; Arora and Melton 2018, amongst others). A range of satellite-based observations of fire have also radically advanced insight into continental-scale fire characteristics and processes, including understory fires (Morton et al. 2013); fire radiative power (Freeborn et al. 2014); and interactions between fire and species distribution (Rogers et al. 2015). Archibald et al. (2013) identified 5 dominant types of fire regimes using remotely sensed observation of fire frequency, radiative power and size. Active fire detection also offers a crucial constraint on land C fire losses, with current observing system such as VIIRS offering new possibilities to detect boreal fires (Waigl et al. 2017).

Beyond monitoring vegetation dynamics and inferring land–atmosphere fluxes, satellite observations have allowed the creation of high-resolution maps of above-ground biomass (AGB) covering large regions such as the pantropics (e.g., Saatchi et al. 2011; Baccini et al. 2012). Pantropical maps were created using allometric equations (e.g., Chave et al. 2014) relating tree height measured by NASA's Geoscience Laser Altimeter System sensor onboard the Ice, Cloud and land Elevation Satellite (Zwally et al. 2002) to AGB. These wall-to-wall maps allow a first-order approximation of remotely sensed deforestation (e.g.,

Hansen et al. 2013) on AGB stocks in the tropics (Harris et al. 2012; Baccini et al. 2012). Recently, maps by Saatchi et al. (2011) and Baccini et al. (2012) have been fused with additional in situ measurements to create a third map, currently considered as the state of the art (Avitabile et al. 2016).

Thurner et al. (2014) created a map for the northern boreal and temperate forests paper using retrievals of Growing Stock Volume from Envisat Advanced Synthetic Aperture Radar (Santoro et al. 2011, 2015), database information about wood density (Chave et al. 2009) and allometric equations. Thurner et al. (2016) used this map in combination with MODIS NPP to evaluate gradients in turnover dynamics across these regions.

While these previous studies relied on a single AGB map to produce estimates of gross emissions from deforestation, more recent studies have produced annually resolved AGB maps which allow tracking the counteracting impact of regrowth and derive net changes of biomass globally. For example, Liu et al. (2015) used the correlation between Vegetation Optical Depth and AGB from Saatchi et al. (2011) to produce annual biomass maps for 1993–2012 at a 0.25° spatial resolution. They concluded to a loss of global AGB driven by a loss of tropical forests partially compensated by gains over boreal, temperate and savannah regions. More recently, Brandt et al. (2018) used a similar approach to describe a net carbon loss across sub-Saharan Africa for the period 2010–2016. Baccini et al. (2017) also concluded that the tropics are a net source of atmospheric CO_2 based on annual maps they constructed from 2003 to 2014.

EO is useful to identify land cover change (e.g., Hansen et al. 2013) and, by extension, intact forest landscapes (Potapov et al. 2008). Potential AGB maps, representative of the hypothetical undisturbed landscape, can be reconstructed using the relationship between AGB (Saatchi et al. 2011; Baccini et al. 2012; Liu et al. 2015) and climate (New et al. 2002) in these intact regions. This approach attributes ~ 1.5% of the recent increase in atmospheric CO_2 to the Amazonian deforestation (Exbrayat and Williams 2015) while climate change has reduced the capacity of these forests to recover (Exbrayat et al. 2017).

International policy efforts such as the Paris agreement on climate change and the Bonn challenge for forest restoration have raised the interest of countries to produce country-scale maps for monitoring and reporting. For example, Rodríguez-Veiga et al. (2016) used local information from the Mexican forestry. Similarly, Xu et al. (2017) produced a biomass map for the Democratic Republic of Congo using additional data which were not available to pantropical maps. Both studies presented measurable increase in mapping quality and uncertainty quantification.

3 Evaluating Terrestrial Ecosystem Models

Land surface models are key components of Earth system models that simulate energy and mass transfers between the land and the atmosphere; hence, these are key components in the prediction of climate variations from short to long time scales (Pitman 2003). EO of the biosphere provides unprecedented means to evaluate vegetation dynamics, carbon fluxes and biomass stocks simulated by land surface models in a temporally and spatially explicit manner. The evaluation strategies have been largely focusing on aspects related to: the timing of seasonal vegetation development and long-term trends in vegetation greenness; the seasonal and spatial variations in photosynthesis patterns; the spatial variations in plant carbon stocks; and EO-derived estimates of carbon turnover times on land.

The representation of phenology in land surface models is a major source of uncertainty for the calculation of energy, water and carbon fluxes (Kala et al. 2014). Many studies have focused on in situ evaluation of modelled LAI (e.g., Richardson et al. 2012; Migliavacca et al. 2012), but long-term EO-derived products have also been used. Zhu et al. (2013) described a systematic overestimation of LAI by 18 Earth system models compared to the GIMMS LAI3 g dataset. This was accompanied by a shift towards earlier peak in LAI in boreal regions. Similarly, Anav et al. (2013) evaluated models participating in the fifth phase of the Coupled Model Intercomparison Project (CMIP5; Taylor et al. 2012) which underpinned the fifth Assessment Report of the Intergovernmental Panel on Climate Change. They described a tendency for CMIP5 models to overestimate LAI, although most models captured LAI trends. The poor performance of models to represent phenology has led several inter-comparison projects to impose EO-derived LAI dataset to all participating models (Huntzinger et al. 2013; Haarsma et al. 2016).

EO-derived products have also been used to assess vegetation productivity simulated by ecosystem models. Kolby Smith et al. (2016) created a long-term NPP dataset based on MODIS NPP algorithm (Running et al. 2004; Zhao and Running 2010) driven by long-term GIMMS FAPAR3g (Zhu et al. 2013). They compared this new dataset with five CMIP5 models which exhibited a much stronger trend of increasing NPP than the EO-based product. They concluded that models' sensitivity to increasing atmospheric CO_2 was too high, probably owing to the lack of representation of nutrient limitation on productivity. Ito et al. (2017) showed that spatial and seasonal variations in GPP simulated by eight ecosystem models were in agreement with the MODIS GPP product. However, they also showed that models failed to simulate GPP anomalies in response to extreme events such as the 1997–1998 El Niño or the eruption of Mount Pinatubo in 1991. Slevin et al. (2017) identified an underestimation of GPP in the tropics when comparing the Joint UK Land Environment Simulator (JULES; Clark et al. 2011) with EO-derived GPP products.

EO-derived products of GPP and NPP now allow skill-based ensemble averaging studies to be applied to ecosystem models. These post-processing procedures have been used in atmospheric sciences for many decades (e.g., Krishnamurti et al. 1999) and can be applied to ecosystem models in a spatially explicit way. Schwalm et al. (2015) applied the Reliability Ensemble Averaging method (Giorgi and Mearns 2002) constrained by FLUXNET MTE-GPP and biomass estimates to ten ecosystem models participating to the MsTMIP (Huntzinger et al. 2013). Exbrayat et al. (2018a) used a similar method to constrain projections of twenty-first-century change in NPP predicted by 30 simulations from the ISIMIP ensemble (Friend et al. 2014; Warszawski et al. 2014). They showed that the uncertainty in global change in NPP could be reduced by two-thirds using a skill-based ensemble averaging while gaining confidence on the sign of the change for more than 80% of the global land surface.

A recent emphasis has been put on the need to move beyond the separate evaluation of pools and fluxes by terrestrial land models. For many years, global models have been initialized using a spin-up procedure from which biomass stocks would emerge as a result of input fluxes and turnover times at steady state (Exbrayat et al. 2014). However, models perform poorly to simulate vegetation carbon stocks in agreement with observation-based products. For example, Fig. 2 presents a comparison of the recent pantropical biomass map from Avitabile et al. (2016) with models from the Inter-Sectoral Impact Model Intercomparison Project (ISIMIP) ensemble (Friend et al. 2014; Warszawski et al. 2014). There is a large uncertainty represented by the inter-model spread while they tend to overestimate biomass stocks in regions of the Americas and Africa located north of 10°N and south 15°S. Friend et al. (2014) clearly demonstrated that the highest disagreement between models

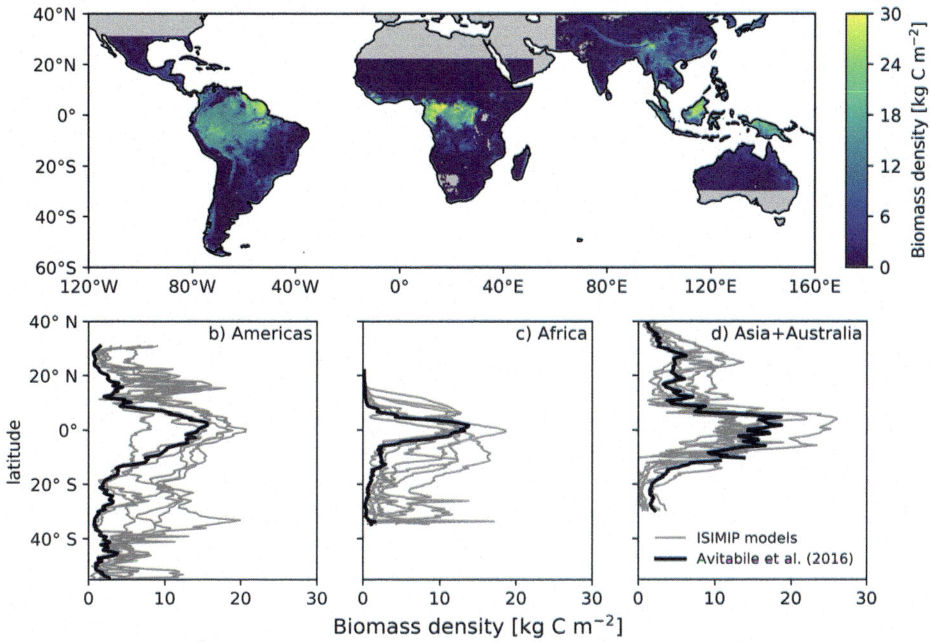

Fig. 2 Biomass density from Avitabile et al. (2016) (top) and comparison of zonal means simulations (bottom, as indicated) from six Inter-Sectoral Impact Model Intercomparison Project (ISIMIP) ecosystem models. Results indicate an overestimation of stocks by models in the Americas and Africa

resides in the internally modelled residence times of carbon which can be inferred from the ratio between observable fluxes and stocks (Friend et al. 2014; Sierra et al. 2017). Under the same future changes in environmental and climate conditions models alternatively predict longer or shorter turnover times of carbon in vegetation. This mismatch reflects a disagreement in the sign of the terrestrial carbon cycle feedback on future changes in climate and atmospheric CO_2. Current EO-based estimates suggest a pervasive control of hydrology on whole ecosystem apparent turnover times of carbon, which are not captured by current Earth system models (Carvalhais et al. 2014). In particular, the spatial patterns of vegetation C turnover times in forests suggest strong climatic controls in mortality patterns associated with drought and heat, but also extreme winter cold temperatures which could expand plant mortality, or reduce it, via reductions in herbivore activity (Thurner et al. 2016). An across-model comparison also revealed that most state-of-the-art global vegetation models do not reflect the direct effects of climate-induced mortality (Thurner et al. 2017), emphasizing the present challenge of understanding mortality induced by climate extreme (Hartmann et al. 2015). Furthermore, recent results have also emphasize the role of land use, in addition to land cover, as a substantial factor for an overall reduction in carbon residence times in terrestrial vegetation (Erb et al. 2016).

In general, all of these works have been emphasizing the mismatch between model and observation-derived ecosystem dynamics and hypothesizing on the missing, or misrepresented, underlying mechanisms that drive carbon dynamics. Downstream, a full body of research has also been focusing on formally integrating these observations into model data assimilation frameworks to maximize information transfer from observation to models.

4 Model Data Integration

While EO can be used to benchmark models, they do not measure all aspects of the terrestrial carbon cycle. Therefore, models are needed to fill gaps, but benchmarking studies reviewed in Sect. 3 have generally pointed to poor performances and systematic biases in forward models. Model data integration aims to synergize data and models through an interactive process. There exist multiple forms of model data integration in which EO has been used to provide global covariates for the extrapolation of in situ data (e.g., Jung et al. 2011), retrieve state variables such as LAI from reflectance through a complex radiative transfer scheme (Lewis et al. 2012), constrain productivity and phenology model parameters in terrestrial Carbon Cycle Data Assimilation Systems (CCDAS; e.g., Knorr et al. 2010) or even provide the initial conditions to detailed forest models (e.g., Rödig et al. 2017, 2018).

The global coverage of EO has allowed the development of a range of "bottom-up" approaches to upscale data-driven in situ models to spatially explicit gridded estimates. One major development in this area has relied on training machine learning algorithm to reproduce local ecosystem fluxes as a function of climate and vegetation properties available from EO (e.g., NDVI, FAPAR). The first example of this approach was reported by Papale and Valentini (2003) who used and Artificial Neural Network trained at 16 European sites to generate continental maps of forest productivity using information about land cover, seasonal temperatures and maximum NDVI from AVHRR. Jung et al. (2009) developed the Model Tree Ensemble (MTE) algorithm to create gridded products of GPP and latent heat fluxes. They first demonstrated the potential for their approach in a synthetic example using LPJmL simulations as training data. Using this method driven by fluxes measured at several FLUXNET sites has allowed the first data-driven description of the distribution of global GPP between biomes and an attribution of dominant regional climate drivers for the period 1998–2005 (Beer et al. 2010). Building on this approach, monthly gridded estimates of GPP extending back to 1982 have been created using GIMMS FAPAR data (Jung et al. 2011). They identified semi-arid and sub-humid as experiencing a high inter-annual variability in productivity due to rainfall variations. These data, used in combination with process-based models, further helped identify the response of savannahs ecosystems to ENSO as the dominant driver of the variability in the land carbon sink (Poulter et al. 2014). Multiple machine learning approaches, relying on various algorithms, have been compared by Tramontana et al. (2016). The evaluation of different approaches was performed as part of the FLUXCOM initiative (http://www.fluxcom.org). The conclusion of this comparison was that machine learning approaches were skilled at reproducing heat and productivity fluxes but may be biased to predict net ecosystem carbon fluxes due to the lack of feedback representation and knowledge of historical disturbance regimes. Nevertheless, bottom-up and top-down approaches estimates of GPP are in good agreement. Figure 3 presents mean annual GPP estimated by FLUXCOM during 2000–2013 and compares it to the MODIS GPP product. Both approaches are partly driven by the same estimates of MODIS FAPAR which yields a high spatial correlation ($r = 0.80$; $p \ll 0.001$). FLUXCOM indicates a mean annual GPP of 124.7 Pg C year^{-1} while MODIS estimates a 136.4 Pg C year^{-1}, a 9% relative difference.

Beyond empirical approaches, the increasing availability of EO has played a key role in the development of more "top-down" terrestrial Carbon Cycle Data Assimilation Systems (CCDAS). Unlike "bottom-up" approaches which consist in extrapolating in situ models, "top-down" CCDAS are centred around using EO, including non-carbon variables

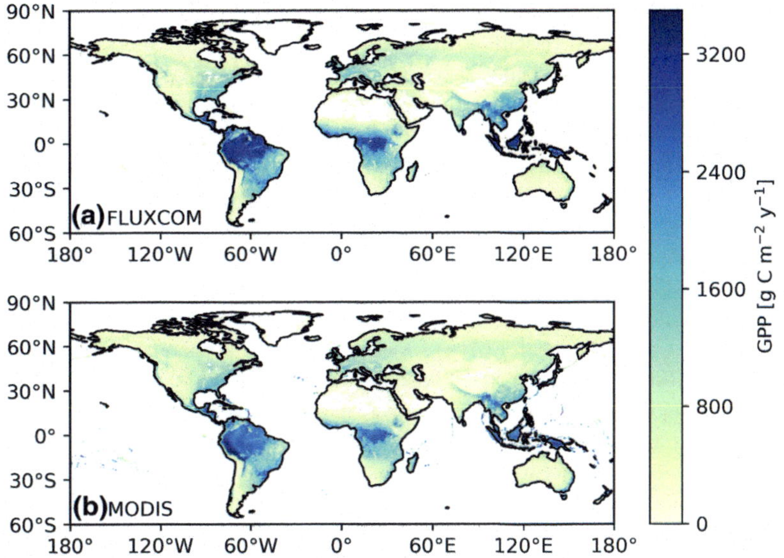

Fig. 3 Mean annual gross primary production (GPP) during 2000–2013 as reported by **a** the FLUXCOM product and **b** MODIS MOD17 GPP product. FLUXCOM corresponds to eddy covariance data upscaled at 0.5° using machine learning. MODIS product is based on a light use efficiency model and was regridded from 30'' (~ 1 km) to 0.5°

(Scholze et al. 2017), to constrain process-based models in a spatially explicit way. One of the first CCDAS was based on the Bethy ecosystem model (Knorr 2000). It has been incrementally improved with additional processes such as dynamic phenology (Knorr et al. 2010). An interesting aspect of CCDAS studies has been to focus on the development and comparison of inversion strategies (e.g., Ziehn et al. 2012) to reduce the computational cost of the assimilation. We refer the reader to a detailed review of the evolution of CCDAS by Kaminski et al. (2013) for more information about this particular framework. One strategy introduced by Peylin et al. (2016) has also been to use a stepwise approach to first constrain parameters related to phenology in the ORCHIDEE model before assimilating fluxes in a subsequent step.

We focus the following paragraphs on examples of new knowledge derived from CCDAS applications and point the readers to the recent review of Scholze et al. (2017) for more technical information about the type of assimilation techniques and EO used. The advantage of EO is that model data integration is performed globally, and CCDAS framework such as CARDAMOM (Bloom et al. 2016) and CASA-GFED (van der Werf et al. 2010) provide compelling methodologies for reconciling land surface and atmospheric constraints on the terrestrial C balance, through which major uncertainties in process representation such as phenology (e.g., Stöckli et al. 2011; Forkel et al. 2014), allocation (Bloom et al. 2016), combustion and emission dynamics (Bloom et al. 2015; Worden et al. 2017a, b) can be explicitly constrained.

Phenology is a poorly represented process, and assimilating reflectance-based EO of NDVI, FAPAR and LAI has allowed in development and validation of new global models. For example, Knorr et al. (2010) assimilated daily FAPAR at seven sites in a generic phenological model. Quaife et al. (2008) demonstrated that assimilating reflectance from the MODIS sensor in the Data Assimilation Linked Ecosystem Carbon model (DALEC;

Williams et al. 2005) led to an improvement in simulated carbon fluxes at a coniferous forest site in Oregon, USA. Stöckli et al. (2011) assimilated 10 years of MODIS LAI and FAPAR data in a phenological model based on the Growing Season Index concept (GSI; Jolly et al. 2005). They identified used the constrained model to produce a 50-year re-analysis of LAI and FAPAR. Forkel et al. (2014) implemented a modified version of the GSI model in the LPJmL dynamic global vegetation model. They retrieved dominant controls of phenology by assimilating 30 years of GIMMS FAPAR, highlighting the codominant role of moisture stress on the variability in phenology (Forkel et al. 2015) which contrasts with classical temperature-based parameterizations. The importance of moisture availability was also found by MacBean et al. (2015) based on the biases in the temperature-driven phenology of the ORCHIDEE model. While these previous studies have relied on plant functional types, Caldararu et al. (2014) successfully fitted a phenological model to pixel-wise MODIS LAI data. Their approach based on carbon optimality concluded that leaf age was also a limiting factor for phenology in evergreen tropical regions.

EO of fire has also been used to constrain emission estimates. Top-down estimates of surface CO emissions amount to a robust constraint on continental-scale fire C emissions: measurements of atmospheric CO—including those from ESA's IASI instrument, NASA's MOPITT, TES and AIRS instruments—have been used to constrain atmospheric chemistry and transport models in data assimilation frameworks (Jiang et al. 2015; Gonzi et al. 2011; Krol et al. 2013; Kopacz et al. 2010, amongst many others). Subsequently, estimates of fire $CO:CO_2$ ratio (Andreae and Merlet 2001; Akagi et al. 2011) have been used to quantify continental-scale fire C fluxes (Gatti et al. 2014; Bowman et al. 2017). However, $CO:CO_2$ has been identified as a potential source of error in extreme fire events (Krol et al. 2013; Bloom et al. 2015), where $CO:CO_2$ values and their uncertainty characteristics are poorly known. Overall, estimates of fire emissions derived from an array of bottom-up and top-down constraints are invaluable for obtaining spatially explicit estimates of fire C fluxes (Bowman et al. 2017; Liu et al. 2017). Ultimately, satellite-based estimation of fire C emissions, characteristics and trends are key to advance process-level understanding of fires as a dynamic component of the Earth system.

Although a number of satellite-derived products (NDVI, LAI, FAPAR, biomass and XCO_2) have been used to constrain both modelled leaf phenology, biomass and net CO_2 fluxes (Kaminski et al. 2013; Forkel et al. 2015; MacBean et al. 2015; Peylin et al. 2016; Bloom et al. 2016), these data only provide indirect information on gross C uptake. For some vegetation types, even ground-based net CO_2 fluxes derived from eddy covariance towers only provide limited capacity in constraining the gross C fluxes (Kuppel et al. 2014). Consequently, SIF products have been used in a variety of ways to assess and improve land surface model (LSM) simulations: (1) to benchmark GPP and SIF temporal dynamics simulated for a range of sites (Lee et al. 2015; Thum et al. 2017); (2) to optimize global-scale GPP estimates from a LSM inter-comparison a posteriori (Parazoo et al. 2014); and to optimize parameters of both fluorescence and photosynthesis models at local to global scales (Zhang et al. 2014; MacBean et al. 2018; Norton et al. 2018; Norton et al., in review). The latter studies have demonstrated considerable potential for SIF to constrain both in situ and global-scale GPP simulations. MacBean et al. (2018) and Norton et al. (under review) show strong reductions in both the spatio-temporal misfit (increased correlation and decreased bias) across vegetation types between modelled and observed GPP and SIF and the simulated global-scale GPP uncertainty. The reduction in GPP uncertainty is a result of constraining both fluorescence, photosynthesis and phenology-related parameters. In many of the above-mentioned modelling studies, an explicit formulation of the relationship between photosynthesis and fluorescence has been developed—largely based on the Soil Canopy Observation Photochemistry and Energy Fluxes (SCOPE) model (van der Tol et al. 2009)—and implemented within each respective LSM (Lee et al. 2015; Thum et al. 2017; Norton et al. 2018; Norton et al., in review). However, SIF has been shown to be linearly correlated with GPP at a range of spatial and temporal scales (Frankenberg et al. 2011; Guanter et al. 2012; Joiner et al. 2014; Yang et al. 2015, 2017; Zhang et al. 2016; Wood et al. 2017). This assumed linear relationship also allows a relatively simple and straightforward means by which modelled GPP and SIF can be compared with, and constrained by, remote sensing SIF estimates at large spatial scales (MacBean et al. 2018).

Continental-scale temporal variability of the terrestrial land sink can be robustly observed through atmospheric CO_2 measurements from satellites—most notably from SCIAMACHY instrument onboard ENVISAT, JAXA's GOSAT and the NASA OCO-2 missions (Buchwitz and Burrows 2004; Yokota et al. 2009; Eldering et al. 2017). Terrestrial CO2 fluxes can be quantitatively retrieved through assimilation of these observations into inverse modelling frameworks (Houweling et al. 2015; Deng et al. 2016), although we note that absolute CO_2 flux estimates are susceptible to a number of model and observation biases (Feng et al. 2016; Miller et al. 2018; Worden et al. 2017b; Basu et al. 2018). Notable insights into terrestrial C cycle processes from satellite-constrained estimates of land C fluxes include multi-year constraints on the Australian C balance (Detmers et al. 2015) regional constraints on the seasonal and inter-annual Amazon C fluxes (Parazoo et al. 2013; Bowman et al. 2017) and Indonesia fire C emissions during the 2015 ENSO (Heymann et al. 2017). Pantropical continental-scale estimates of inter-annual CO_2 flux variations by Liu et al. (2017) demonstrated the synergistic capacity of GOSAT and OCO-2 CO2 measurements—along with ancillary constraints from solar-induced fluorescence—to disentangle the processes regulating the temporal variability of the terrestrial C sink.

While previous studies have focused on fluxes, reconciling stocks is the key as these are responsible for uncertainty in residence times (Carvalhais et al. 2014; Friend et al. 2014). Bloom et al. (2016) have used pantropical biomass estimates from Saatchi et al. (2011) to constrain global retrievals of carbon allocation and residence times in the CARDAMOM framework. Figure 4 shows an updated version of these data based on the assimilation of biomass estimates from Avitabile et al. (2016) and MODIS LAI for the period 2000–2015. These simulations, limited to the area covered by Avitabile et al. (2016),

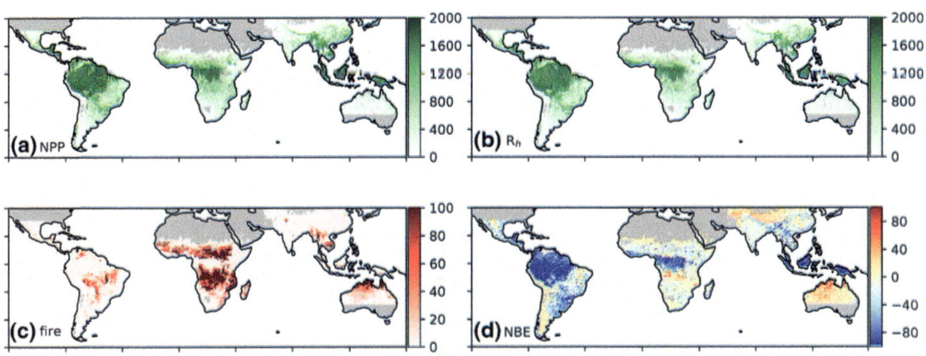

Fig. 4 CARDAMOM (CARbon DAta MOdel fraMework) retrievals of land–atmosphere C fluxes, averaged over 2000–2015. All fluxes are in g C m^{-2}. In d, net biome exchange (NBE) is calculated as $NBE = -NPP + R_h + fire$; hence, < 0 values correspond to a sink (in blue) and > 0 correspond to a source of carbon (in red)

provide EO-constrained estimation of NPP (43.3 Pg C $year^{-1}$), heterotrophic respiraton (Rh; 40.2 Pg C $year^{-1}$), fire emissions (1.2 Pg C $year^{-1}$) and Net Biome Exchange (− 1.7 Pg C $year^{-1}$, corresponding to a sink) as reported in Exbrayat et al. (2018b).

Furthermore, an accurate representation of stocks in ecosystem models is required to robustly estimate emissions related to land use change. Li et al. (2017) used estimates of current biomass compiled by Carvalhais et al. (2014) to constrain centennial emissions from land use change in an ensemble of nine models. While global numbers of cumulative emissions from land use change were similar between the unconstrained and constrained models, regional differences appeared. For example, the data-constrained estimates yielded larger emissions from land use change in the tropics and smaller in temperate regions, compared to the unconstrained estimates. In a more recent study, Lienert and Joos (2018) also used biomass data from Carvalhais et al. (2014) to constrain emissions from land use change using alternative representation of emissions due to net and/or gross land use transitions.

Studies presented in previous paragraphs have focused on using EO data to constrain fluxes and state variables in conceptual models. However, remotely sensed information of vegetation structure can be connected to highly detailed forest models to provide mechanistic estimates of forest biomass and productivity (Shugart et al. 2015; Knapp et al. 2018a, b). Forest structure is indeed an important element to describe the state of forests. Precise estimates of forest structure need to consider small-scale variations resulting from local disturbances, on the one hand, and require large-scale information on the state of the forest that can be detected by remote sensing, on the other hand (Rödig et al. 2017). Local forest models can simulate and analyse different kinds of local disturbances and thus small-scale changes in forest structures more accurately than global ecosystem models. Remote sensing has the potential to provide global high-resolution measurements of the structure of a forest (e.g., forest height by LIDAR or interferometric radar measurements).

As an example, Rödig et al. (2017) used remote sensing data with a resolution of 1 km^2 (i.e. canopy height map derived from ICESat) to establish large-scale applications of a local forest gap model (i.e. FORMIND, Fischer et al. 2016). Forest gap models (Shugart et al. 2018) simulate forest succession at the individual tree level. The advantage of using a local forest model at the large scale is that it brings along information on many different forest attributes (e.g., productivity, carbon sequestration, water fluxes) in a very fine resolution. In combination with remote sensing, this enables the derivation of high-resolution maps of carbon stocks and fluxes—which was conducted for the whole Amazon (Rödig et al. 2017, 2018). By this approach, it was possible to simulate each tree in the Amazon. Finally growth of more than 410 billion trees was analysed. According to this study, forests in the Amazon store high amounts of above-ground biomass (76 Gt of carbon) and are an important sink of 0.56 Gt C a^{-1} under current conditions (Rödig et al. 2018).

5 Outlook

EO has made an essential contribution to our understanding of the terrestrial carbon cycle since the 1980s. It ranges from the continuous monitoring of vegetation activity through NDVI, FAPAR and LAI to providing wall-to-wall constraints for model-based estimates of land–atmosphere carbon fluxes. Nowadays, multiple dedicated missions and services, such as the Sentinel satellites of the European Union's Copernicus programme, provide almost real-time observations with a high level of quality. For example, the ESA TROPOMI instrument of Sentinel-5P provides CO measurements with an unprecedented spatio-temporal coverage (Borsdorff et al. 2018). In the next few years, multiple sensors will be launched to complement the existing constellation of Sentinels and provide coincident observations of several aspects of the biosphere (Stavros et al. 2017).

NASA's Global Ecosystem Dynamics Investigation (GEDI) mission, a LIDAR system on board the International Space Station (ISS), will provide a global coverage of canopy height and foliage vertical profiles. It will provide updated and more detailed structural information for integration with forest models (e.g., Rödig et al. 2017, 2018). ESA's 7th Earth Explorer BIOMASS (Le Toan et al. 2011) will provide repeated measurements of tropical biomass with a P-band synthetic aperture radar. In contrast to currently available biomass maps, temporally resolved information losses from deforestation and gain from regrowth will be especially useful to reduce the uncertainty and correct bias in CCDAS framework (Smallman et al. 2017).

Recent studies have demonstrated the potential for satellite CO_2 observations to constrain land–atmosphere exchange (Liu et al. 2017). NASA's upcoming OCO-3 will replace OCO-2 on board the ISS, while the geostationary GeoCARB will focus on the Americas. Both systems, and the dedicated ESA's Fluorescence Explorer (FLEX) mission, will provide measurements of SIF which has a great potential to constrain models of ecosystem productivity (MacBean et al. 2018), especially following the implementation of mechanistic representation of leaf physiology in CCDAS (Norton et al. 2018).

Finally, the carbon cycle is tightly linked to the energy and water cycles semi-arid areas in particular have been pointed as key ecosystems to understand the global land carbon sink (Poulter et al. 2014). Therefore, the development of new non-carbon EO and their assimilation in CCDAS frameworks plays a major role in simulating the carbon cycle (Scholze et al. 2017). NASA's ECOSTRESS will measure evapotranspiration (Stavros et al. 2017) which will be used to obtain estimates of water-use efficiency, the ratio of productivity to evapotranspiration, which will be useful to drive process-based models of the biosphere.

Overall, the next few years will see an increase in the amount of observing systems with a ever-increasing spatial resolution and higher frequency. One of the key challenges for the modelling community is to build systems able to ingest all this information in an efficient way to provide high confidence retrievals of the terrestrial carbon cycle.

Acknowledgements This review stemmed from the workshop "Space-based Measurement of Forest Properties for Carbon Cycle Research" at the International Space Science Institute in Bern during November 2017. Contribution from AAB was carried out at the Jet Propulsion Laboratory, California Institute of Technology, under a contract with the National Aeronautics and Space Administration. The first and last authors were supported by the Natural Environment Research Council through the National Centre for Earth Observation, contract number PR140015, and the Newton Fund, through CSSP Brazil.

References

Akagi SK, Yokelson RJ, Wiedinmyer C et al (2011) Emission factors for open and domestic biomass burning for use in atmospheric models. Atmos Chem Phys 11:4039–4072. https://doi.org/10.5194/acp-11-4039-2011

Anav A, Friedlingstein P, Kidston M et al (2013) Evaluating the land and ocean components of the global carbon cycle in the CMIP5 earth system models. J Clim 26:6801–6843. https://doi.org/10.1175/JCLI-D-12-00417.1

Andela N, Morton DC, Giglio L et al (2017) A human-driven decline in global burned area. Science 356:1356–1362. https://doi.org/10.1126/science.aal4108

Andreae MO, Merlet P (2001) Emission of trace gases and aerosols from biomass burning. Glob Biogeochem Cycles 15:955–966. https://doi.org/10.1029/2000GB001382
Archibald S, Lehmann CER, Gómez-Dans JL, Bradstock RA (2013) Defining pyromes and global syndromes of fire regimes. Proc Natl Acad Sci USA 110:6442–7. https://doi.org/10.1073/pnas.1211466110
Arneth A, Sitch S, Pongratz J et al (2017) Historical carbon dioxide emissions caused by land-use changes are possibly larger than assumed. Nat Geosci 10:79–84. https://doi.org/10.1038/ngeo2882
Arora VK, Melton JR (2018) Reduction in global area burned and wildfire emissions since 1930s enhances carbon uptake by land. Nat Commun 9:1326. https://doi.org/10.1038/s41467-018-03838-0
Arora VK, Boer GJ, Friedlingstein P et al (2013) Carbon-concentration and carbon-climate feedbacks in CMIP5 earth system models. J Clim 26:5289–5314. https://doi.org/10.1175/JCLI-D-12-00494.1
Avitabile V, Herold M, Heuvelink GBM et al (2016) An integrated pan-tropical biomass map using multiple reference datasets. Glob Change Biol 22:1406–1420. https://doi.org/10.1111/gcb.13139
Baccini A, Goetz SJ, Walker WS et al (2012) Estimated carbon dioxide emissions from tropical deforestation improved by carbon-density maps. Nat Clim Change 2:182–185. https://doi.org/10.1038/nclimate1354
Baccini A, Walker W, Carvalho L et al (2017) Tropical forests are a net carbon source based on aboveground measurements of gain and loss. Science (80) 358:230–234. https://doi.org/10.1126/science.aam5962
Baldocchi D, Falge E, Gu LH et al (2001) FLUXNET: a new tool to study the temporal and spatial variability of ecosystem-scale carbon dioxide, water vapor, and energy flux densities. Bull Am Meteorol Soc 82:2415–2434. https://doi.org/10.1175/1520-0477(2001)082%3c2415:FANTTS%3e2.3.CO;2
Baret F, Weiss M, Lacaze R et al (2013) GEOV1: LAI and FAPAR essential climate variables and FCOVER global time series capitalizing over existing products. Part 1: principles of development and production. Remote Sens Environ 137:299–309. https://doi.org/10.1016/J.RSE.2012.12.027
Bastos A, Running SW, Gouveia C, Trigo RM (2013) The global NPP dependence on ENSO: La Niña and the extraordinary year of 2011. J Geophys Res Biogeosci 118:1247–1255. https://doi.org/10.1002/jgrg.20100
Bastos A, Gouveia CM, Trigo RM, Running SW (2014) Analysing the spatio-temporal impacts of the 2003 and 2010 extreme heatwaves on plant productivity in Europe. Biogeosciences 11:3421–3435. https://doi.org/10.5194/bg-11-3421-2014
Bastos A, Ciais P, Park T et al (2017) Was the extreme Northern Hemisphere greening in 2015 predictable? Environ Res Lett 12:44016. https://doi.org/10.1088/1748-9326/aa67b5
Basu S, Baker DF, Chevallier F et al (2018) The impact of transport model differences on CO_2 surface flux estimates from OCO-2 retrievals of column average CO_2. Atmos Chem Phys 18:7189–7215. https://doi.org/10.5194/acp-18-7189-2018
Beer C, Reichstein M, Tomelleri E et al (2010) Terrestrial gross carbon dioxide uptake: global distribution and covariation with climate. Science 329:834–838. https://doi.org/10.1126/science.1184984
Bloom AA, Worden J, Jiang Z et al (2015) Remote-sensing constraints on South America fire traits by Bayesian fusion of atmospheric and surface data. Geophys Res Lett 42:1268–1274. https://doi.org/10.1002/2014GL062584
Bloom AA, Exbrayat J-F, van der Velde IR, Feng L, Williams M (2016) The decadal state of the terrestrial carbon cycle: global retrievals of terrestrial carbon allocation, pools, and residence times. Proc Natl Acad Sci 113(5):1285–1290. https://doi.org/10.1073/pnas.1515160113
Borsdorff T, Aan de Brugh J, Hu H et al (2018) Measuring carbon monoxide with TROPOMI: first results and a comparison with ECMWF-IFS analysis data. Geophys Res Lett 45:2826–2832. https://doi.org/10.1002/2018GL077045
Bowman KW, Liu J, Bloom AA et al (2017) Global and Brazilian carbon response to El Niño Modoki 2011–2010. Earth Space Sci 4:637–660. https://doi.org/10.1002/2016EA000204
Brandt M, Wigneron J-P, Chave J et al (2018) Satellite passive microwaves reveal recent climate-induced carbon losses in African drylands. Nat Ecol Evol 2:827–835. https://doi.org/10.1038/s41559-018-0530-6
Broich M, Huete A, Tulbure MG et al (2014) Land surface phenological response to decadal climate variability across Australia using satellite remote sensing. Biogeosciences 11:5181–5198. https://doi.org/10.5194/bg-11-5181-2014
Buchwitz M, Burrows JP (2004) Retrieval of CH_4, CO, and CO_2 total column amounts from SCIAMACHY near-infrared nadir spectra: retrieval algorithm and first results. In: Schaefer K, Comeron A, Carleer MR, Picard RH (eds) Proceedings of SPIE, p 375
Buitenwerf R, Rose L, Higgins SI (2015) Three decades of multi-dimensional change in global leaf phenology. Nat Clim Change 5:364–368. https://doi.org/10.1038/nclimate2533

Caldararu S, Purves DW, Palmer PI (2014) Phenology as a strategy for carbon optimality: a global model. Biogeosciences 11:763–778. https://doi.org/10.5194/bg-11-763-2014
Canadell JG, Le Quéré C, Raupach MR et al (2007) Contributions to accelerating atmospheric CO_2 growth from economic activity, carbon intensity, and efficiency of natural sinks. Proc Natl Acad Sci USA 104:18866–18870. https://doi.org/10.1073/pnas.0702737104
Carvalhais N, Forkel M, Khomik M et al (2014) Global covariation of carbon turnover times with climate in terrestrial ecosystems. Nature 514:213–217. https://doi.org/10.1038/nature13731
Chave J, Coomes D, Jansen S et al (2009) Towards a worldwide wood economics spectrum. Ecol Lett 12:351–366. https://doi.org/10.1111/j.1461-0248.2009.01285.x
Chave J, Réjou-Méchain M, Búrquez A et al (2014) Improved allometric models to estimate the aboveground biomass of tropical trees. Glob Change Biol 20:3177–3190. https://doi.org/10.1111/gcb.12629
Clark DB, Mercado LM, Sitch S et al (2011) The joint UK land environment simulator (JULES), model description—part 2: carbon fluxes and vegetation dynamics. Geosci Model Dev 4:701–722. https://doi.org/10.5194/gmd-4-701-2011
Dannenberg MP, Wise EK, Janko M et al (2018) Atmospheric teleconnection influence on North American land surface phenology. Environ Res Lett 13:34029. https://doi.org/10.1088/1748-9326/aaa85a
Deng F, Jones DBA, O'Dell CW et al (2016) Combining GOSAT × CO_2 observations over land and ocean to improve regional CO_2 flux estimates. J Geophys Res Atmos 121:1896–1913. https://doi.org/10.1002/2015JD024157
Detmers RG, Hasekamp O, Aben I et al (2015) Anomalous carbon uptake in Australia as seen by GOSAT. Geophys Res Lett 42:8177–8184. https://doi.org/10.1002/2015GL065161
Eldering A, Wennberg PO, Crisp D et al (2017) The orbiting carbon observatory-2 early science investigations of regional carbon dioxide fluxes. Science 358:eaam5745. https://doi.org/10.1126/science.aam5745
Erb K-H, Fetzel T, Plutzar C et al (2016) Biomass turnover time in terrestrial ecosystems halved by land use. Nat Geosci 9:674–678. https://doi.org/10.1038/ngeo2782
Exbrayat J-F, Williams M (2015) Quantifying the net contribution of the historical Amazonian deforestation to climate change. Geophys Res Lett 42:2968–2976. https://doi.org/10.1002/2015GL063497
Exbrayat J-F, Pitman AJ, Abramowitz G (2014) Response of microbial decomposition to spin-up explains CMIP5 soil carbon range until 2100. Geosci Model Dev 7:2683–2692. https://doi.org/10.5194/gmd-7-2683-2014
Exbrayat J-F, Liu YY, Williams M (2017) Impact of deforestation and climate on the Amazon Basin's above-ground biomass during 1993–2012. Sci Rep 7:15615. https://doi.org/10.1038/s41598-017-15788-6
Exbrayat J-F, Anthony Bloom A, Falloon P et al (2018a) Reliability ensemble averaging of 21st century projections of terrestrial net primary productivity reduces global and regional uncertainties. Earth Syst Dyn 9:1. https://doi.org/10.5194/esd-9-153-2018
Exbrayat J-F, Smallman TL, Anthony Bloom A, Hutley LB, Williams M (2018b) Inverse determination of the influence of fire on vegetation carbon turnover in the pantropics. Glob Biogeochem Cycles 32(12):1776–1789. https://doi.org/10.1029/2018GB005925
Fasullo JT, Boening C, Landerer FW, Nerem RS (2013) Australia's unique influence on global sea level in 2010–2011. Geophys Res Lett 40:4368–4373. https://doi.org/10.1002/grl.50834
Feng L, Palmer PI, Parker RJ et al (2016) Estimates of European uptake of CO_2 inferred from GOSAT × CO_2 retrievals: sensitivity to measurement bias inside and outside Europe. Atmos Chem Phys 16:1289–1302. https://doi.org/10.5194/acp-16-1289-2016
Field CB, Randerson JT, Malmström CM (1995) Global net primary production: combining ecology and remote sensing. Remote Sens Environ 51:74–88. https://doi.org/10.1016/0034-4257(94)00066-V
Fischer R, Bohn F, Dantas de Paula M et al (2016) Lessons learned from applying a forest gap model to understand ecosystem and carbon dynamics of complex tropical forests. Ecol Modell 326:124–133. https://doi.org/10.1016/j.ecolmodel.2015.11.018
Flannigan MD, Haar THV (1986) Forest fire monitoring using NOAA satellite AVHRR. Can J For Res 16:975–982. https://doi.org/10.1139/x86-171
Forkel M, Carvalhais N, Schaphoff S et al (2014) Identifying environmental controls on vegetation greenness phenology through model-data integration. Biogeosciences 11:7025–7050. https://doi.org/10.5194/bg-11-7025-2014
Forkel M, Migliavacca M, Thonicke K et al (2015) Codominant water control on global interannual variability and trends in land surface phenology and greenness. Glob Change Biol 21:3414–3435. https://doi.org/10.1111/gcb.12950

Forkel M, Carvalhais N, Rodenbeck C et al (2016) Enhanced seasonal CO_2 exchange caused by amplified plant productivity in northern ecosystems. Science 351:696–699. https://doi.org/10.1126/science.aac4971

Frankenberg C, Fisher JB, Worden J et al (2011) New global observations of the terrestrial carbon cycle from GOSAT: patterns of plant fluorescence with gross primary productivity. Geophys Res Lett 1:1. https://doi.org/10.1029/2011gl048738

Freeborn PH, Wooster MJ, Roy DP, Cochrane MA (2014) Quantification of MODIS fire radiative power (FRP) measurement uncertainty for use in satellite-based active fire characterization and biomass burning estimation. Geophys Res Lett 41:1988–1994. https://doi.org/10.1002/2013GL059086

Friedlingstein P, Cox P, Betts R et al (2006) Climate-carbon cycle feedback analysis: results from the C4MIP model intercomparison. J Clim 19:3337–3353. https://doi.org/10.1175/JCLI3800.1

Friend AD, Lucht W, Rademacher TT et al (2014) Carbon residence time dominates uncertainty in terrestrial vegetation responses to future climate and atmospheric CO_2. Proc Natl Acad Sci USA 111:3280–3285. https://doi.org/10.1073/pnas.1222477110

Gatti LV, Gloor M, Miller JB et al (2014) Drought sensitivity of Amazonian carbon balance revealed by atmospheric measurements. Nature 506:76–80. https://doi.org/10.1038/nature12957

Giglio L, Randerson JT, van der Werf GR (2013) Analysis of daily, monthly, and annual burned area using the fourth-generation global fire emissions database (GFED4). J Geophys Res Biogeosci 118:317–328. https://doi.org/10.1002/jgrg.20042

Giorgi F, Mearns LO (2002) Calculation of average, uncertainty range, and reliability of regional climate changes from AOGCM simulations via the "reliability ensemble averaging" (REA) method. J Clim 15:1141–1158. https://doi.org/10.1175/1520-0442(2002)015%3c1141:COAURA%3e2.0.CO;2

Gonzi S, Feng L, Palmer PI (2011) Seasonal cycle of emissions of CO inferred from MOPITT profiles of CO: sensitivity to pyroconvection and profile retrieval assumptions. Geophys Res Lett 38:1–11. https://doi.org/10.1029/2011gl046789

Guanter L, Frankenberg C, Dudhia A et al (2012) Retrieval and global assessment of terrestrial chlorophyll fluorescence from GOSAT space measurements. Remote Sens Environ 121:236–251. https://doi.org/10.1016/j.rse.2012.02.006

Haarsma RJ, Roberts MJ, Vidale PL et al (2016) High resolution model intercomparison project (HighResMIP v1.0) for CMIP6. Geosci Model Dev 9:4185–4208. https://doi.org/10.5194/gmd-9-4185-2016

Hansen MC, Potapov PV, Moore R et al (2013) High-resolution global maps of 21st-century forest cover change. Science 342:850–853. https://doi.org/10.1126/science.1244693

Harris NL, Brown S, Hagen SC et al (2012) Baseline map of carbon emissions from deforestation in tropical regions. Science 336:1573–1576. https://doi.org/10.1126/science.1217962

Hartmann H, Adams HD, Anderegg WRL et al (2015) Research frontiers in drought-induced tree mortality: crossing scales and disciplines. New Phytol 205:965–969. https://doi.org/10.1111/nph.13246

Heymann J, Reuter M, Buchwitz M et al (2017) CO_2 emission of Indonesian fires in 2015 estimated from satellite-derived atmospheric CO_2 concentrations. Geophys Res Lett 45:1621. https://doi.org/10.1002/2016gl072042

Houweling S, Baker D, Basu S et al (2015) An intercomparison of inverse models for estimating sources and sinks of CO_2 using GOSAT measurements. J Geophys Res Atmos 120:5253–5266. https://doi.org/10.1002/2014JD022962

Huete A, Didan K, Miura T et al (2002) Overview of the radiometric and biophysical performance of the MODIS vegetation indices. Remote Sens Environ 83:195–213. https://doi.org/10.1016/S0034-4257(02)00096-2

Huete AR, Didan K, Shimabukuro YE et al (2006) Amazon rainforests green-up with sunlight in dry season. Geophys Res Lett 33:L06405. https://doi.org/10.1029/2005GL025583

Huntzinger DN, Schwalm C, Michalak AM et al (2013) The North American carbon program multi-scale synthesis and terrestrial model intercomparison project ? Part 1: overview and experimental design. Geosci Model Dev 6:2121–2133. https://doi.org/10.5194/gmd-6-2121-2013

Ito A, Nishina K, Reyer CPO et al (2017) Photosynthetic productivity and its efficiencies in ISIMIP2a biome models: benchmarking for impact assessment studies. Environ Res Lett 12:085001. https://doi.org/10.1088/1748-9326/aa7a19

Jiang Z, Jones DBA, Worden HM, Henze DK (2015) Sensitivity of top-down CO source estimates to the modeled vertical structure in atmospheric CO. Atmos Chem Phys 15:1521–1537. https://doi.org/10.5194/acp-15-1521-2015

Joiner J, Guanter L, Lindstrot R et al (2013) Global monitoring of terrestrial chlorophyll fluorescence from moderate-spectral-resolution near-infrared satellite measurements: methodology, simulations, and application to GOME-2. Atmos Meas Tech 6:2803–2823. https://doi.org/10.5194/amt-6-2803-2013

Joiner J, Yoshida Y, Vasilkov AP et al (2014) The seasonal cycle of satellite chlorophyll fluorescence observations and its relationship to vegetation phenology and ecosystem atmosphere carbon exchange. Remote Sens Environ 152:375–391. https://doi.org/10.1016/j.rse.2014.06.022

Joiner J, Yoshida Y, Guanter L, Middleton EM (2016) New methods for the retrieval of chlorophyll red fluorescence from hyperspectral satellite instruments: simulations and application to GOME-2 and SCIAMACHY. Atmos Meas Tech 9:3939–3967. https://doi.org/10.5194/amt-9-3939-2016

Jolly WM, Nemani R, Running SW (2005) A generalized, bioclimatic index to predict foliar phenology in response to climate. Glob Change Biol 11:619–632. https://doi.org/10.1111/j.1365-2486.2005.00930.x

Jung M, Reichstein M, Bondeau A (2009) Towards global empirical upscaling of FLUXNET eddy covariance observations: validation of a model tree ensemble approach using a biosphere model. Biogeosciences 6:2001–2013. https://doi.org/10.5194/bg-6-2001-2009

Jung M, Reichstein M, Margolis HA et al (2011) Global patterns of land-atmosphere fluxes of carbon dioxide, latent heat, and sensible heat derived from eddy covariance, satellite, and meteorological observations. J Geophys Res 116:G00J07. https://doi.org/10.1029/2010jg001566

Kala J, Decker M, Exbrayat J-F et al (2014) Influence of leaf area index prescriptions on simulations of heat, moisture, and carbon fluxes. J Hydrometeorol 15:489–503. https://doi.org/10.1175/JHM-D-13-063.1

Kaminski T, Knorr W, Schürmann G et al (2013) The BETHY/JSBACH carbon cycle data assimilation system: experiences and challenges. J Geophys Res Biogeosci 118:1414. https://doi.org/10.1002/jgrg.20118

Knapp N, Fischer R, Huth A (2018a) Linking lidar and forest modeling to assess biomass estimation across scales and disturbance states. Remote Sens Environ 205:199–209. https://doi.org/10.1016/j.rse.2017.11.018

Knapp N, Huth A, Kugler F et al (2018b) Model-assisted estimation of tropical forest biomass change: a comparison of approaches. Remote Sens 10:731. https://doi.org/10.3390/rs10050731

Knorr W (2000) Annual and interannual CO_2 exchanges of the terrestrial biosphere: process-based simulations and uncertainties. Glob Ecol Biogeogr 9:225–252. https://doi.org/10.1046/j.1365-2699.2000.00159.x

Knorr W, Kaminski T, Scholze M et al (2010) Carbon cycle data assimilation with a generic phenology model. J Geophys Res 115:G04017. https://doi.org/10.1029/2009JG001119

Knyazikhin Y, Martonchik JV, Myneni RB et al (1998) Synergistic algorithm for estimating vegetation canopy leaf area index and fraction of absorbed photosynthetically active radiation from MODIS and MISR data. J Geophys Res Atmos 103:32257–32275. https://doi.org/10.1029/98JD02462

Köhler P, Joiner J (2015) A linear method for the retrieval of sun-induced chlorophyll fluorescence from GOME-2 and SCIAMACHY data. Atmos Meas Tech 8:2589–2608. https://doi.org/10.5194/amt-8-2589-2015

Kolby Smith W, Reed SC, Cleveland CC et al (2016) Large divergence of satellite and earth system model estimates of global terrestrial CO_2 fertilization. Nat Clim Change 6:306–310. https://doi.org/10.1038/nclimate2879

Kopacz M, Jacob DJ, Fisher JA et al (2010) Global estimates of CO sources with high resolution by adjoint inversion of multiple satellite datasets (MOPITT, AIRS, SCIAMACHY, TES). Atmos Chem Phys 10:855–876. https://doi.org/10.5194/acp-10-855-2010

Krishnamurti TN, Kishtawal CM, LaRow TE et al (1999) Improved weather and seasonal climate forecasts from multimodel superensemble. Science 285:1548–1550. https://doi.org/10.1126/science.285.5433.1548

Krol M, Peters W, Hooghiemstra P et al (2013) How much CO was emitted by the 2010 fires around Moscow? Atmos Chem Phys 13:4737–4747. https://doi.org/10.5194/acp-13-4737-2013

Kuppel S, Peylin P, Maignan F et al (2014) Model-data fusion across ecosystems: from multisite optimizations to global simulations. Geosci Model Dev 7:2581–2597. https://doi.org/10.5194/gmd-7-2581-2014

Le Quéré C, Andrew RM, Friedlingstein P et al (2018) Global carbon budget 2017. Earth Syst Sci Data 10:405–448. https://doi.org/10.5194/essd-10-405-2018

Le Toan T, Quegan S, Davidson MWJWJ et al (2011) The BIOMASS mission: mapping global forest biomass to better understand the terrestrial carbon cycle. Remote Sens Environ 115:2850–2860. https://doi.org/10.1016/j.rse.2011.03.020

Lee J-E, Berry JA, van der Tol C et al (2015) Simulations of chlorophyll fluorescence incorporated into the community land model version 4. Glob Change Biol 21:3469–3477. https://doi.org/10.1111/gcb.12948

Lewis P, Gómez-Dans J, Kaminski T et al (2012) An earth observation land data assimilation system (EO-LDAS). Remote Sens Environ 120:219–235. https://doi.org/10.1016/j.rse.2011.12.027

Li W, Ciais P, Peng S et al (2017) Land-use and land-cover change carbon emissions between 1901 and 2012 constrained by biomass observations. Biogeosciences 14:5053–5067. https://doi.org/10.5194/bg-14-5053-2017

Lienert S, Joos F (2018) A Bayesian ensemble data assimilation to constrain model parameters and land-use carbon emissions. Biogeosciences 15:2909–2930. https://doi.org/10.5194/bg-15-2909-2018

Liu YY, van Dijk AIJM, de Jeu RAM et al (2015) Recent reversal in loss of global terrestrial biomass. Nat Clim Change 5:470–474. https://doi.org/10.1038/nclimate2581

Liu J, Bowman KW, Schimel DS et al (2017) Contrasting carbon cycle responses of the tropical continents to the 2015–2016 El Niño. Science eaam358:5690. https://doi.org/10.1126/science.aam5690

MacBean N, Maignan F, Peylin P et al (2015) Using satellite data to improve the leaf phenology of a global terrestrial biosphere model. Biogeosciences 12:7185–7208. https://doi.org/10.5194/bg-12-7185-2015

MacBean N, Maignan F, Bacour C et al (2018) Strong constraint on modelled global carbon uptake using solar-induced chlorophyll fluorescence data. Sci Rep. https://doi.org/10.1038/s41598-018-20024-w

Migliavacca M, Sonnentag O, Keenan TF et al (2012) On the uncertainty of phenological responses to climate change, and implications for a terrestrial biosphere model. Biogeosciences 9:2063–2083. https://doi.org/10.5194/bg-9-2063-2012

Miller SM, Michalak AM, Yadav V, Tadić JM (2018) Characterizing biospheric carbon balance using CO_2 observations from the OCO-2 satellite. Atmos Chem Phys 18:6785–6799. https://doi.org/10.5194/acp-18-6785-2018

Morton DC, Le Page Y, DeFries R et al (2013) Understorey fire frequency and the fate of burned forests in southern Amazonia. Philos Trans R Soc B Biol Sci 368:20120163. https://doi.org/10.1098/rstb.2012.0163

Myneni RB, Hall FG, Sellers PJ, Marshak AL (1995) The interpretation of spectral vegetation indexes. IEEE Trans Geosci Remote Sens 33:481–486. https://doi.org/10.1109/36.377948

Myneni RB, Keeling CD, Tucker CJ et al (1997) Increased plant growth in the northern high latitudes from 1981 to 1991. Nature 386:698–702. https://doi.org/10.1038/386698a0

Myneni RB, Hoffman S, Knyazikhin Y et al (2002) Global products of vegetation leaf area and fraction absorbed PAR from year one of MODIS data. Remote Sens Environ 83:214–231. https://doi.org/10.1016/S0034-4257(02)00074-3

Myneni RB, Yang W, Nemani RR et al (2007) Large seasonal swings in leaf area of Amazon rainforests. Proc Natl Acad Sci 104:4820–4823. https://doi.org/10.1073/pnas.0611338104

New M, Lister D, Hulme M, Makin I (2002) A high-resolution data set of surface climate over global land areas. Clim Res 21:1–25. https://doi.org/10.3354/cr021001

Norton AJ, Rayner PJ, Koffi EN, Scholze M (2018a) Assimilating solar-induced chlorophyll fluorescence into the terrestrial biosphere model BETHY-SCOPE v1.0: model description and information content. Geosci Model Dev 11:1517–1536. https://doi.org/10.5194/gmd-11-1517-2018

Norton AJ, Rayner PJ, Koffi EN et al (2018b) Estimating global gross primary productivity using chlorophyll fluorescence and a data assimilation system with the BETHY-SCOPE model. Biogeosciences 12:835. https://doi.org/10.5194/bg-2018-270

Papale D, Valentini R (2003) A new assessment of European forests carbon exchanges by eddy fluxes and artificial neural network spatialization. Glob Change Biol 9:525–535. https://doi.org/10.1046/j.1365-2486.2003.00609.x

Parazoo NC, Bowman K, Frankenberg C et al (2013) Interpreting seasonal changes in the carbon balance of southern Amazonia using measurements of XCO_2 and chlorophyll fluorescence from GOSAT. Geophys Res Lett 40:2829–2833. https://doi.org/10.1002/grl.50452

Parazoo NC, Bowman K, Fisher JB et al (2014) Terrestrial gross primary production inferred from satellite fluorescence and vegetation models. Glob Change Biol 20:3103–3121. https://doi.org/10.1111/gcb.12652

Peylin P, Bacour C, MacBean N et al (2016) A new stepwise carbon cycle data assimilation system using multiple data streams to constrain the simulated land surface carbon cycle. Geosci Model Dev 9:3321–3346. https://doi.org/10.5194/gmd-9-3321-2016

Pinzon JE, Tucker CJ (2014) A non-stationary 1981–2012 AVHRR NDVI3g time series. Remote Sens 6:6929–6960. https://doi.org/10.3390/rs6086929

Pitman AJ (2003) The evolution of, and revolution in, land surface schemes designed for climate models. Int J Climatol 23:479–510. https://doi.org/10.1002/joc.893

Porcar-Castell A, Tyystjärvi E, Atherton J et al (2014) Linking chlorophyll a fluorescence to photosynthesis for remote sensing applications: mechanisms and challenges. J Exp Bot 65:4065–4095. https://doi.org/10.1093/jxb/eru191

Potapov P, Yaroshenko A, Turubanova S et al (2008) Mapping the world's intact forest landscapes by remote sensing. Ecol Soc 13:51. https://doi.org/10.5751/es-02670-130251

Potter CS, Randerson JT, Field CB et al (1993) Terrestrial ecosystem production: a process model based on global satellite and surface data. Glob Biogeochem Cycles 7:811–841. https://doi.org/10.1029/93GB02725
Poulter B, Frank D, Ciais P et al (2014) Contribution of semi-arid ecosystems to interannual variability of the global carbon cycle. Nature 509:600–603. https://doi.org/10.1038/nature13376
Prince SD, Goward SN (1995) Global primary production: a remote sensing approach. J Biogeogr 22:815. https://doi.org/10.2307/2845983
Quaife T, Lewis P, De Kauwe M et al (2008) Assimilating canopy reflectance data into an ecosystem model with an Ensemble Kalman Filter. Remote Sens Environ 112:1347. https://doi.org/10.1016/j.rse.2007.05.020
Richardson AD, Anderson RS, Arain MA et al (2012) Terrestrial biosphere models need better representation of vegetation phenology: results from the North American Carbon Program Site Synthesis. Glob Change Biol 18:566–584. https://doi.org/10.1111/j.1365-2486.2011.02562.x
Rödig E, Cuntz M, Heinke J et al (2017) Spatial heterogeneity of biomass and forest structure of the Amazon rain forest: linking remote sensing, forest modelling and field inventory. Glob Ecol Biogeogr 26:1292–1302. https://doi.org/10.1111/geb.12639
Rödig E, Cuntz M, Rammig A et al (2018) The importance of forest structure for carbon fluxes of the Amazon rainforest. Environ Res Lett 13:054013. https://doi.org/10.1088/1748-9326/aabc61
Rodríguez-Veiga P, Saatchi S, Tansey K, Balzter H (2016) Magnitude, spatial distribution and uncertainty of forest biomass stocks in Mexico. Remote Sens Environ 183:265–281
Rogers BM, Soja AJ, Goulden ML, Randerson JT (2015) Influence of tree species on continental differences in boreal fires and climate feedbacks. Nat Geosci 8:228–234. https://doi.org/10.1038/ngeo2352
Running SW, Nemani RR, Heinsch FA et al (2004) A continuous satellite-derived measure of global terrestrial primary production. Bioscience 54:547. https://doi.org/10.1641/0006-3568(2004)054%5b0547:ACSMOG%5d2.0.CO;2
Saatchi SS, Harris NL, Brown S et al (2011) Benchmark map of forest carbon stocks in tropical regions across three continents. Proc Natl Acad Sci USA 108:9899–9904. https://doi.org/10.1073/pnas.1019576108
Santoro M, Beer C, Cartus O et al (2011) Retrieval of growing stock volume in boreal forest using hyper-temporal series of Envisat ASAR ScanSAR backscatter measurements. Remote Sens Environ 115:490–507. https://doi.org/10.1016/j.rse.2010.09.018
Santoro M, Beaudoin A, Beer C et al (2015) Forest growing stock volume of the northern hemisphere: spatially explicit estimates for 2010 derived from Envisat ASAR. Remote Sens Environ 168:316–334. https://doi.org/10.1016/j.rse.2015.07.005
Scholze M, Buchwitz M, Dorigo W et al (2017) Reviews and syntheses: systematic earth observations for use in terrestrial carbon cycle data assimilation systems. Biogeosciences 14:3401–3429
Schwalm CR, Huntzinger DN, Fisher JB et al (2015) Toward "optimal" integration of terrestrial biosphere models. Geophys Res Lett 42:4418–4428. https://doi.org/10.1002/2015GL064002
Shugart HH, Asner GP, Fischer R et al (2015) Computer and remote-sensing infrastructure to enhance large-scale testing of individual-based forest models. Front Ecol Environ 13:503–511. https://doi.org/10.1890/140327
Shugart HH, Wang B, Fischer R et al (2018) Gap models and their individual-based relatives in the assessment of the consequences of global change. Environ Res Lett 13:033001. https://doi.org/10.1088/1748-9326/aaaacc
Sierra CA, Müller M, Metzler H et al (2017) The muddle of ages, turnover, transit, and residence times in the carbon cycle. Glob Change Biol 23:1763–1773. https://doi.org/10.1111/gcb.13556
Sitch S, Friedlingstein P, Gruber N et al (2015) Recent trends and drivers of regional sources and sinks of carbon dioxide. Biogeosciences 12:653–679. https://doi.org/10.5194/bg-12-653-2015
Slevin D, Tett SFB, Exbrayat J-F et al (2017) Global evaluation of gross primary productivity in the JULES land surface model v3.4.1. Geosci Model Dev 10:2651–2670. https://doi.org/10.5194/gmd-10-2651-2017
Smallman TL, Exbrayat J-F, Mencuccini M et al (2017) Assimilation of repeated woody biomass observations constrains decadal ecosystem carbon cycle uncertainty in aggrading forests. J Geophys Res Biogeosci 122:528–545. https://doi.org/10.1002/2016JG003520
Stavros EN, Schimel D, Pavlick R et al (2017) ISS observations offer insights into plant function. Nat Ecol Evol 1:0194. https://doi.org/10.1038/s41559-017-0194
Stöckli R, Rutishauser T, Baker I et al (2011) A global reanalysis of vegetation phenology. J Geophys Res 116:G03020. https://doi.org/10.1029/2010JG001545

Sun Y, Frankenberg C, Jung M et al (2018) Overview of solar-induced chlorophyll fluorescence (SIF) from the orbiting carbon observatory-2: retrieval, cross-mission comparison, and global monitoring for GPP. Remote Sens Environ 209:808–823. https://doi.org/10.1016/j.rse.2018.02.016
Taylor KE, Stouffer RJ, Meehl GA (2012) An overview of CMIP5 and the experiment design. Bull Am Meteorol Soc 93:485–498. https://doi.org/10.1175/BAMS-D-11-00094.1
Thum T, Zaehle S, Köhler P et al (2017) Modelling sun-induced fluorescence and photosynthesis with a land surface model at local and regional scales in northern Europe. Biogeosciences 14:1969–1987. https://doi.org/10.5194/bg-14-1969-2017
Thurner M, Beer C, Santoro M et al (2014) Carbon stock and density of northern boreal and temperate forests. Glob Ecol Biogeogr 23:297–310. https://doi.org/10.1111/geb.12125
Thurner M, Beer C, Carvalhais N et al (2016) Large-scale variation in boreal and temperate forest carbon turnover rate related to climate. Geophys Res Lett 43:4576–4585. https://doi.org/10.1002/2016GL068794
Thurner M, Beer C, Ciais P et al (2017) Evaluation of climate-related carbon turnover processes in global vegetation models for boreal and temperate forests. Glob Change Biol 23:3076–3091. https://doi.org/10.1111/gcb.13660
Tramontana G, Jung M, Schwalm CR et al (2016) Predicting carbon dioxide and energy fluxes across global FLUXNET sites with regression algorithms. Biogeosciences 13:4291–4313. https://doi.org/10.5194/bg-13-4291-2016
van der Tol C, Verhoef W, Rosema A (2009) A model for chlorophyll fluorescence and photosynthesis at leaf scale. Agric For Meteorol 149:96. https://doi.org/10.1016/j.agrformet.2008.07.007
van der Werf GR, Morton DC, DeFries RS et al (2009) CO_2 emissions from forest loss. Nat Geosci 2:737–738. https://doi.org/10.1038/ngeo671
van der Werf GR, Randerson JT, Giglio L et al (2010) Global fire emissions and the contribution of deforestation, savanna, forest, agricultural, and peat fires (1997–2009). Atmos Chem Phys 10:11707–11735. https://doi.org/10.5194/acp-10-11707-2010
Waigl CF, Stuefer M, Prakash A, Ichoku C (2017) Detecting high and low-intensity fires in Alaska using VIIRS I-band data: an improved operational approach for high latitudes. Remote Sens Environ 199:389–400. https://doi.org/10.1016/j.rse.2017.07.003
Warszawski L, Frieler K, Huber V et al (2014) The inter-sectoral impact model intercomparison project (ISI-MIP): project framework. Proc Natl Acad Sci USA 111:3228–3232. https://doi.org/10.1073/pnas.1312330110
Williams M, Schwarz PA, Law BE et al (2005) An improved analysis of forest carbon dynamics using data assimilation. Glob Change Biol 11:89–105. https://doi.org/10.1111/j.1365-2486.2004.00891.x
Williams M, Richardson AD, Reichstein M et al (2009) Improving land surface models with FLUXNET data. Biogeosciences 6:1341–1359. https://doi.org/10.5194/bg-6-1341-2009
Wood JD, Griffis TJ, Baker JM et al (2017) Multiscale analyses of solar-induced florescence and gross primary production. Geophys Res Lett 53:785–818. https://doi.org/10.1002/2016gl070775
Worden JR, Bloom AA, Pandey S et al (2017a) Reduced biomass burning emissions reconcile conflicting estimates of the post-2006 atmospheric methane budget. Nat Commun 8:2227. https://doi.org/10.1038/s41467-017-02246-0
Worden JR, Doran G, Kulawik S et al (2017b) Evaluation and attribution of OCO-2 XCO_2 uncertainties. Atmos Meas Tech 10:2759–2771. https://doi.org/10.5194/amt-10-2759-2017
Xu L, Saatchi SS, Shapiro A et al (2017) Spatial distribution of carbon stored in forests of the Democratic Republic of Congo. Sci Rep 7:15030. https://doi.org/10.1038/s41598-017-15050-z
Yang X, Tang J, Mustard JF et al (2015) Solar-induced chlorophyll fluorescence that correlates with canopy photosynthesis on diurnal and seasonal scales in a temperate deciduous forest. Geophys Res Lett 42:2977–2987. https://doi.org/10.1002/2015GL063201
Yang H, Yang X, Zhang Y et al (2017) Chlorophyll fluorescence tracks seasonal variations of photosynthesis from leaf to canopy in a temperate forest. Glob Change Biol 23:2874–2886. https://doi.org/10.1111/gcb.13590
Yokota T, Yoshida Y, Eguchi N et al (2009) Global concentrations of CO_2 and CH_4 retrieved from GOSAT: first preliminary results. SOLA 5:160–163. https://doi.org/10.2151/sola.2009-041
Zhang Y, Guanter L, Berry JA et al (2014) Estimation of vegetation photosynthetic capacity from space-based measurements of chlorophyll fluorescence for terrestrial biosphere models. Glob Change Biol 20:3727–3742. https://doi.org/10.1111/gcb.12664
Zhang Y, Xiao X, Jin C et al (2016) Consistency between sun-induced chlorophyll fluorescence and gross primary production of vegetation in North America. Remote Sens Environ 183:154–169. https://doi.org/10.1016/j.rse.2016.05.015

Zhao M, Running SW (2010) Drought-induced reduction in global. Science (80) 329:940–943. https://doi.org/10.1126/science.1192666
Zhu Z, Bi J, Pan Y et al (2013) Global data sets of vegetation leaf area index (LAI)3g and fraction of photosynthetically active radiation (FPAR)3g derived from global inventory modeling and mapping studies (GIMMS) normalized difference vegetation index (NDVI3G) for the period 1981 to 2011. Remote Sens 5:927–948. https://doi.org/10.3390/rs5020927
Zhu Z, Piao S, Myneni RB et al (2016) Greening of the earth and its drivers. Nat Clim Change 6:791–795. https://doi.org/10.1038/nclimate3004
Zwally HJ, Schutz B, Abdalati W et al (2002) ICESat's laser measurements of polar ice, atmosphere, ocean, and land. J Geodyn 34:405–445. https://doi.org/10.1016/S0264-3707(02)00042-X
Ziehn T, Scholze M, Knorr W (2012) On the capability of Monte Carlo and adjoint inversion techniques to derive posterior parameter uncertainties in terrestrial ecosystem models. Glob Biogeochem Cycles https://doi.org/10.1029/2011GB004185

Publisher's Note Springer Nature remains neutral with regard to jurisdictional claims in published maps and institutional affiliations.

Affiliations

Jean-François Exbrayat[1] · A. Anthony Bloom[2] · Nuno Carvalhais[3,6] · Rico Fischer[4] · Andreas Huth[4,7,8] · Natasha MacBean[5] · Mathew Williams[1]

1 National Centre for Earth Observation and School of GeoSciences, University of Edinburgh, Crew Building, Alexander Crum Brown Road, Edinburgh EH9 3FF, UK

2 Jet Propulsion Laboratory, California Institute of Technology, Pasadena, CA 91109, USA

3 Max Planck Institute for Biogeochemistry, Department Biogeochemical Integration, Hans-Knoell-Str. 10, 07745 Jena, Germany

4 Helmholtz Centre for Environmental Research – UFZ, Department of Ecological Modelling, Permoserstraße 15, 04318 Leipzig, Germany

5 Department of Geography, Indiana University, 701 E Kirkwood Avenue, Bloomington, IN 47405, USA

6 CENSE, Departamento de Ciências e Engenharia do Ambiente, Faculdade de Ciências e Tecnologia, Universidade NOVA de Lisboa, Campus de Caparica, 2829-516 Caparica, Portugal

7 Institute of Environmental Systems Research, University of Osnabrück, Barbarastraße 12, 49076 Osnabrück, Germany

8 German Centre for Integrative Biodiversity Research (iDiv) Halle-Jena-Leipzig, Deutscher Platz 5e, 04103 Leipzig, Germany

Surveys in Geophysics (2019) 40:757–778
https://doi.org/10.1007/s10712-019-09510-6

The Role and Need for Space-Based Forest Biomass-Related Measurements in Environmental Management and Policy

Martin Herold, et al. ***[full author details at the end of the article]***

Received: 16 July 2018 / Accepted: 14 January 2019 / Published online: 11 February 2019

Abstract
The achievement of international goals and national commitments related to forest conservation and management, climate change, and sustainable development requires credible, accurate, and reliable monitoring of stocks and changes in forest biomass and carbon. Most prominently, the Paris Agreement on Climate Change and the United Nations' Sustainable Development Goals in particular require data on biomass to monitor progress. Unprecedented opportunities to provide forest biomass data are created by a series of upcoming space-based missions, many of which provide open data targeted at large areas and better spatial resolution biomass monitoring than has previously been achieved. We assess various policy needs for biomass data and recommend a long-term collaborative effort among forest biomass data producers and users to meet these needs. A gap remains, however, between what can be achieved in the research domain and what is required to support policy making and meet reporting requirements. There is no single biomass dataset that serves all users in terms of definition and type of biomass measurement, geographic area, and uncertainty requirements, and whether there is need for the most recent up-to-date biomass estimate or a long-term biomass trend. The research and user communities should embrace the potential strength of the multitude of upcoming missions in combination to provide for these varying needs and to ensure continuity for long-term data provision which one-off research missions cannot provide. International coordination bodies such as Global Forest Observations Initiative (GFOI), Committee on Earth Observation Satellites (CEOS), and Global Observation of Forest Cover and Land Dynamics (GOFC-GOLD) will be integral in addressing these issues in a way that fulfils these needs in a timely fashion. Further coordination work should particularly look into how space-based data can be better linked with field reference data sources such as forest plot networks, and there is also a need to ensure that reference data cover a range of forest types, management regimes, and disturbance regimes worldwide.

Keywords Remote sensing · Climate change · Carbon stocks · Reducing emissions from deforestation and forest degradation (REDD) · Sustainable Development Goals (SDGs)

1 Introduction

Vegetation biomass and its dynamics are key components of the global carbon cycle, and central to provision of food and fibre worldwide. Living biomass in the terrestrial biosphere contains 450–650 PgC (Prentice et al. 2001), which is approximately 30% of the terrestrial carbon pool, since dead organic matter in litter and soils contains 1500–2400 PgC (Batjes 1996). The land sink, defined as uptake of carbon in mainly biomass and soils on land, was estimated to be 3.0 ± 0.8 PgCyr^{-1} between 2005 and 2014. To put this into perspective, this is similar to the growth in carbon in the atmosphere which was 4.4 ± 0.1 GtCyr^{-1} (Le Quéré et al. 2015). The link between vegetation biomass and the carbon cycle (Houghton et al. 2009) makes management of forest biomass relevant to the challenge of limiting global warming to 2 °C, an internationally agreed target (UNFCCC 2015).

The change in quantity of biomass impacts the climate in three ways. First, the land sector currently represents the largest negative emissions potential (Griscom et al. 2017), mainly through sequestration of carbon in vegetation but also soils [with options such as carbon capture and storage being only currently operational at relatively small scales and with high marginal abatement costs (Irlam 2017)]. Second, disturbances (including fire), biomass losses (for example, forest degradation), and decomposition of the biomass result in carbon dioxide and other greenhouse gas (GHG) emissions. The anthropogenic impact on biomass is significant and represents approximately 23.8% of the global potential net primary productivity which is 15.6 PgCyr^{-1} (Haberl et al. 2007). Third, the presence of woody biomass influences climate by modifying water and energy exchanges with the atmosphere, thus influencing local hydroclimates (Lorenz and Pitman 2014; Zomer et al. 2016; Ellison et al. 2017). Vegetative biomass is also a key component of many of the world's ecosystems and represents habitats which support biodiversity and provide resources for human use. In addition, many communities rely on the resources found in forests and other ecosystems for food, fuel, and fibre, all of which are biomass components (Hermans-Neumann et al. 2016). Related to this, the estimation of biomass is important for quantifying and understanding the impacts of dynamic land use or land-cover events such as encroachment of vegetation and removal of biomass. Additionally, this information is useful for estimating the likelihood and impact of fire, for monitoring plantation dynamics and forest management practices, and for assessing resources for wood products and bioenergy.

Accurate information on biomass distributions and on the spatial and temporal dynamics of biomass change is therefore essential to plan adaptation and mitigation actions in the land use sector. Progress on domestic policies in the land use sector can be assessed with biomass data, which must meet the reporting needs of international policies, for example the Paris Climate Agreement (UNFCCC 2015). These include the Nationally Determined Contributions and regular global stocktakes which countries contribute to, and which represent a country's proposed contribution to achieving the goals of the United Nations Framework Convention on Climate Change (UNFCCC).

Knowledge of the distribution of biomass across the globe is limited, and consistent data on changes over time at both the global and regional levels are even less available. At the same time, there is a proliferation of biomass mapping efforts based on airborne and space-based data. In this paper, we focus estimates based on space-based data, which are often calibrated and validated using field data, including remote sensing such as terrestrial lidar and also airborne data. From a space-based perspective, both optical and radar (Synthetic Aperture Radar (SAR)) data can be useful for biomass mapping. Radar-based measurements, however, have larger sensitivity to forest biomass than optical systems, in particular, when they are combined with Light Detection and Ranging (lidar) (Asner et al. 2011; Mitchell et al. 2017). It is important to clarify that remote sensing cannot provide direct biomass measurements, but provides information from which biomass can be estimated. Long wavelength (L-band) radar and lidar sensors have been commonly used to estimate forest height and structure and aboveground biomass (AGB). Multiple large-geographical-scale AGB maps with coarse spatial resolution (>500 m) have been produced using space-based lidar data from the Geoscience Laser Altimeter System (GLAS) (Baccini et al. 2012; Saatchi et al. 2011). Biomass maps at 30 m or finer resolution have also been produced; however, space-based maps typically have low accuracy because of different forest structures (e.g., heights, allocations of AGB) and varying environmental conditions (e.g., inconsistent estimates between frozen and unfrozen conditions) (Lucas et al. 2015). An additional challenge is that existing continental-to-global-scale maps are based on data from sensors which are no longer operational (GCOS 2016). Such maps remain one-time efforts lacking long-term continuity and are made available as independent products by the scientific community. These products mainly provide a snapshot of biomass at one moment in time (or are produced from composites of various timeframes) and do not provide information on changes over time. This does not fulfil user needs related to information on biomass changes, and the proliferation of maps confuses end users over which map to use for specific purposes and why.

Efforts including JAXA's L-band radar missions (JERS-1, ALOS, ALOS-2) and NASA's Geoscience Laser Altimeter System (GLAS) mission onboard the Ice, Cloud, and land Elevation Satellite (ICESat) have for the past two decades constituted the main testing grounds for space-based methodology development for AGB estimation. There are now concrete plans, however, for a new generation space-based missions using long wavelength L- and P-band radar and lidar targeted at forest AGB mapping and estimation specifically, including many which have an open data policy (Fig. 1). In this figure, the lifetime of radar satellites/space-based missions where not fixed/known has been estimated to be 5 years, although may be much longer, and advanced non-commercial missions only were included. Where the end date was not known, the date is in grey. Tandem-L is a phase A study and thus is not confirmed. These missions overlap in terms of their objectives and their intended times of operation and thus have potential to close the gap for data and information needed to stimulate activities and report progress on many forest-related policy and management applications.

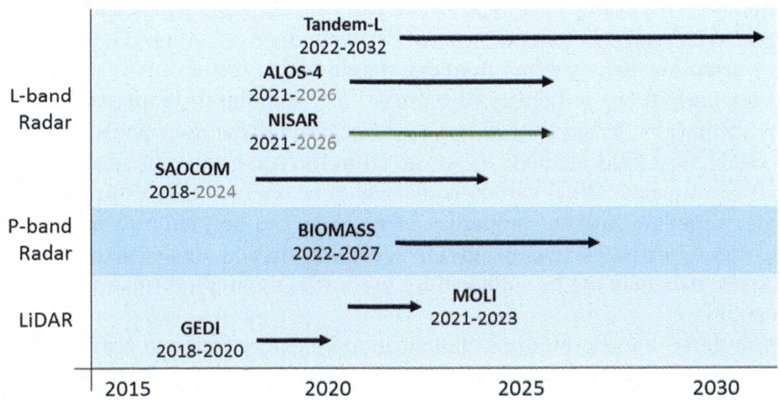

Fig. 1 Near-future space-based missions with an open data policy, where one of the key mission goals is biomass monitoring. ALOS (Advanced Land Observation Satellite); NISAR (NASA-ISRO Synthetic Aperture Radar); SAOCOM (Satélite Argentino de Observación Con Microondas) (A, 2018, and B, 2019); GEDI (Global Ecosystems Dynamics Investigation Lidar); MOLI (Multi-footprint Observation Lidar and Imager)

With the future space-based biomass monitoring on the horizon, there is need to summarize and consolidate the role of and needs for biomass information based on data from these missions. To this end, our assessment:

1. Identifies and describes the requirements of biomass data for key policy and forest management processes;
2. Synthesizes the requirements and assesses the key gaps for policy (and forest management processes), considering the availability and experiences with large scale biomass products;
3. Derives recommendations which aim to bridge the gaps identified by better aligning various space-based biomass mapping efforts to serve these policy and management processes.

2 Requirements

2.1 Defining Biomass

Different definitions of biomass are used for different applications. Biomass of terrestrial ecosystems is recognized in major documents including IPCC (2003, 2006), Tompo et al. (2010), Baede et al. (2007), and FAO (2015) as broadly the mass of live and dead organic matter. Biomass can be divided into AGB and belowground biomass (BGB), and live and dead biomass; however, classification schemes vary (GlobBiomass 2017). BGB is typically defined as living root biomass, although fine roots (for example, <2 mm) are sometimes excluded (IPCC 2003). AGB is defined as all living biomass above the soil (IPCC 2003). In other definitions, deadwood and other organic material (litter) which are lying above the soil are included in the definition of AGB, but deadwood and litter which are in the soil are not (GlobBiomass 2017). Material laying above the soil may be in various states of decomposition, and the Intergovernmental Panel on Climate Change (IPCC) classification guidance requires the live and dead biomass pools to be counted separately, which can cause some difficulties, for example, stumps can be part of both living and dead wood.

Biomass is measured in units of mass of dry weight, or of carbon (GlobBiomass 2017), which is conventionally converted from dry weight with a factor of 0.47–0.5 (IPCC 2006). Biomass regularly refers to density of biomass, e.g., amount of biomass per unit area (in this paper, hereafter referred to as biomass). Biomass in the various pools can be estimated through established field methods, however, from the perspective of space data, typically only AGB is estimated. Other carbon pools cannot be operationally estimated from space-based data, although different properties of biomass can be predicted using information from different space-based sensors, giving some information about other carbon pools. In general, space data may not be sufficient for users who require information about the other biomass pools.

Figure 2 shows a forest biomass classification system, which can also be used in other ecosystems. Often the classes are related to the reporting needs of different policies and can vary for different observations/products in terms of including live and/or dead biomass, or the division of various vegetation compartments (woody, branches, and leaves). A clear and transparent definition of biomass should be provided with each measurement/product, to ensure that multiple types of measurements and estimations are used consistently and correctly.

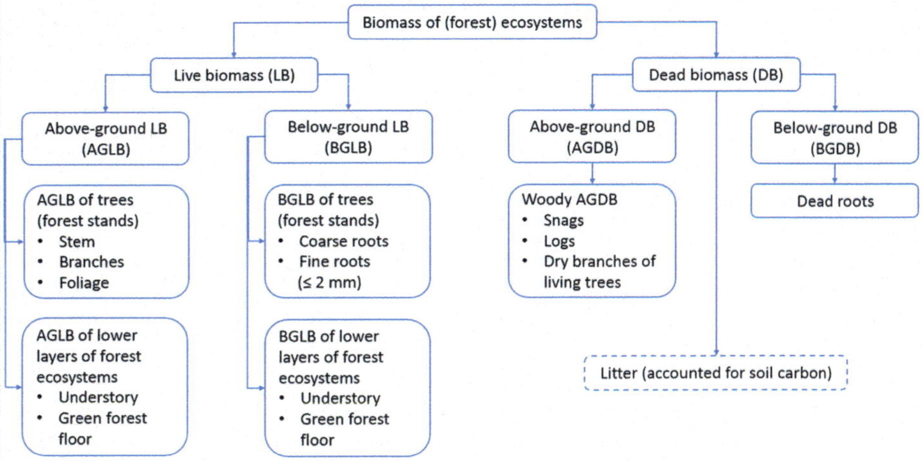

Fig. 2 Classification scheme for biomass in forest ecosystems adapted from GlobBiomass (2017)

2.2 Users and Their Needs

Multiple international policies that require biomass information have been developed and adopted in recent years. Multilateral agreements include performance-based incentive systems to curb trends in forest loss in the tropics and to stimulate reforestation and forest restoration, and to enhance the goods and services provided by forests. Most prominently, international negotiations related to climate change led to increased interest in monitoring forest biomass using space-based data (Kumar and Mutanga 2017). Voluntary international processes and their targets related to forests will require biomass data to track the achievement of their goals. These processes include the United Nations Forum on Forests, the Aichi targets under the UN convention on Biological Diversity which includes at least halving the loss of all natural habitats, the 2011 Bonn Challenge to restore 150 million hectares of degraded land by 2020, the 2014 New York Declaration with several targets of reduction in loss of natural forest, the land degradation neutrality target under the UN Convention to Combat Desertification, the UNFCCC NDCs, and the related Sustainable Development Goals (SDGs). Other agreements such as the Committee on Food Security could also benefit from biomass information that could be related to crop and pasture productivity. We explore five key national and international policies, commitments, and agreements in more detail.

2.2.1 Estimating Aboveground Biomass as an Essential Climate Variable

Essential Climate Variables (ECVs) are observable physical, chemical, or biological variables which can be important to characterize the Earth's climate system (GCOS 2010). Users of ECV products include global, regional, and national groups such as climate and integrated assessment modellers, World Meteorological Organization, national meteorological services, and others that require consistent products over large areas. These ECVs have been identified by the Climate Change Convention (UNFCCC) under its agenda item on research and systematic observations (under the leadership of the Global Climate Observing System (GCOS)) and include AGB (ECV T12). Other ECVs are also impacted by changes to ECV T12, including Albedo (ECV T8), Land Cover (ECV T9), FAPAR (Fraction of Absorbed Photosynthetically Active Radiation, ECV T10), LAI (Leaf Area Index, ECV T11), and fire disturbance (ECV T13) (GCOS 2010). To monitor ECV T12, GCOS proposed the development of gridded AGB datasets (GCOS 2015). Specifically GCOS highlights the further need for actions to:

1. Encourage inter-agency collaboration on developing methods to combine biomass estimates from current and upcoming missions (GCOS Action: 52);
2. Encourage inter-agency collaboration to develop validation methodologies (GCOS action 53)
3. Develop a set of validation sites covering the major forest types, particularly in the tropics (GCOS action 54);
4. Promote access to well-calibrated and validated regional- and national-scale biomass maps including uncertainty assessment (GCOS action 55);
5. Improve access to high-quality forest inventories, especially in the tropics, which can be used for research purposes and Reducing emissions from deforestation and forest degradation plus (REDD+) (GCOS action 56)

GCOS has recently updated the more specific requirements for estimating AGB as ECV [see also Global Observing Systems Information Center (GOSIC) (GCOS 2016; WMO 2018)]. In the context of ECVs, AGB is generally defined as the mass of live organic matter above the soil in terrestrial vegetation, thus excluding roots, litter, and dead wood (GCOS 2016). ECV AGB is quantified in mass of dry weight in metric tons. Biomass is quantified in g m^{-2} (dry matter) or multipliers (e.g., Mg ha^{-1}). There are multiple uncertainty metrics for biomass estimates, which include the relative and absolute systematic deviation and confidence interval or root-mean-square error (RMSE) for the overall biomass estimate and for each biomass class/range which can be estimated from reference data of better quality. RMSE alone is not a strong indicator of uncertainty since it mixes systematic deviation and precision. Systematic deviation is often the most significant error and varies among biomass ranges. Ideally, an error distribution as a function of biomass should be provided, but is rarely achieved in practice. As a minimum, a comparison of the ECV product with independent (in situ) reference data should provide uncertainty related to systematic deviation and precision among multiple biomass class/ranges in per cent (for relative) and Mg ha^{-1} (for absolute).

GCOS additionally asks for consideration of stability, which is to reflect the change in uncertainty over time (GCOS 2016). The metric and units for stability are the same as for uncertainty characterization and include the relative and absolute error assessments and confidence interval or RMSE, overall and by biomass class/range estimated using multi-date reference data of better quality. The stability requirements are more important than for overall uncertainty since the aim for multi-date ECV data is to provide information on biomass changes, and stability needs to be greater to actually be assessed. Since biomass varies in space, when moving to more detailed spatial resolutions, this variability quickly becomes large. This makes it very hard to estimate accurately when using multiple observation sources (which may have different spatial and temporal resolutions), especially for natural and tropical forests. The current GCOS suggestion is a horizontal resolution of 0.25 ha (50 m×50 m resolution) outside the (sub-) tropics and about 1 ha (100 m×100 m) for tropical forests. ECV biomass products should move beyond just being one-time products and ideally be provided annually or at least every 5 years to capture the most important changes.

2.2.2 National GHG Reporting Needs in the Context of the UNFCCC

Biomass measurements are of particular significance for policies related to the UNFCCC, because countries must monitor emissions related to gains and loss of biomass, and must report on these regularly (Herold and Johns 2007). Data and information on forest biomass stocks are not only required for international information/data needs, but also for national needs to support national policy formulation and mitigation activities.

Guidelines for reporting emissions and removals on the national level are provided by the IPCC Good Practice Guidance (GPG) and guidelines (GL), which countries use when reporting through their GHG inventories or other reporting avenues to the UNFCCC. A refinement of the 2006 IPCC GL (IPCC 2006) is ongoing with a planned release for 2019 and will be the first update of the Agriculture, Forestry and Other Land Use (AFOLU) guidelines since 2006. Updates will include additional information on tropical and subtropical forest biomass factors, as a result of recent efforts by REDD+ countries, and more information on the role of space-based data in biomass monitoring. More specifically, the guidelines will include a new section introducing the use of biomass estimates from maps estimated from space-based data. The use of biomass maps is increasingly important, since these wall-to-wall datasets have the potential to complement plot-based biomass measurements available through national forest inventories (NFIs). The characteristics and usefulness of biomass maps produced using space data for national GHG inventories depend on multiple factors:

1. The definitions for forest and biomass or AGB used to produce the map and how this definition relates to the one used in the national GHG inventory;
2. The type of space-based data sources in terms of spatial resolution, temporal coverage, and the degree to which the signal responds, or is sensitive to AGB;
3. The method used to construct the map. Methods can range from simple interpolation of field estimates of biomass using spatial covariates to more complex modelling of AGB using field estimates and observed space-based data signals (Goetz et al. 2009; Avitabile et al. 2011);
4. The availability and reliability of biomass-related field data, which are of better quality than the map and are needed to produce and validate the biomass map;
5. The degree to which map uncertainty is characterized and the manner in which it is used to assess systematic deviation and precision for large-area estimates in support of national GHG inventories.

With these factors in mind, biomass maps can improve the stratification of field-based biomass inventories, increase data in under-sampled or inaccessible areas, and serve as an independent data source for verification purposes (provided that the field data were not used to calibrate the biomass maps used).

In general, net carbon emissions from land use change are estimated by multiplying activity data (area of change) by an emission factor or removal factor (the carbon stock change per unit of change) (IPCC 2006). In this respect, the use of biomass maps for the estimation of carbon emissions at IPCC Tier 2 and Tier 3 levels can be achieved in three ways. First, biomass maps can provide the base to estimate carbon stocks. Such analyses require consistency among the activity data and biomass maps concerning definitions, geolocation, and spatial and temporal characteristics. The use of regionally aggregated carbon stocks (e.g., using average estimates for different forest types) helps to reduce inherent pixel-level uncertainties in biomass map data for national-scale estimations. Countries have used this approach to increase data density in areas under-sampled by field inventories. Second, biomass change and carbon emissions can be estimated directly from multi-date biomass maps. This approach provides an assessment of carbon stock changes in AGB from land use change and, in particular, also includes changes within forests remaining forests such as degradation and regrowth, management and harvest, and natural disturbances. This method requires consistent and well-calibrated biomass maps using field measured and space-based data to accurately estimate biomass changes and a quality requirement that has so far not been achieved for national GHG inventories. Improvements in both the field estimates of biomass change and remote sensing technologies in the coming years could lead to such approaches becoming efficient and accurate for GHG inventory purposes. Third, biomass maps can be integrated with remote sensing-assisted time series of land use change and/or with IPCC Tier 3 models to estimate emissions. This way, the biomass map data can be linked to land use to better reflect the complexity of forest-related carbon fluxes. A critical element of this type of application is the consistency among the various data sources and models concerning definitions (e.g., forest, biomass pools) and spatial and temporal data characteristics. Map unit prediction uncertainties in biomass maps propagate to larger area estimates and can lead to substantial uncertainties in national emissions estimation if not properly considered, particularly in relation to the effect of spatial autocorrelation. The application of these three approaches requires maps which are well-calibrated for national circumstances (McRoberts et al. 2019). Many available large-area biomass maps might not be consistent with national definitions of forest and/or biomass pools and often exhibit large systematic deviation in the estimation of carbon stock and changes for national and local assessments (Avitabile et al. 2016).

Incorporating biomass maps in national GHG inventory efforts requires a long-term perspective, such as establishing a data protocol to assure accessible data in the future. Comparability across timescales is essential to meet the IPCC guidance for consistency. Developing countries in particular have been making significant progress in using space-based data for providing activity data and national forest inventories for estimating emission factors (Romijn et al. 2015). These increasing capacities should help the potential uptake of novel biomass mapping and estimation possibilities by countries but so far few examples exist; additional national-scale efforts are needed to demonstrate that potential.

2.2.3 Result-Based Schemes Including Multilateral, Bilateral, and Project-Based Voluntary Initiatives

Efforts to reduce emissions from deforestation and degradation, sustainably manage forests, and conserve and enhance forest carbon stocks (REDD+) in developing countries have been moving into accessing result-based payments (third REDD+ phase), which requires robust and credible systems for measuring, reporting, and verification (MRV). REDD+ efforts have been re-enforced as one of the mitigation foci of the Paris Agreement (UNFCCC 2015; Turnhout et al. 2017). The result-based nature of REDD + requires that GHG emissions and removals related to REDD+ activities are benchmarked against a reference level to estimate their impact in units of tCO_2 equivalent. In many cases, emissions are estimated using activity data or land use change area (e.g., forest cover gain or loss) and an emissions factor, which is essentially the carbon stock or biomass of the forest. Biomass maps offer opportunities to not just look at loss or gain in the context of forests, but to look at change which can occur to varying degrees (for example, in the case of degradation or regrowth). Reporting under REDD+ frameworks is very much related to national GHG inventories; however, the objectives and requirements vary across frameworks, as outlined in Table 1.

REDD+ readiness efforts have triggered significant investments in improving developing countries' technical and institutional capacity for forest monitoring, including the capacity to utilize space-based data and remote sensing technologies which are seen as cost-effective methods. These investments and interest have spurred the forest observation community to develop new data sources through targeted research and methods (see, for example, http://www.gfoi.org/rd/), provide justification for space missions (including those for forest biomass), and develop improved guidance (GOFC-GOLD 2016).

2.2.4 Sustainable Development Goals

The SDGs, adopted by the UN in 2015, have 17 goals aimed at sustainably transforming our world (UN 2016). There are 232 targets, each with one or more indicators which must be monitored (UN 2017). New geospatial data and technologies are expected to be crucial for a number of the indicators. In particular, biomass data are essential for monitoring Goal 13 (Climate action) and Goal 15 (Life on land) (Romijn et al. 2016). Specifically, indicators 15.2.1 *Forest cover under sustainable forest management,* 15.2.2 *Net permanent forest loss,* and 15.3.1 *Percentage of land that is degraded over total land area* require forest biomass data. Several other targets could potentially use biomass data, for example goal 2.3 related to doubling agricultural productivity by 2030 could utilize biomass data for *indicator 2.3.2 Total Factor Productivity.* Likewise, indicator 7.2.1 *Renewable energy share in the total final energy consumption* could also include information on biomass for bioenergy in plantations. SDGs are typically set per country; however, spatial differences within a country can be more adequately monitored using space-based data, thus giving information on the extent to which SDG targets are being met at the sub-national level. The time frame for annual reporting on the SDG indicators is 2015-2030.

2.2.5 Local Forest and Land Management

Forest monitoring at the local level can serve many purposes, including to monitor tree growth and wood production in commercial operations including bioenergy production, to monitor illegal activities, such as illegal logging and illegal charcoal production, to provide information for the monitoring and reporting requirements related to sustainable timber certification, and to provide data on land use which can inform sustainable landscape management planning. Sustainable management plans may include fire prevention plans to inform forest laws and strategies. Balancing different utilities that forests can provide—both market goods and services and non-market goods—require, however, more comprehensive data that go beyond pure biomass information. In addition, biomass data are typically required at relatively small (sub-national) spatial scales, and with fine resolution, to capture details and subtle changes in the landscape. For this reason, airborne and field measured biomass data are often typically more suitable than space-based data. However, space-based information could be a useful data source if it can provide timely data at

Table 1 Various REDD+-related reporting fora, their purpose, and considerations of uncertainties

Context	Purpose	Requirements/consideration of uncertainties
REDD+-related submissions and National GHG inventories to UNFCCC	Regular reporting required at the national level, REDD+ reports, in developing countries in the context of result-based finance, and biennial update reports (BURs) for all countries	Key GHG source categories should be reported on Tier 2 level (using national emissions factors) in the national GHG Inventory. REDD+ FREL/FRLs, and REDD+ Annexes to the BURs are assessed through specific technical assessments and reviews by UNFCCC roster of experts. The stepwise approach allows continuous improvements including uncertainty
Green Climate Fund (GCF)	Finance for forest transformation for climate change mitigation and adaptation. Result-based payments for REDD+	In the case of result-based payments for REDD+, these are based on the UNFCCC Technical Assessment and Technical Reviews and a further assessment through a score card where uncertainty is part of the scoring system to allocate payments
Forest Carbon Partnership Facility's Carbon Fund (FCPF CF)	Pilot market transactions for REDD+	An IPCC Tier 2 is required to estimate emissions and removals. Uncertainty needs to be assessed and reported. Verification of emission reductions by an independent third party. Potential discount based on overall uncertainty of reported and verified emission reductions
Other initiatives (e.g. Verified Carbon Standard)	Issuance of emission reduction units for compliance and/or non-compliance carbon market	Varies by standard, but uncertainties should be assessed and considered. Independent verification is often part of the process. Often a preference for conservative approaches

fine resolutions. Space-based data can also be combined with field or airborne data, from NFIs for example, to upscale the data across larger areas. Currently, several space-based monitoring systems provide early warnings on forest disturbance, such as illegal activities (Hansen et al. 2016); this information can be combined with biomass maps to determine whether forest loss is occurring in high biomass forests, which are typically high biodiversity areas, and are therefore valuable habitats which be prioritized for protection.

2.3 Transparency and Broader Stakeholder Engagements

Enhancing transparency is fundamental to realize the bottom-up nature of the Paris Agreement, and should be understood as catalyst for action by providing, in the context of biomass data, open and consistent time series, transparent definitions, and assumptions and methodologies that will enhance the credibility and reliability of land use sector mitigation activities (Carter et al. 2018a) in both developing and developed countries. Transparency is already one of the principles of the IPCC guidance for GHG inventories together with accuracy, completeness, comparability, and consistency (UNFCCC 2003). Open and transparent estimations are also required in cases where results have to be assessed and verified. This includes publicly available input data sources, and the full transparency of methods, including algorithms. This builds public trust in reported results and encourages cooperation among multiple stakeholders including the collaborative design of interventions.

To better understand the user needs, a survey was undertaken with the aim of assessing stakeholders' data needs when monitoring GHG emissions in the land use sector (Romijn et al. 2018). A total of 557 answered the questionnaire and provided information on the type of organization for whom they were working. The survey results show that current open and freely available biomass datasets are not able to fulfil all stakeholder needs completely. Users were found to require detailed documentation regarding the scope and usability of the data (97% of all survey respondents), data sources which are comparable with alternatives, uncertainty estimates for evaluating mitigation options (94% of all survey respondents), more region-specific data with greater accuracy and detail for sub-national application, and regular updates and continuity for establishing consistent time series (83% of all respondents) (Romijn et al. 2018). Specifically, on biomass data needs, the survey asked different user groups their desired different levels of accuracy in terms of IPCC Tiers (Fig. 3). Responses are divided into groups, and ALL relates to all respondents, including the other groups, such as "local stakeholders", "donor organizations", "intergovernmental organizations", and others. Developing country governments include Non-Annex 1 governments, and developed country governments include Annex-1 governments according to the UNFCCC definition. "Research" includes universities and research institutes, and non-governmental organizations (NGO) include International (I) NGOs. Tier 1 emission factors are default values for broad continental forest types, e.g., from IPCC Emission Factor Data Base, Tier 2 emission factors are country specific data for key factors, e.g., from field inventories, permanent plots, and Tier 3 emission factors are based on detailed national inventory of key C stocks, comprehensive field sampling repeated at regular time intervals, soils data, and use of locally calibrated models. The government groups all selected Tier 2 or Tier 3, with Annex 1 (developed) country experts selecting Tier 3 (74%), even more so than the research community. The NGO community selected Tier 1 information (20%). On average, 43% of respondents selected Tier 3, while 47% of respondents selected Tier 2.

Multiple datasets are available which can be used by stakeholders to estimate biomass (Fig. 4). The datasets in Fig. 4 are IPCC (2003, 2006), Baccini et al. (2012), Saatchi et al. (2011), Harris et al. (2012). The bars show the proportion of respondents who were aware of the dataset and the proportion that both used it and found it "useful" or "very useful" when responding to the question "Was this dataset useful for your research and/or business purposes?" Other response options were "somewhat useful", "not very useful", and "not at all useful". Several differences were found in the awareness and usefulness of these datasets. While almost all stakeholders are aware and use the IPCC guidance, the knowledge and uptake of biomass maps were much less. This emphasized the current status that available maps are not used in GHG reporting by countries. There is a significant gap between stakeholders being aware of certain data sources and whether they actually found them useful for their purpose. This highlights the large divide between what is available and what is used. NGO respondents were more likely to be aware and users of these biomass datasets than other user groups.

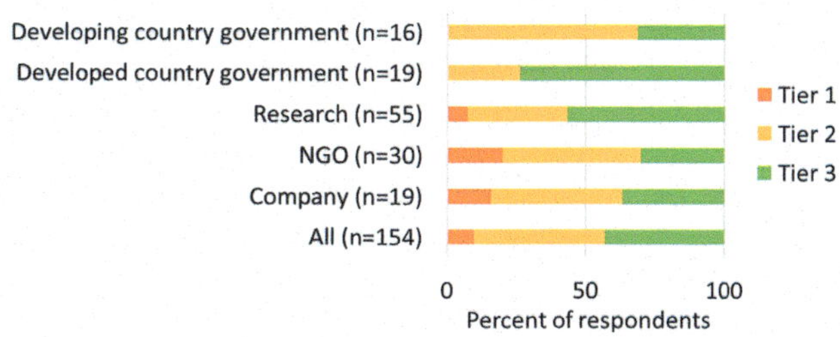

Fig. 3 Desired level of IPCC Tiers. Amended from Romijn et al. (2018)

3 Synthesis of User Needs and Gaps

A summary of the key initiatives, user groups, and requirements for biomass mapping in Table 2 emphasizes that biomass data derived from space-based data are useful for multiple purposes from national to international reporting related to forest change, to monitoring climate variables and sustainability goals. Some stakeholders are already using biomass maps along with other biomass data sources for their monitoring and reporting needs and find these useful. In light of these requirements, several gaps remain, which might be potentially filled by future space-based missions and the availability of those data. These gaps are described, and opportunities discussed, before recommendations for the future of biomass mapping are made.

The gaps and barriers for the use of biomass estimates from biomass maps by environmental policy and management (Table 2) can be summarized as follows:

Insufficient Level of Detail: Different users related to different policy and management processes require different levels of detail in terms of:

- *Spatial Resolution*: from spatially explicit/stand-level information of less than one hectare to many kilometres for global vegetation models;
- *Geographic Extent*: ranging from a forest management unit to global scale;
- *Temporal Coverage*: some users prefer a particular year; some want data that are most recent (near-real-time information) or available for longer time series;
- *Definition*: there is variability in needs in terms of which vegetation components are considered and whether dead biomass is included or not.

Given that most available regional and global biomass maps are limited in their spatial resolution and temporal coverage, most of them have not been found suitable for many of the applications we have assessed here.

Biomass Maps Versus Change: Most regional and global biomass maps exist as one-time products. While there is value in and need for high-quality biomass stocks data, many applications require biomass change information. Although case studies exist (Næsset et al. 2013, 2015), it is only recently that reports of estimation of biomass change over large

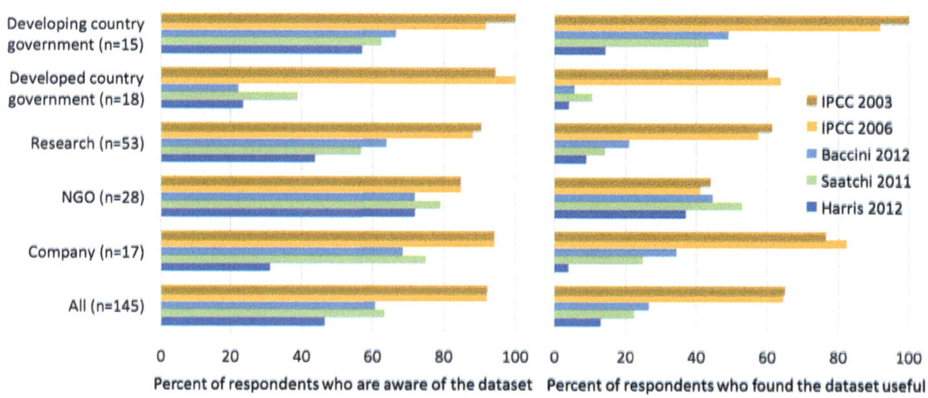

Fig. 4 The most frequently selected five datasets which the questionnaire respondents were aware of when asked "Are you aware of this dataset" and "Was this dataset useful for your research and/or business purposes?"

areas from space-based data are starting to appear in the scientific literature (Baccini et al. 2017).

Conflicting Information: Urgency by researchers and scientists to improve space-based monitoring capabilities has naturally led to a diversification of methods, with positive implications in terms of innovation, and demonstrating the potential. Different datasets are available and often provide inconsistent information which confuses users. Identifying the sources of differences in these maps is difficult, due to lack of detailed descriptions/transparency, different definitions, temporal coverage, and also the lack of information on the uncertainty of the maps (Avitabile et al. 2016; Böttcher et al. 2017). Enhanced user's guidance on which data sources are most suitable for which purposes could be included in the upcoming 2006 IPCC GL refinement. The user should be provided with documentation of the product to understand which one to use for a certain purpose and why.

Limitations in the Complementary Use of Field/Airborne and Space Data: Biomass monitoring still suffers from the conceptual and technical divide between space-based data and traditional means of gathering biomass data (i.e. collecting data in the field). High-quality in situ data are needed for calibrating and validating space-based biomass maps, and the quality of this calibration or validation data must be higher than that of the map (Stehman 2009; GFOI 2016). But lack of in situ data in certain locations, use of different definitions of forests and biomass, and uncertainties in both field and space-based data have led to limited complementarity. Current biomass mapping from space is essentially disconnected from plot-based national forest inventory efforts (Böttcher et al. 2017).

Lack of Capacity of Key Stakeholders: One of the main barriers to progress in general and relating in particular to the uptake of space-based biomass derived data for biomass monitoring is the capacity of users such as country experts and non-technical experts to access and use the data. A study of countries' capacities to measure, monitor, and report on REDD+ found many to have low capacity, particularly in Africa (Romijn et al. 2012). Despite efforts to increase capacities, a capacity gap remains in many developing countries (Romijn et al. 2015) which continue to limit the uptake of new technologies. The use of biomass maps by key stakeholders is also hindered by the lack of capacity to understand the methods that were used to produce them.

Gaps Between Theoretical Opportunities and Policy and Management Praxis: An often-overlooked limitation of biomass and forest monitoring capabilities arises from a gap between what is needed by users, what has been successfully demonstrated in a research environment, and what can be developed and maintained in an operational environment, such as a country's National Forest Monitoring System. The technologies offered may not meet the user requirements or be practical for operational implementation, and an analysis of user requirements should be undertaken to ensure that what is offered can be adopted successfully. In particular, investments to develop new data sources and methods are significant and often do not consider how such novel approaches can be scaled geographically and then reliably maintained at a reasonable cost by users. This problem is often especially acute in countries with low technical capacity. Unclear institutional arrangements and turnover of technical staff can create problems for responding to new requirements and technical opportunities and for sustainability of monitoring efforts (Ochieng et al. 2016).

Uncertainty Characterization is Missing or Not Suited: Currently global maps are available, but using these datasets to estimate biomass at national scale does not meet the level of uncertainty required for REDD+ (Mitchell et al. 2017). Some do not provide adequate uncertainty estimates so these have to be estimated by the users, which can be difficult, and this process relies on assumptions and thus is uncertain in itself (Carter et al. 2018b). To add to the challenge, most biomass mapping accuracy varies by forest type and biomass

Table 2 Summary of key initiatives, user groups, and requirements for biomass mapping

Policy initiative	Key user groups	Biomass data needs	Considerations on uncertainty
IPCC/GCOS	Global climate data users and global assessments (i.e. IPCC WG 1 and WG3)	AGB as one of the Essential Climate Variables (ECVs) Optimal: regular global forest biomass monitoring at 50–100 m resolution at annual intervals Focus on long-term consistency as input to global and regional climate and vegetation models	Uncertainty and stability of ECV products are to be assessed and reported for relative and absolute systematic deviation and confidence interval or RMSE, overall and by biomass class/range estimated using reference data of better quality Coordinated international activities for uncertainty characterization and reference data collection are required
UNFCCC national reporting, transparency framework and global stocktake	National GHG inventory experts and UNFCCC roster of experts for technical assessments and review	Estimation of carbon emissions and removals from forest changes; also including non-forest areas with significant woody biomass (i.e. agriculture/grasslands) Reporting done by countries and all countries need to report regularly (at least a summary of the GHG inventory biennially) Various options for integrating biomass maps in national GHG inventories Greatest need in tropical countries where availability of plot data and sustainability of NFIs remains an issue Regular global stocktake by UNFCCC may require data for reconciling country reporting to provide global estimates of progress	IPCC guidance: estimates should be unbiased (on the national level) and uncertainties reduced as much as is practical IPCC Tiers: national data needed for Tier 2 Uncertainty estimates required on national estimate Continuous improvement process as part of national GHG inventories Need for consistent (global) data sources (including uncertainty characterization) for technical assessment and global stocktake

Table 2 (continued)

Policy initiative	Key user groups	Biomass data needs	Considerations on uncertainty
REDD + frameworks	National REDD + agencies, forest monitoring experts, REDD + donors/investors and independent assessment bodies	Estimation of emission factors and forest carbon stock changes for establishing forest reference level and REDD + results to access result-based finance Assessment of potential forest mitigation activities Requirements can vary according to the financial initiative (i.e. GCF, FCPF CF, bilateral organizations, voluntary markets; see Table 1) Independent data sources that verification teams can use for comparison	Uncertainty requirements vary by performance scheme Consistency, accuracy, and transparency needs to be sufficient to be convincing for technical assessments and verification of the results of REDD+ activities Stepwise improvements possible and encouraged Discounts in payments due to high uncertainty or conservative adjustments possible
Sustainable development goals	UN Stats, national statistical office responsible for SDG indicator reporting	Biomass changes as one sub-indicator of SDG 15.3.1 but also important input for others that are required for all countries Detailed requirements are still developing but reporting is to be done by countries on annual basis 2015–2030	Not yet known but reporting of SDG indicator should allow for the assessment of trends 2015–2030
Local forest and land management	National forest institutions, forest industry, certification schemes	Management plans Forest laws and strategies Fire prevention plans Bioenergy plans Etc	Emphasis on stand-level/spatially explicit information provision Quantification and understanding of uncertainties should be incorporated in management decisions
Stakeholder engagement and transparency	Societal actors and broader set of forest-related stakeholders, including local communities	Free and open biomass data accessible to non-technical stakeholders Biomass sometimes used for independent assessments, as a simple proxy for forest health, intactness and habitat quality, and for forest resources assessments, etc	Transparent information and uncertainty characterization required by all stakeholder groups Uncertainty requirements vary by stakeholder group

range. Biomass mapping typically performs poorly in high biomass and dry regions of the tropics, and uncertainties are particularly large there (Avitabile et al. 2016). Users have specific requirements on what they expect from uncertainty characterization in terms of geographic scope (i.e. global vs. country), type of biomass estimate needed (i.e. total or average per country or forest type), and the uncertainty measure required. Such user-targeted uncertainty characterization is essentially lacking from all large-area biomass mapping products (Böttcher et al. 2017).

3.1 Summary and Recommendations

The availability of new space-based data targeted at forest biomass mapping can be a game changer in terms of the level of detail and uncertainty that will become available for monitoring applications in the short term. At the same time, there have never been as many, and as explicit requests from environmental policy and resource management for quality biomass data. The question then is whether such novel data sources will also be a game changer when it comes to filling the existing data and information gaps on forest biomass for these applications. In light of the assessment of user needs here, we structure our recommendations in two main blocks, one related to the user engagement, and one for a coordinated and collaborative data provision process.

3.1.1 Enhancing User Engagement

There is a need to regularly gather, synthesize, and communicate user requirements of forest biomass data and space-based AGB data products. Products which better meet user needs will support forest monitoring and wider land use management applications and improve environmental policy to enable countries to help meet international climate and sustainable goals. There is no doubt that the advent of new space-based biomass missions will significantly improve the spatially and timely provision of forest biomass data worldwide in future. While research and investments in biomass monitoring technology advance, there is still little information on how it can be leveraged through the value chain of land use planning, enforcement, and policy making to improve outcomes for forests and sustainable development. The greatest advancements in forest monitoring technology often happen in agencies, laboratories, companies, or institutions in the developed world, without appropriate guidance and best practices which guide users (in particular those in developing countries) to adopt and use this technology. Consideration of institutional and financial requirements to scale, operationalize, and maintain the new technology in the context of specific applications should be undertaken and be an essential part of preparatory activities.

Further effort is required to stimulate uptake of biomass data by key stakeholders. While the integration of biomass mapping products in climate and vegetation models is already progressing (i.e. ESA Biomass CCI: http://cci.esa.int/), there has been little progress in demonstrating how biomass mapping from space can be integrated in national forest monitoring and national GHG inventories efforts. International institutions and organizations (such as IPCC, The Food and Agriculture Organization of the United Nations (FAO), The Global Forest Observations Initiative (GFOI), Global Observation for Forest Cover and Land Dynamics (GOFC-GOLD), Committee on Earth Observation Satellites (CEOS)) should stimulate and implement demonstration activities that then can lead to experiences and improved guidance, training materials, and capacity building efforts. Once there are successful demonstrations with users, it becomes much more straightforward to include them in ongoing capacity development processes. The 2019 refinement of the 2006 IPCC GL is expected to provide some initial guidance.

Biomass maps should be developed as a co-creation of both producers and users. Particular focus is needed to match estimates from biomass maps produced using space data with field-based biomass estimates and to harmonize these biomass estimates based on national definitions. The plot size of the data should also be considered since, for example, sub-national statistics or approximated geolocation of small (i.e. 0.1 ha) plots are not suitable for direct comparison and integration with space-based data. This requires a sustained dialogue and collaboration between the remote sensing community and the national forest inventory institutions, which is essential to combine, integrate, and reconcile biomass estimates obtained with different approaches (e.g., sample-based vs. model-based or mapping approach) and driven by different data sources, either field data or remotely sensed data which include space-based sources. Moreover, this dialogue has to be complemented by the implementation of data sharing policies that allow the access of these field-based datasets to users in general and specifically the remote sensing community, and allows in turn access to the national forest inventory institutions of remotely sensed data.

Uncertainty information is required by all stakeholders interested in using biomass maps (Romijn et al. 2018). While the data producing community aims to provide uncertainty estimates [i.e. through the CEOS WGCV efforts (Duncanson et al. 2019)], it is important to note that different applications have specific requirements on what they expect from biomass and uncertainty characterization in terms of:

- Geographic scope which ranges from global scale (i.e. for use in climate and vegetation models) or national level estimates since it is most relevant to country GHG inventories;
- The final biomass estimate needed by the users that can be the total or average biomass per country or forest type or a unit-specific value;
- The uncertainty estimates required, which can include relative and absolute systematic deviation and confidence interval or RMSE. They can be provided for the overall estimate or by biomass class/range;
- Whether a biomass estimate is needed for one time (i.e. one year) or for multiple years, and whether it includes the estimation of biomass changes;
- Compliance with IPCC good practice guidelines: (1) "neither over- nor underestimates so far as can be judged," and (2) "uncertainties are reduced as far as is practicable" (IPCC 2006).

This suggests that one biomass product and uncertainty characterization approach will not work for all users—so any effort of assessing biomass products should be co-developed by both producers and users. Often user organizations have their own forest and biomass data that they would like to integrate with biomass map data. Such efforts are, thus, not only aiming at the technical credibility of a biomass map product but also at creating ownership and saliency that are often most important for ensuring user uptake. For example, a biomass map can only become part of IPCC Tier 2 or Tier 3 emission estimation if integrated with field data acquired in the country. It is also often the user alone that can fully judge what implications estimated biomass and uncertainties have on their application. This information can be fed back to the producer on where to improve.

An increasing democratization of forest information can bring about a transformation in the forest sector because it will be easier to hold stakeholders (i.e. a state or the private sector) accountable for their commitments, like those related to REDD+ (Böttcher et al. 2017). Enhancing transparency is one fundamental step to make the bottom-up nature of the Paris Climate Agreement work in practice, and to facilitate open exchange of data and information, and more participatory monitoring and impact assessments involving the private sector, civil society and government agencies (Romijn et al. 2018). The open-source nature of many space-based data and products is leading the way, and this should be expanded to include other forest-related data sources. National Forest Agencies and international partners should develop protocols/agreement to make more accessible field plot data with precise geolocation to the research communities.

3.1.2 Promoting Collaborative and Sustained Data Provision

There is a need for international forest monitoring and data coordination groups (GFOI, GOFC-GOLD, and CEOS) to represent the needs of the data users and encourage data providers/programme funding organizations to work collaboratively to provide complementary, relevant, transparent, and validated biomass data and information.

Although the mission of space-based data providers is to provide stand-alone information on biomass, and there have been some advances in terms of data sharing (Albinet et al. 2019; Duncanson et al. 2019), further collaborations are needed. Different maps which are produced independently can cause confusion, and guidance is needed if the user is to adopt the appropriate map(s). In the previous section, we have already outlined the need to work with the user from the beginning. Such a partnership, however, requires a coordinated and collaborative data provision process to be successful. The various biomass-focussed space missions (see Fig. 1) have been thus far planned individually, but the one sensor/one map approach will not be the best solution for many of the applications we have presented here. The actual strength of these missions, which has not been fully exploited, is in their combination to increase accuracy, detail, and geographic coverage for biomass estimation, and to develop an increasingly longer time series that is certain and consistent enough to track biomass changes. Space agencies and organizations providing such data are encouraged to take action for an integrated suite of products taking advantage of having multiple measurements (radar, lidar) and creating consistent time series from multiple sensors. Historical data sources used for biomass mapping to expand existing time series should also be considered in this integrated suite, to aim for the timely delivery of new biomass and biomass change data. To achieve this, an increasing investment and collaboration in R&D that looks into the combined potential of these missions is needed to advance the technical underpinnings and to transition from viable prototypes to operational systems within a timeframe consistent with ambitious targets set by policy frameworks.

The coordination and collaboration activities advocated here can only work if the space-based data sources are available, free, and open. At the time of writing, all missions described in Fig. 1, have an open data policy. Previous open data space missions, such as the Landsat program and the EU/ESA Copernicus missions, have demonstrated the tremendous value of freely available earth observation data for international and national environmental policy and management applications.

Almost, all of the upcoming space-based biomass mapping efforts are planned as one-off research missions. We encourage taking steps now to operationalize biomass maps produced using space data (e.g. as part of EU Copernicus programme) building upon the experiences and progress made for ensuring seamless continuity and consistency of biomass mapping from space after currently planned missions end. A long-term vision for data availability is necessary for all users needing sustained observation of biomass changes (i.e. for climate users and reporting of the SDGs).

Any biomass maps produced from space data need to be underpinned by adequate field reference data for calibration and validation. In general, any calibration/validation has to rely on a coordinated effort to provide consistent protocols, and to collect and share reference data. Space agencies often have limitations in supporting large in situ data collection efforts for forest biomass. There is need to find mechanisms to increase support and partnerships with forest plot networks and implementing agencies that both can provide reference data that covers a range of forest types and disturbance regimes worldwide (see Phillips et al. 2019). Field surveys need to be planned to allow for better integration with space-based data (Næsset et al. 2015), in particular with NFI data that tend to use small plots, and are often not acquired with space-data integration in mind.

High-quality field data are needed for both calibration and validation, and the requirements for each can vary. Calibration data should be able to provide enough detail in order to allow the inter-comparison and inter-calibration of multi-source space-based data (i.e. from various radar wavelengths and lidar). Such data can increasingly be provided from terrestrial laser scanning (Disney et al. 2019) and lidar-based drone measurements (Kellner et al. 2019), including a comprehensive error analysis (McRoberts et al. 2019). Reference data for product validation should be able to support uncertainty analysis targeted for specific user needs. The product validation procedures by CEOS (Duncanson et al. 2019) are important but tend to focus on the producer perspective. The engagement with different users is essential in product uncertainty analysis (see Phillips et al. 2019).

In summary, the key challenge is to avoid a widening gap between what is evolving and demonstrated in research and in the context of space-based missions, and the diverse demands of users, particularly in terms of the need for operational systems. This can only be achieved through a long-term collaborative effort among producers and users. Such efforts are currently most advanced for climate-related biomass mapping, which can lead the way to advanced guidance and practices for the use of forest biomass monitoring data outside of climate applications. Further work should particularly look into how new biomass monitoring technologies can be integrated in the processes and institutions of land use planning, enforcement, and policy making to achieve forest and climate goals.

Acknowledgements This research has been funded by the European Space Agency (ESA) as part of the CCI Biomass project of the Climate Change Initiative (CCI) (ESA ESRIN/Contract No. 4000123662), and also by FAO, GFOI, GOFC-GOLD, and Wageningen University. We thank those involved in the EC supported H2020 REDD+Copernicus project and those who contributed to GFOI Expert Workshops (www.gfoi.org/rd). M.J. Sanz was supported by the Spanish Government through María de Maeztu excellence accreditation MDM-2017-0714.

References

Albinet C, Whitehurst AS, Jewell LA, et al (2019) A joint ESA-NASA multim-mission algorithm and analysis platform for BIOMASS, NISAR and GEDI. Surv Geophys (In review)

Asner GP, Clark JK, Mascaro J, Garc GAG (2011) High-resolution mapping of forest carbon stocks in the Colombian Amazon. Biogeosciences 9(7):2683–2696. https://doi.org/10.1186/1750-0680-6-7

Avitabile V, Herold M, Henry M, Schmullius C (2011) Mapping biomass with remote sensing: a comparison of methods for the case study of Uganda. Carbon Balance Manag 6:1–14. https://doi.org/10.1186/1750-0680-6-7

Avitabile V, Herold M, Heuvelink GBM et al (2016) An integrated pan-tropical biomass map using multiple reference datasets. Glob Change Biol 22:1406–1420. https://doi.org/10.1111/gcb.13139

Baccini A, Goetz SJ, Walker WS et al (2012) Estimated carbon dioxide emissions from tropical deforestation improved by carbon-density maps. Nat Clim Change 2:182–185. https://doi.org/10.1038/nclimate1354

Baccini A, Walker W, Carvalho L et al (2017) Tropical forests are a net carbon source based on aboveground measurements of gain and loss. Science 358:230–234. https://doi.org/10.1126/science.aam5962

Baede A, van der Linden P, Verbruggen A (2007) Annex to IPCC fourth assessment report. Switzerland, Geneva

Batjes NH (1996) Total carbon and nitrogen in the soils of the world. Eur J Soil Sci 47:151–163

Böttcher H, Herrmann LM, Herold M, et al (2017) Independent Monitoring: Building trust and consensus around GHG data for increased accountability of mitigation in the land use sector

Carter S, Arts B, Giller KE et al (2018a) Climate-smart land use requires local solutions, transdisciplinary research, policy coherence and transparency. Carbon Manag. https://doi.org/10.1080/17583004.2018.1457907

Carter S, Herold M, Avitabile V et al (2018b) Agriculture-driven deforestation in the tropics from 1990–2015: emissions, trends and uncertainties. Environ Res Lett 13:014002. https://doi.org/10.1088/1748-9326/aa9ea4

Disney MI, Burt A, Calders K, et al (2019) Innovations in ground and airborne technologies as reference and for training and validation: terrestrial laser scanning (TLS). Surv Geophys (In review)

Duncanson L, Armston J, Disney M, et al (2019) The importance of global land product validation: towards a standardized protocol for aboveground biomass. Surv Geophys (In review)

Ellison D, Morris CE, Locatelli B et al (2017) Trees, forests and water: cool insights for a hot world. Glob Environ Change 43:51–61. https://doi.org/10.1016/j.gloenvcha.2017.01.002

FAO (2015) FRA 2015 terms and definitions. Rome, Italy

GCOS (2010) Implementation plan for the global observing system for climate change in support of the UNFCCC. WMO, IOC, UNEP, ICS, Geneva, Switzerland

GCOS (2015) Staus of the global observing system for climate. Full report. GCOS-195. World Meteorological Organization, Geneva, Switzerland

GCOS (2016) The global observing system for climate: implementation needs. GCOS-200. World Meteorological Organization, Geneva, Switzerland

GFOI (2016) Integration of remote-sensing and ground-based observations for estimation of emissions and removals of greenhouse gases in forests: methods and guidance from the global forest observations initiative

GlobBiomass (2017) Due GlobBiomass final report

Goetz SJ, Baccini A, Laporte NT et al (2009) Mapping and monitoring carbon stocks with satellite observations: a comparison of methods. Carbon Balance Manag 4:2. https://doi.org/10.1186/1750-0680-4-2

GOFC-GOLD (2016) A sourcebook of methods and procedures for monitoring and reporting anthropogenic greenhouse gas emissions and removals associated with deforestation, gains and losses of carbon stocks in forests remaining forests, and forestation report version. COP22-1

Griscom BW, Adams J, Ellis PW et al (2017) Natural climate solutions. PNAS 114:11645–11650. https://doi.org/10.1073/pnas.1710465114

Haberl H, Erb KH, Krausmann F et al (2007) Quantifying and mapping the human appropriation of net primary production in earth's terrestrial ecosystems. PNAS 104:12942–12947. https://doi.org/10.1073/pnas.0704243104

Hansen MC, Krylov A, Tyukavina A et al (2016) Humid tropical forest disturbance alerts using landsat data. Environ Res Lett 11:34008. https://doi.org/10.1088/1748-9326/11/3/034008

Harris NL, Brown S, Hagen SC et al (2012) Baseline map of carbon emissions from deforestation in tropical regions. Science 336:1573–1576. https://doi.org/10.1126/science.1217962

Hermans-Neumann K, Gerstner K, Geijzendorffer IR et al (2016) Why do forest products become less available? A pan-tropical comparison of drivers of forest- resource degradation. Environ Res Lett 11:125010. https://doi.org/10.1088/1748-9326/11/12/125010

Herold M, Johns T (2007) Linking requirements with capabilities for deforestation monitoring in the context of the UNFCCC-REDD process. Environ Res Lett 2:045025. https://doi.org/10.1088/1748-9326/2/4/045025

Houghton RA, Hall F, Goetz SJ (2009) Importance of biomass in the global carbon cycle. J Geophys Res 114:G00E03. https://doi.org/10.1029/2009JG000935

IPCC (2003) Good practice guidance for land use, land use change and forestry. Kanagawa, Japan

IPCC (2006) 2006 IPCC guidelines for national greenhouse gas inventories. In: Eggleston HS, Buendia L, Miwa K, Ngara T, Tanabe K (eds) National greenhouse gas inventories programme. IGES, Japan

Irlam L (2017) Global costs of carbon capture and storage—2017 update. Melbourne, Australia

Kellner JR, Armston JD, Birrer M, et al (2019) New opportunities for forest remote sensing through ultra-high-density drone lidar. Surv Geophys (In review)

Kumar L, Mutanga O (2017) Remote sensing of above-ground biomass. Remote Sens 9:1–8. https://doi.org/10.3390/rs9090935

Le Quéré C, Moriarty R, Andrew RM et al (2015) Global carbon budget 2015. Earth Syst Sci Data 7:349–396. https://doi.org/10.5194/essd-7-349-2015

Lorenz R, Pitman AJ (2014) Effect of land-atmosphere coupling strength on impacts from Amazonian deforestation. Geophys Res Lett 41:5987–5995. https://doi.org/10.1002/2014GL061017

Lucas RM, Mitchell AL, Armston J (2015) Measurement of forest above-ground biomass using active and passive remote sensing at large (subnational to global) scales. Curr For Rep 1:162–177. https://doi.org/10.1007/s40725-015-0021-9

McRoberts R, Næsset E, Liknes GC, Chen Q, Walters BF, Saatchi S, Herold M (2019) Using a finer resolution biomass map to assess the accuracy of a regional map-based estimate of forest biomass. Surv Geophys. https://doi.org/10.1007/s10712-019-09507-1

Mitchell AL, Rosenqvist A, Mora B (2017) Current remote sensing approaches to monitoring forest degradation in support of countries measurement, reporting and verification (MRV) systems for REDD+. Carbon Balance Manag 12:1–22. https://doi.org/10.1186/s13021-017-0078-9

Næsset E, Bollandsås OM, Gobakken T et al (2013) Model-assisted estimation of change in forest biomass over an 11 year period in a sample survey supported by airborne LiDAR: a case study with post-stratification to provide "activity data". Remote Sens Environ 128:299–314. https://doi.org/10.1016/j.rse.2012.10.008

Næsset E, Bollandsås OM, Gobakken T et al (2015) The effects of field plot size on model-assisted estimation of aboveground biomass change using multitemporal interferometric SAR and airborne laser scanning data. Remote Sens Environ 168:252–264. https://doi.org/10.1016/j.rse.2015.07.002

Ochieng RM, Visseren-hamakers IJ, Arts B et al (2016) Institutional effectiveness of REDD + MRV: countries progress in implementing technical guidelines and good governance requirements. Environ Sci Policy 61:42–52. https://doi.org/10.1016/j.envsci.2016.03.018

Phillips OL, Sullivan MJ, Baker TR, et al (2019) Species matter: multi-scalar impacts of wood density on tropical forest biomass. Surv Geophys (In review)

Prentice IC, Farquhar GD, Fasham MJR et al (2001) The carbon cycle and atmospheric carbon dioxide. In: Houghton JT, Ding Y, Griggs DJ et al (eds) Climate change 2001: the scientific basis. Contribution of Working Group I to the third assessment report of the intergovernmental panel on climate change. Cambridge University Press, Cambridge

Romijn E, Herold M, Kooistra L et al (2012) Assessing capacities of non-annex I countries for national forest monitoring in the context of REDD+. Environ Sci Policy 19–20:33–48. https://doi.org/10.1016/j.envsci.2012.01.005

Romijn E, Lantican CB, Herold M et al (2015) Assessing change in national forest monitoring capacities of 99 tropical countries. For Ecol Manag 352:109–123. https://doi.org/10.1016/j.foreco.2015.06.003

Romijn E, Herold M, Mora B, et al (2016) Monitoring progress towards sustainable development goals the role of land monitoring. http://www.gofcgold.wur.nl/documents/newsletter/Sustainable_Development_Goals-infobrief.pdf. Accessed 11 Dec 2018

Romijn E, De Sy V, Herold M et al (2018) Independent data for transparent monitoring of greenhouse gas emissions from the land use sector: what do stakeholders think and need? Environ Sci Policy 85:101–112. https://doi.org/10.1016/j.envsci.2018.03.016

Saatchi SS, Harris NL, Brown S et al (2011) Benchmark map of forest carbon stocks in tropical regions across three continents. Proc Natl Acad Sci USA 108:9899–9904. https://doi.org/10.1073/pnas.1019576108

Stehman S (2009) Sampling designs for accuracy assessment of land cover. J Remote Sens 30:5243–5272

Tompo E, Gschwantner T, Lawrence M, McRoberts R (2010) National forest inventories. Pathways for common reporting. Springer, Berlin
Turnhout E, Gupta A, Weatherley-Singh J, Vijge MJ, De Koning J, Visseren-Hamakers IJ, Herold M, Lederer M (2017) Envisioning REDD+ in a post-Paris era: between evolving expectations and current practice. Wiley Interdiscip Rev Clim Change 8(1):e425
UN (2016) UN sustainable development goals. https://sustainabledevelopment.un.org/?menu=1300. Accessed 22 Dec 2016
UN (2017) Revised list of global sustainable development goal indicators. Report of the inter-agency and expert group on sustainable development goal indicators (E/CN.3/2017/2), Annex III. 1–26
UNFCCC (2003) Report of the conference of the parties on its eighth session: 17/CP.8. New Delhi, India
UNFCCC (2015) The Paris agreement FCCC/CP/2015/L.9/Rev.1. Paris, France
WMO (2018) Essential climate variables: global climate observing system. https://gcos.wmo.int/en/essential-climate-variables
Zomer RJ, Neufeldt H, Xu J et al (2016) Global tree cover and biomass carbon on agricultural land: the contribution of agroforestry to global and national carbon budgets. Sci Rep 6:29987. https://doi.org/10.1038/srep29987

Affiliations

Martin Herold[1] · Sarah Carter[1] · Valerio Avitabile[2] · Andrés B. Espejo[3] · Inge Jonckheere[4] · Richard Lucas[5,6] · Ronald E. McRoberts[7] · Erik Næsset[8] · Joanne Nightingale[9] · Rachael Petersen[10] · Johannes Reiche[1] · Erika Romijn[1] · Ake Rosenqvist[11] · Danaë M. A. Rozendaal[1] · Frank Martin Seifert[12] · María J. Sanz[13] · Veronique De Sy[1]

✉ Sarah Carter
sarah.carter@wur.nl

1 Laboratory of Geo-Information Science and Remote Sensing, Wageningen University and Research, Droevendaalsesteeg 3, 6708 PB Wageningen, The Netherlands

2 Joint Research Centre, European Commission, Via E. Fermi 2749, 21027 Ispra, Italy

3 Climate Change Group, The World Bank, 1818 H Street, NW, Washington, DC 20433, USA

4 Food and Agriculture Organization of the United Nations, UN-REDD Programme, REDD+ Team / NFM Group, Viale delle Terme di Caracalla, 00193 Rome, Italy

5 Earth Observation and Ecosystem Dynamics Research Group, Department of Geography and Earth Sciences (DGES), Aberystwyth University, Aberystwyth, Ceredigion SY23 3DB, UK

6 Centre for Ecosystem Science, School of Biological Earth and Environmental Sciences (BEES), The University of New South Wales, High Street, Kensington, NSW 2052, Australia

7 Northern Research Station, U.S. Forest Service, Saint Paul, MN 55108, USA

8 Faculty of Environmental Sciences and Natural Resource Management, Norwegian University of Life Sciences, P.O. Box 5003, 1432 Ås, Norway

9 National Physical Laboratory, Hampton Road, Teddington, Middlesex TW11 0LW, UK

10 World Resources Institute, Washington, DC 20002, USA

11 Solo Earth Observation (soloEO), Tokyo 104-0054, Japan

12 ESA-ESRIN, Via Galileo Galilei, Frascati, Italy

13 BC3 Basque Centre for Climate Change, Sede Building 1, 1st Floor, Scientific Campus, University of the Basque Country, 48940 Leioa, Spain

Surveys in Geophysics (2019) 40:779–801
https://doi.org/10.1007/s10712-019-09539-7

The Status of Technologies to Measure Forest Biomass and Structural Properties: State of the Art in SAR Tomography of Tropical Forests

Stefano Tebaldini[1] · Dinh Ho Tong Minh[2] · Mauro Mariotti d'Alessandro[1] · Ludovic Villard[3] · Thuy Le Toan[3] · Jerome Chave[4]

Received: 19 June 2018 / Accepted: 11 April 2019 / Published online: 16 May 2019

Abstract
Synthetic aperture radar (SAR) tomography (TomoSAR) is an emerging technology to image the 3D structure of the illuminated media. TomoSAR exploits the key feature of microwaves to penetrate into vegetation, snow, and ice, hence providing the possibility to see features that are hidden to optical and hyper-spectral systems. The research on the use of P-band waves, in particular, has been largely propelled since 2007 in experimental studies supporting the future spaceborne Mission BIOMASS, to be launched in 2022 with the aim of mapping forest aboveground biomass (AGB) accurately and globally. The results obtained in the frame of these studies demonstrated that TomoSAR can be used for accurate retrieval of geophysical variables such as forest height and terrain topography and, especially in the case of dense tropical forests, to provide a more direct link to AGB. This paper aims at providing the reader with a comprehensive understanding of TomoSAR and its application for remote sensing of forested areas, with special attention to the case of tropical forests. We will introduce the basic physical principles behind TomoSAR, present the most relevant experimental results of the last decade, and discuss the potentials of BIOMASS tomography.

Keywords Remote sensing · Forestry · Synthetic aperture radar (SAR) · Tomography · Microwaves · Aboveground biomass · Forest height · Terrain topography · BIOMASS

✉ Stefano Tebaldini
stefano.tebaldini@polimi.it

[1] Dipartimento di Elettronica, Informazione e Bioingegneria (DEIB), Politecnico di Milano, Via Ponzio 34/5, 20133 Milan, Italy

[2] UMR TETIS, IRSTEA, University of Montpellier, 34093 Montpellier, France

[3] Centre d'Etudes Spatiales de la Biosphère, CNRS-CNES-Université Paul Sabatier-IRD, Toulouse Cedex 9, France

[4] Laboratoire Evolution et Diversité Biologique, UMR5174, CNRS, Université Paul Sabatier, IRD, Toulouse Cedex 9, France

1 Introduction

Synthetic aperture radar (SAR) imagery is nowadays a most relevant technology for remote sensing of the natural environment, as witnessed by the increasing number of spaceborne SARs and their improving performance (Moreira 2014). Indeed, SAR systems provide a powerful and unique combination of features relevant to remote sensing, such as large spatial coverage, resolution of the order of few meters, and the possibility to operate largely independent of weather conditions and solar illumination (Curlander and McDonough 1991). Another most relevant feature, which is peculiar to microwave systems, is the capability to probe the interior part of the illuminated media. Indeed, microwaves, especially at long wavelengths, can penetrate for meters, or even tens of meters, into natural media that are non-transparent at optical frequencies, as it is the case for vegetation, snow, ice, and sand. This feature makes SAR data sensitive to the vertical structure of those media, hence providing access to features that are hidden to optical and hyper-spectral sensors, see for example Mariotti d'Alessandro et al. (2013), Tebaldini et al. (2016a, b), Rekioua et al. (2017), Paillou et al. (2003). The downside is that microwave scattering from distributed media may be quite complex, involving a number of different mechanisms through which the wave is backscattered. Considering forested areas, which are the focus of this paper, the radar signal is determined by direct scattering from elements within the vegetation canopy, from the underlying terrain, as well as from multiple scattering resulting from the waves bouncing off the ground in the direction of the radar after being scattered downward by the tree canopy and trunks (Treuhaft and Siqueira 2000; Mariotti d'Alessandro et al. 2013; Papathanassiou and Cloude 2001; Smith-Jonforsen et al. 2005; Lin and Sarabandi 1992). As a result, SAR data analysis has traditionally been carried out based on mathematical models that provide the best trade-off between the variety of phenomena captured by the model and the possibility to produce robust estimates of forest parameters through model inversion; see, for example, Freeman (2007), Treuhaft and Siqueira (2000).

The introduction of SAR Tomography (TomoSAR) techniques has opened up the way to a completely new approach to look at SAR data. A TomoSAR survey is typically obtained by repeatedly flying a radar to acquire multiple SAR images of the same area from different trajectories, as shown in the left panel of Fig. 1. These data are then focused via digital signal processing algorithms to produce a collection of voxels that represent the backscattered energy in three dimensions, thus providing the possibility to see how the vertical structure of the vegetation interacts with the wave (Reigber and Moreira 2000; Mariotti d'Alessandro and Tebaldini 2012). An example is shown in the right panel of Fig. 1, where a 3D representation of a tropical forest at the Paracou Ecological Research Station, French Guiana, is provided by displaying four horizontal tomographic sections.

The use of TomoSAR for investigating forest structure has been under analysis for more than a decade, based on theoretical studies and on the analysis of real data from airborne campaigns. Research on the use of P-band waves (wavelength $\approx$ 70 cm), in particular, has accelerated since 2007 in view of the future P-band spaceborne Mission BIOMASS, which will be launched by the European Space Agency (ESA) in 2022 with the aim of providing AGB accurately and globally (ESA 2012). Early research was generally aimed at retrieving information about the vertical structure of the vegetation, to complement traditional radiometric and interferometric measurements, (Tebaldini 2009; Mariotti d'Alessandro and Tebaldini 2012; Frey and Meyer 2011a, b). Boosted by the first encouraging results, successive studies demonstrated that P-band TomoSAR could effectively be used to derive an accurate characterization of forest structural properties and of the interaction with radar

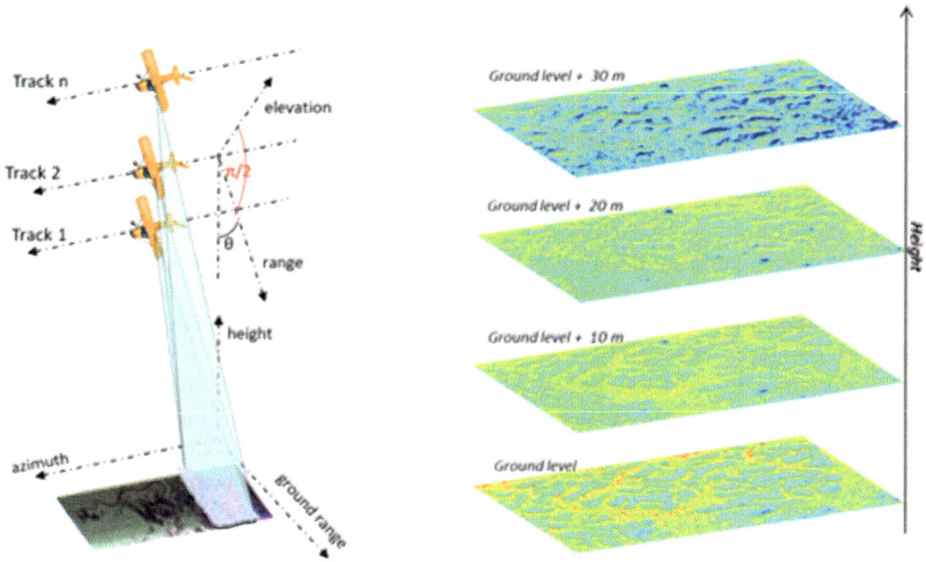

Fig. 1 Left: TomoSAR illumination. Right: tomographic sections. Data: TropiSAR, P-band, Paracou Research Station. Data acquisition by ONERA

waves. The works in Frey and Meyer (2011a), Tebaldini and Rocca (2012), and Mariotti d'Alessandro et al. (2013), for example, first showed the variation in wave polarization inside the vegetation layer, which could then be used as a fingerprint to detect and quantify the arising of double-bounce scattering from ground–trunk interactions, see in particular (Mariotti d'Alessandro et al. 2013). TomoSAR was also demonstrated to be a most valuable tool to retrieve forest canopy height (Tebaldini and Rocca 2012; Ho Tong Minh et al. 2016) and, by virtue of the penetration capabilities of P-band waves, sub-canopy terrain topography (Tebaldini 2009; Gatti et al. 2011; Mariotti d'Alessandro and Tebaldini 2018a, b). A most important result concerning the application of P-band TomoSAR in tropical forests is the one published in Ho Tong Minh et al. (2014a). In that paper, it was first shown by analyzing a tropical site in French Guiana that tomographic intensity at the height of the 'main canopy' in a tropical forest provides a much higher correlation to forest AGB than traditional 2D SAR intensity. It was later shown that model parameterization at one site could be used to predict AGB based on TomoSAR at two sites in French Guiana in Ho Tong Minh et al. (2016), and below, we extend this to three more forest sites in Gabon.

Along with P-band, research on Tomography was carried out at L-band (wavelength $\approx$ 25 cm) as well, motivated by the proposal of the L-band bistatic SAR systems Tandem-L and SAOCOM-CS (Moreira et al. 2015; ESA 2015). Most noticeable results are those concerning separation of ground and volume scattering (Pardini and Papathanassiou 2017; Tebaldini and Rocca 2012), the impact of weather changes (Pardini et al. 2014), and a study on AGB retrieval from vertical structure parameters extracted using TomoSAR (Toraño Caicoya et al. 2015). Most recently, a study has showed that L-band TomoSAR provides greatly improved correlation to forest AGB in boreal forests (Blomberg et al. 2018). It is worth noting that most research at L-band has thus far been focused on temperate and boreal forests, under the general assumption that L-band waves can hardly penetrate to the ground in dense tropical forests. Yet, in two recent studies carried out in Gabon, it was clearly observed that

L-band TomoSAR can actually characterize the full vertical structure of tropical forests (La Valle et al. 2017; Pardini et al. 2018).

The aim of this paper is to provide the reader with a comprehensive understanding of TomoSAR and its application for remote sensing of forested areas, by introducing the basic physical principles behind TomoSAR, the most relevant experimental results obtained in the last decade, and the potential for spaceborne applications. In the exposition, we will mostly focus on the case of P-band TomoSAR and tropical forests. This choice is due to two reasons. In the first place, it is in tropical environments that the use of Tomography appears today to make the most significant difference with respect to conventional SAR and interferometric SAR (InSAR) methods. The second reason is the upcoming BIOMASS Mission, which will implement tomographic imaging for the first 14 months of its lifetime, with optimized performance on equatorial areas (ESA 2012).

This paper is structured as follows. The basic principles required to understand tomographic imaging are introduced and discussed in Sect. 2. Section 3 is intended to provide a brief introduction to forest scattering from forested areas. The use of TomoSAR of the remote sensing of forested areas is discussed in Sects. 4, 5, and 6, which focuses on imaging of the forest structure, forest biomass, and the retrieval of forest height and terrain topography. A discussion of the potentials of tomographic imaging using the future spaceborne mission BIOMASS is provided in Sect. 7. Conclusions are drawn in Sect. 8.

2 Synthetic Aperture Radar Tomography

The expression synthetic aperture radar (SAR) tomography (TomoSAR) is generally used to indicate a microwave imaging technology to focus the illuminated scatterers in 3D space by processing data from multiple SAR acquisitions (Reigber and Moreira 2000). SAR tomographic imaging has been receiving increasing attention in the last years by different research groups, including application fields such as 3D urban scenes, snow, ice sheets, glaciers, and of course forested areas (Reigber and Moreira 2000; Tebaldini 2009; Frey and Meier 2011a; Mariotti d'Alessandro and Tebaldini 2012; Ho Tong Minh et al. 2014a; Banda et al. 2016; Frey et al. 2016; Rekioua et al. 2017; Tebaldini et al. 2016b; Yitayew et al. 2017).

The rationale of TomoSAR is easily understood by considering it as an extension of SAR imaging from 2D to 3D. Conventional SAR systems transmit short pulses and receive the echoes backscattered by the illuminated targets along a single flight line, i.e., a 1D synthetic aperture in the SAR jargon. In this way, the received signal can be focused through digital signal processing techniques to produce a 2D image of the illuminated area, where targets are resolved in the range/azimuth plane (Curlander and McDonough 1991). TomoSAR imaging is based on the collection of multiple flight lines, i.e., a 2D synthetic aperture. This allows focusing the received signal not only in the range/azimuth plane, as in conventional 2D SAR imaging, but also in elevation. A sketch of this concept is shown in Fig. 2, where TomoSAR voxels are represented by the light blue ellipses.

The geometrical resolution in range and azimuth direction is the same as conventional 2D SAR, that is:

$$\Delta r = \frac{c}{2B} \tag{1}$$

$$\Delta x = \frac{\lambda R}{2L_s} \tag{2}$$

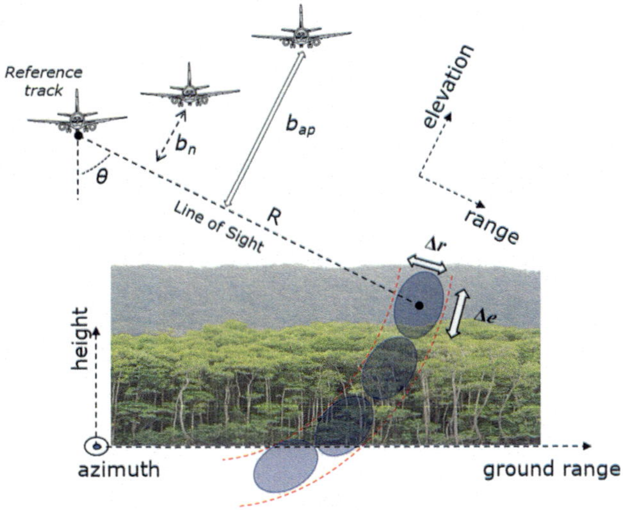

Fig. 2 TomoSAR geometry and tomographic voxels in the height/ground range plane. The interferometric baseline b_n is defined as the projection of the vector connecting each sensor to the one chosen as common reference onto the normal to the Line of Sight of the radar, i.e., the direction connecting the reference sensor to the target. Baseline aperture b_{ap} is the total span of the available interferometric baselines

where r and x indicate range and azimuth, respectively, c is the wave velocity in vacuum, B is the pulse bandwidth, λ is the carrier wavelength, L_s is the 1D synthetic aperture length (in azimuth), and R is the stand-off distance from the imaged target.

Resolution in elevation depends on the total length of the synthetic aperture in elevation, usually referred to as baseline aperture (Bamler and Hartl 1998), given by the formula:

$$\Delta e = \frac{\lambda R}{2b_{\mathrm{ap}}} \tag{3}$$

where e indicates elevation and b_{ap} is the total baseline aperture.

Vertical resolution is then obtained by projecting the voxel along the vertical direction. In the case where range resolution is significantly finer than resolution in elevation ($\Delta r << \Delta e$), vertical resolution is expressed as

$$\Delta z = \Delta e \cdot \sin(\theta) = \frac{\lambda R}{2\Delta b} \cdot \sin(\theta) \tag{4}$$

where θ is the incidence angle (Tebaldini et al. 2016a, b). A real-world example is provided in Fig. 3, which shows two tomographic sections that represent the tropical forest at La Lopé, Gabon, as seen at P-band. In both panels, the color scale is proportional to signal intensity (blue = low, red = high), and the superimposed white line indicates terrain topography from Lidar measurements. The two panels were obtained by processing a stack of ten P-band airborne SAR images acquired by DLR during the ESA campaign AFRISAR (AfriSAR 2017). In this case, vertical resolution ranges from approximately 8–15 m from near to far range.

2.1 Models and Algorithms for SAR Tomography

A focused SAR image can be generally linked to the vertical distribution of the illuminated scatterers through expressions that account for the distance between scattering elements and the flight trajectory. Assuming approximately parallel trajectories flown

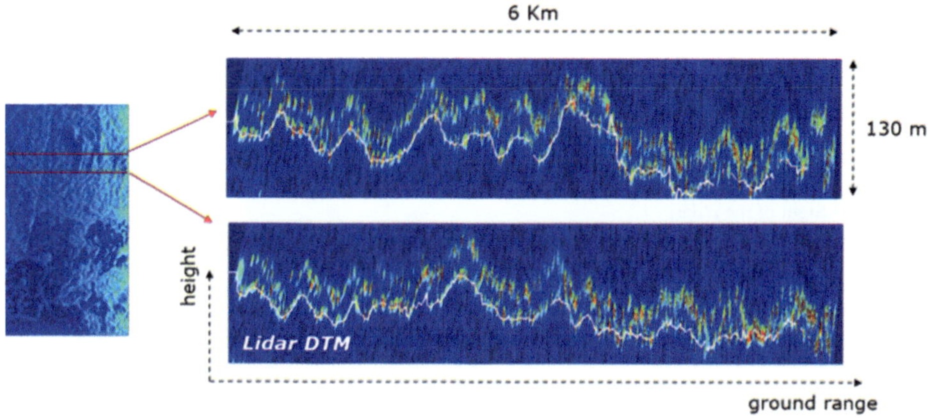

Fig. 3 Two tomographic vertical sections of the tropical forest at La Lopé, Gabon. The small image on the left represent a SAR image of the same area, where the two red lines indicate the position of the two tomographic sections on the right. The color scale is proportional to signal intensity (blue = low, red = high). The superimposed white line indicates terrain topography, as obtained by Lidar measurements. Data: AfriSAR at P-band. Data acquisition by DLR. Tomographic processing by PoliMi

along the x-axis, SAR data can be expressed by the 2D model (Bamler and Hartl 1998; Tebaldini et al. 2016a, b):

$$I_n = \int s(y, z) \cdot \exp\left(j\frac{4\pi}{\lambda}R_n(y, z)\right)\mathrm{d}y\mathrm{d}z \tag{5}$$

where I_n is a complex-valued pixel from the SAR image acquired along the nth trajectory, $s(y,z)$ represent individual scattering elements in the height/ground range plane, R_n is the distance from the scattering element at (y,z) to the radar sensor in the nth trajectory. The integral in Eq. (5) is limited to the circular crown of radius R, thickness Δr, and centered on the nth trajectory, as shown by the red-dashed lines in Fig. 2 (Bamler and Hartl 1998; Tebaldini et al. 2016a, b). A convenient approximation to Eq. (5) is obtained by expanding the expression of R_n at the first order about a reference position, see in particular (Tebaldini et al. 2016a, b), yielding:

$$I_n = \int P(e) \cdot \exp\left(j\frac{4\pi}{\lambda}\frac{b_n}{R} \cdot e\right)\mathrm{d}e \tag{6}$$

where e is elevation, $P(e)$ is the projection of $s(y,z)$ along the elevation axis, and b_n is the interferometric baseline, as shown in Fig. 2. Equation (6) provides an immediate comprehension on the basic principle of SAR Tomography. Indeed, by defining the elevation wavenumber as (Tebaldini et al. 2016a, b):

$$K_n = \frac{4\pi}{\lambda R}b_n \tag{7}$$

it can immediately be seen that Eq. (6) states that the SAR data I_n are linked to the distribution of scattering elements along elevation $P(e)$ via a Fourier transform. It then follows that $P(e)$ can be simply retrieved by taking the (inverse) Fourier transform of SAR data with respect to the interferometric baseline, thus adding one dimension w.r.t. conventional SAR imaging (Reigber and Moreira 2000; Tebaldini 2009; Tebaldini et al. 2016a, b).

Following standard arguments from Fourier analysis, Eqs. (6) and (7) also provide an easy way to assess the performance of tomographic imaging depending on the distribution of available interferometric baselines b_n. In the first place, resolution in elevation is obtained by taking the inverse of the range of elevation wavenumber (divided by 2π), leading to Eq. (3). Beside resolution, the other fundamental parameter for assessing the quality of tomographic is height of ambiguity, z_{amb}, which represents the distance from a target along the vertical direction at which artifacts appear due to finite sampling in the wavenumber domain. By assuming a uniform baseline distribution of the form $b_n = n \cdot \Delta b$, one gets that wavenumbers are sampled by $\Delta K = \frac{4\pi}{\lambda R}\Delta b$, and hence the ambiguous height interval is

$$z_{\text{amb}} = \frac{\lambda R}{2\Delta b} \cdot \sin(\theta) \tag{8}$$

The fundamental requirement for correct tomographic imaging of forested area is that the height of ambiguity z_{amb} is larger than forest height. In designing a tomographic survey, it is good practice to choose a baseline sampling Δb such that the height of ambiguity is roughly twice that of forest height. The number of passes is then easily obtained by taking the ratio between Eqs. (8) and (4), or equivalently between baseline sampling Δb and baseline aperture b_{ap}.

A large variety of approaches for tomographic processing is found in the literature. The most general and accurate approaches to tomographic imaging are obtained as 3D time domain back projection, which allow taking into account strongly irregular trajectories and very large baselines (Frey et al. 2009; Ponce et al. 2014; Tebaldini et al. 2016a, b). In most cases, however, tomographic imaging can be carried out by decoupling focusing in the range–azimuth plane from focusing in elevation. This approach allows casting tomographic processing in terms of a one-dimensional problem, as depicted in Eq. (6), resulting in a substantial advantage in terms of computational burden and enabling the employment of a large variety of techniques from spectral analysis (Gini et al. 2002; Budillon et al. 2011; Zhu and Bamler 2012; Aguilera et al. 2013; Huang et al. 2017). A most interesting aspect of these techniques is that they provide super-resolution capabilities, that is, the capability to resolve targets at a finer resolution than the one expressed in (4). Unfortunately, super-resolution is achieved at the expense of radiometric accuracy, which prevents the application of super-resolution techniques in a general context (Gini et al. 2002; Pardini and Papathanassiou 2017). Finally, a fundamental requirement to enable tomographic focusing is that the knowledge about the position of the radar along the trajectory in all flights is accurate enough to predict variations in the distance travelled by the wave to within an accuracy much better than the system wavelength. This accuracy is seldom met by current navigational systems concerning the location of one flight line with respect to another. As a result, SAR images are affected by space-varying phase disturbances, commonly referred to as phase screens, which produce blurring (Tebaldini and Monti Guarnieri 2010; Tebaldini et al. 2016a, b). For this reason, a preprocessing phase calibration step is quite often required before tomographic focusing (Tebaldini and Monti Guarnieri 2010; Tebaldini et al. 2016a, b; Azcueta and Tebaldini 2017; Mancon et al. 2017).

3 An Introduction to Forest Scattering in the Microwave Regime

A most relevant feature of low-frequency microwaves is the capability to penetrate through the whole vegetation layer down to the underlying ground. As the wave travels through the vegetation, it interacts with leaves, branches, tree trunks, and finally the terrain, generating new waves that are scattered in all directions and further interact with the vegetation and the terrain. As a result, the radar signal can be modeled as the result of different scattering mechanisms (SM). Consistent with a large part of the literature, we will assume here four SMs, as pictorially depicted in Fig. 4 (Sarabandi 1992; Ulaby et al. 1998; Cloude and Pottier 1997; Freeman and Durden 1998; Treuhaft and Siqueira 2000; Papathanassiou and Cloude 2001; Smith-Jonforsen et al. 2005).

(i) Backscattering from the tree canopies. This SM is the result of direct backscattering from woody elements within the vegetation layer. Accordingly, it provides the most direct information about the vertical structure of the vegetation. The resulting signal is depolarized and hence presents with varying intensity in all polarimetric channels.
(ii) Backscattering from the terrain. This SM gives rise to a strongly polarized signal, for which the intensity is much larger in like-polarized returns than in cross-polarized ones. Especially at P-band, terrain scattering is weak compared to forest scattering and can usually be neglected.
(iii) Double-bounce scattering from trunk–ground interactions. This SM occurs as a result of the two specular reflections of the wave onto the tree trunks and the terrain, or vice versa. After the second reflection, the signal is conveyed back to the radar at varying intensity, depending on the tree characteristics and on topographic slope. Intensity is maximal when terrain topography is flat, whereas it tends to vanish on both positive and negative slopes (Smith-Jonforsen et al. 2005). It may be shown through geometrical arguments that this SM may be regarded as an equivalent point-like scatterer located at the tree trunk base. Hence, it appears at the terrain level in tomographic images. Another peculiar trait of trunk–ground scattering is found in the phase difference between HH and VV polarizations, which typically assumes values ranging from about 90° to 180°, depending on the Fresnel coefficients of the terrain and the tree trunks (Freeman and Durden 1998). Assuming flat topography, trunk–ground scattering is not expected to contribute at HV polarization (Freeman and Durden 1998).

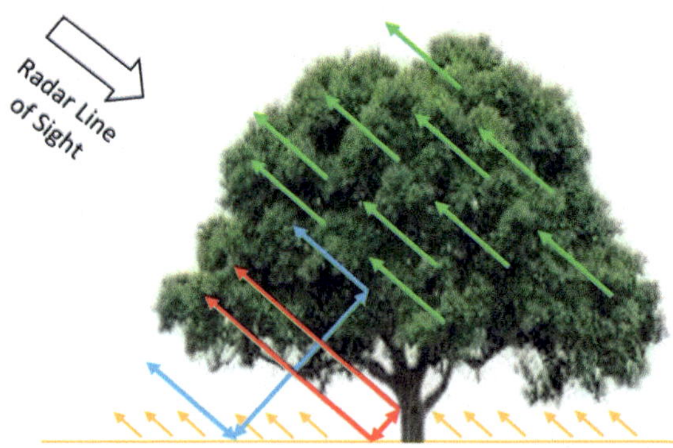

Fig. 4 Pictorial view of the four main scattering mechanisms arising in forested areas. Light green: canopy backscattering. Yellow: terrain backscattering. Red: trunk–ground double-bounce scattering. Blue: canopy–ground double-bounce scattering

(iv) Double-bounce scattering from canopy–ground interactions. This SM results from the waves bouncing off the ground in the direction of the radar after being scattered downward by vegetation elements within the canopy (woody elements at P-band), or vice versa. Canopy–ground scattering appears at the terrain level in tomographic images, as discussed in the case of trunk–ground scattering. The resulting signal is depolarized and hence presents with varying intensity in all polarimetric channels.

4 Sensitivity to Forest Structure

The first tomographic campaign carried out in the frame of BIOMASS studies is BIOSAR 2007, which took place at the Remningstorp forest site, in Southern Sweden (BIOSAR 2008). Prevailing tree species are Norway spruce, Scots pine and birch. The dominant soil type is till with a field layer, when present, of blueberry and narrow-thinned grass. Tree heights are on the order of 20 m, with emergent trees up to 30 m. The topography is fairly flat, terrain elevation above sea level ranging between 120 and 145 m. The acquisition campaign was carried out by DLR from March to May 2007 and comprises 14 fully polarimetric P-band SAR images. The horizontal baseline spacing is approximately 10 m, resulting in a maximum horizontal baseline of approximately 80 m and a vertical resolution ranging from approximately 10 m to 40 m from near to far range.

Two tomographic vertical sections of the Remningstorp forest are shown in the right-hand part of Fig. 5. The two vertical sections are referred to the same area, but were obtained by processing different polarimetric channels. In both panels, the color scale is proportional to signal intensity (blue = low, red = high), and the superimposed lines indicates terrain topography (black) and canopy height (green) obtained from Lidar measurements. Looking at the vertical section at HH (top panel), the first feature that immediately catches the eye is that the highest levels of signal intensity (red areas) are found at terrain level, whereas the signal from the forest canopy is nearly undetectable.

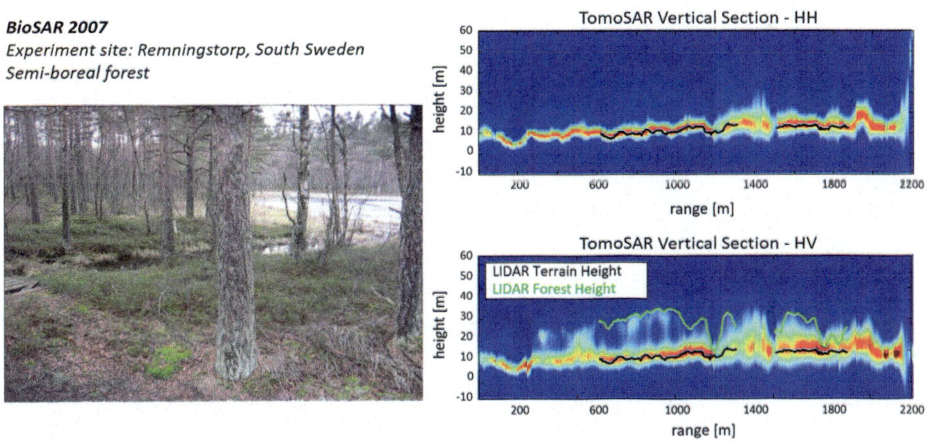

Fig. 5 Tomographic vertical sections of the semi-boreal forest at Remningstorp, Southern Sweden. The color scale is proportional to signal intensity (blue = low, red = high). For visualization purposes, both panels were normalized such that the sum along each column is unitary. The superimposed lines indicates terrain topography and forest height, as obtained by Lidar measurements. Data: BioSAR 2007, P-band. Data acquisition by DLR

Consistently, with the discussion in Sect. 3, this behavior clearly indicates that the signal in like-polarized channels is dominated by trunk–ground scattering, as also further confirmed by polarimetric analysis in Tebaldini (2009, 2010). The most surprising result, however, is that a similar behavior is observed at HV (bottom panel of Fig. 5), for which the brightest areas are again found at the ground level, and the signal scattered from the forest canopies is only barely detectable, indicating that double-bounce scattering is present at HV as well (Tebaldini 2010).

The subsequent tomographic campaign on boreal forest was BIOSAR (2008), flown in Northern Sweden at P-band and L-band (BIOSAR 2009). Tomographic data from BIOSAR (2008) confirmed the intuitions of BIOSAR 2007, revealing the presence of double-bounce scattering and allowing to study their dependence on topographic slope (Tebaldini and Rocca 2012).

The first tomographic campaign focused on tropical areas was TropiSAR, which was flown in summer 2009 at the two tropical sites of Paracou and Nouragues, French Guiana (TropiSAR 2011). The Paracou site was the first to be investigated using tomography. The forest at this site is classified as a lowland moist forest, with approximately 140–200 tree species per hectare. The tree top height reaches 45 m, with an average canopy height of about 30 m. The acquisition campaign was carried out by ONERA. The tomographic dataset at Paracou and comprises six fully polarimetric P-band SAR images. Vertical resolution is approximately 20 m.

The right-hand panels of Fig. 6 reports an HH and an HV tomographic vertical section of the forest in Paracou. For visualization purposes, both panels were resampled such that the terrain level is 0. It is immediate to note how different these sections look as compared to those from Remningstorp. Indeed, scattering from the forest canopies is well detectable at both polarizations. At HV, one can even see that the brightest areas are often found at the canopy level, rather than on the ground. Yet, it can also be appreciated that intensity does not vanish in correspondence with the terrain level, demonstrating that, by and large, the wave is actually penetrating through the whole vegetation

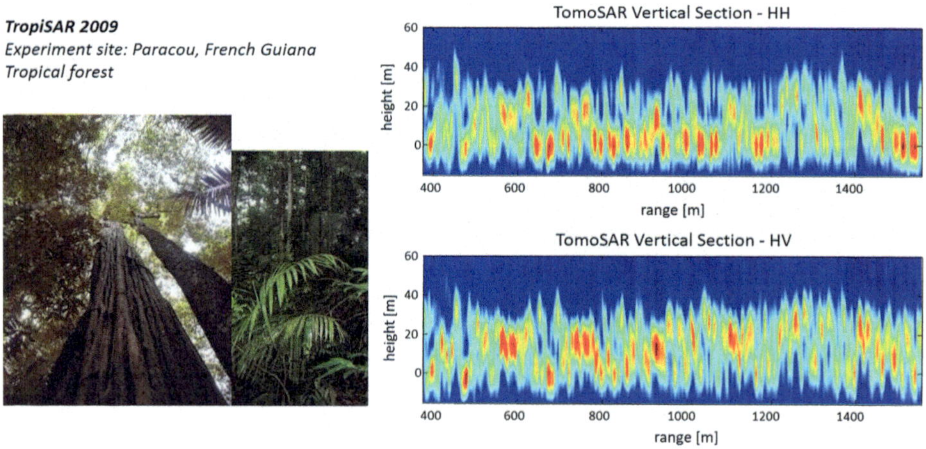

Fig. 6 Tomographic vertical sections of the semi-boreal forest at Paracou, French Guiana. The color scale is proportional to signal intensity (blue = low, red = high). For visualization purposes, both panels were normalized such that the sum along each column is unitary and resampled such that terrain level is 0 m. Data: TropiSAR 2009, P-band. Data acquisition by ONERA

layer down to the ground. An in-depth analysis is reported in Mariotti D'Alessandro et al. (2013).

In conclusion, although the results in this section have only been discussed qualitatively, they demonstrate a most important point: SAR tomography is highly sensitive to forest structure. An in-depth assessment of the role of SAR tomography in the remote sensing of forested areas will be presented in the next sections.

5 The Link to Forest Aboveground Biomass

Aboveground biomass (AGB, in Mg of dry matter per hectare or in t/ha) is the mass of vegetation standing aboveground. It is a key quantity as it constitutes an important ecosystem service, but is also a major store of carbon in the biosphere. The first works that studied the link between SAR tomography and AGB in a quantitative manner are those by Ho Tong Minh et al. (2014a, 2016). Both papers are based on an analysis of the correlation between forest AGB available from in situ surveys and the intensity of tomographic horizontal sections corresponding to different heights w.r.t. to the terrain. The analyzed datasets are the ones collected by ONERA during the TropiSAR campaign (TropiSAR 2011). The Paracou site was investigated in Ho Tong Minh et al. (2014a), whereas the Nouragues site was studied in Ho Tong Minh et al. (2016).

The Paracou site is located in a lowland tropical rain forest near Sinnamary. Terrain elevation is between 5 and 50 m, and mean annual temperature is 26 °C, with an annual range of 1–1.5 °C. The landscape is characterized by a relatively flat terrain, which is dissected by narrow streams. As mentioned in the last section, the forest in Paracou is classified as a lowland moist forest. The tree flora at Paracou exceeds 550 woody species attaining 2 cm diameter at breast height (DBH) have been described in Molino and Sabatier (2001), and a single hectare of forest may harbor 140–200 tree species. Top-of-canopy height reaches up to 45 m with the average value around 30 m. The Nouragues Ecological Research Station is located 120 km south of Cayenne, French Guiana, and was established in 1986. This area is a protected natural reserve characterized by a lowland moist tropical rainforest (Sabatier and Prévost 1988; Van Der Meer and Bongers 1996). Recent floristic censuses have recorded over 660 species of trees above 10 cm in trunk diameter (DBH) in a 12-ha plot. The landscape is a succession of small hills, between 60 and 120 m asl covered by a pristine forest. One prominent feature of the landscape is the presence of a granitic hill, called inselberg, with no vegetation at the top. Top-of-canopy height reaches up to 55 m with the average value around 35 m.

In situ forest AGB measurements were available in Paracou from 16 permanent plots established since 1984, and in Nouragues from two large- and long-term permanent plots established in 1992–1994 and regularly surveyed to the present (Ho Tong Minh et al. 2016). At both test sites, plots were subdivided in 100×100 m subplots (1 ha), resulting in 85 plots in Paracou and 22 in Nouragues (Ho Tong Minh et al. 2016).

Tomographic data consisted of fully polarimetric P-band SAR images acquired on 14 August 2009 in Nouragues (5 flight tracks) and 10 days later in Paracou (6 flight tracks). Importantly, the tomographic flight lines were displaced in a vertical plane rather than in a horizontal plane, which helped limiting spatial variations in vertical resolution across the scene swath (Dubois-Fernandez et al. 2012). This allowed both forest sites to be imaged at an approximately constant vertical resolution of 20 m without the need for super-resolution imaging techniques, thus preserving radiometric accuracy (Mariotti d'Alessandro et al. 2013; Ho Tong Minh et al. 2016).

A synthesis of the analysis carried out in Ho Tong Minh et al. (2014a, 2016) is presented in Fig. 7. The main results are summarized as follows:

- 2D SAR intensity is poorly correlated with AGB.
- Tomographic intensity at 0 m is poorly and negatively correlated with AGB.
- Tomographic intensity at 15 m is poorly correlated with AGB.
- Tomographic intensity at 30 m is highly correlated with AGB. The observed sensitivity is $\approx$ 50 Mg/ha per dB.

Interestingly, the relation between TomoSAR intensity at 30 m and AGB becomes increasingly accurate by aggregating plots at a larger scale and produces a correlation coefficient of 0.97 for plot sizes of about 6 ha.

Based on these results, an inversion procedure was introduced in Ho Tong Minh et al. (2014a) to retrieve AGB by assuming linear dependency on HV tomographic intensity at 30 m. The inversion was trained by using few 1-ha samples and validated with the remaining, resulting in a final total AGB error (RMSE) lower than 10% at 1 ha resolution This approach was later extended in Ho Tong Minh et al. (2016), where a cross-validation was assessed using training plots from Nouragues and validation plots from Paracou, and vice

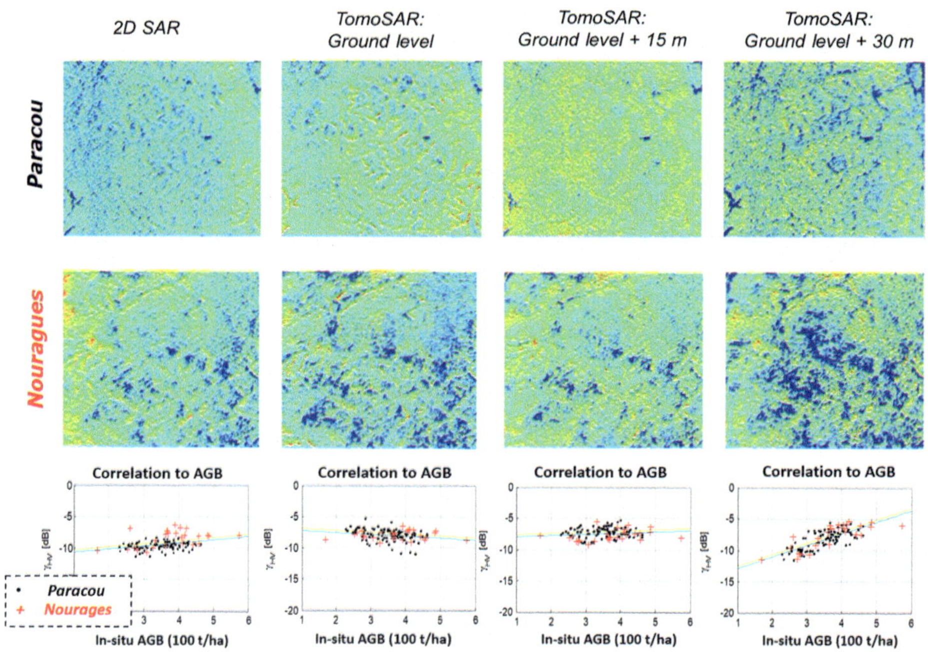

Fig. 7 Correlation between SAR and TomoSAR intensities and AGB at Paracou and Nouragues, French Guiana. Top two rows: 2D SAR intensity in dB (leftmost panel) and TomoSAR intensities in dB from ground level, 15 m and 30 m above ground level for Paracou and Nouragues. The same color scale is used for all panels. Polarization is HV in all panels. Bottom row: scatterplots between intensity and AGB (as derived from in situ observations). The size of each plot is approximately 1 ha. Figures taken from Ho Tong Minh et al. (2016)

versa, resulting in a total AGB error at the two sites around 16–18% at 1 ha resolution, as shown in Fig. 8.

After TropiSAR, the next campaign focused on tropical forests was AfriSAR, which was carried out in Gabon in 2015 and 2016 (AfriSAR 2017). The campaign was shared between ONERA (dry season, July 2015) and DLR (wet season, February 2016) and included tomographic acquisitions at a vertical resolution of about 10 m to 15 m at the forest sites of Lopé, Mondah, Mabounié, and Rabi. These four forest sites are characterized by different physical forest structure types, biomass levels, growth stages, and different levels and kinds of disturbance. All sites contain ground data (permanent tree inventories) together with aerial Lidar scanning of the regions of interest from which reference biomass data have been generated (Labrière et al. 2018). Lopé is located 250 km east of the Libreville airport, and it is characterized by a mosaic between forests and savannas. Forest AGB ranges between a few tons per hectare (in the case of open woody savannas) and up 600t/ha. Two tomographic vertical sections of the Lopé forest sites are shown in Sect. 2, Fig. 3. Mondah is located 25 km north of the Libreville airport. It is a relatively young forest with high variability of density, including disturbed areas due to the proximity to the city. Tree height can also be higher than 50 m. Mabounié is located 180 km southeast of the Libreville airport. The landscape is mostly forested (including swamps and temporarily flooded areas), and most areas are rather hilly (altitude ranges between 25 and 230 m asl). Due to the presence of rare earths, mining exploration took place during the last decades, and many degraded areas are still visible. Rabi is located 260 km south from the Libreville airport. The area of interest contains a 25-ha permanent plot maintained by the Smithsonian Institute, for which extensive ground measurements are available. Next to this plot, an oil extraction area is present around which many degraded areas can be found.

Tomographic data have been generated using data acquired by ONERA at the sites of Lopé, Mondah, and Rabi, and correlated to in situ AGB following the same methodology as in Ho Tong Minh (2014a). The results from AfriSAR are shown in Fig. 9, where we

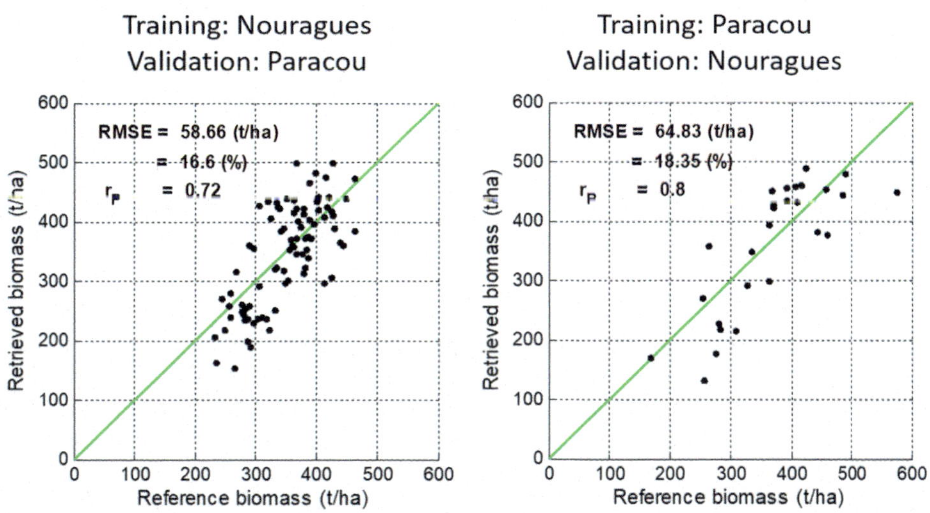

Fig. 8 TomoSAR biomass retrieval result based on cross-validations: comparison of retrieved AGB and in situ AGB. Left: Training in Nouragues and validation in Paracou. Right: Training in Paracou and Validation in Nouragues. Figures taken from Ho Tong Minh et al. (2016)

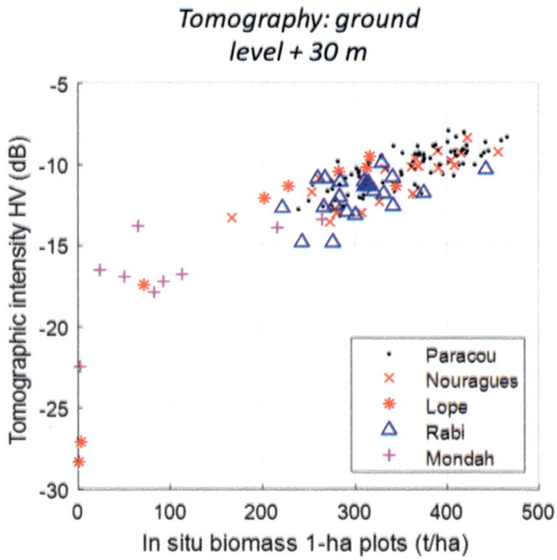

Fig. 9 Plot of TomoSAR intensity at 30 m against AGB at the five tropical forest sites in South America and Equatorial Africa

overlay those from TropiSAR for reference. Crucially, the results are observed to be consistent with those obtained in French Guiana, preserving a similar sensitivity and accuracy for AGB values larger than 200 t/ha.

Accordingly, all the results obtained so far at five tropical forest sites in South America and Equatorial Africa clearly indicate that tomographic intensity at 30 m is dramatically more correlated with AGB than 2D SAR intensity. The observed sensitivity was found to be about 50 Mg/ha per dB across the range of AGB values from about 200 to 500 tons/ha.

Understanding the reasons behind this empirical result is of primary relevance to achieve a more complete understanding of forest scattering at P-band. This, in turn, is of crucial importance in the context of the BIOMASS mission. A better understanding is expected to result in better algorithms for AGB retrieval and possibly limits as much as possible the need for external data for calibration. In the remainder of this section, we discuss two possible reasons, of both ecological and electromagnetic nature, that could help understand the physics underlying the observations:

- 30 m is a biophysically relevant height in dense tropical forests.
- Ground scattering acts as a noise factor on 2D SAR intensity, limiting its sensitivity to AGB. This disturbing factor is most efficiently canceled out in tomographic intensity at 30 m, which would explain the dramatic increase in sensitivity.

5.1 The ecological reason: the role of the 30 m layer

Considering the vertical resolution of the data considered in this paper, tomographic intensity at 30 m accounts approximately for scatterers in the layer from 20 to 40 m above the terrain. Accordingly, the question is whether there is any biophysical reason connecting this layer to total AGB.

To answer this question, we assume a simple structural model of tropical rain forests, which accounts for five layers: the overstorey, the main canopy, the understory, the shrub layer, and the forest floor. This structure is classically observed in aerial Lidar scanning, and even more precisely using terrestrial Lidar scanning. The overstorey refers to the crowns of emergent trees which are above the rest of the canopy (above 40 m). The canopy is the dense ceiling of closely packed trees and their branches centered at about 30 m, while the understory denotes more widely spaced, smaller tree species and young individuals that form a broken layer below the canopy (below 20 m). The shrub layer is characterized by shrubby species and young trees that grow only 2–6 m off the forest floor.

The canopy layer is the principal site for the interchange of heat, water vapor, and atmospheric gases. Under the canopy, there is little direct sunlight due to the extinction of the light through the canopy layer. For these reasons, it is expected that the layer from 20 to 40 m contains a major part of the leaves and a large proportion of woody elements, including trunks and most of the branches (primary, secondary, and higher order) that contribute to the total AGB. Still, the question remains whether the fraction of biomass contained in the 30-m layer is actually representative of the total AGB.

A first answer to this question was given in Ho Tong Minh et al. (2014a) by assuming a forest structure as derived from the TROLL model (Chave 1999; Maréchaux and Chave 2017), which is a spatially explicit forest growth simulator designed to study structural, successional and spatial patterns in natural tropical forests. The model includes competition for light, treefall gap formation and recruitment, which are the critical phenomena in the morphology of tropical forests. The parameters of the model for the species groups have been determined using field data in French Guiana. As a result, an area of 400×400 m^2 was generated, from which biomass between 20 and 40 m was extracted and compared to total AGB. Simulations showed that biomass contained in the 20–40 m layer is about 40% of the total AGB, and that it is strongly correlated ($rp = 0.92$) to total AGB over the whole range of AGB from 250 to 700 t/ha (Ho Tong Minh et al. 2014a).

Most interestingly, this result is confirmed by the recent reanalysis of high-resolution discrete-return airborne Lidar data at nine tropical sites in South America (Meyer 2018), which analyzes the correlation between AGB and the area occupied at different heights by large trees. Correlation (R^2) was found to be maximum at a height of 27–30 m at all the nine study sites, (Meyer 2018). In conclusion, both ecological modeling and empirical Lidar measurements support the idea that the 30-m layer do actually play a special role in tropical forests, as the fraction of biomass included in it provides a reliable proxy to total AGB.

5.2 The EM reason: the role of ground scattering

The behavior of ground scattering in a tropical forest was thoroughly investigated in Mariotti d'Alessandro et al. (2013) based on the tomographic P-band dataset acquired by ONERA at the forest site of Paracou, French Guiana.

The analysis in that paper focused on the variation in backscattered intensity and copolar phase (i.e., the phase between HH and VV) w.r.t. ground range slope, considering both 2D SAR intensity and Tomographic data at ground level, see Fig. 10.

Figure 10 shows that both the intensity and the copolar phase are modulated by ground range slope. In particular, in flat areas, intensity is seen to increase and the copolar phase approaches $-180°$, which is a clear indication of the occurrence of double-bounce scattering from ground–trunk interactions (Mariotti d'Alessandro et al. 2013). With tomographic data, the variation in intensity associated with double-bounce scattering was accurately estimated as about 5 dB (see the bottom panels of Fig. 10).

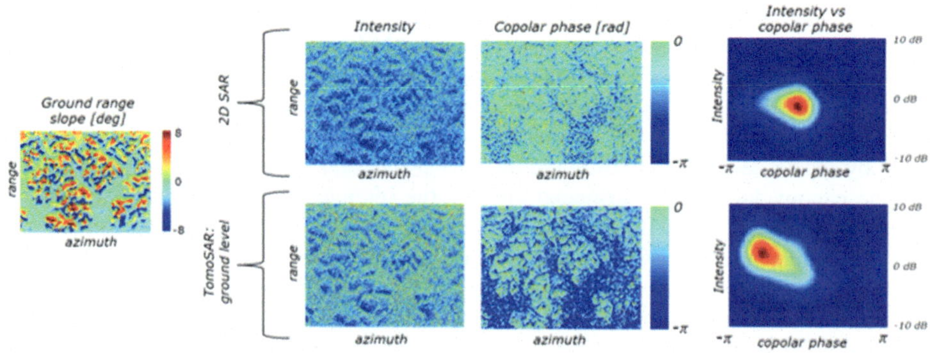

Fig. 10 Top row: backscattered intensity, copolar phase, and 2D histogram relating backscattered intensity and copolar phase for the original (i.e., non-tomographic) SLC data. Bottom row: backscattered intensity, copolar phase, and 2D histogram relating backscattered intensity and copolar phase for tomographic data corresponding to ground level. Leftmost panel: ground range slope: positive-valued pixels indicate the terrain is tilted toward the radar. Figure taken from Mariotti d'Alessandro et al. (2013)

Further analysis based on physical optics showed that the characteristic parameter that rules ground–trunk scattering is not only tree height, but also the length of the base available for ground reflections (Mariotti d'Alessandro et al. 2013), which indicates that this phenomenon is connected to the presence of nearby trees, understory, and undulating topography. The immediate conclusion that can be drawn from this analysis is that even in a dense tropical forest, double-bounce scattering from ground–trunk interactions is relevant on flat areas. Another conclusion is that ground–trunk scattering in tropical forests is strongly connected to local topography, and also to local forest features, such as tree density, and density of the understory. Moreover, although not considered in Mariotti d'Alessandro et al. (2013), soil moisture and canopy–ground scattering are expected to play a role, as predicted by EM models (Truong-Loi et al. 2015).

These conclusions are consistent with the results in Ho Tong Minh (2014a, 2016), showing that ground scattering is poorly and negatively correlated to AGB. Indeed, a slight decay of ground scattering with increasing AGB can be explained by assuming that total wave attenuation increases with AGB. However, as noted above, ground scattering is also determined by several other factors, which necessarily weakens the correlation to AGB.

6 Forest Height and Terrain Topography

The retrieval of canopy height using SAR tomography has been considered since the early experiments in 2007. The basic principle enabling the use of SAR tomography for this task is immediately understood by considering tomographic vertical sections, such as those in Figs. 3, 5, and 6. Indeed, wave scattering from forested areas is bound to occur between the terrain and the top of the canopy. Hence, canopy height can be retrieved, at least in principle, by tracing the upper envelope in tomographic sections. In practice, the upper envelope is extracted after interpolating the tomographic sections in ground coordinates with respect to the DTM, that is, by setting the terrain level at $z=0$ m, see for example (Tebaldini and Rocca 2012). An advantage of this approach is that it is less computationally expensive than model-based inversion and can be applied in the absence of a specific model of the forest vertical structure. In the recent literature, this approach has been successfully applied on boreal and tropical forests, using data from BIOSAR 2008 and TropiSAR. Performance was assessed based on a pixel-to-pixel comparison against Lidar measurements and resulted in an accuracy of about 3 m at a resolution of 60×60 m^2 in the case of boreal forests (Tebaldini and Rocca 2012) and about 2.5 m at a resolution of 60×60 m^2 in the case of tropical forests (Ho Tong Minh et al. 2016). The achievement of a better accuracy in the latter case is most likely due to the better vertical resolution of TropiSAR data w.r.t. BIOSAR 2008. In Fig. 11, we report a recent example of the application of this approach to the case of tropical forests using data from AfriSAR. Performance assessment of the TomoSAR height retrieval algorithm is part of an ongoing study.

In addition to forest height, SAR tomography is also being investigated to retrieve a digital terrain model. This task was first considered for the case of boreal forests in (Tebaldini 2010), where terrain topography was retrieved by assuming a parametric model. In that paper, accuracy was assessed to be about 2 m by comparison against Lidar measurements of ground elevation (Tebaldini 2010). In dense tropical areas, terrain topography retrieval was first considered in Mariotti D'Alessandro et al. (2013). In this case, retrieval was obtained by jointly processing tomographic data at different polarizations, according to the procedure introduced in Tebaldini (2009). For terrain topography, accurate assessment was not considered in Mariotti d'Alessandro and Tebaldini (2012). However, that terrain localization was accurate enough to enable the later developments in Mariotti d'Alessandro et al. (2013), Ho Tong Minh et al. (2014a, 2016). A recent assessment of the accuracy achievable in tropical areas based on data from the AfriSAR campaign is reported in Pardini et al. (2018), Mariotti d'Alessandro and Tebaldini (2018a, 2019). The results indicate a final accuracy comparable to that of Lidar system, supporting the idea that SAR tomography will be useful in the future as an alternative to Lidar mapping. An example is shown in Fig. 12.

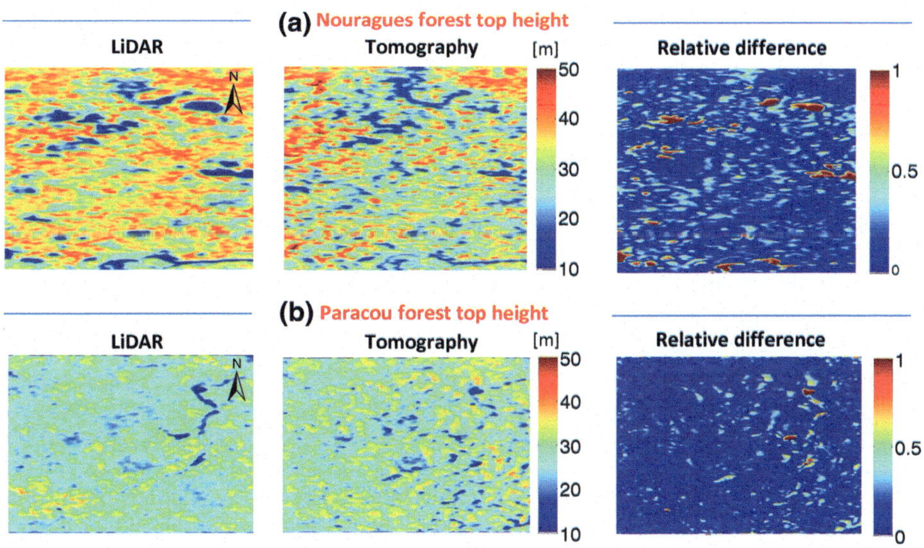

Fig. 11 Retrieval of forest height at Nouragues and Paracou. Left panels: forest height from Lidar survey. Middle panels: forest height by tracing the envelope of tomographic intensities. Right panels: relative error $\frac{|H_{\mathrm{Tomo}} - H_{\mathrm{Lidar}}|}{H_{\mathrm{Lidar}}}$. Data: TropiSAR. Data acquisition by ONERA. Figures taken from Ho Tong Minh et al. (2016)

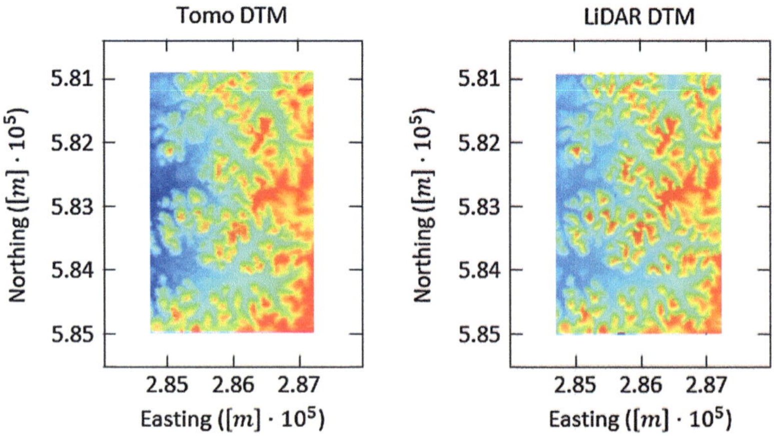

Fig. 12 Retrieval of sub-canopy terrain topography. Left: DTM by SAR tomography. Right: Digital Terrain Model (DLR) from Lidar survey. Data: TropiSAR. Data acquisition by ONERA

7 BIOMASS Tomography

BIOMASS was selected in 2013 to be ESA's 7th Earth Explorer Core Mission. The mission primary objective is to generate accurate maps of forest biomass and height at a global scale. BIOMASS will be implemented as a fully polarimetric SAR operating at P-band, taking advantage of the understorey penetration capabilities of P-band wavelengths (ESA 2012). Launch date is currently planned in 2022. The satellite will operate in two different observation phases, referred to as *tomographic* and *interferometric* phase, respectively. The tomographic phase will operate during the first 14 months of the mission lifetime. In this phase, the satellite will be orbited to provide seven consecutive acquisitions per site from slightly different points of view, hence enabling TomoSAR imaging. Geographical coverage during the tomographic phase will be global, and the seven passes will be acquired with a time lag of 3 days from one another. In the subsequent interferometric phase, the satellite orbits will be modified to achieve faster coverage. This phase will produce three consecutive acquisitions per site at a revisit time of 3 days, enabling AGB and forest height retrieval by SAR Polarimetry and Polarimetric Interferometry (ESA 2012).

The achievement of radiometrically and geometrically accurate tomographic products in the context of BIOMASS is a challenging task. Indeed, one has to account for several potentially damaging factors w.r.t. the airborne case, including degraded signal-to-noise ratio, coarser spatial resolution, propagation through the ionosphere, and temporal decorrelation effects due to the fact that the seven acquisitions needed to implement SAR tomography will be collected in 18 days.

The impact of system range and azimuth resolution has drawn attention since early experiments. Indeed, airborne campaign data provide a resolution on the order of 1 m, whereas BIOMASS will generate data at a resolution of about 10 m in azimuth and only 25 m in range, due to the 6 MHz bandwidth limit imposed by ITU regulations (ESA 2012). This disparity in spatial resolution has stimulated the development of techniques to simulate BIOMASS acquisitions based on campaign data. The feasibility of tomographic imaging at BIOMASS resolution was first demonstrated in Tebaldini and Rocca (2012), where it was demonstrated that forest height in boreal forests could be retrieved to within an

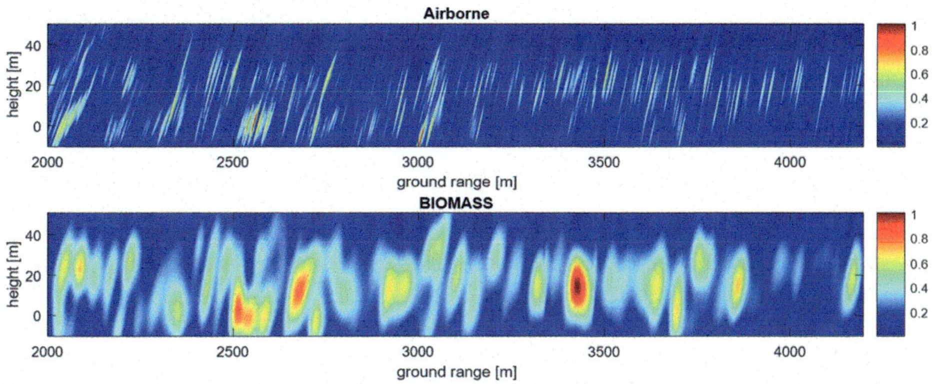

Fig. 13 Airborne tomography (top panel) and BIOMASS tomography from synthetic data derived from real airborne data (bottom panel). Data: TropiSAR P-band. Data acquisition by ONERA

accuracy better than 4 m based on BIOMASS (simulated) data. The next relevant study on this subject is the one published in Ho Tong Minh et al. (2015a), based on simulated BIOMASS acquisitions on tropical forests derived from the TropiSAR campaign. In that paper, the correlation between in situ AGB and tomographic intensity at 30 m was observed to be about 0.84 for plot sizes of about 6 ha when using BIOMASS data, whereas it was 0.97 using airborne data at full resolution (see Sect. 5), hence providing evidence that use of tomography is expected to play a key role in the context of a spaceborne mission as well. Figure 13 shows an example of a tomographic vertical derived from synthetic BIOMASS data derived from airborne data (Mariotti d'Alessandro and Tebaldini 2018b).

The impact of temporal decorrelation was thoroughly studied in the frame of the long-term campaign TropiSCAT. The TropiSCAT campaign was implemented as a P-band fully polarimetric ground-based campaign, installed at the forest site of Paracou, French Guiana. The goal of the TropiSCAT campaign was to analyze the vertical distribution of temporal decorrelation in tropical forests. The system consists of 20 antennas installed on the 55-m high Guyaflux tower, each of which can be used as a transmitter or a receiver to form an equivalent monostatic vertical array of 15 elements for each polarization. The system was operated to collect tomographic data every 15 min for an overall period of about 1 year (Ho Tong Minh et al. 2013, 2014b). The results published in Ho Tong et al. (2014b, 2015b) indicated that the degradation of tomographic imaging due to temporal decorrelation is acceptable as long as the time lag between two consecutive acquisitions is 4 days or less. A recent study considered the emulation of BIOMASS tomography by mixing acquisitions from TropiSCAT gathered every 3 days, including sunny and rainy days. The results showed a total radiometric error due to temporal decorrelation of 1–1.5 dB, which would entail a biomass retrieval error around about 20% or better at spatial scales on the order of 6 ha (Bai et al. 2018).

8 Conclusions

SAR tomography is an increasingly studied technique to image the 3D structure of natural media, such as snow, ice, and vegetation. In the context of forest remote sensing, a most important finding is that tomographic intensity is significantly better correlated to forest AGB in tropical forests than conventional 2D SAR intensity, as it was observed in experimental data from two tropical sites in South America and three in equatorial Africa. Two possible reasons were considered to provide physical support to this finding, of both biophysical and electromagnetic nature: (i) 30 m is a biophysically relevant height in dense tropical forests, and (ii) ground scattering acts as a disturbing factor that needs to be removed. The following conclusions were drawn:

- The hypothesis that 30 m is a biophysically relevant height in dense tropical forests is strongly supported by ecological modeling (Chave 1999) and by a reanalysis of Lidar data across Amazonia as published in Meyer (2018).
- Ground scattering (which includes terrain scattering and double scattering from trunk–ground and canopy–ground interactions) appears to be determined by a complex set of factors other than forest biomass, including local topography, tree density, understorey, and soil moisture. For this reason, it appears unlikely that ground scattering can be directly related to AGB in an operational context, at least in the absence of specific knowledge about local terrain and vegetation features.

In summary, tomography appears to bring the most complete information about AGB in tropical forests by virtue of its ability to single out the returns from different layers within the vegetation while rejecting ground scattering.

Besides forest biomass, SAR tomography was also demonstrated to be a powerful tool for mapping canopy height and sub-canopy terrain topography, as shown by studies conducted at both boreal and tropical sites. Based on results from the recent literature, forest height can be retrieved by SAR tomography to within an accuracy better than 3 m in tropical forests, as validated through a pixel-to-pixel comparison against Lidar data. Retrieval of terrain topography under forests has been shown to be possible in both boreal forests and dense tropical forests, supporting the idea that SAR tomography might be used in a recent future as an alternative to Lidar mapping.

Tomography analyses of forested areas will be implemented for the first time from space during the tomographic phase of the BIOMASS mission. The tomographic phase of BIOMASS will last for the first 14 months of mission lifetime, providing global geographical coverage and enabling tomographic imaging with seven passes acquired with a time lag of 3 days from one another. Notwithstanding many limitations w.r.t. airborne tomography, mostly arising from the coarser resolution and increased temporal decorrelation, studies based on simulated BIOMASS data derived from airborne campaign indicate that accurate tomographic imaging is feasible, and support the idea that forest AGB in tropical could be retrieved to within a 20% accuracy at spatial scales on the order of 6 ha.

Acknowledgements All of the results presented within this paper were obtained in the frame of studies funded by the European Space Agency (ESA) in support of the BIOMASS mission, and we acknowledge ESA for the support it gave to the research on SAR Tomography over the last decade. This paper stemmed from the most fruitful *Forest Properties Workshop* organized in November 2017 in Bern (CH) by the International Space Science Institute (ISSI), which we wish to warmly acknowledge for this initiative. We also acknowledge various funding sources including CNES (France, TOSCA) and an "Investissement d'Avenir" program managed by Agence Nationale de la Recherche (CEBA, Ref. ANR-10-LABX-25-01).

References

AFRISAR (2017) Technical assistance for the deployment of airborne SAR and geophysical measurements during the AFRISAR experiment. Final Report to ESA, 2017

Aguilera E, Nannini M, Reigber A (2013) Wavelet-based compressed sensing for SAR tomography of forested areas. IEEE Trans Geosci Remote Sens 51(12):5283–5295

Azcueta M, Tebaldini S (2017) Non-cooperative bistatic SAR clock drift compensation for tomographic acquisitions. Remote Sensing 9(11):1087

Bai Y, Tebaldini S, Minh DHT, Yang W (2018) An empirical study on the impact of changing weather conditions on repeat-pass SAR tomography. IEEE J Sel Top Appl Earth Obs Remote Sens 11(10):3505–3511

Bamler R, Hartl P (1998) Synthetic aperture radar interferometry-inverse problems 14:R1–R54

Banda F, Dall J, Tebaldini S (2016) Single and multipolarimetric P-Band SAR tomography of subsurface ice structure. IEEE Trans Geosci Remote Sens 54(5):2832–2845

BIOSAR (2008) Technical assistance for the deployment of airborne SAR and geophysical measurements during the BIOSAR 2007 experiment, Final Report to ESA, 2008

BIOSAR (2009) Technical assistance for the deployment of airborne SAR and geophysical measurements during the BIOSAR 2008 experiment. Final Report to ESA, 2009

Blomberg E, Ferro-Famil L, Soja MJ, Ulander LMH, Tebaldini S (2018) Forest biomass retrieval from L-band SAR using tomographic ground backscatter removal. IEEE Geosci Remote Sens Lett https://doi.org/10.1109/lgrs.2018.2819884

Budillon A, Evangelista A, Schirinzi G (2011) Three-dimensional SAR focusing from multipass signals using compressive sampling. IEEE Trans Geosci Remote Sens 49(1):488–499

Chave J (1999) Study of structural, successional and spatial patterns in tropical rain forests using TROLL, a spatially explicit forest model. Ecol Modell 124(2–3):233–254

Cloude SR, Pottier E (1997) An entropy based classification scheme for land applications of polarimetric SAR. IEEE trans Geosci Rem Sens 35(1):68–78

Curlander JC, McDonough RN (1991) Synthetic aperture radar. Wiley Interscience, New York

Dubois-Fernandez PC et al (2012) The TropiSAR Airborne Campaign in French Guiana: objectives, description, and observed temporal behavior of the backscatter signal. IEEE trans Geosci Rem Sens 50(8):3228–3241. https://doi.org/10.1109/TGRS.2011.2180728

ESA (2012) Report for mission selection: biomass. ESA SP-1324/1 (3 volume series), European Space Agency, Noordwijk

ESA (2015) ESA, SAOCOM-CS mission science document, ESA, EOP-SM/2764/. ESA, Noordwijk

Freeman A (2007) Fitting a two-component scattering model to polarimetric SAR data from forests. IEEE Trans Geosci Remote Sens 45(8):2583–2592

Freeman A, Durden SL (1998) A three-component scattering model for polarimetric SAR data. IEEE Trans Geosci Remote Sens 36(3):963–973

Frey O, Meier E (2011a) Analyzing tomographic SAR data of a forest with respect to frequency, polarization, and focusing technique. IEEE Trans Geosci Remote Sens 49(10):3648–3659

Frey O, Meier E (2011b) 3-D time-domain SAR imaging of a forest using airborne multibaseline data at L- and P-bands. IEEE Trans Geosci Remote Sens 49(10):3660–3664

Frey O, Magnard C, Ruegg M, Meier E (2009) Focusing of airborne synthetic aperture radar data from highly nonlinear flight tracks. IEEE Trans Geosci Remote Sens 47(6):1844–1858

Frey O, Werner CL, Caduff R, Wiesmann A (2016) A time series of SAR tomographic profiles of a snowpack. In: Proceedings of EUSAR 2016: 11th European conference on synthetic aperture radar, Hamburg, Germany, 2016, pp 1–5

Gatti G, Tebaldini S, Mariotti d'Alessandro M, Rocca F (2011) ALGAE: A Fast Algebraic Estimation of Interferogram Phase Offsets in Space-Varying Geometries. IEEE trans Geosci Rem Sens 49(6):2343–2353

Gini F, Lombardini F, Montanari M (2002) Layover solution in multibaseline SAR interferometry. IEEE Trans Aerosp Electron Syst 38(4):1344–1356

Ho Tong Minh D, Tebaldini S, Rocca F, Koleck T, Borderies P, Albinet C, Villard L, Hamadi A, Le Toan T (2013) Ground-based array for tomographic imaging of the tropical forest in p-band. IEEE Trans Geosci Remote Sens 51(8):4460–4472

Ho Tong Minh D, Le Toan T, Rocca F, Tebaldini S, D'Alessandro MM, Villard L (2014a) Relating P-band synthetic aperture radar tomography to tropical forest biomass. IEEE Trans Geosci Remote Sens 52(2):967–979

Ho Tong Minh D, Tebaldini S, Rocca F, Le Toan T, Borderies P, Koleck T, Albinet C, Hamadi A, Villard L (2014b) Vertical structure of p-band temporal decorrelation at the Paracou forest: results from tropiscat. IEEE Geosci Remote Sens Lett 11(8):1438–1442

Ho Tong Minh D, Tebaldini S, Rocca F, Le Toan T, Villard L, Dubois-Fernandez PC (2015a) Capabilities of BIOMASS tomography for investigating tropical forests. IEEE Trans Geosci Remote Sens 53(2):965–975

Ho Tong Minh D, Tebaldini S, Rocca F, Le Toan T (2015b) The impact of temporal decorrelation on BIOMASS tomography of tropical forests. IEEE Geosci Remote Sens Lett 12(6):1297–1301

Ho Tong Minh D, Le Toan T, Rocca F, Tebaldini S, Villard L, Réjou-Méchain M, Phillips OL, Feldpausch TR, Dubois-Fernandez P, Scipal K, Chave J (2016) SAR tomography for the retrieval of forest biomass and height: cross-validation at two tropical forest sites in French Guiana. Remote Sens Environ 175:138–147

Huang Y, Levy-Vehel J, Ferro-Famil L, Reigber A (2017) Three-dimensional imaging of objects concealed below a forest canopy using SAR tomography at L-band and wavelet-based sparse estimation. IEEE Geosci Remote Sens Lett 14(9):1454–1458

Labrière N, Tao S, Chave J, Scipal K, Le Toan T, Abernethy K, Alonso A, Barbier N, Bissiengou P, Casal T, Davies SJ, Ferraz A, Hérault B, Jaouen G, Jeffery KJ, Kenfack D, Korte L, Lewis SL, Malhi Y, Memiaghe HR, Poulsen JR, Réjou-Méchain M, Villard L, Vincent G, White LJT, Saatchi S (2018) In situ reference datasets from the TropiSAR and AfriSAR campaigns in support of upcoming spaceborne biomass missions. JSTARS 2018 (in press)

Lavalle M, Hawkins B, Hensley S (2017) Tomographic imaging with UAVSAR: current status and new results from the 2016 AfriSAR campaign. In: 2017 IEEE international geoscience and remote sensing symposium (IGARSS)

Lin Y-C, Sarabandi K (1992) Electromagnetic scattering model for a tree trunk above a tilted ground plane. IEEE Trans Geosci Remote Sens 33(4):1063–1070

Mancon S, Monti Guarnieri A, Giudici D, Tebaldini S (2017) On the phase calibration by multisquint analysis in TOPSAR and stripmap interferometry. IEEE Trans Geosci Remote Sens 55(1):134–147

Maréchaux I, Chave J (2017) An individual-based forest model to jointly simulate carbon and tree diversity in Amazonia: description and applications. Ecol Monogr 87:632–664. https://doi.org/10.1002/ecm.1271

Mariotti d'Alessandro M, Tebaldini S (2018a) Retrieval of terrain topography in tropical forests using P-band SAR tomography, IGARSS 2018, Valencia, 2018

Mariotti d'Alessandro M, Tebaldini S (2018b) Cross sensor simulation of tomographic SAR data. In: IGARSS 2018–2018 IEEE international geoscience and remote sensing symposium, Valencia, 2018, pp 8695–8698

Mariotti d'Alessandro M, Tebaldini S (2019) Digital terrain model retrieval in tropical forests through P-band SAR tomography, on IEEE Transactions on Geoscience and Remote Sensing, Early Access

Mariotti d'Alessandro M, Tebaldini S (2012) Phenomenology of P-band scattering from a tropical forest through three-dimensional SAR tomography. IEEE Geosci Remote Sens Lett 9(3):442–446

Mariotti D'Alessandro M, Tebaldini S, Rocca F (2013) Phenomenology of ground scattering in a tropical forest through polarimetric synthetic aperture radar tomography. IEEE Trans Geosci Remote Sens 51(8):4430–4437

Meyer V et al. (2018) Canopy area of large trees explains aboveground biomass variations across nine neotropical forest landscapes. Biogeosci Discuss. https://doi.org/10.5194/bg-2017-547. https://www.biogeosciences-discuss.net/bg-2017-547/bg-2017-547.pdf

Molino JF, Sabatier D (2001) Tree diversity in tropical rain forests: a validation of the intermediate disturbance hypothesis. Science 294:17021704

Moreira A (2014) A golden age for spaceborne SAR systems. 2014 20th international conference on microwaves, radar and wireless communications (MIKON). Gdansk 2014:1–4. https://doi.org/10.1109/MIKON.2014.6899903

Moreira A et al (2015) Tandem-L: a highly innovative bistatic SAR mission for global observation of dynamic processes on the earth's surface. IEEE Geosci Remote Sens Mag 3(2):8–23

Paillou P, Grandjean G, Baghdadi N, Heggy E, August-Bernex Th, Achache J (2003) Sub-surface imaging in central-southern Egypt using low frequency radar: bir Safsaf revisited. IEEE Trans Geosci Remote Sens 41(7):1672–1684

Papathanassiou KP, Cloude SR (2001) Single baseline polarimetric SAR interferometry. IEEE Trans Geosci Remote Sens 39(11):2352–2363

Pardini M, Papathanassiou K (2017) On the estimation of ground and volume polarimetric covariances in forest scenarios with SAR tomography. IEEE Geosci Remote Sens Lett 14(10):1860–1864

Pardini M, Cantini A, Lombardini F, Papathanassiou K (2014) 3-D structure of forests: first analysis of tomogram changes due to weather and seasonal effects at L-band. EUSAR 2014. In: 10th European conference on synthetic aperture radar, Berlin, Germany, 2014, pp 1–4

Pardini M, Tello M, Cazcarra-Bes V, Papathanassiou KP, Hajnsek I (2018) L- and p-band 3-d SAR reflectivity profiles versus Lidar waveforms: the afrisar case. IEEE J Sel Top Appl Earth Obs Remote Sens 11(10):3386–3401

Ponce O, Prats-Iraola P, Scheiber R, Reigber A, Moreira A, Aguilera E (2014) Polarimetric 3-D reconstruction from multicircular SAR at P-band. IEEE Geosci Remote Sens Lett 11(4):803–807

Reigber A, Moreira A (2000) First demonstration of airborne SAR tomography using multibaseline L-band data. IEEE Trans Geosci Remote Sens Remote Sens 38(5):2142–2152

Rekioua B, Davy M, Ferro-Famil L, Tebaldini S (2017) Snowpack permittivity profile retrieval from tomographic SAR data. C R Phys 18(1):57–65

Sabatier D, Prévost MF (1988) Quelques données sur la composition floristique et la diversité des peuplements forestiers de Guyane Française. Bois et Forêts des Tropiques 219:31–55

Sarabandi K (1992) Scattering from dielectric structures above impedance surfaces and resistive sheets. IEEE Trans Antennas Propag 40(1):67–78

Smith-Jonforsen G, Ulander L, Luo X (2005) Low vhf-band backscatter from coniferous forests on sloping terrain. IEEE Trans Geosci Remote Sens 43(10):2246–2260

Tebaldini S (2009) Algebraic synthesis of forest scenarios from multibaseline PolInSAR data. IEEE Trans Geosci Remote Sens 47(12):4132–4142

Tebaldini S (2010) Single and multipolarimetric SAR tomography of forested areas: a parametric approach. IEEE trans Geosci Rem Sens 48(5):2375–2387

Tebaldini S, Monti Guarnieri A (2010) On the role of phase stability in SAR multibaseline applications. IEEE Trans Geosci Remote Sens 48(7):2953–2966

Tebaldini S, Rocca F (2012) Multibaseline polarimetric SAR tomography of a boreal forest at P- and L-bands. IEEE Trans Geosci Remote Sens 50(1):232–246

Tebaldini S, Rocca F, Mariotti D'Alessandro M, Ferro-Famil L (2016a) Phase calibration of airborne tomographic SAR data via phase center double localization. IEEE Trans Geosci Remote Sens 54(3):1775–1792

Tebaldini S, Nagler T, Rott H, Heilig A (2016b) Imaging the internal structure of an alpine glacier via L-band airborne SAR tomography. IEEE Trans Geosci Remote Sens 54(12):7197–7209

Toraño Caicoya A, Pardini M, Hajnsek I, Papathanassiou K (2015) Forest above-ground biomass estimation from vertical reflectivity profiles at L-band. IEEE Geosci Remote Sens Lett 12(12):2379–2383

Treuhaft RN, Siqueira PR (2000) Vertical structure of vegetated land surfaces from interferometric and polarimetric radar. Radio Sci 35(1):141–177. https://doi.org/10.1029/1999RS900108

TROPISAR (2011) Technical assistance for the deployment of airborne SAR and geophysical measurements during the TROPISAR 2009 experiment. Final Report to ESA, 2011

Truong-Loï ML, Saatchi S, Jaruwatanadilok S (2015) Soil moisture estimation under tropical forests using UHF radar polarimetry. IEEE Trans Geosci Remote Sens 53(4):1718–1727

Ulaby F, McDonald K, Sarabandi K, Dobson M (1998) Michigan microwave canopy scattering models (mimics). In: International geoscience and remote sensing symposium, 1988. IGARSS'88. Remote sensing: moving toward the 21st century. vol 2, pp 1009–1009

Van Der Meer PJ, Bongers F (1996) Patterns of tree-fall and branch-fall in a tropical rain forest in French Guiana. J Ecol 84(1):19–29

Yitayew TG, Ferro-Famil L, Eltoft T, Tebaldini S (2017) Lake and fjord ice imaging using a multifrequency ground-based tomographic SAR system. IEEE J Sel Top Appl Earth Obs Remote Sens 10(10):4457–4468

Zhu XX, Bamler R (2012) Super-resolution power and robustness of compressive sensing for spectral estimation with application to spaceborne tomographic SAR. IEEE Trans Geosci Remote Sens 50(1):247–258

Surveys in Geophysics (2019) 40:803–837
https://doi.org/10.1007/s10712-019-09553-9

Early Lessons on Combining Lidar and Multi-baseline SAR Measurements for Forest Structure Characterization

Matteo Pardini[1] · John Armston[2] · Wenlu Qi[2] · Seung Kuk Lee[2,3] · Marivi Tello[1] · Victor Cazcarra Bes[1] · Changhyun Choi[1] · Konstantinos P. Papathanassiou[1] · Ralph O. Dubayah[2] · Lola E. Fatoyinbo[3]

Received: 10 December 2018 / Accepted: 21 June 2019 / Published online: 9 July 2019

Abstract
The estimation and monitoring of 3D forest structure at large scales strongly rely on the use of remote sensing techniques. Today, two of them are able to provide 3D forest structure estimates: lidar and synthetic aperture radar (SAR) configurations. The differences in wavelength, imaging geometry, and technical implementation make the measurements provided by the two configurations different and, when it comes to the sensitivity to individual 3D forest structure components, complementary. Accordingly, the potential of combining lidar and SAR measurements toward an improved 3D forest structure estimation has been recognised from the very beginning. However, until today there is no established framework for this combination. This paper attempts to review differences, commonalities, and complementarities of lidar and SAR measurements. First, vertical lidar reflectance and SAR reflectivity profiles at different wavelengths are compared in different forest types. Then, current perspectives on their combination for the generation of enhanced structure products are discussed. Two promising frameworks for combining lidar and SAR measurements are reviewed. The first one is a model-based framework where lidar-derived parameters are used to initialize SAR scattering models, and relies on both the validity of the models and on the physical equivalence of the used lidar and SAR parameters. The second one is a structure-based framework based on the ability of lidar and SAR measurements to express physical forest structure by means of appropriate indices. These indices can then be used to establish a link between the two kind of measurements. The review is supported by experimental results achieved using space- and airborne data acquired in recent relevant mission and campaigns.

Keywords Forest · Structure · Height · Lidar · Synthetic-aperture radar (SAR) · Full waveforms · Polarimetric SAR interferometry (Pol-InSAR) · SAR tomography (TomoSAR)

✉ Matteo Pardini
matteo.pardini@dlr.de

1 Microwave and Radar Institute, German Aerospace Center, Wessling 82234, Germany

2 Department of Geographical Sciences, University of Maryland, College Park, MD 20742, USA

3 Biospheric Sciences Laboratory, NASA Goddard Space Flight Center, Greenbelt, MD 20771, USA

1 Introduction

The development and functionality of forests are widely reflected in their biophysical structure. According to a commonly used definition, forest (canopy) structure is "the organisation in space and time, including the position, extent, quantity, type and connectivity, of the aboveground components of vegetation" (Parker 1995; Lefsky et al. 1999; Spies 1998; Harding et al. 2001; Snyder 2010). Thus, it expresses forest state, functionality, biodiversity and evolution (Hall et al. 2011; Brokaw and Lent 1999; McElhinny et al. 2005). Accordingly, forest structure is an indicator of forest successional stage and development as well as sustainability and habitability, and is therefore an important parameter for assessing forest productivity (Bohn and Huth 2017), biomass and biodiversity (Bergen et al. 2009; Goetz et al. 2007; Turner et al. 2003). Forest structure changes are associated with dynamic processes such as growth, regeneration, decay, and natural or anthropogenic disturbance. Knowledge about such processes is important for modeling the function and development of forest ecosystems, and for developing accurate and robust forest biomass estimators (Frolking et al. 2009). Mapping forest structure is therefore critical for understanding the history, function, and future of forest ecosystems.

Traditionally, forest structure characterisation relies on sampling at local scales by means of either field inventory plots or more recently terrestrial laser scanning techniques able to catch the 3D arrangement of vegetation compartments. However, any extrapolation to larger scales is limited by the ability of these measurements to represent larger areas. Moreover, the fact that the temporal continuity of such plot measurements is in many cases difficult to be established limits even more their ability to characterise forest structure change. In this context, remote sensing techniques offer the potential to overcome, at least partially, these limitations (Hall et al. 2011; Bergen et al. 2009; Goetz et al. 2007; Turner et al. 2003; Frolking et al. 2009). Today, only two remote sensing techniques can provide 3D information and contribute to the characterisation of forest structure at large scales (Hall et al. 2011): lidar and synthetic-aperture radar (SAR).

Lidar systems are active configurations usually operating in the infrared or visible region of the electromagnetic spectrum at wavelengths in the nm range. They transmit laser pulses in nadir-looking geometry which "illuminate" a given footprint on the ground whose diameter is typically in the order of decimeters for small-footprint lidar up to tens of meters for large-footprint configurations. The transmitted pulses are reflected by the vegetation elements within the footprint back to the sensor, where they are usually "incoherently" recorded, i.e., only the energy (amplitude) of the reflected light is detected as a function of the signal runtime. A single transmitted laser pulse can be reflected by different vegetation elements located at different heights within the footprint, leading to multiple reflections and an extended distribution of light energy that returns to the sensor. The distribution of light energy that returns to the sensor is known as waveform and is directly related to the 3D distribution of the intercepted vegetation surfaces within the footprint (Dubayah and Drake 2000).

SAR configurations also transmit electromagnetic pulses, but differently from the lidar sensors in the microwave region of the electromagnetic spectrum (wavelengths in the order of cm to dm) in a side-looking geometry. The transmitted pulses interact with Earth's surface, and only a portion of them is scattered back to the sensor. The backscattered pulses are "coherently" recorded by the receiver, i.e., both their amplitude and phase are measured. The amplitude and phase of the backscattered signal depend on the geometric and dielectric properties of the scatterers and their distribution within the antenna footprint, i.e., the illuminated area/volume. With increasing wavelength (decreasing frequency) radar pulses penetrate more and more into and through vegetation layers, and interact with vegetation elements located at different heights within the forest volume and with the underlying ground. However, a single SAR image, even if it results from the interaction of the transmitted pulse(s) with the whole 3D forest structure, does not allow a reconstruction of the 3D distribution of scatterers. For this, a set of SAR images acquired under (slightly) different angular directions (i.e., incidence angles) is required in the context of interferometric (InSAR) and tomographic SAR measurements (Moreira et al. 2013). The side-looking geometry complicates additionally the interpretation of a single SAR image, but at the same time, it allows the separation of the scatterers in the ground range direction and the realisation of wide swath widths up to hundreds of kilometers. Spaceborne SAR configurations of the latest generation can therefore realize revisit times between 1 and 2 weeks with resolution on ground between 1 m and 10 m.

Accordingly, the combined use of lidar and SAR data has the potential to enhance the quality of forest structure characterization by improving the accuracy of physical forest structure parameter estimates, (e.g., forest height and biomass). The combination of SAR and lidar data becomes even more relevant in view of a multitude of SAR and lidar spaceborne missions that are curently operational or planned to be launched and operated in the coming years to meet a number of forest-related objectives. Lidar missions like NASA's GEDI (Dubayah 2015) and IceSAT-2 (Neuenschwander and Magruder 2018) or JAXA's MOLI (Mitsuhashi et al. 2018) and SAR missions like ESA's BIOMASS (Le Toan et al. 2011) NASA's NISAR (Rosen et al. 2017) and/or DLR's TanDEM-X (Krieger et al. 2013) and Tandem-L (Moreira et al. 2015) are expected to operate and acquire data simultaneously or contiguously in time. For this, the understanding of the complementarities and possible synergies between lidar and SAR (i.e., polarimetric SAR interferometry, Pol-InSAR, and/or SAR tomography, TomoSAR) measurements is required.

The similarity of information content of lidar and SAR measurements arises from their common sensitivity to the size and location of vegetation elements in the forest volume. The difference, thus complementarity, arises from the different wavelengths and the different acquisition geometries (i.e., nadir-looking vs. side-looking) defining the interaction of the transmitted pulses with the forest elements. For lidar, the penetration into and through the canopy layer is supported by the nadir-looking geometry that facilitates the penetration through vegetation gaps, while longer wavelengths allow SAR pulses to penetrate into and through canopy layers even in side-looking geometries. The structure information revealed by lidar measurements primarily implies the geometry of trunk and branch structure forming and supporting the canopy. On the other hand, the relative importance of the different tree components (e.g., leaves, branches, trunks) in SAR measurements depends on their vertical and horizontal distribution and the wavelength. Further, in contrast to lidar, SAR measurements are also sensitive to the dielectric properties of the forest which depend strongly on the amount and distribution of water across the vegetation elements (Hall et al. 2011; Moreira et al. 2013).

The difference in information content is one dimension that can be exploited for the development of lidar-SAR fusion algorithms. In this case, lidar and SAR data are combined in a commonly valid scattering or allometric model. One example is the direct use of lidar measurements or lidar-derived parameters to support and/or improve model-based inversions of forest structure parameters from SAR data. The use of lidar data allows to employ more complex physical models and to decouple the geometric from the dielectric information of the SAR data. This could be particularly important for instance to support smaller SAR observations spaces repeated in time to detect structural or dielectric changes. This concept has been demonstrated to estimate top height from X-band SAR interferometric coherence measurements by means of the Random-Volume-over-Ground (RVoG) model using the lidar ground topography (Kugler et al. 2014). Similarly, the use of lidar ground and top forest height in the RVoG model allows the estimation of the attenuation of SAR pulses through the vegetation again from interferometric coherences (Praks et al. 2012; Pardini et al. 2016; Qi and Dubayah 2016). Furthermore, higher-dimensional frameworks for the extraction of structure information from radar reflectivity profiles can benefit from the use of lidar vegetation profiles for parameterisation (Brolly et al. 2016). However, these approaches rely not only on the validity of the model(s) used to combine the two data sets, but also require that lidar and SAR configurations share a set of physically equivalent parameters. However, this is not always the case as, especially with increasing difference in the lidar and radar wavelengths, the differences in the scattering and propagation make lidar and SAR pulses interact differently with the physical vegetation elements. The loss of physical equivalence between SAR and lidar parameters makes the lidar-based parameterisation of SAR models lose significance (Treuhaft et al. 2009; Brolly et al. 2016; Pardini et al. 2018a, b).

The second dimension that can be used to combine lidar and radar data is their complementarity in terms of acquisition and measurement characteristics driven by the individual technologies and acquisition strategies. With particular reference to spaceborne implementations, lidar measurements are typically acquired by means of footprint samples along rather narrow stripes so that large scale coverage and short revisit times become challenging to achieve. In contrast, SAR measurements are continuous with high spatial resolution, and can be implemented by means of wide swaths that allow large scale coverage and short revisit times (Bamler and Hartl 1998; Moreira et al. 2013). At the same time, lidar configurations allow direct measurement of vegetation reflectance profiles with a fairly high vertical resolution, while SAR systems measure backscattering amplitudes and/or interferometric coherences at different polarizations (Pol-InSAR) or radar reflectivity profiles (TomoSAR) with a lower vertical resolution, which are not always straightforward to be interpreted in terms of physical forest structure.

In this case, the combination of lidar and SAR measurements strongly depends on the ability to establish a (physical or statistical) link between them at the different scales/resolutions performed. This link can be established by means of statistical regression(s), and several examples for such an approach have been reported in the literature, for example aiming at biomass estimation by combining lidar heights and SAR backscatter and/or interferometric parameters [see, e.g., (Kellndorfer et al. 2010; Sun et al. 2011; Tsui et al. 2013; Fatoyinbo and Simard 2013; Kaasalainen et al. 2015)]. On the other hand, lidar and SAR measurements can be linked more systematically by means of a scattering model or by means of common structure-dependent indices. The choice between the two frameworks strongly depends on the individual information content. The applicability of the *model-based* framework relies on the common sensitivity to geometric parameters. For instance, the lidar ground height has been used to bridge the spatial (Qi and Dubayah 2016, here together with forest heights at lidar footprint locations) or the temporal (Persson et al. 2017) dissimilarities of lidar and SAR measurements to obtain continuous estimates of forest height and of its changes over large scales. On the other hand, the *structure-based* framework aims at establishing a correspondence between lidar and Pol-InSAR/TomoSAR measurements by means of their capability in reflecting physical 3D forest structure through appropriate indices without necessarily exchanging parameters (Cazcarra-Bes et al. 2017; Tello et al. 2018; Pardini et al. 2018a). In this way, not only the commonalities, but also the complementarities of the different configurations and wavelengths can be exploited. Although promising, the latter framework is in a very early stage of development, considering also that the interpretation of 3D SAR reflectivity in terms of physical forest structure attributes is not as advanced as for lidar profiles.

The objective of this paper is to review (i) the differences, thus indicating the complementarities, in terms of information content of lidar and SAR measurements at different wavelengths, and (ii) the model-based and the structure-based frameworks in which their combination appears promising. Section 2 presents from a theoretical point of view how lidar and SAR measurements are linked to forest structure, and the factors affecting this relationship. The complementarity between the information content of lidar and SAR measurements is then addressed in Sect. 3 by using reflectance and reflectivity profiles in different forest types. The dependency of the SAR profiles on wavelength, dielectric properties, 3D structure, and temporal changes is shown. For this, real airborne SAR (DLR E-SAR and F-SAR platforms) and lidar data acquired over the temperate Traunstein forest (South of Germany) are used. As an example of a tropical forest site lidar and SAR results over the ForestGEO (Smithsonian Institution Forest Global Earth Observatories) plot in Rabi (Gabon) are also shown. The model-based framework is reviewed in Sect. 4 with a case study concerning forest height estimation. With reference to a spaceborne implementation, one InSAR coherence is used to fill the spatial gaps of a lidar acquisition that, in turn, parameterises the RVoG model. Different parameterisation strategies are compared. Concerning the structure-based framework, Sect. 5 reviews a case study on the Rabi plot in which indices expressing horizontal and vertical heterogeneity of structure are calculated from lidar waveforms and TomoSAR profiles, in correspondence to structure indices with the same meaning already established in forestry and ecology. Finally, Sect. 6 draws the conclusions.

2 Basics of Lidar and SAR Measurements of Forest Structure

2.1 Lidar Measurements

Lidar measurements rely on the transmission of laser pulses of finite length towards a scene. The pulses are intercepted, attenuated and reflected (i.e., backscattered) by the branches and leaves of the vegetation canopy, and by the ground, and return back to the sensor. The receiver records the reflected distribution of (light) energy (i.e., the waveform) as a function of time. The time a pulse takes after being transmitted to return to the sensor depends on the range and the scattering properties of intercepted objects within the illuminated part of the scene (i.e., the footprint) as well as the distance between the laser source and the receiver, and the actual atmospheric conditions. The range between the sensor and an intercepted object is calculated as $r_i = c(t_i - t_0)/2$, where c is the speed of light, t_0 is the timing of the transmitted pulse and t_i is the timing of a peak or discretised range bin of the received waveform. Accordingly, the received waveform depends directly on the 3D distribution of the intercepted vegetation elements within the footprint (Dubayah and Drake 2000).

The reflected energy from a vegetation layer of thickness (i.e., height) H_V located over a reference elevation z_G—ignoring for simplicity the propagation through the atmosphere and assuming a random distribution of scattering elements and penetration until the ground level z_G—can generally be modelled as

$$P(x, y, z) = P_0 \int_{z_G}^{z_G+H_V} \mathrm{d}z \iint \beta(x, y, z) \cdot \mathrm{e}^{-\tau(z-z_G)} \mathrm{d}y\mathrm{d}x, \tag{1}$$

where P_0 is the total returned power, where $\beta(x, y, z)$ is the volumetric reflection of the vegetation layer, τ the volumetric extinction coefficient that accounts for the two-way attenuation within the vegetation layer.

Vegetation lidar instruments use in general wavelengths of 1064 nm or 1550 nm and are designed to maximise the measurement signal-to-noise ratio considering atmospheric attenuation, eye-safey, and the reflectance of ground and vegetation elements at those wavelengths. The way the distribution of light energy that returns to the sensor is received and recorded categorises lidar instruments as discrete return or waveform:

1. "Discrete return" lidar systems record only individual (discrete) peaks (i.e., time-stamped ranges triggered real-time) of the waveform. They identify peaks and record a point at each peak location in the waveform curve. These individual or discrete points are called returns. A discrete system may record 1–4 (and sometimes more) returns from each laser pulse.
2. "Full-waveform" lidar systems record the distribution of transmitted and returned light energy. Accordingly, (full) waveform lidar data contain more information compared to discrete return lidar systems, but are more complex to process. Waveform lidar systems sample the returned waveform at a higher frequency and record the energy returned over equal time intervals.

Discrete return lidars are typically designed for fine-scale topographic airborne mapping, therefore their footprint is typically kept to 0.1 to 2 m by employing a small beam divergence angle. However, large-footprint (> 5 m) waveform lidar have several advantages over small footprint ones for mapping vegetation (Dubayah and Drake 2000). First, with a footprint comparable to the average crown diameter (10–25 m), there is a higher probability to contain both ground and tree top at the same time in a waveform. Second, larger footprints can cover wider areas at a lower cost. Large-footprint full-waveform lidar systems from airborne platforms such as the NASA Laser Vegetation, and Ice Sensor (LVIS; Blair et al. 1999) are being used not only for many vegetation studies, but also for calibration and validation of structure parameter extraction algorithms for instruments on spaceborne platforms, like the GEDI (Global Ecosystem Dynamics Investigation).

Central to the use of lidar measurements for estimating vegetation structure parameters is the separation of reflections (returns) from canopy and ground surfaces. Simple waveforms are common from bare ground and complex waveforms from heterogeneous forest canopies. In addition to the characteristics of canopy structure, waveform shapes are affected by ground slope, pulse width and the spatial distribution of wavefront energy (Dubayah et al. 2000). The first and last modes along the profile of a recorded waveform are typically associated with the highest and lowest reflecting surfaces within the footprint, respectively. Identification of these modes within the waveform is required to precisely geolocate the corresponding reflecting surfaces and calculate statistical quantile metrics from the integrated waveform between these ranging points. These statistical quantiles are

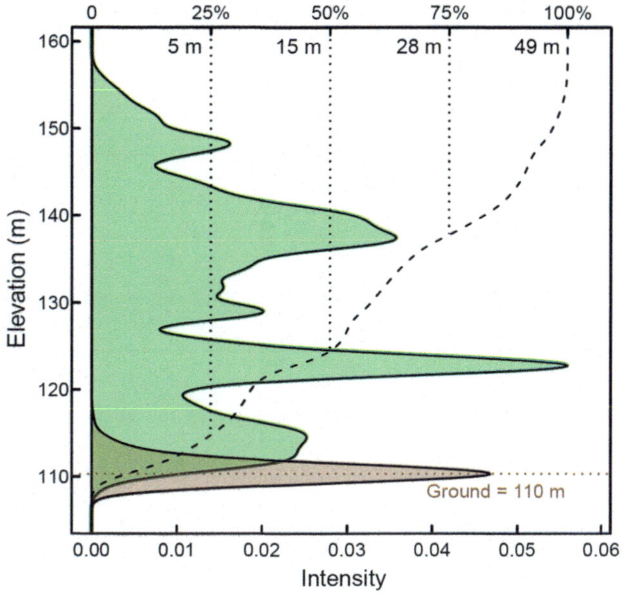

Fig. 1 Example of a lidar waveform simulated from airborne discrete return lidar data acquired over a complex rainforest canopy at La Selva Biological Station, Costa Rica, showing the energy returned from the ground surface (brown) and the energy returned from the canopy (green) as a function of elevation. The reference ground elevation is shown along with the 25th, 50th, 75th, and 100th percentile relative height (RH) metrics extracted from the cumulated energy profile (dashed line)

typically referred to as relative height (RH) metrics (sketched in Fig. 1), which are often used in models to predict aboveground biomass density (Dubayah et al. 2010). The derivation of vertical structure of vegetation canopies from lidar waveform data, though, depends on the knowledge of the relationship between lidar waveforms and the spatial structure and optical properties of vegetation canopies.

In addition to "lidar perceived" statistical metrics, large-footprint waveform lidars enable the estimation of biophysical metrics, as they are a direct measurement of the vertical distribution of the intercepted surfaces between the canopy top and the ground. In the absence of multiple scattering, a received waveform is the product of the projected area of canopy/background materials along the path of the laser pulse, their single scattering albedo, and phase function. The projected area of canopy/background materials is often expressed as the canopy gap probability (P_{gap}), which is a parameter that is fundamental to linking lidar measurements, vertical and horizontal canopy structure, and the radiation regime of a plant canopy (Armston et al. 2013; Ni-Meister et al. 2001). The directional gap probability, P_{gap} (θ), is defined as the probability of a beam of infinitesimal width at zenith angle θ to the local normal, being directly transmitted through a canopy. The vertical distribution of lidar-derived P_{gap} is indirectly related to the vertical distribution of plant area, enabling the estimation of Plant Area Index (PAI), vegetation cover and their vertical profiles (Ni-Meister et al. 2001; Tang et al. 2012).

Surface elevation, height, and vertical profiles of PAI and vegetation cover are all structure parameters that are being generated from GEDI waveform lidar measurements. However, in contrast to airborne scanning systems such as LVIS, spaceborne lidar instruments such as NASA's ICESat/GLAS and GEDI are profiling instruments and may only measure a fraction of the Earth's surface. For example, during GEDI's 2-year mission life, waveform measurements are made over the Earth's surface between 51.6°N and 51.6°S, resulting in approximately 10 billion observations of the Earth's land surface in absence of clouds. The sampling pattern is optimised to maximise the geographic coverage of these observations. In the case of GEDI, shown in Fig. 2,

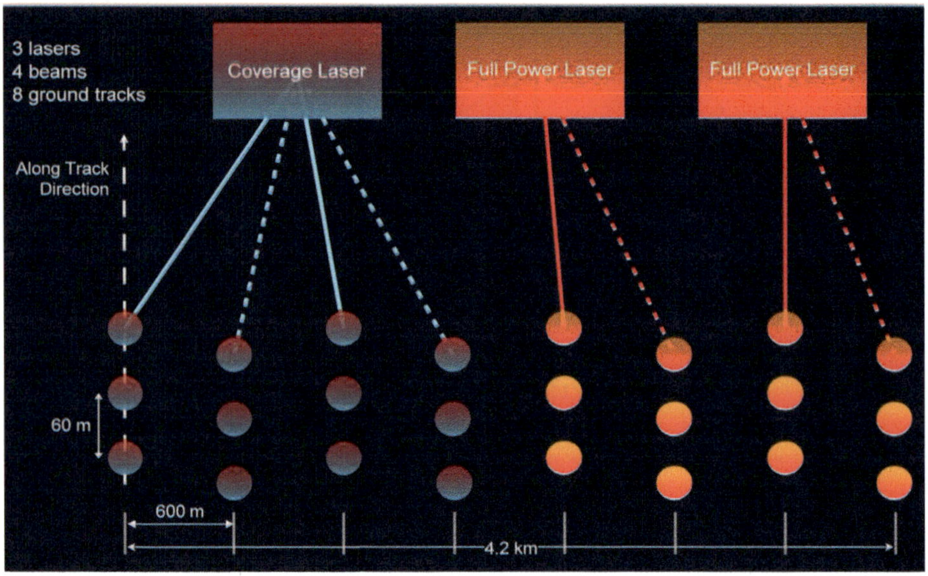

Fig. 2 The sampling pattern of the NASA Global Ecosystems Dynamics Investigation (GEDI) mission, illustrating the arrangement of ground tracks generated from the full power and coverage laser beams. Vertical and horizontal spacing of footprints are not to scale

three lasers (two full power and a coverage one) generate four beams that are dithered to allow data collection along 8 tracks separated by 600 m in across-track direction. This way, a total across-track width of 4.2 km is covered by a single GEDI overflight. Along each track, footprints with a nominal diameter of 25 m are illuminated on ground and separated by 60 m in the along-track direction.

Lidar acquisitions are limited by the presence of clouds and dense atmospheric haze: in these cases, the signal is often attenuated before it reaches the ground. In addition, the estimation of the sub-canopy topography, vegetation height and the interpretation of the waveform in terms of physical structure is affected directly by multiple scattering, sloped terrains and off-nadir pointing (Yang et al. 2011). Multiple scattering occurring in optically thick media distort the waveform making their interpretation more difficult. Additionally, off-nadir pointing and/or a sloped terrain may also deform the vertical waveform shape with respect to a nadir pointing and/or flat terrain. Sloped terrains typically extend the ground return and make it closer to, or even overlapping with, the vegetation returns. This effect becomes more critical at larger footprint sizes. For an off-nadir pointing angle, a larger amount of light energy tends to remain trapped in the upper canopy layers. If penetration to the ground occurs, the ground peak can disappear even on a flat terrain, similarly to the sloped terrain and nadir pointing case. However, in an off-nadir pointing lidar the path length is tilted with respect to the tree growth direction, and this has to be accounted for in the height estimation process. Finally, it is worth remarking that spaceborne lidar sensor are further limited due to the sampling on ground, that leads to a gapped coverage and hampers high spatial resolution (e.g., 1 ha) mapping of forest structure characteristics. In this case, high-resolution mapping can be achieved by combining the lidar data with those from different sensors (e.g., radar or optical).

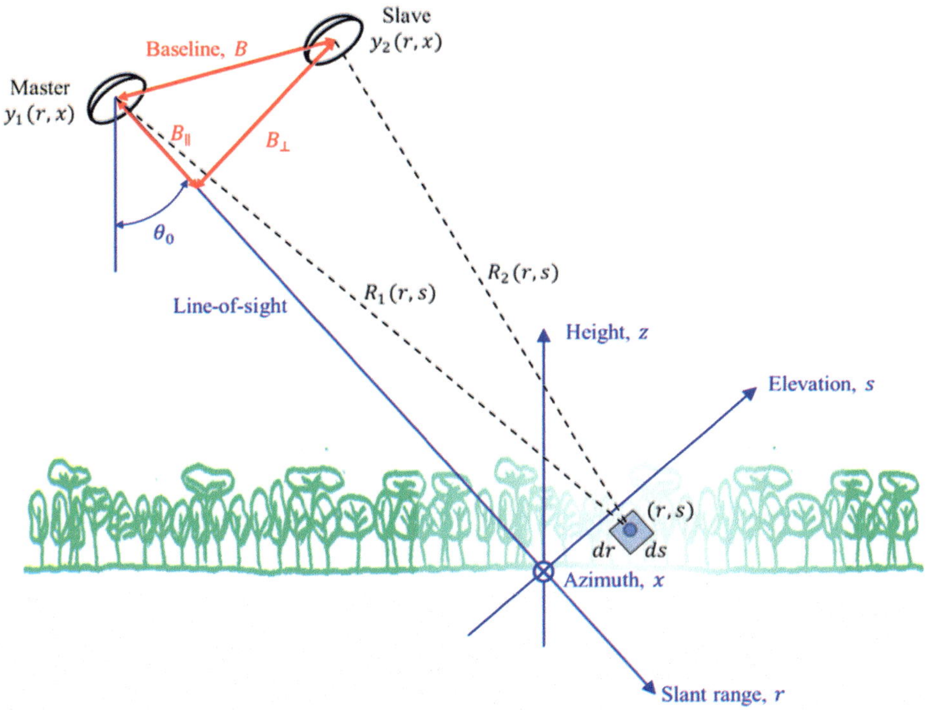

Fig. 3 SAR and (Pol-)InSAR acquisition geometry in the basic single baseline configuration. The azimuth axis is orthogonal to the plane of the page. Angles and distances are not to scale

2.2 SAR Measurements

SAR systems are installed on moving platforms, typically airborne or spaceborne, and transmit electromagnetic pulses in side-looking geometry (see Fig. 3). The radiation backscattered from the illuminated footprint is recorded continuously. The illuminated width of the footprint, i.e. the swath width, extends for 1–20 km in the airborne case and for 30–500 km in the spaceborne case, with typical resolution between 1 and 10 m (Moreira et al. 2013).

The received signals are coherently processed in order to focus 2D SAR (amplitude and phase) single-look complex (SLC) images in the slant range (r, parallel to the line of sight)—azimuth (x, parallel to the flight direction) plane. According to the Born approximation, the backscattered radiation is the linear superposition of the contributions arising from the individual scatterers (i.e., the physical vegetation elements). Multiple scattering phenomena from their mutual interactions are neglected. As a consequence, the SLC complex amplitude for a fixed polarization channel at the generic range–azimuth (r', x') coordinates can be expressed as a volume integral (Bamler and Hartl 1998; Fornaro et al. 2003):

$$y(r', x') = \iiint h(r' - r, x' - x) \cdot \xi(r, x, s) \cdot e^{-j\frac{4\pi}{\lambda}R(r,s)} \mathrm{d}r\mathrm{d}x\mathrm{d}s, \tag{2}$$

where s is the elevation coordinate, $R(r, s)$ is the distance between the antenna and the elementary volume defined by $\mathrm{d}r\mathrm{d}x\mathrm{d}s$, λ is the wavelength, and $h(r, x)$ is the (end-to-end) system point-spread function after focusing which depends on the range and azimuth resolutions (Bamler and Hartl 1998). $\xi(r,x,s)$ is the (unknown) 3D complex radar reflectivity, resulting from the interaction between the transmitted pulse and the scatterer. From (2), it is apparent that a SAR image alone contains backscatter amplitude information that depends on the full 3D forest structure, but it does not provide structure reconstruction capability as a consequence of the integration in s.

The sensitivity to the vertical dimension can be recovered in an InSAR configuration constituted by two SAR images $y_1(r',x')$ and $y_2(r',x')$ acquired under a (slight) angular diversity (incidence angle) from tracks or orbits separated in space by a baseline of length B. With reference to Fig. 3, InSAR can localize a scatterer by measuring the phase difference in the SLCs induced by the difference of the distances from the two SAR platform positions to the scatterer. The sensitivity (i.e., the derivative) of the InSAR phase difference with respect to the vertical height corresponds to the so-called vertical wavenumber, which is given by (Reigber and Moreira 2000; Papathanassiou and Cloude 2001):

$$k_Z := \frac{4\pi}{\lambda r}\frac{B_\perp}{\sin\theta_0}, \tag{3}$$

where $B_\perp$ is the orthogonal baseline, i.e., the length of the projection of the InSAR baseline in the direction orthogonal to the line of sight, and θ_0 is the incidence angle. As any phase value, the interferometric phase is ambiguous by integer multiples of 2π. As a consequence, the related height measurements are ambiguous over a height interval HoA $= 2\pi/k_Z$.

The scattering from natural media, including forests, is typically regarded as a stochastic process. Although Gaussianity may not hold depending on wavelength and/or resolution, the use of second-order statistics has been proven to be valuable for the extraction of scatterer information. Thus, the complex coherence is considered and it is defined as:

$$\gamma_{1,2}(k_Z) := \frac{E\{y_1(r',x')y_2^*(r',x')\}}{\sqrt{E\{|y_1(r',x')|^2\}E\{|y_2(r',x')|^2\}}}, \tag{4}$$

where $E\{\cdot\}$ indicates the statistical expectation operator, and the asterisk indicates the complex conjugate. The dependence of the coherence on k_Z has been explicitly indicated at the left-hand side of (4), while range and azimuth have been dropped for simplicity. The estimation of $\gamma_{1,2}(k_Z)$ in (4) would require several realizations of the underlying stochastic process, which are generally not available in reality. Therefore, assuming spatial ergodicity, reliable measurements of the statistical expectations in (4) are typically obtained by averaging neighboring pixels within a range–azimuth (multilook) cell.

By following all the algebraic development (see Appendix 1), and by considering the presence of noise, the complex coherence (3) becomes (Zebker and Villasenor 1992; Bamler and Hartl 1998):

$$\gamma_{1,2}(k_Z) = \gamma_{\mathrm{Sys}}\gamma_{\mathrm{SNR}}\gamma_S(k_Z)\gamma_{\mathrm{vol}}(k_Z). \tag{5}$$

In (5), γ_{Sys} includes a wide range of decorrelation effects induced by the SAR system and processing inaccuracies (e.g., quantization, co-registration, range and azimuth ambiguities, …), while γ_{SNR} is the decorrelation induced by the presence of noise, and therefore by a finite signal-to-noise ratio. $\gamma_S(k_Z)$ is a (real-valued) decorrelation term that depends on $h(r,x)$ (Zebker and Villasenor 1992; Bamler and Hartl 1998). The former two terms can be compensated by exploiting the available SAR system characterisation, while $\gamma_S(k_Z)$ is compensated by applying the so-called spectral shift filtering (Bamler and Hartl 1998). The remaining term $\gamma_{\text{vol}}(k_Z)$ is the volume decorrelation, and expresses the spectral decorrelation properties of the 3D reflectivity:

$$\gamma_{\text{vol}}\left(k_Z\right) := \mathrm{e}^{jk_Z z_G} \frac{\int_0^{H_V} F_{\text{vol}}(z)\mathrm{e}^{jk_Z z}\mathrm{d}z}{\int_0^{H_V} F_{\text{vol}}(z)\mathrm{d}z}. \tag{6}$$

In (6), $F_{\text{vol}}(z)$ is the spatial density of backscattered power along height (i.e., a projection of $\xi(r, x, s)$, see Appendix 1) for the fixed range cell, called also vertical reflectivity profile. z_G is a reference volume bottom height [see also (1)] and H_V is the volume extension above z_G. If the wavelength allows penetration down to the underlying ground, then z_G is the ground height with respect to the global or local height considered as a reference.

The vertical reflectivity $F_{\text{vol}}(z)$ reflects the physical (geometric) 3D distribution of vegetation elements through scattering and attenuation processes that depend on the SAR wavelength, incidence angle, polarisation, and the dielectric properties of the canopy layers:

- The wavelength defines directly the scattering of and the propagation through the vegetation. Longer wavelengths (e.g., P- and L-bands) are more sensitive to larger vegetation elements. At the same time, a (high) penetration through them provides propagation and sensitivity down to the ground. The sensitivity to smaller elements, which can still be relevant from an ecological point of view, increases with decreasing wavelength (from S- up to X-band). However, the attenuation of the vegetation increases. Hence, at shorter wavelengths the propagation through (even small) vegetation gaps contributes to the penetration to a larger extent than at longer wavelengths.
- For a fixed slant range, wavefronts at steeper (i.e., closer to nadir) incidence angles have a wider projection on the horizontal plane, thus $F_{\text{vol}}(z)$ results more sensitive to the projection of the horizontal distribution of the effective scatterers. In contrast, shallower incidence angles (i.e., far from nadir), increase the sensitivity to the vertical distribution. At the same time, the incidence angle defines the length of the propagation path through the vegetation volume, hence the (two-way) attenuation of the transmitted wave as a function of height. Steeper incidence angles are less attenuated than shallower ones for the same stand characteristics. Not only this, but steeper incidence angles facilitate penetration through gaps at shorter wavelengths.
- The orientation of the scattering vegetation elements as a function of height affects the related distribution of the backscattered power in the different SAR polarisation channels. Therefore, the relative power levels of scattering components at different heights in $F_{\text{vol}}(z)$ can be changed by changing polarisation channel.
- The dielectric properties of the vegetation elements affect their scattering and attenuation, which depend on the water content of the tree tissuesand on its distribution. Dielectric changes can occur due to both environmental (e.g., rains, droughts, snowfalls…) and seasonal changes, and $F_{\text{vol}}(z)$ can change accordingly although the extent and type of change can be wavelength-dependent (Pardini and Papathanassiou 2018; Bai et al 2018; Kugler et al 2014).

Equation (6) shows that the volume coherence $\gamma_{\text{vol}}(k_Z)$ and the vertical reflectivity profile $F_{\text{vol}}(z)$ form a Fourier pair, i.e., an InSAR acquisition corresponds to one Fourier component of the reflectivity profile at a spatial frequency equal to the vertical wavenumber. One single-baseline InSAR measurement makes it possible to extract the height of the backscattering phase center within the volume, i.e., the Digital Elevation Model

(DEM), corresponding (at first order) to the "center of mass" of $F_{vol}(z)$. It is calculated as $z_{DEM}(k_Z) = \varphi_{vol,u}(k_Z)/k_Z$, with $\varphi_{vol,u}(k_Z)$ being the phase of the volume coherence after unwrapping. In order to be able to estimate a larger number of parameters, one can increase the observation space by acquiring interferometric pairs in different polarisation channels, and/or by acquiring multiple images from multiple track/orbit displacements (multi-baseline acquisition). Two different approaches can then be followed:

(A1) The first one is to parameterise $F_{vol}(z)$ in terms of geometrical and scattering properties and to use then the set of measured $\gamma_{vol}(k_Z)$ to estimate the individual model parameters, as it is done with Pol-InSAR inversions (Papathanassiou and Cloude 2001). The scattering model, whose validity is in general frequency-dependent, is essential for the significance and accuracy of the estimated parameters. The model should contain enough physical structure to interpret the interferometric measurements, and at the same time, it must be simple in terms of number of parameters in order to be determinable with the available (in general limited) number of measured coherences.

(A2) The second approach is the most direct one, and attempts to invert the Fourier relationship (6) provided that a set of measurements of $\gamma_{vol}(k_Z)$ is available at suitable k_Z, according to the TomoSAR imaging principle (Reigber and Moreira 2000). This approach is more expensive in terms of acquisitions needed, but it has the advantage that no assumption on the shape of $F_{vol}(z)$ is required, allowing the reconstruction of arbitrary vertical reflectivity profiles. The smallest available HoA (i.e., the largest k_Z) is the TomoSAR vertical resolution (Rayleigh limit).

Pol-InSAR model-based inversion approaches have been demonstrated, e.g., for forest height estimation in boreal, temperate and tropical forests at multiple frequencies from P- up to X-band, considering both airborne and spaceborne platforms (Papathanassiou and Cloude 2001; Garestier et al. 2008; Neumann et al 2010; Lee et al. 2013; Kugler et al. 2014; Lavalle and Hensley 2015; Simard and Denbina 2018). The TomoSAR reconstruction of $F_{vol}(z)$ has been demonstrated in airborne experiments across different forest types (Reigber and Moreira 2000; Frey and Meyer 2011; Tebaldini and Rocca 2012; Mariotti et al. 2012; Pardini et al. 2018a, b). Several algorithmic solutions have been found in the spectral estimation theory (Gini et al. 2002; Frey and Meyer 2011; Aguilera et al. 2013) to improve the imaging performance against the typical low number and the irregular distribution of the flight tracks/orbits. Among the several alternatives, the model-free Capon spectral estimator (Lombardini and Reigber 2003) is widely employed, and can provide vertical resolution beyond the Rayleigh limit (Cazcarra et al. 2019).

As a final remark, a fundamental prerequisite for a successful estimation of 3D structure parameters from SAR acquisitions is the stationarity of $F_{vol}(z)$ within the total acquisition time. Changes can be induced by, e.g., the movement of the vegetation under the action of wind, or changes in the dielectric properties, and introduce an additional (wavelength-dependent) temporal decorrelation factor in (4). If not properly accounted for, the inversion problem may become ambiguous if not unfeasible (Lee et al. 2013; Lavalle and Hensley 2015). On the other hand, (bistatic) single-pass (Pol-)InSAR acquisitions provide temporal decorrelation-free measurements of $\gamma_{vol}(k_Z)$, although dielectric changes may still occur between acquisitions. In this frame, a full separation between geometric and dielectric contributions within the same vertical reflectivity profile or its parameters would only be possible with single-pass multistatic TomoSAR constellations.

3 A Comparison Between Lidar and SAR Profiles at Different Wavelengths

In this Section, the dependency of the TomoSAR vertical reflectivity profiles on wavelength, dielectric properties, and structure types is discussed by comparing profiles extracted in the temperate Traunstein (Germany) and the tropical Rabi (Gabon) forest sites. Details on the used lidar and SAR data sets are reported in Appendix 2. TomoSAR vertical profiles have been estimated in the HV channel in square range–azimuth multilook cells measuring 20 m × 20 m, corresponding to the profile resolution in the horizontal range–azimuth plane. The profiles have been obtained by means of the Capon spectral estimator (Lombardini and Reigber 2003). Since lidar waveforms were available only in Rabi, lidar profiles have been generated in Traunstein as the histogram of all the recorded ALS returns as a function of height.

The profiles in Fig. 4 have been calculated in the Traunstein forest, and refer to different growth stages ("transition", "close-to-nature", "mature" and "young") which are typical of this forest site. The profiles in Fig. 5 have been estimated in the Rabi forest in correspondence of a ForestGEO inventory plot (see Appendix 2).

The ability of the different wavelengths of penetrating into and through the canopy layers can be assessed directly by detecting the presence and measuring the power of the ground scattering in the profiles. In both Traunstein and Rabi, the ground scattering is well visible in almost all the P-band profiles (see Figs. 4a, 5a). A larger variation of ground power across forest stands occurs at L-band due to the higher sensitivity to the spatial variations of the attenuation levels induced by the changes of both canopy density and water content (see Figs. 4b, 5b). In Traunstein, the L-band profiles in Fig. 4c have been obtained with data acquired around 3.5 years after the ones used to generate the profiles in Fig. 4b, and show a change of ground power. This can be due to a change of canopy attenuation caused by a change of both canopy density, as a result of forest management, and dielectric properties, as a result of seasonality-induced redistributions of water content. The effect of forest management is particularly visible in the "transition" stands, in which old taller trees are cut to allow the growth of the younger shorter ones below. This is recognisable also in the sparsity of the lidar returns in Fig. 4e. The effect of seasonal dielectric changes is likely visible in the forest stand beyond a horizontal distance of 2 km (Autumn 2009 vs. Spring 2013). Indeed, an increase of ground power occurs in the presence of the thick canopy cover shown by the lidar profiles. Finally, the X-band profiles (Fig. 4d) are the ones characterized by the lowest ground contribution due to the largest attenuation of the shortest X-band wavelength. However, the ground scattering becomes dominant in the "transition" stands. Here, the sparser canopy facilitates penetration through its largest gaps even in the slanted SAR geometry.

In both Traunstein and Rabi, P- and L-band waves exhibit similar penetration, but different sensitivity to the physical canopy elements. In Traunstein, the P-band canopy contributions are weaker than the ground ones in many stands in contrast to L-band, similarly to other temperate and boreal forests (Frey and Meyer 2011; Tebaldini and Rocca 2012). However, there are cases in which the P- and L-band profiles are more similar (e.g., the forest stand beyond a horizontal distance of 2 km). The spatial variation of the canopy backscattering distribution (which in general occurs at very small scales) is reflected in different ways by the wavelengths. At P-band, the visibility of the canopy layers changes in space faster than at L-band. At L-band, canopy scattering contributions are either concentrated closer to the canopy top, or spread along the

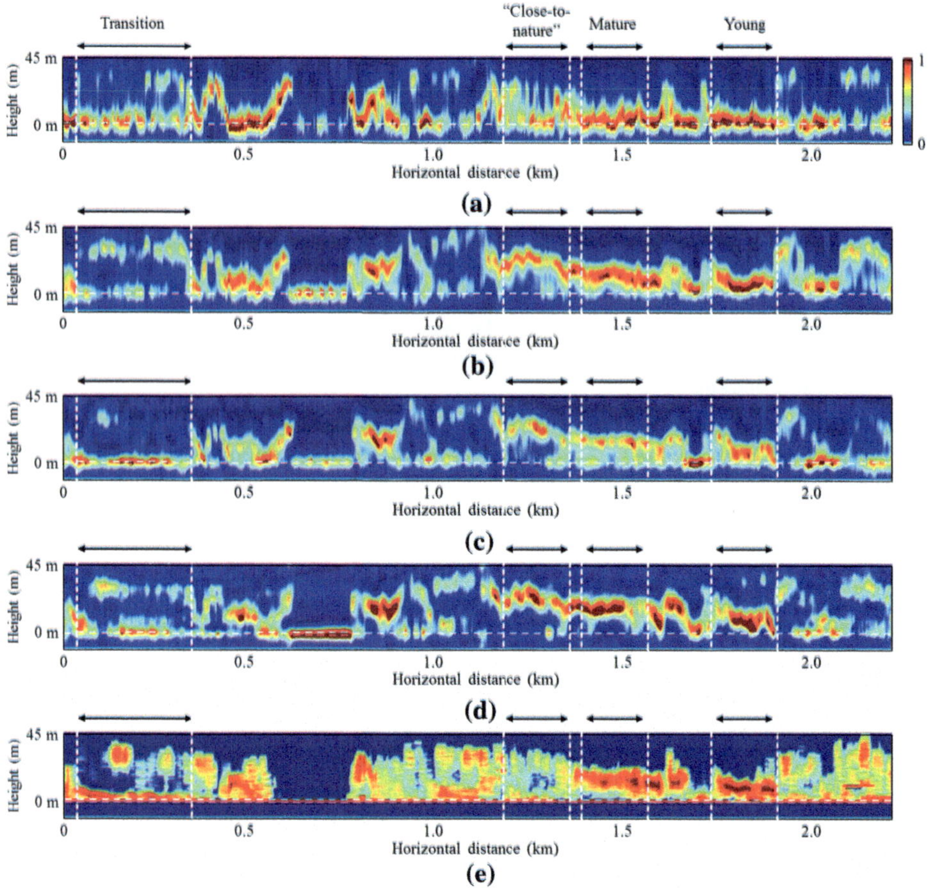

Fig. 4 Traunstein forest (south of Germany). SAR vertical reflectivity profiles and lidar profiles (histograms of the lidar returns as a function of height) over a representative azimuth transect at constant range, HV polarization channel. The height axis is relative to the ground height (0 m), indicated with the horizontal white dashed line. The vertical white dashed lines delimit stands of interest in different growth stages. SAR profile intensities are normalised by the total backscattered power. **a** P-band (2009); **b** L-band (2009); **c** L-band (2013); **d** X-band (2013); **e** lidar profiles

whole volume extension. Seasonal dielectric changes affect not only the distribution of the power at the ground, but also within the canopy layers (e.g., the forest stand beyond a horizontal distance of 2 km). Due to the larger attenuation, at X-band the canopy scattering contributions tend to be closer to the canopy top than at the other wavelengths. In contrast to Traunstein, P- and L-band profiles tend to be more similar in Rabi. However, large differences can occur, as shown for the stand corresponding to the horizontal distances between 600 and 800 m (delimited by the vertical dashed lines in Fig. 5). Here, stronger canopy contributions are close to the canopy top at P-band, while they are distributed along height at L-band. The distribution of the physical size of the involved scatterers might again explain this difference: the bigger ones might be close to the top, while smaller ones (which are semi-transparent at P-band) might be distributed down

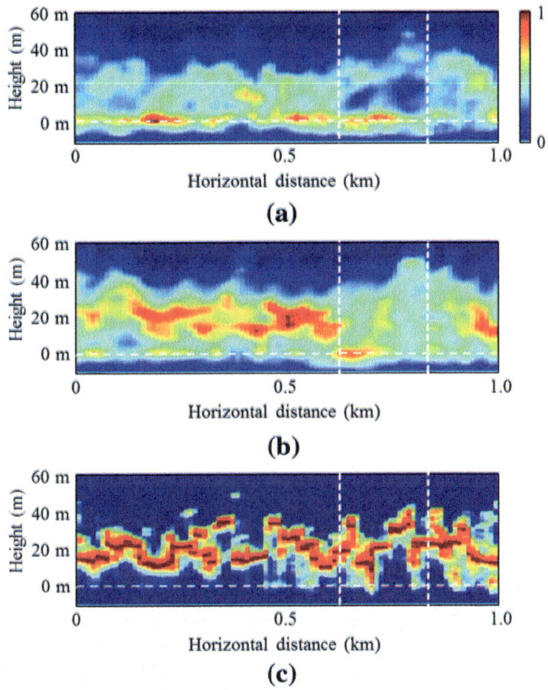

Fig. 5 Rabi ForestGEO plot (Gabon). Vertical profiles along a North–South transect. The height axis is relative to the ground height (0 m), indicated with the white dashed line. SAR profile intensities are normalised by the total backscattered power. **a** P-band and **b** L-band TomoSAR vertical reflectivity profiles, **c** LVIS waveforms

to the ground. On the contrary, the most powerful contributions in the lidar profiles are close to the canopy top due to a reduced penetration.

The profiles in Fig. 6 are extracted from Fig. 4 (Traunstein forest), and can be used to gain some additional insights about the capability of each wavelength to reflect 3D structure properties. Each plot shows 10 profiles extracted every 10 m in one of the four growth stages highlighted in Fig. 4:

- In the "transition" stands, all wavelengths show similar vertical profiles, with relevant vegetation backscattering contributions up to 10 m and between 20 and 40 m. However, the relative profile amplitudes and their spatial variation across profile coordinates are wavelength and time-dependent. The contributions close to the canopy top appear in 2009 more heterogeneous at P- than at L-band. This might be due to electromagnetic semitransparency at P-band, affecting any interpretation in terms of horizontal heterogeneity. As a result of the sparsity induced at the canopy top by management activities, in 2013 the upper canopy scattering contributions become weak at L-band, but stronger at X-band. In addition, the sparsity of the vegetation elements makes X-band penetrate well down to the ground, and the related reflectivity profiles reflect closely the geometric distribution of the lidar returns.
- The "close-to-nature" stands are characterized by the largest heterogeneity in terms of tree species and heights. The resulting structural complexity is reflected by the lidar profiles, which are composed by returns at all heights. In parallel, all the SAR frequencies show this heterogeneity, but at different extents. P-band shows still large penetration capabilities, and the profiles contain larger or even dominant scattering contributions close to ground. These contributions are present also in the L-band

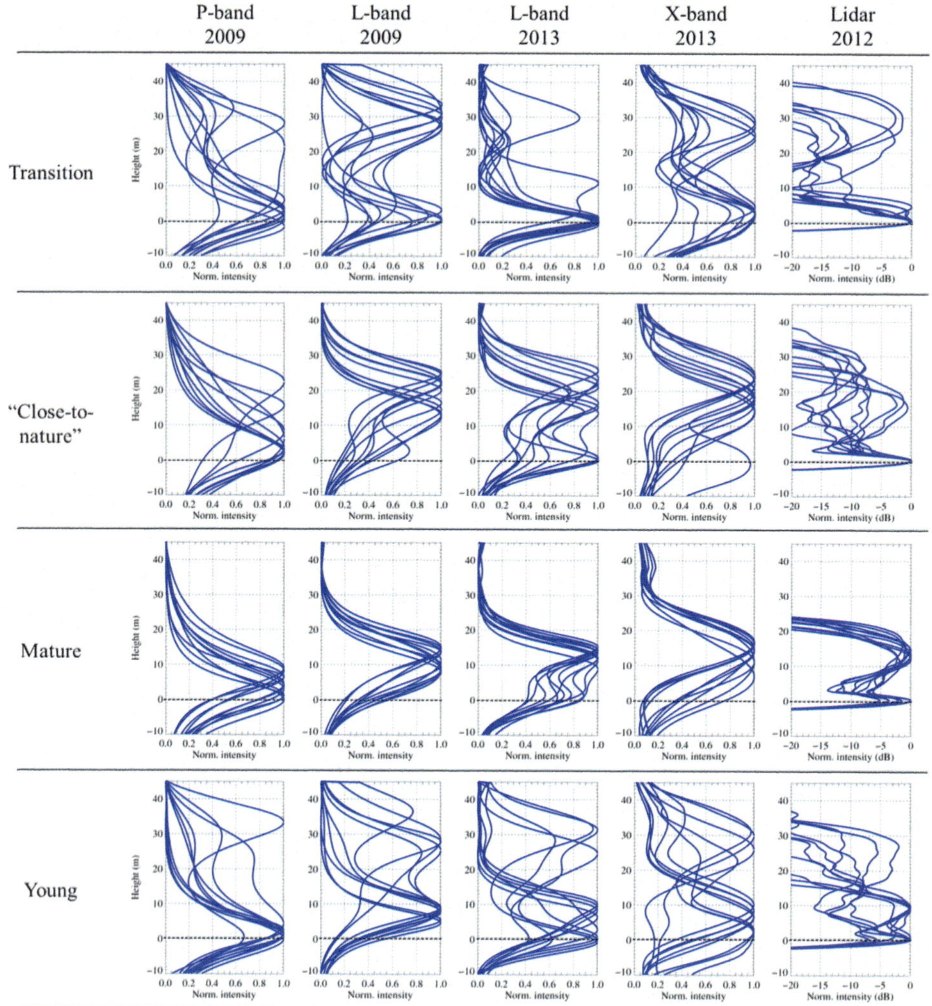

Fig. 6 Traunstein forest (south of Germany). Examples of lidar and TomoSAR profiles (HV channel) in the forest stands delimited in Fig. 4. The height axis is relative to the ground height (0 m), indicated with the horizontal black dashed line. For the sake of visualization the profiles have been normalized by their maximum

profiles together with additional stronger ones in correspondence of the upper canopy layers. The difference between the L-band profiles in 2009 and 2013 is probably caused by the difference in TomoSAR vertical resolution (better in 2013) and dielectric differences at a lower extent, as no management action is documented in these stands. At X-band, the reduced penetration makes the profiles sensitive to the upper canopy variations between 15 m and 30 m.

- The "mature" stands are constituted by (tall) trees uniformly distributed in space. This uniformity characterizes the lidar and SAR profiles at all wavelengths. The P-band profiles show the least sensitivity to the canopy top, having their main peak located closer to the ground. Similarly to the "close-to-nature" stands, the change of

the L-band profiles from 2009 to 2013 is mostly due to an improvement in the vertical resolution and to dielectric changes. In 2013, beyond the difference in attenuation, the L- and X-band profiles convey similar structural (i.e., geometric) information.
- Finally, the "young" stands are composed of mainly a layer of short trees with low density and some sparser taller trees. The shorter trees are visible at all the frequencies, while the taller trees are less visible at P-band. In 2013, the L-band profiles show a significant spatial variability of the canopy layer close to the ground corresponding to the shorter trees. Such variability is not shown either by the lidar or the X-band profiles. The spatial distribution of the X-band profiles is very similar to the lidar ones, and once more this could indicate their increased sensitivity to the geometric properties of the canopy.

To summarize, these examples have shown that, together with lidar profiles, SAR profiles at all wavelengths contain information about 3D structure. P- and L-band can penetrate through the canopy layers down to the ground even in dense tropical forest environments, and can complement lidar profiles in mapping vertical heterogeneity. At P-band, vegetation elements can be semitransparent in some stands, and the fast and strong spatial changes of attenuation might make the canopy appear more heterogeneous horizontally. At L-band, the scattering from smaller vegetation elements becomes more significant, and different degrees of vertical heterogeneity could be more distinguishable than, e.g., at P-band. The complementarity between these two frequencies seems to depend on the imaged forest structures. The L-band examples clearly show also the dependency of the reflectivity profiles on the dielectric properties, which might affect the interpretation of the profiles and their spatial changes in terms of physical structure. Finally, the X-band profiles resemble more closely the geometric distribution of the canopy layers in the lidar profiles, at least in those sparser stands where penetration until the ground occurs. This might be an indication that penetration is facilitated by vegetation gaps along the SAR line of sight. In the absence of these, the resulting lower X-band penetration increases the sensitivity to the top canopy spatial variations, therefore complementing and improving the capability of longer wavelengths in the characterization of horizontal heterogeneity.

4 The Lidar-SAR Model-Based Framework for Forest Height Estimation

In this section, a case study is considered in which the common sensitivity of lidar and SAR measurements to geometric structure parameters is exploited in the parameterisation of the Random-Volume-over-Ground Pol-InSAR backscattering model using sparsely sampled lidar full waveforms for obtaining accurate, continuous and high-resolution forest (top) height estimates.

4.1 The Random-Volume-Over-Ground Model

The Random-Volume-over-Ground (RVoG) model describes the variation of the reflectivity profiles with the polarization channel $\boldsymbol{\omega}$ (Papathanassiou and Cloude 2001). First of all, it is assumed that $F_{\mathrm{vol}}(z)$ is the sum of the ground-only and the volume-only vertical reflectivity profiles (two-layer model). Next, it is assumed that the backscattered power at the ground level is the only component of $F_{\mathrm{vol}}(z)$ that changes with polarisation. In other words, the volume-only profile is the same across all polarimetric channels except for a scalar factor. Additionally, the ground-only profile is modelled by a Dirac-δ function (Papathanassiou and Cloude 2001). As a consequence:

$$\gamma_{\mathrm{vol}}\left(k_Z, \boldsymbol{\omega}\right) = \mathrm{e}^{jk_Z z_G} \frac{\gamma_V\left(k_Z\right) + \mu(\boldsymbol{\omega})}{1 + \mu(\boldsymbol{\omega})} \tag{7}$$

where the dependency of the individual parameters on the polarisation channel $\boldsymbol{\omega}$ has been explicitly indicated. $\gamma_V(k_Z)$ is the (polarisation-independent) volume-only coherence and $\mu(\boldsymbol{\omega})$ is the ground-to-volume amplitude ratio. $\gamma_V(k_Z)$ is obtained from the volume-only vertical reflectivity profile $F_V(z)$ with a relationship formally identical to (6). The dependency of $\gamma_V(k_Z)$ on the forest top height H_V is made explicit by modeling $F_V(z)$. A widely and very successful model is an exponential distribution of scatterers, i.e.

$$F_V(z) = \mathrm{e}^{-2(H_V - z)\tau/\cos\theta_0} \quad \text{for } 0 \le z \le H_V \tag{8}$$

where defines the attenuation rate of the profile, which is a function of both density of scatterers and their dielectric constant. The exponential profile tends to fit better at shorter wavelengths. At longer wavelengths, or more in general in the cases in which more effective scatterers are located closer to the ground, the exponential decay in (6) may not be valid. Different parameterisations might be used instead [see, e.g., (Garestier et al. 2008)].

4.2 Case Study: Fusion of GEDI Waveforms and TanDEM-X Coherences

The GEDI lidar acquires footprints according to the spatial sampling pattern illustrated in Fig. 2. The use of single-pass bistatic X-band InSAR coherences acquired during the DLR's TanDEM-X mission has been proposed to bridge the spatial gaps between the footprints in order to obtain continuous and high-resolution forest mapping, and the inversion of the RVoG model (7, 8) is a possible solution. Only one TanDEM-X coherence in one polarimetric channel is most typically available for inversion. On the other hand, fixed a polarimetric channel, the RVoG model (7) presents 4 unknown parameters, i.e., H_V, τ, $\mu(\boldsymbol{\omega})$ and z_G. As one complex coherence can account for only two real parameters, the inversion of a full RVoG model results underdetermined. Additionally, the X-band penetration is not guaranteed in all forest types, and the use of an external z_G is needed to avoid any penetration-dependent height estimation bias (Kugler et al. 2014). The spatially sparse set of GEDI footprints can then be used to initialize some of the RVoG parameters, thus making the inversion determined (Qi and Dubayah 2016, 2017) and correcting penetration-induced biases. Depending on the model assumptions, the following strategies can be adopted:

(S1) The sparse GEDI ground topography values are interpolated on the TanDEM-X spatial grid of the complex InSAR coherences. Given the interpolated topography, the RVoG model is fitted to the complex coherences to estimate H_V and τ assuming $\mu = 0$ (no ground contribution).

(S2) Assuming $\mu = 0$, the GEDI ground topography and top height are used in the RVoG model to estimate extinction values from TanDEM-X complex coherences in correspondence of the footprints. Then, both topography and extinction are interpolated over the TanDEM-X grid. Finally, a continuous and high-resolution estimate of H_V is obtained by fitting the RVoG to the coherences using the interpolated topography and extinction.

(S3) As in (S2), but allowing $\mu \neq 0$, i.e., values of μ and τ are jointly estimated in correspondence of the footprints, interpolated and used in the inversion of the remaining parameters.

The interpolation of the three parameters (i.e., ground topography, extinction and ground-to-volume ratio) estimated in correspondence of the footprints can be carried out, e.g., by means of a simple kriging interpolation (Qi et al. 2017; Oliver and Webster 1990), in which it is assumed that samples are correlated according to a negative exponential spatial correlation function. Alternatively, the TanDEM-X high-resolution DEM can be used to obtain a ground topography with higher resolution than the one that a simple interpolation would provide (Lee et al. 2018). A wavelet transformation is used to decompose the TanDEM-X DEM into three high spatial frequency components (describing local changes of topography) and a low-frequency one, according to the methodology outlined in (De Grandi et al. 2016). The low-frequency part of the TanDEM-X DEM is substituted by the GEDI interpolated ground topography, and a GEDI-TanDEM-X combined ground topography is obtained by inverting the wavelet transformation.

Real data results have been generated over the two tropical forest sites of La Selva (Costa Rica) and Lopé (Gabon) by using TanDEM-X HH acquisitions and simulated GEDI acquisition patterns (Qi and Dubayah 2016, 2017). Details about the radar and lidar data sets are reported in Appendix 2. Over La Selva, the lidar data were processed to simulate GEDI full waveforms following the method in Blair and Hofton (1999) with measurement noise added following Hancock et al. (2011). GEDI acquisition tracks were simulated also over the Lopé site. In both cases, the nominal lidar sampling after 2 years of mission has been considered. In the experiments, TanDEM-X interferometric coherences have been calculated over 25 m×25 m (ground range) multilook cell approximately corresponding to the GEDI footprint diameter (25 m). Therefore the TanDEM-X estimates of H_V have been obtained with 25 m resolution. To reduce statistical variations, the estimates have been averaged on a threefold bigger cell. As a consequence, the final resolution of the height map is 75 m. Figure 7 shows the estimated TanDEM-X height maps and their validation against the lidar top height (2D histograms) for different configurations of the model parameters over La Selva. The adoption of the strategy (S1) (Fig. 7b) improves both the estimation RMSE and the bias with respect to the case in which the TanDEM-X absolute coherence values alone are used to estimate H_V with $\tau = \tau_0$ fixed a priori (Fig. 7a). However, the overall estimation performance is not satisfactory, due to the still large bias (around 4 m). A noticeable improvement is obtained by applying (S2), as the bias reduces to around 0.5 m (Fig. 7c). The use of (S3) keeps almost unaltered the already very small bias and further reduces the RMSE to 4.3 m (13%) (Fig. 7d). Such estimation performance is definitely better than just interpolating the GEDI top heights over the TanDEM-X grid. Moreover, the residual RMSE can be reduced by averaging further the final estimates to a coarser resolution.

The combination strategy (S1) with the refined wavelet-based interpolation of the ground topography has been applied to the Lopé acquisition. The mean value of the interpolation error between the original topography and the interpolated one (Fig. 8a) is 0.8 m and the standard deviation is around 8 m. The forest height (Fig. 8b) could be estimated with a correlation coefficient of 0.81 and an RMSE of 5.47 m (i.e., around 10%) with respect to the lidar RH95 heights (Fig. 8c). This performance is still comparable to the one obtained for the forest height inverted using the original lidar fully sampled topography (correlation coefficient of 0.92 and RMSE of 3.58 m).

As a final remark, the effectiveness of the parameterisation of SAR interferometric models by means of lidar waveform parameters is influenced by the difference of wavelengths, the

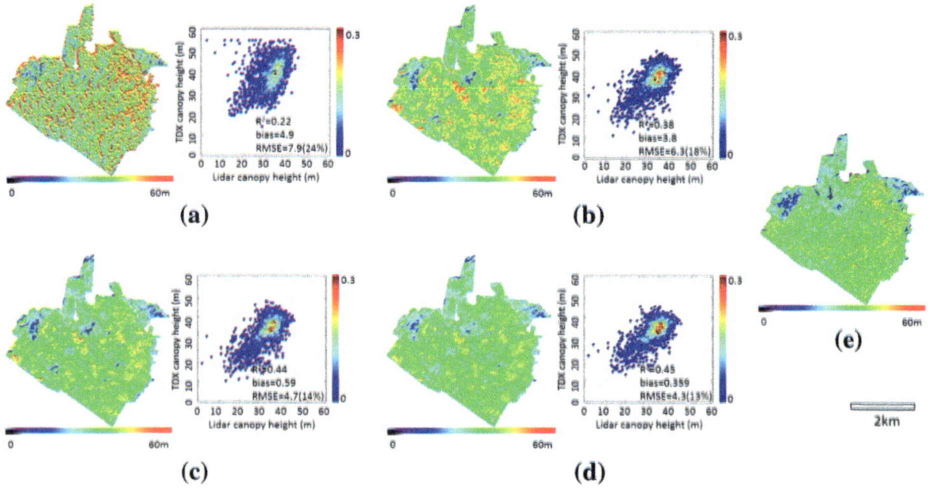

Fig. 7 La Selva Biological Station. TanDEM-X forest top height maps before and after combination with GEDI simulated data and 2D histograms against lidar canopy height **a** Inversion from TanDEM-X data only; **b** TanDEM-X inversion using the interpolated GEDI terrain height; **c** TanDEM-X inversion using the GEDI interpolated terrain height and σ, for $\mu = 0$; **d** TanDEM-X inversion using the GEDI interpolated terrain height, σ and μ; **e** reference lidar top height

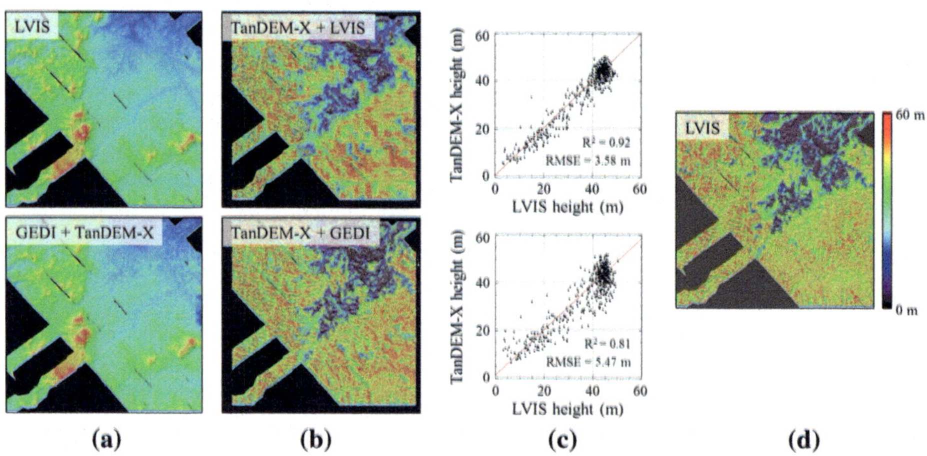

Fig. 8 Lopé test site. **a** LVIS terrain height (upper) and terrain height resulting from the simulated GEDI-TanDEM-X combination; **b** TanDEM-X top height inversion results from the LVIS terrain height (upper) and the one resulting from the GEDI-TanDEM-X combination (bottom); **c** validation plots against LVIS RH95; **e** reference LVIS RH95 canopy height

difference in the looking geometry, and the resolution of the parameterisation itself. Concerning the resolution of the parameterisation, in the example above τ and μ are radar reflectivity parameters, which depend on the (relatively) fast spatial variations of dielectric properties and physical arrangement of the vegetation elements. In general, any mismatch between the resolution of an external parameterisation and the resolution of the TanDEM-X height inversion causes a loss of estimation performance. A constant value of τ makes the inversion of forest height feasible even from single-baseline coherences without external parameterisations, but can lead to an unsatisfactory estimation performance. The interpolation of the values of τ and μ retrieved at the footprint locations improves the estimation performance, but it can track only a low-resolution component of these variations depending on the spatial sampling rate provided by the footprints. A way to reduce the consequential mismatch between the interpolated parameters and the SAR data could be to reduce the resolution of both the lidar parameterisation (for instance by aggregating multiple footprints) and at the same time of the inversion (multilook cell) from the interferometric coherences, for instance as noted by Brolly et al. (2016).

5 The Lidar-SAR Structure-Based Framework

The applicability of the lidar-SAR model-based framework presented in Sect. 4 becomes limited with increasing difference of the wavelength involved. This is a consequence of the fact that different wavelengths "see" different effective scatterers, as shown by the examples in Sect. 3. A direct relationship between lidar profiles and SAR vertical reflectivity profiles becomes difficult to be established at the increase of the wavelength difference, resolution and look angle (Brolly et al. 2016; Pardini et al. 2018a, b). In the structure-based framework, lidar and SAR measurements are used to calculate one or more structure indices independently, i.e., without exchanging a parametrization of the underlying backscattering models. The structure indices considered here are defined with the aim of describing (components of) physical structure, and they use features extracted from the vertical reflectance/reflectivity profiles that reflect physical heterogeneity within a certain scale. In this Section, we review meaningful structure indices both from field data and lidar and SAR profiles, and we show in an example with real data how they can be used to relate the structure information content of the profiles.

5.1 Structure from Ground Measurements

Forest structural information is often quantified by means of scalar indices whose definition depends on the specific application. Basing on field inventory data, structure indices are defined at plot level by means of single-tree attributes such as diameter, height, basal area, canopy dimension, species composition, and stand density. A number of indices have been proposed in the literature, like, e.g., the aggregation index (Clark and Evans 1954), the diametric differentiation index, the mingling index, the contagion index, or the complex stand index (Pastorella and Paletto 2013). However, there is not yet an overall and unequivocal measure able to express forest structure in terms of forestry and ecology appropriate for a wide variety of application, spatial scales, and forest types (Zenner and Hibbs 2000; Pommerening 2002; Pretzsch 2009; del Rio et al. 2016). Despite this difficulty, there is a common understanding that the structural heterogeneity in both the horizontal and the vertical dimensions are two complementary aspects that should be considered.

One way to quantify horizontal structure or heterogeneity is to use the concept of density. High density means low heterogeneity, and vice versa. The stand density index SDI (Reineke 1933) (measured in trees per hectare) links space utilization to tree size. It is closely related to basal area and by definition relates the stand density with the equivalent density of a stand with a quadratic mean diameter of 25 cm:

$$\mathrm{SDI} = N_S \left(\frac{D_g}{25} \right)^{1.605} \tag{9}$$

where N_S is the number of stems per hectare and D_g is the quadratic mean diameter at breast height (*dbh*) in cm within the scale at which SDI is calculated. A horizontal (dimensionless) structure index can be defined as

$$\mathrm{HS}_{\mathrm{field}} := 1 - \mathrm{SDI}_{\mathrm{norm}}, \tag{10}$$

where $\mathrm{SDI}_{\mathrm{norm}}$ is the normalized SDI within an area. $\mathrm{HS}_{\mathrm{field}}$ varies between 0 and 1, indicating the minimum and maximum horizontal heterogeneity, respectively. Additional refinements can be applied, e.g., considering the taller trees to better characterize canopy closure.

Quantifying vertical structure is often treated as quantifying the degree of heterogeneity in the vertical direction. In this sense, vertical structure has been expressed by means of indices such as the Gini coefficient, the Shannon index or the standard deviation of single tree heights (Liang et al. 2007; Barbeito et al. 2009; Bohn and Huth 2017). However, tree height measurements are often not performed so that these indices cannot be calculated directly from inventory data. An alternative way to express tree size variability is to use the diameter at breast height (*dbh*), which is a standard inventory measurement and can be related to tree height through models. Therefore, a vertical structure index can be defined as (McElhinny et al. 2005):

$$\mathrm{VS}_{\mathrm{field}} := \mathrm{std}(\{\mathrm{dbh}\}), \tag{11}$$

in cm, where $\{\mathrm{dbh}\}$ is the ensemble of diameter at breast height of all the trees included in a stand, given in cm. Without loss of generality, the vertical structure index can be normalized to its maximum within the area under consideration, becoming a dimensionless index as the horizontal one.

5.2 Structure from Lidar and TomoSAR Profiles

The definition of indices from lidar and TomoSAR vertical profiles with the same structural meaning as the ones from field inventory data in Sect. 5.1 is challenged by the fact that the resolution of conventional remote sensing configurations (especially SAR) does not allow single trees to be distinguished directly. An additional complication in the interpretation of the profiles in term of geometric structure is caused by the specific dependencies of their information content (see Sect. 2). Consequently, the direct translation of the indices in Sect. 5.2 is not possible, and the use of physically significant profile features is desired to express the same heterogeneity. Recently, TomoSAR profiles have been interpreted in terms of forest structure by considering the distribution of their peaks. Even if the correspondence between the 3D distribution of peaks and the variability in the distribution of trees may not be given at all frequencies and/or spatial scales, it is supported in a number of models and experiments (Lin and Sarabandi 1999; Frey and Meyer 2011; Thirion et al. 2006; Brolly and Woodhouse 2013; Whitehurst et al. 2013; Cazcarra-Bes et al. 2017; Tello et al. 2018).

Canopy height variation above the ground height within a certain scale can be considered as a proxy for horizontal structure (Couteron et al. 2005; Carabajal and Harding 2006; Neumann et al. 2012; De Grandi et al. 2016). Here, this idea is applied to the top peaks (i.e.,

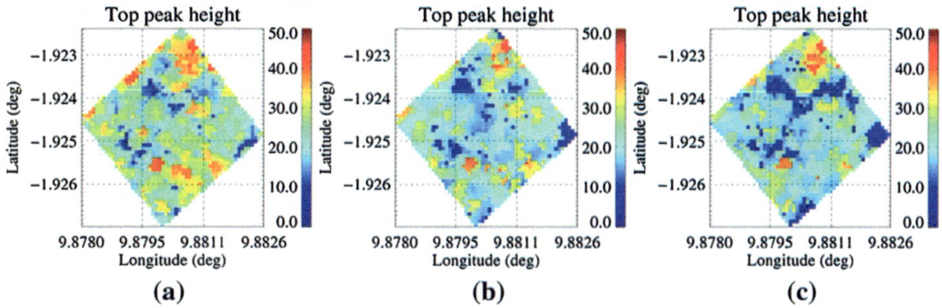

Fig. 9 Rabi ForestGEO plot. Top peak heights maps extracted from **a** LVIS waveforms, **b** L-band and **c** P-band vertical reflectivity profiles

the peaks at the highest height) of the waveforms and the vertical reflectivity profiles. Let $Z_{\mathrm{TOP}} = \{z_{T1}, z_{T2}, \ldots, z_{TM}\}$ be the ensemble of the M unique top peak heights within the structure window. The horizontal structure index HS can be calculated as:

$$\mathrm{HS} := \mathrm{var}\{Z_{\mathrm{TOP}}\}, \tag{12}$$

where $\mathrm{var}\{\cdot\}$ denotes the variance of a set of values. Along this line, a way to quantify vertical structure is to consider the ensemble $Z = \{z_1, z_2, \ldots, z_N\}$ of all the unique peak heights (excluding the ground peak, if present) between the ground and the canopy top. Then, a vertical structure index can be calculated as:

$$\mathrm{VS} := N\mathrm{var}\{Z\}. \tag{13}$$

Similarly to $\mathrm{HS}_{\mathrm{field}}$ and $\mathrm{VS}_{\mathrm{field}}$, HS and VS can be normalized to their maximum value over an area. The increase in their values toward 1 corresponds to increasing heterogeneity.

5.3 Real Data Examples

The structure indices defined in the previous Sections have been applied to the field inventory measurements LVIS lidar and L- and P-band TomoSAR (Capon spectral estimator) profiles extracted over the ForestGEO plot in the Rabi test site (see Appendix 2). The profiles have horizontal resolution 20 m × 20 m and have been used to calculate the structure indices within a window measuring 100 m × 100 m. This means aggregating information from 25 statistically independent profile cells. A smaller window would reduce the statistical validity of the number of independent cells, while a larger window could provide biased values of the structure indicators as different structure types would be mixed together, and therefore lose ecological significance.

Figure 9 shows the maps of the top peak heights of LVIS profiles, and the L- and P-band TomoSAR profiles (Capon spectral estimator). Due to penetration differences, the L- and P-band top peak heights are in general lower than the LVIS ones, as shown by the profiles in Fig. 5. Nevertheless, there is a limited number of cases in which the L-band top peak height can be even higher than the LVIS one, and they are more likely focusing artifacts (e.g., sidelobes) coming from very high decorrelations. The generated $\mathrm{HS}_{\mathrm{field}}$ and HS maps are shown in Fig. 10. All of them can well locate heterogeneous areas in the North and the South parts of the plot, and some discrepancies are visible in the remaining areas. The temporal difference (around 4 years) between the field measurements and the lidar and SAR

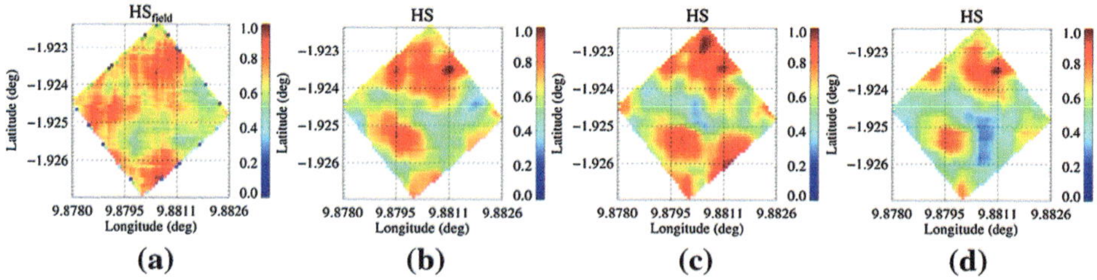

Fig. 10 Rabi ForestGEO plot. Maps of **a** HS_{field}, and HS calculated from **b** LVIS waveforms, **c** L-band and **d** P-band vertical reflectivity profiles

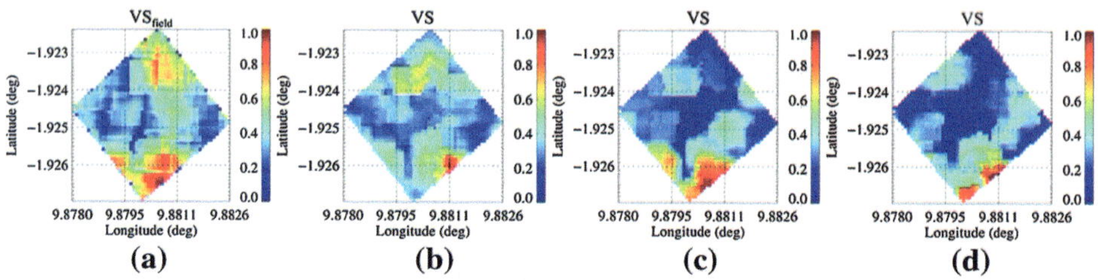

Fig. 11 Rabi ForestGEO plot. Maps of **a** VS_{field}, and VS calculated from **b** LVIS waveforms, **c** L-band and **d** P-band vertical reflectivity profiles

acquisitions may also affect the comparison between field and remote sensing indices. The LVIS and the L-

band HS maps are the most similar ones, although in some stands the L-band profiles lead to higher values of HS as a consequence of the mentioned artifacts. At P-band, many stands result in more homogeneity as a consequence of the lower sensitivity to the top canopy variations.

Figure 11 shows the VS_{field} and the VS maps. All maps highlight an area of high vertical structure at the southeastern part of the site, which seems to be the mean feature in the plot. More to the North, another high structure area is indicated in the map derived from the field data. This area is also visible in the LVIS map and in the L-band map, while it becomes hardly visible in the P-band one. The correlation plot of Fig. 12 clearly shows that L-band reconstructs the wider range and distribution of structure indices obtained from the field data. The profiles of Fig. 5 offer some insights to understand these differences. In the LVIS profiles, the majority of peaks appear within 20 to 45 m, while peaks below 20 m are rather seldom. On the other hand, the TomoSAR peaks at L- as well as at P-band are distributed across the whole canopy extent down to the ground. The peaks from the P-band TomoSAR profiles are particularly numerous closer to the ground level. Occasionally, gaps in the canopy layer become visible at P-band which are not noticeable at L-band or in the LVIS profiles. Therefore, LVIS profiles are more sensitive than TomoSAR reflectivity profiles to variations mostly concentrated in the highest part of the canopy and can better predict VS_{field} in those areas. However, TomoSAR long-wavelength profiles are more sensitive to variations of vertical structure when these are driven by the presence of sub-canopy elements.

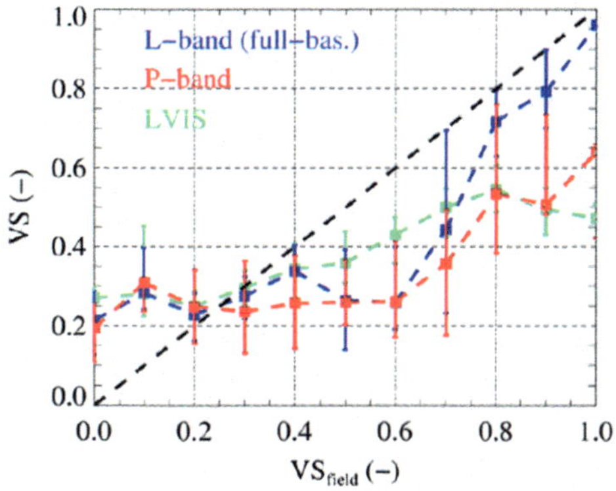

Fig. 12 Rabi, ForestGEO plot. Comparison between VS and VS_{field}. The square symbols indicate (conditioned) mean values, while the vertical bars delimit the VS interval where 75% of values are found

6 Conclusions

In this paper, first, a comparison between lidar waveforms and TomoSAR reflectivity profiles at multiple wavelengths has been attempted by means of real data in a temperate and a tropical test site. It has been shown that the wavelength affects directly penetration through the vegetation volume down to the ground. While lidar penetration occurs through vegetation gaps and is facilitated by the nadir-looking geometry, SAR penetration typically reduces from P- up to X-band, being however further affected by forest type, density, and dielectric properties (attenuation induced by water content). At the same time, it has been shown, even when the penetration is comparable, the measurements provided by lidar and by SAR systems at different wavelengths are sensitive to different components of forest structure, and therefore all of them can complement each other for structure characterization. This comparison has then be used as a basis to gain some early insights on the capabilities of two most promising combination frameworks, i.e., the model-based and the structure-based frameworks, based on two recent case studies.

The common sensitivity to geometric parameters drives the parameterisation of SAR backscattering models by means of lidar measurements in the model-based framework. A fusion algorithm could then be based on the complementarity of type and/or spatial distribution and resolution of the measurements. With reference to forest height estimation, some of the experimented model-based combination methodologies have been presented by means of a case study developed in the context of the NASA GEDI lidar mission. The TanDEM-X coherences can be used to continuously extend the coverage of the top heights measured by GEDI on the sparsely sampled waveform footprints. The RVoG model has been shown to be simple and representative enough to allow accurate parameterisation and inversion. Depending on the SAR wavelength and configurations, additional parameters can be accommodated (Simard and Denbina 2018). In general, (i) the choice of an appropriate backscattering model and its dimensionality, (ii) the choice a meaningful lidar parameterisation, and (iii) its interpolation from the lidar footprint to the continuous SAR coverage are three critical steps that contribute to the effectiveness of the combination.

The structure-based framework can directly accommodate both commonalities and complementarities in the information content of lidar and SAR measurements, bridging the limitations of the model-based framework in the cases in which the physical equivalence between lidar and SAR parameters is not given. An example has been shown in which indices expressing horizontal and vertical structure (heterogeneity) have been applied to lidar waveforms and TomoSAR vertical reflectivity profiles, in correspondence to structure indices established in ecology and forestry with ground inventory measurements. Lidar and P- and L-band wavelengths fully complement each other in both the horizontal and vertical structure if significant penetration differences exist and/or structural differences are driven by element sizes. For sure, the sensitivity of lidar waveforms to vertical structure variations mostly close to the canopy top is complemented by the higher sensitivity of long-wavelength TomoSAR profiles to variations closer to the ground. The definition of ecologically significant structure indices is of crucial importance. Nevertheless, the structure space defined by the two indices (whose independence is not guaranteed) seems to be able to relate lidar and SAR measurements and field measurements, therefore providing a biophysical interpretation. Limitations and ambiguities of this framework are still under investigation, remembering that indices might only describe an incomplete "subspace" of the whole structure information.

It is worth noting that a combination framework that unifies both the model-based and the structure-based ones might also be developed. For instance, the structure-based framework could be used to identify the spatial regions of validity of a certain parameterisation, thus defining models and supporting inversion strategies in the model-based framework.

A critical factor affecting both frameworks is the scale at which structure parameters and indices are retrieved. Indeed, it affects both the meaningfulness of the parameterisation with respect to the final inversion performance and the structure indices with respect to the underlying stand or forest. Highresolution sensors allow to explore multiple scales, representing an additional degree of freedom for structure characterisation to be exploited in the combination between lidar and SAR, and SAR at multiple wavelengths. Existing models, algorithms and indices should be assessed also in this perspective, and new ones developed, if necessary. At the present stage, addressing these issues and understanding the underlying relationships is a unique challenge that still needs to be mastered toward the development of effective fusion techniques for forest structure observation, and to optimize system and acquisitions configurations.

Acknowledgements This review originates from the workshop "Space-based Measurement of Forest Properties for Carbon Cycle Research" held at the International Space Science Institute in Bern in November 2017. We thank P. Biber and M. Heym (Technische Universität München, TUM, Munich, Germany), and A. Huth and R. Fischer (Helmholtz Zentrum für Umweltforschung, UFZ, Leipzig, Germany) for inputs and discussions on forest structure and structure metrics. Part of this work was supported by the Helmholtz Alliance Remote Sensing and Earth System Dynamics funded by the Initiative and networking Fund of the Helmholtz Association. Laser Vegetation and Ice Sensor (LVIS) data sets were provided by the team in the Laser Remote Sensing Branch at NASA Goddard Space Flight Center with support from the University of Maryland, College Park. Finally, we thank the two anonymous reviewers that helped to improve the manuscript with their constructive comments.

Appendix 1: Derivation of the Relationship Between InSAR Volume Coherences and Vertical Reflectivity Profile

The purpose of this Appendix is to summarize the algebraic steps that lead to the Fourier relationship between the InSAR coherence and the vertical reflectivity profile starting from the volume integral (2).

First of all, it is commonly assumed that the (end-to-end) system point-spread function after focusing is separable in the range and azimuth dimensions, i.e.

$$h(r,x) = h_r(r) \cdot h_x(x) \tag{14}$$

being $h_r(r)$ and $h_x(x)$ the range and azimuth system point-spread functions, usually approximated as ideal rectangular functions with width equal to the respective resolution. The formulation in (14) implies that the viewing geometry is independent of the azimuth coordinate,[1] therefore the integral on dx can be solved, and (1) becomes:

$$y(r') = \iint h_r(r' - r) \cdot \bar{\xi}(r, s) \cdot e^{-j\frac{4\pi}{\lambda}R(r,s)} \mathrm{d}r\mathrm{d}s, \tag{15}$$

$\bar{\xi}(r, s)$ being the integral along azimuth of the complex reflectivity weighted by the azimuth point-spread function. The dependency on the azimuth has been dropped for notational simplicity. From (15), a single SAR image provides a single tomographic projection of the 3D reflectivity filtered by the range response (Bamler and Hartl 1998). The interpretation of the convolution integral in (15) is facilitated in the domain of the 2D range-elevation spatial frequencies after a Fourier transform. Indeed, the SAR image spectrum is a 2D slice out of the complete 3D reflectivity spectrum (Bamler and Hartl 1998).

In general, an additional 2D spectral slice with minimum overlap increases the amount of information available on the reflectivity spectrum toward its reconstruction. Conventional InSAR configurations realize this by varying the complex exponential in (15), i.e., by collecting one additional SAR image in spatial (incidence angle) diversity. First of all, the two images are co-registered, i.e., both images are resampled onto a common grid so that the co-registered pixels refer to a common position in the 3D space. After co-registration, and in absence of system (including noise) and processing non-idealities, the SLC complex amplitudes for the same pixel become:

$$\begin{aligned} y_1(r') &= \iint h_r(r' - r) \cdot \bar{\xi}(r, s) \cdot e^{-j\frac{4\pi}{\lambda}R_1(r,s)} \mathrm{d}r\mathrm{d}s, \\ y_2(r') &= \iint h_r(r' - r) \cdot \bar{\xi}(r, s) \cdot e^{-j\frac{4\pi}{\lambda}R_2(r,s)} \mathrm{d}r\mathrm{d}s, \end{aligned} \tag{16}$$

under the additional assumption that $\bar{\xi}(r, s)$ does not change with the difference of incidence angle induced by the track/orbit displacement and with the (possible) time difference between the two acquisition. The interferogram is defined as

$$\varphi(r') = \angle y_1(r') y_2^*(r'), \tag{17}$$

which for an ideal deterministic point scatterer located at (r', s') becomes:

$$\varphi(r') = \frac{4\pi}{\lambda}\left[R_2(r', s') - R_1(r', s')\right] \tag{18}$$

Let $B_{\|}$ and $B_{\perp}$ be the spatial track/orbit displacements in the direction parallel and orthogonal to the line of sight, respectively. Therefore, one can write:

$$R_2(r, s) = \sqrt{\left(r - B_{\|}\right)^2 + \left(s - B_{\perp}\right)^2}. \tag{19}$$

Some processing steps and approximations can be applied in order to reach a convenient formulation, which are detailed in Bamler and Hartl (1998) and Fornaro et al (2003). In

[1] The dependence on the azimuth direction can be readily included. Here it has not been considered for simplicity.

these operations, $y_1(r')$ is taken as a phase reference, and for this reason it is called master; $y_2(r')$ is then called slave. The small variation of look angle between the two acquisitions allows to retain the paraxial approximation. By developing (19), a residual s^2-dependent phase contribution is originated by the curvature of the wavefront which can be included in the complex reflectivity, as its (second-order) amplitude distribution is normally of interest. Alternatively, it can simply be neglected if the distance between the sensor and the scatterers is large enough (as it typically is) to make the plane-wave approximation hold. Finally, the phase of the slave image is "flattened" with respect to a reference height by using the interferometric phase (18). In this case, the reference height can be chosen constant for the entire scene, or varying locally according to a reference external DEM. After these manipulations, by further assuming $B_{\|} \ll r$, it results:

$$\begin{aligned} y_1(r') &= \iint h_r(r' - r) \cdot \xi'(r, z) \mathrm{d}r\mathrm{d}z, \\ y_2(r') &\cong \iint h_r(r' - r) \cdot \xi'(r, z) \cdot e^{jk_z z} \mathrm{d}r\mathrm{d}z. \end{aligned} \tag{20}$$

In (20), a change of variable from elevation to vertical height z has been performed, with $s = z/\sin\theta_0$. $\xi'(r, z)$ is the final complex reflectivity after the change of variables and the inclusion of phase residuals ψ, i.e., $\xi'(r, z) = \bar{\xi}(r, z/\sin\theta_0)\mathrm{e}^{j\psi}/\sin\theta_0$, being θ_0 the local incidence angle. k_Z is the so-called vertical wavenumber, defined in (3).

Being composed of distributed scatterers, the scattering from forest scenarios is better described by its statistics, particularly by the complex coherence defined in (4). This results in:

$$\begin{aligned} E\{y_1(r')\} &= E\{y_2(r')\} = 0, \\ E\left\{|y_1(r')|^2\right\} &= E\left\{|y_2(r')|^2\right\} = P_y, \\ E\{y_1(r')y_2^*(r')\} &= P_y\gamma_{1,2}(k_Z), \end{aligned} \tag{21}$$

and using (20):

$$E\{y_1(r')y_2^*(r')\} = \iint |h_r(r' - r)|^2 \cdot F_\xi(r, z) \cdot \mathrm{e}^{jk_z z} \mathrm{d}r\mathrm{d}z. \tag{22}$$

$F_\xi(r, z)$ represents the spatial density of backscattered power in the height-range plane. It is interesting to consider the case of surface scattering at a constant height $z = z_0$. Its complex reflectivity can be written as a Dirac-δ with power P_S. After appropriate algebraic manipulations, it provides the following interferometric coherence:

$$\gamma_{1,2}(k_Z) = \frac{E\{y_1(r')y_2^*(r')\}}{P_S} = \gamma_S(k_Z)\mathrm{e}^{jk_Z z_0}. \tag{23}$$

Therefore, the coherence phase is proportional to the surface of the height. $\gamma_S(k_Z)$ is a positive (real-valued) geometric decorrelation term that depends on the range point-spread function.[2] However, forest scatterers are distributed within a volume, that can be seen as the coherent superposition of the number of surfaces with different total powers. As a consequence, it simply follows:

$$E\{y_1(r')y_2^*(r')\} = \gamma_S(k_Z)\sum_n P_S(z_n)e^{jk_Z z_n}. \tag{24}$$

In the most general case, by letting the number of surfaces tend to infinity, the sum in (24) becomes an integral, and it is readily obtained

$$\gamma_{1,2}(k_Z) = \gamma_S(k_Z)\gamma_{\text{vol}}(k_Z), \tag{25}$$

where:

$$\gamma_{\text{vol}}(k_Z) := e^{jk_Z z_G}\frac{\int_0^{H_V} F_{vol}(z)e^{jk_Z z}\mathrm{d}z}{\int_0^{H_V} F_{\text{vol}}(z)\mathrm{d}z}, \tag{26}$$

which provides the Fourier relationship between vertical reflectivity profile and volume coherence anticipated in Sect. 2.1.

Appendix 2: Descriptions of Test Sites and Data Sets

Traunstein Forest (South of Germany)

This temperate forest site is located close to the city of Traunstein in the South of Germany. Most forest stands are structurally complex due to species richness and management practices, with a majority of conifer stands (around 70%). The forest top height can reach 40 m, and mean biomass level is about 200 t/h, significantly higher than other managed forests in the same ecological zone. In this test site, TomoSAR data sets at different frequencies have been collected by means of the German Aerospace Center (DLR) E-SAR and F-SAR airborne platforms since 2008. The data sets used in this paper were acquired at P- and L-band in 2009, and at X-band in 2013 with approximately the same flight tracks and viewing (master) geometry. Due to the 4-year time difference between the P- and X-band data sets, an additional L-band TomoSAR data set of 2013 has been considered. In this way, two direct comparisons can be performed, i.e., P- versus L-band in 2009, and L- versus X-band in 2013. All acquisitions share the same (master) geometry. While the acquisitions at P- and L-band are in repeat-pass mode, the one at X-band is in repeated single-pass mode (with changing k_Z for different passes) to avoid temporal decorrelation (Pardini et al. 2018b). The most relevant acquisition parameters are reported in Table 1. On November 18, 2012, discrete return lidar data have been acquired as well. During the total 4-year time, structural changes occurred in some stands due to forest management activities. Additionally, all acquisitions are in leaf-off conditions, but differences of water content and of its distribution in the tree trunk (Gates 1991) could have affected the SAR backscattering.

[2] If the dependence of the problem on the azimuth coordinate is taken into account, a similar term would arise also in azimuth. However, the related decorrelation is normally lower than the range one, thus it is considered negligible.

Table 1 Summary of parameters characterizing the airborne SAR acquisitions over the Traunstein and Rabi forests

Test site	Traunstein (Germany)	Traunstein (Germany)	Traunstein (Germany)	Traunstein (Germany)	Rabi (Gabon)	Rabi (Gabon)
Platform	E-SAR	E-SAR	F-SAR	F-SAR	F-SAR	F-SAR
Frequency	P-band	L-band	L-band	X-band	P-band	L-band
Wavelength (m)	0.85	0.23	0.23	0.031	0.69	0.23
Acquisition date	05/11/2009	28/10/2009	04/04/2013	22/03/2013	07/02/2016	
Resolution range × azimuth (m)	6.64 × 2.4	2.12 × 1.2	1.28 × 0.6	0.96 × 0.5	3.84 × 2	1.92 × 0.64
Number of tracks	6	4	5	6	10	10
Total acquisition time[a]	~1 h	< 1 h	~1 h	~1 h	~2 h	~2 h
Vertical resolution, m[b]	20	20	10	20	15	5

[a]Time needed to fly all the acquisition tracks

[b]Average Rayleigh TomoSAR vertical resolution (defined in Sect. 2.2) for the considered transect

Rabi and Lopé (Gabon)

The Rabi forest site is located at the southwestern Gamba Complex of Gabon, and consists of a diverse mix of upland and wet-forest with a mean tree height of about 40 m on a fairly flat topography. Within this forest site, a ForestGEO (Smithsonian Institution Forest Global Earth Observatories) plot was established between 2010 and 2012 in which the position and the diameter at breast height (*dbh*) of more than 175,000 trees belonging to 340 species with $dbh \geq 1$ cm were inventoried across 25 ha. P- and L-band TomoSAR acquisitions were carried out with the F-SAR airborne platform during the AfriSAR campaign in 2016 (see Table 1 for the relevant acquisition parameters) (Pardini et al. 2018a). During the same campaign, the Rabi site has been covered also by NASA LVIS lidar waveforms. For AfriSAR, LVIS operated at 7315 m altitude, with a wavelength equal to 1064 nm, ~20 m footprint diameter and 10 m footprint spacing along latitude and longitude. Together with the waveforms, ground topography (Digital Terrain Model, DTM) and relative height metrics have been processed.

The Lopé site lies within the Lopé National Park, and it consists of a mosaic of savannah and (dense) forest, with varying species richness and tree density across stands. The maximum tree height exceeds 50 m in many stands. The forest (above ground) biomass ranges between 10 t/ha in savanna areas up to ~400 t/ha in mature forest stands. The terrain is hilly, with many local slopes steeper than 20°. This test site was covered during the AfriSAR campaign by both airborne SAR and LVIS lidar acquisitions similarly to the Rabi site. For the experiments in Sect. 4, a TanDEM-X bistatic single-pass single polarisation (HH) interferometric acquisition has been considered (see Table 2) that was performed a few weeks apart from the LVIS acquisitions.

La Selva (Costa Rica)

La Selva Biological Station (LSBS) forest site is located in northeastern Costa Rica. The site is a protected low-land region of tropical rain forest, covering an area of 1600 ha. The area contains a mixture of old-growth, secondary and selectively logged forests with height up to 60 m. The mean biomass of old-growth forest, which is the major components of total LSBS biomass, is around 169 t/ha (Clark et al. 2011). Airborne discrete return lidar data were collected over LSBS in September and October 2009. In 2011, the LSBS was covered by a dual polarisation (HH/VV) TanDEM-X bistatic acquisition on December 5, whose characteristic parameters are reported in Table 2.

Table 2 Summary of parameters characterizing the TanDEM-X SAR acquisitions over Lopé and La Selva

Test site	Lopè (Gabon)	La Selva (Costa Rica)
Acquisition date	14/12/2015	05/11/2011
Acquisition mode	Bistatic	
Polarization	HH	HH/VV
Resolution range × azimuth (m)	1.7 × 3.3	1.7 × 6.6
incidence angle (°)	45	38
HoA (m)	80	70

References

Aguilera E, Nannini M, Reigber A (2013) Wavelet-based compressed sensing for SAR tomography of forested areas. IEEE Trans Geosci Remote Sens 51(12):5283–5295
Armston J, Disney M, Lewis P, Scarth P, Phinn S, Lucas R, Bunting P, Goodwin N (2013) Direct retrieval of canopy gap probability using airborne waveform lidar. Remote Sens Environ 134:24–38
Bai Y, Tebaldini S, Minh DHT, Yang W (2018) An empirical study on the impact of changing weather conditions on repeat-pass SAR tomography. J Sel Top Appl Earth Obs Remote Sens 11(10):3505–3511
Bamler R, Hartl P (1998) Synthetic aperture radar interferometry. Inverse Probl 14:R1–R54
Barbeito I, Montes F, Canellas I (2009) Evaluating the behavior of vertical structure indices in Scots pine forests. Ann For Sci 66(710):1–10
Bergen KM, Goetz SJ, Dubayah RO, Henebry GM, Hunsaker CT, Imhoff ML, Nelson RF, Parker GG, Radeloff VC (2009) Remote sensing of vegetation 3-D structure for biodiversity and habitat: review and implications for lidar and radar spaceborne missions. J Geophys Res 114:1–13
Blair JB, Hofton MA (1999) Modeling laser altimeter return waveforms over complex vegetation using high-resolution elevation data. Geophys Res Lett 26:2509–2512
Blair J, Rabine D, Hofton M (1999) The laser vegetation imaging sensor: a medium altitude, digitization-only, airborne laser altimeter for mapping vegetation and topography. ISPR J Photogramm Remote Sens 54(2/3):115–122
Bohn FJ, Huth A (2017) The importance of forest structure to biodiversity-productivity relationships. R Soc Open Sci 4(1):160521
Brokaw N, Lent R (1999) Vertical structure. In: Maintaining biodiversity in forest ecosystems. Cambridge University Press, Cambridge, UK
Brolly M, Woodhouse IH (2013) Vertical backscatter profile of forests predicted by a macroecological plant model. Int J Remote Sens 34(4):1026–1040
Brolly M, Simard M, Tang H, Dubayah R (2016) A lidar-radar framework to assess the impact of vertical forest structure on interferometric coherence. IEEE J Sel Top Appl Earth Obs Remote Sens 9:5830–5841
Carabajal CC, Harding DJ (2006) SRTM C-band and ICESat laser altimetry elevation comparisons as a function of tree cover and relief. Photogramm Eng Remote Sens 72(3):287–298
Cazcarra Bes V, Tello M, Fischer R, Heym M, Papathanassiou K (2017) Monitoring forest structure dynamics by means of L-band SAR tomography. Remote Sens 9(12):1229
Cazcarra Bes V, Pardini M, Tello M, Papathanassiou K (2019) Comparison of tomographic SAR reflectivity reconstruction algorithms for forest applications at L-band. IEEE Trans Geosci Rem Sensing (Accepted for publication)
Clark P, Evans C (1954) Distance to nearest neighbor as a measure of spatial relationships in populations. Ecology 35(4):445–452
Clark ML, Roberts DA, Ewel JJ, Clark DB (2011) Estimation of tropical rain forest aboveground biomass with small-footprint lidar and hyperspectral sensors. Remote Sens Environ 115:2931–2942
Couteron P, Pelissier R, Nicolini EA, Paget D (2005) Predicting tropical forest stand structure parameters from Fourier transform of very high-resolution remotely sensed canopy images. J Appl Ecol 42:1121–1128
De Grandi EC, Mitchard E, Hoekman D (2016) Wavelet based analysis of TanDEM-X and LiDAR DEMs across a tropical vegetation heterogeneita gradient driven by fire distrurbance in Indonesia. Remote Sens 8:641–667
Del Rio M (2016) Characterization of structure, dynamics, and productivity of mixed species stands: review and prespectives. Eur J For Res 135(1):23–49
Dubayah RO (2015) The global ecosystem dynamics investigation (GEDI) lidar. In: Proceedings of 1st BIOMASS science workshop
Dubayah RO, Drake JB (2000) Lidar remote sensing for forestry. J For 98(6):44–46
Dubayah RO, Sheldon SL, Clark DM, Hofton MA, Blair JB, Hurtt GC, Chazdon RL (2010) Estimation of tropical forest height and biomass dynamics using lidar remote sensing at La Selva, Costa Rica. J Geophys Res 115:1–17
Fatoyinbo TE, Simard M (2013) Height and biomass of mangroves in Africa from ICESat/GLAS and SRTM. Int J Remote Sens 34(2):668–681
Fornaro G, Serafino F, Soldovieri F (2003) Three-dimensional focusing with multipass SAR data. IEEE Trans Geosci Remote Sens 41(3):507–517
Frey O, Meyer E (2011) Analyzing tomographic SAR data of a forest with respect to frequency, polarization, and focusing techniques. IEEE Trans Geosci Remote Sens 49(10):3648–3659

Frolking S, Palace MW, Clark DB, Chambers JQ, Shugart HH, Hurtt GC (2009) Forest disturbance and recovery: a general review in the context of spaceborne remote sensing of impacts on aboveground biomass and canopy structure. J Geophys Res: Biogeosci 114:1–27

Garestier F, Dubois-Fernandez P, Champion I (2008) Forest height inversion using high resolution P-band Pol-InSAR data. IEEE Trans Geosci Remote Sens 46(10):3544–3559

Gates DM (1991) Water relations of forest trees. IEEE Trans Geosci Remote Sens 29(6):836–842

Gini F, Lombardini F, Montanari M (2002) Layover solution in multibaseline SAR interferometry. IEEE Trans Aerosp Electron Syst 38(4):1344–1356

Goetz S, Steinberg D, Dubayah RO, Blair B (2007) Laser remote sensing of canopy habitat heterogeneity as a predictor of bird species richness in an eastern temperate forest. Remote Sens Environ 108:254–263

Hall FG, Bergen K, Blair JB, Dubayah R, Houghton R, Hurtt G, Kellndorfer J, Lefsky M, Ranson J, Saatchi S, Shugart HH, Wickland D (2011) Characterizing 3D vegetation structure from space: mission requirements. Remote Sens Environ 115:2753–2775

Hancock S, Disney M, Muller JP, Lewis P, Foster M (2011) A threshold insensitive method for locating the forest canopy top with waveform lidar. Remote Sens Environ 115:3286–3297

Harding DJ, Lefsky MA, Parker GG, Blair JB (2001) Laser altimeter canopy height profiles: methods and validation for closed-canopy, broadleaf forest. Remote Sens Environ 76(3):283–297

Kaasalainen S, Holopainen M, Karjalainen M, Vastaranta M, Kankare V, Karila K, Osmanoglu B (2015) Combining lidar and synthetic aperture radar data to estimate forest biomass: status and prospects. Forests 6:252–270

Kellndorfer JM, Walker WS, LaPoint E, Kirsch K, Bishop J, Fiske G (2010) Statistical fusion of lidar, InSAR, and optical remote sensing data for forest stand height characterization: a regional-scale method based on LVIS, SRTM, Landsat ETM+, and ancillary data sets. J Geophys Res 115:G00E08

Krieger G, Zink M, Bachmann M, Bräutigam B, Schulze D, Martone M, Rizzoli P, Steinbrecher U, Antony JW, De Zan F, Hajnsek Papathanassiou K, Kugler F, Rodriguez Cassola M, Younis M, Baumgartner S, López Dekker P, Prats P, Moreira A (2013) TanDEM-X: a radar interferometer with two formation-flying satellites. Acta Astr 89:83–98

Kugler F, Schulze D, Hajnsek I, Pretzsch H, Papathanassiou K (2014) TanDEM-X Pol-InSAR performance for forest height estimation. IEEE Trans Geosci Remote Sens 52(10):6404–6422

Lavalle M, Hensley S (2015) Extraction of structural and dynamic properties of forests from polarimetric-interferometric SAR data affected by temporal decorrelation. IEEE Trans Geosci Remote Sens 53(9):4752–4767

Le Toan T, Quegan S, Davidson MJW, Baltzer H, Paillou P, Papathanassiou K, Plummer S, Rocca F, Saatchi S, Shugart S, Ulander L (2011) The BIOMASS mission: mapping global forest biomass to better understand the terrestrial carbon cycle. Remote Sens Environ 115:2850–2860

Lee SK, Kugler F, Papathanassiou K, Hajnsek I (2013) Quantification of temporal decorrelation effects at L-band for polarimetric SAR interferometry applications. IEEE J Sel Top Appl Earth Obs Remote Sens 6(3):1351–1367

Lee SK, Fatoyinbo T, Qi W, Hancock S, Armston J, Dubayah R (2018) GEDI and TanDEM-X fusion for 3D forest structure parameter retrieval. In: Proceedings of international symposium of geoscience and remote sensing (IGARSS)

Lefsky MA, Harding DJ, Cohen WB, Parker GG, Shugart HH (1999) Surface lidar remote sensing of basal area and biomass in deciduous forests of eastern Maryland, USA. Remote Sens Environ 67(1):83–98

Liang J, Buongiorno J, Monserud RA, Kruger EL, Zhou M (2007) Effects of diversity of tree species and size on forest basal area growth, recruitment, and mortality. Forest Ecol Manag 243(1):116–127

Lin Y, Sarabandi K (1999) A Monte Carlo coherent scattering model for forest canopies using fractal-generated trees. IEEE Trans Geosci Remote Sens 37(1):440–451

Lombardini F, Reigber A (2003) Adaptive spectral estimation for multibaseline SAR tomography with airborne L-band data. In: Proceedings of international symposium of geoscience and remote sensing (IGARSS)

Mariotti d'Alessandro M, Tebaldini S, Rocca F (2012) Phenomenology of ground scattering in a tropical forest through polarimetric synthetic aperture radar tomography. IEEE Trans Geosci Remote Sens 51(8):4430–4437

McElhinny C, Gibbons P, Brack C, Bauhus J (2005) Forest and woodland stand structural complexity: its definition and measurement. For Ecol Manag 218(1–3):1–24

Mitsuhashi R, Murooka J, Sakaizawa D, Imai T, Kimura T, Hayashi M, Mizutani K, Sawada Y, Endo T, Kajiwara K, Honda Y, Asai K (2018). The development of vegetation Lidar mission 'MOLI'. In: Proceedings of forest SAT

Moreira A, Prats-Iraola P, Younis M, Krieger G, Hajnsek I, Papathanassiou KP (2013) A tutorial on synthetic aperture radar. IEEE Geosci Remote Sens Mag 1(1):6–43

Moreira A, Krieger G, Hajnsek I, Papathanassiou K, Younis M, Lopez Dekker P, Huber S, Villano M, Pardini M, Eineder M, De Zan F, Parizzi A (2015) Tandem-L: a highly innovative bistatic SAR mission for global observation of dynamic processes on the Earth's surface. IEEE Geosci Remote Sens Mag 3(2):8–23

Neuenschwander A, Magruder L (2018) Mapping global forests using data from NASA's ICESat-2 Mission. In: Proceedings of Forest SAT

Neumann M, Ferro-Famil L, Reigber A (2010) Estimation of forest structure, ground, and canopy layer characteristics from multibaseline polarimetric interferometric SAR data. IEEE Trans Geosci Remote Sens 48(3):1086–1104

Neumann M, Saatchi S, Clark DB (2012) Quantifying spatial and temporal dynamics of tropical forest structure using high resolution airborne lidar. In: Proceedings of international symposium of geoscience and remote sensing (IGARSS)

Ni-Meister W, Jupp DLB, Dubayah R (2001) Modeling lidar waveforms in heterogeneous and discrete canopies. IEEE Trans Geosci Remote Sens 39(9):1943–1958

Oliver MA, Webster R (1990) Kriging: a method of interpolation for geographical information systems. Int J Geogr Inf Syst 4:313–332

Papathanassiou K, Cloude S (2001) Single-baseline polarimetric SAR interferometry. IEEE Trans Geosci Remote Sens 39(11):2352–2363

Pardini M, Papathanassiou K (2018) TomoSAR vertical profiles at multiple frequencies: an experimental analysis. In: Proceedings of European SAR conference (EUSAR)

Pardini M, Qi W, Dubayah R, Papathanassiou KP (2016) Exploiting TanDEM-X Pol-InSAR data for forest structure observation and potential synergies with NASA's Global Ecosystem Dynamics Investigation Löidar (GEDI) mission. In: Proceedings of European SAR conference (EUSAR)

Pardini M, Tello M, Cazcarra-Bes V, Papathanassiou K, Hajnsek I (2018a) L- and P-Band 3-D SAR reflectivity profiles versus lidar waveforms: the AfriSAR case. IEEE J Sel Top Appl Earth Obs Remote Sens 11(10):3386–3401

Pardini M, Papathanassiou K, Lombardini F (2018b) Impact of dielectric changes on L-band 3-D SAR reflectivity profiles of forest volumes. IEEE Trans Geosci Remote Sens 56(12):7324–7337

Parker GG (1995) Structure and microclimate of forest canopies. In: Lowman M, Nadkarni N (eds) Forest canopies: a review of research on a biological frontier. Academic press, New York, pp 73–106

Pastorella F, Paletto A (2013) Stand structure indices as tools to support forest management: an application in Trentino forests (ItalyO). J For Sci 59(4):159–168

Persson HJ, Ollson H, Soja MJ, Ulander LMH, Fransson JES (2017) Experiences from large-scale forest mapping of Sweden using TanDEM-X data. Remote Sens 9(12):1253

Pommerening A (2002) Approaches to quantifying forest structures. Forestry 75(3):305–324

Praks J, Antropov O, Hallikainen M (2012) LIDAR-aided SAR interferometry studies in Boreal forest: scattering phase center and extinction coefficient at X- and L-band. IEEE Trans Geosci Remote Sens 50(10):3831–3843

Pretzsch H (2009) Forest dynamics, growth and yield. Springer, Berlin

Qi W, Dubayah R (2016) Combining Tandem-X InSAR and simulated GEDI lidar observations for forest structure mapping. Remote Sens Environ 187:253–266

Qi W, Dubayah R (2017) Forest structure modelling of a coniferous forest using TanDEM-X InSAR and simulated GEDI lidar data. In: Proceedings of IEEE international symposium of geoscience and remote sensing (IGARSS)

Reigber A, Moreira A (2000) First demonstration of airborne SAR tomography using multibaseline L-band data. IEEE Trans Geosci Remote Sens 38(5):2142–2152

Reineke R (1933) Perfecting a stand-density index for even aged forests. J Agric Res 46(7):627–638

Rosen P, Kim Y, Kumar R, Misra T, Bhan R, Raju Sagi T (2017) Global persistent SAR sampling with the NASA-ISRO SAR (NISAR) mission. In: Proceedings of IEEE radar conference

Simard M, Denbina M (2018) An assessment of temporal decorrelation compensation methods for forest canopy height estimation using airborne L-band same-day repeat-pass polarimetric SAR interferometry. IEEE J Sel Top Appl Earth Obs Remote Sens 11(1):95–111

Snyder M (2010) What is forest stand structure and how is it measured? North Woodl 64:15

Spies TA (1998) Forest structure: a key to the ecosystem. Northw Sci 72:34–36

Sun G, Ranson KJ, Guo Z, Zhang Z, Montesano P, Kimes D (2011) Forest biomass mapping from lidar and radar synergies. Remote Sens Environ 115:2906–2916

Tang H, Dubayah R, Swatantran A, Hofton M, Sheldon S, Clark DB, Blair B (2012) Retrieval of vertical LAI profiles over tropical rain forests using waveform lidar at La Selva, Costa Rica. Remote Sens Environ 124:242–250

Tebaldini S, Rocca F (2012) Multibaseline polarimetric SAR tomography of a boreal forest at P- and L-bands. IEEE Trans Geosci Remote Sens 50(1):232–246

Tello M, Cazcarra-Bes V, Pardini M, Papathanassiou K (2018) Forest structure characterization from SAR tomography at L-band. IEEE J Sel Top Appl Earth Obs Remote Sens 11(10):3402–3414

Thirion L, Colin E, Dahon C (2006) Capabilities of a forest coherent scattering model applied to radiometry, interferometry, and polarimetry at P- and L-Band. IEEE Trans Geosci Remote Sens 44(4):849–862

Treuhaft RN, Chapman BD, dos Santos JR, Gonçalves FG, Dutra LV, Graça PMLA, Drake JB (2009) Vegetation profiles in tropical forests from multibaseline interferometric synthetic aperture radar, field, and lidar measurements. J Geophys Res 114:D23110

Tsui OW, Coops NC, Wulder MA, Marshall P (2013) Integrating airborne LiDAR and space-borne radar via multivariate kriging to estimate above-ground biomass. Remote Sens Environ 139:340–352

Turner W, Spector S, Gardiner N, Fladeland M, Sterling E, Steininger M (2003) Remote sensing for biodiversity science and conservation. Trend Ecol Evol 18:306–314

Whitehurst AS, Swatantran A, Blair JB, Hofton MA, Dubayah R (2013) Characterization of canopy layering in forested ecosystems using full waveform lidar. Remote Sens 5(4):2014–2036

Yang W, Ni-Meister W, Lee S (2011) Assessment of the impacts of surface topography, off-nadir pointing, and vegetation structure on vegetation structure on vegetation lidar waveforms using an extended geometric optical and radiative transfer model. Remote Sens Environ 115:2810–2822

Zebker HA, Villasenor J (1992) Decorrelation in interferometric radar echoes. IEEE Trans Geosci Remote Sens 30(5):950–959

Zenner EK, Hibbs D (2000) A new method for modelling the heterogeneity of forest structure. For Ecol Manag 129:75–87

Surveys in Geophysics (2019) 40:839–862
https://doi.org/10.1007/s10712-019-09533-z

Recent Advances in Forest Observation with Visual Interpretation of Very High-Resolution Imagery

Dmitry Schepaschenko[1] · Linda See[1] · Myroslava Lesiv[1] · Jean-François Bastin[2,3] · Danilo Mollicone[2] · Nandin-Erdene Tsendbazar[4] · Lucy Bastin[5,6] · Ian McCallum[1] · Juan Carlos Laso Bayas[1] · Artem Baklanov[1,7] · Christoph Perger[1] · Martina Dürauer[1] · Steffen Fritz[1]

Received: 29 June 2018 / Accepted: 12 April 2019 / Published online: 11 May 2019

Abstract
The land area covered by freely available very high-resolution (VHR) imagery has grown dramatically over recent years, which has considerable relevance for forest observation and monitoring. For example, it is possible to recognize and extract a number of features related to forest type, forest management, degradation and disturbance using VHR imagery. Moreover, time series of medium-to-high-resolution imagery such as MODIS, Landsat or Sentinel has allowed for monitoring of parameters related to forest cover change. Although automatic classification is used regularly to monitor forests using medium-resolution imagery, VHR imagery and changes in web-based technology have opened up new possibilities for the role of visual interpretation in forest observation. Visual interpretation of VHR is typically employed to provide training and/or validation data for other remote sensing-based techniques or to derive statistics directly on forest cover/forest cover change over large regions. Hence, this paper reviews the state of the art in tools designed for visual interpretation of VHR, including Geo-Wiki, LACO-Wiki and Collect Earth as well as issues related to interpretation of VHR imagery and approaches to quality assurance. We have also listed a number of success stories where visual interpretation plays a crucial role, including a global forest mask harmonized with FAO FRA country statistics; estimation of dryland forest area; quantification of deforestation; national reporting to the UNFCCC; and drivers of forest change.

Keywords Forest cover · Biomass · Forest monitoring · Remote sensing · Satellite imagery · Visual interpretation

✉ Dmitry Schepaschenko
schepd@iiasa.ac.at

Extended author information available on the last page of the article

1 Introduction

Remote sensing plays a critical role in the estimation of forest parameters, and in the monitoring of disturbances and changes in forest cover. Different types of satellite sensors (i.e., optical, hyperspectral, LiDAR and radar at varying spatial and temporal resolutions) play different and complementary roles in forest monitoring. Time series of moderate resolution imagery (MODIS, MERIS, etc.) have been used extensively to produce land cover, forest cover and forest-type maps (Defourny et al. 2006; Friedl et al. 2010), forest cover change, tree density (DiMiceli et al. 2011), vegetation indices, gross and net primary production (Running et al. 2004) and forest disturbances (Justice et al. 2002). Radar and LiDAR data have also been used successfully to estimate biomass and canopy height (Baccini et al. 2008; Saatchi et al. 2011; Simard et al. 2011; Thurner et al. 2014; Santoro et al. 2015). With the opening up of the Landsat archive in 2008 (Wulder et al. 2012) including a time series of more than 40 years, it has become possible to monitor long-term changes in forests at a higher resolution (Sexton et al. 2013; Hansen et al. 2013). The open data policy of the European Space Agency (ESA) with respect to the Sentinel satellites means more frequent coverage of the Earth at a higher resolution than Landsat, which is particularly relevant for forest areas dominated by clouds. Sentinel 1 can also complement the monitoring of forest structures using radar data at a higher spatial and temporal resolution than other radar products such as ENVISAT ASAR.

More recently, with changes in Web 2.0 technology (Hudson-Smith et al. 2009) and the development of applications such as Google Earth and Microsoft Bing Maps, very high-resolution (VHR) satellite imagery can be viewed over many parts of the world. Moreover, the land area now covered by VHR imagery has also grown dramatically over recent years (Lesiv et al. 2018b). This opens up the possibility to visually identify land cover and land use features, as well as the structure of forests. For the purpose of this paper, we define VHR imagery as having a spatial resolution of less than 2 m, while high-resolution (HR) imagery refers to the resolutions of Landsat (30 m) or Sentinel 2 (10 m). Forest parameters of interest that can be identified from VHR imagery include: the type of land cover/land use; identification of forest/non-forest areas; whether forest areas are homogeneous or heterogeneous; forest cover fragmentation; forest disturbances, degradation and change over time; identification of low, medium and high values of biomass; whether forests are young or old; differentiation between natural forests and plantation; forest and tree crops; forest phenology (evergreen or deciduous); and leaf types (broad leaf or needle leaf).

Although the automatic processing of remote sensing data for forest monitoring remains the standard procedure, visual interpretation of VHR imagery for reference data collection has a number of advantages over in situ data collection and can even be viewed as a bridge between remote sensing and in situ approaches. For example, it can be used to collect large amounts of training data for automatic classification algorithms. A probability-based sample design can be implemented so that the visually interpreted data can be used for the validation of maps produced using remote sensing. Areas with high uncertainty can then be identified, and additional reference data were collected to further improve automatic classification. Visual interpretation can also be used for deriving forest statistics directly from satellite imagery (Bastin et al. 2017a). Finally, visual interpretation can be used to identify drivers of change, e.g., automatically generated maps of forest loss can be further translated to areas of clearcutting, burnt areas, disease dieback, shifting cultivation, etc. (Ontikov et al. 2016; Curtis et al. 2018). Note that we do not explicitly discuss applications of biodiversity here because extracting meaningful biodiversity information from remotely sensed imagery in many forest contexts requires reliable and geographically representative information on the spectral and temporal signatures of species, and these traits must be combined with further assumptions about phylogenetic dissimilarity. Although there are candidate methods available, they are still highly experimental and therefore uncertain.

In this review paper, we focus on the state of the art in visual interpretation of satellite imagery, recognizing the continuing role of HR imagery but focusing on advances that are possible from VHR imagery, and highlight their unique and complementary role in remote sensing and in situ data collection for the observation and monitoring of forests. We first focus on the types of satellite imagery available and then provide an overview of the tools that have been developed for visual interpretation. The quality of visual interpretation is a key concern for users so we review recent work undertaken in this area. Finally, we highlight some successful case studies that have employed satellite imagery for forest observation.

2 Availability and Utilization of Imagery in Forest Observation

2.1 Availability and Distribution of Imagery

As a result of advancements in Earth Observation (EO), HR satellite data are being made openly available, e.g., free access to Landsat data by the United States Geological Survey (USGS) since 2008. Landsat data have played a significant role in monitoring of the Earth, particularly in forest observation studies, as more than 40 years of satellite data with consistent spectral bands are available. The data are distributed through various platforms such as the USGS Earth Explorer and NASA's Earth Data (USGS 2018; NASA 2018) as well as Google Earth Engine (Gorelick et al. 2017). Complementary to Landsat, ESA's Sentinel 2 mission was launched in 2015 (Drusch et al. 2012). Sentinel 2 data (10–60 m), along with other Sentinel mission data, are freely accessible to the public via the Copernicus Open Access Hub (Copernicus 2018).

In addition, VHR satellite data offer further possibilities for forest observation. Multitemporal and multispectral VHR data are acquired by passive sensors (e.g., QuickBird, WorldView-2, GeoEye, Pleiades) (Solano-Correa et al. 2018) and are distributed commercially via DigitalGlobe, Geo-Airbus and Planet (Airbus 2018; Harris 2018; Planet 2018). However, some of this imagery is openly available or available upon request via the USGS Earth Explorer and ESA's Earth Online portal (USGS 2018; ESA 2018). In addition to the data providers, there are multiple platforms that allow users to view VHR data, e.g., Microsoft Bing Maps and Google Earth. The latter was released in 2005 and allows seamless viewing and exploration of medium-resolution, HR and VHR satellite imagery globally (Sheppard and Cizek 2009; Bey et al. 2016). By 2011, Google Earth had been downloaded over one billion times (Google 2018). The appeal of these tools is evident, not only for private users but also for scientists, policy-makers and stakeholders in tackling environmental and planning issues (Butler 2006). Building upon these platforms, several free and open-source software applications have been developed to facilitate the collection and analysis of land and forest cover characteristics including: Geo-Wiki, the GLCF Labeling Tool, LACO-Wiki and TimeSync (Bey et al. 2016. In addition to using VHR imagery, these software applications use archives of Landsat and MODIS imagery to display automatically generated time series of vegetation index profiles (Bey et al. 2016). A recent assessment has been made of the spatial and temporal availability of VHR data in Google Earth and Microsoft Bing Maps (Lesiv et al. 2018b), including suitability for forest monitoring.

VHR and LiDAR imageries are also available from light-weight unmanned aerial vehicles (UAV), which opens up further possibilities for detailed land and forest monitoring (Zahawi et al. 2015; Paz 2017). UAVs are low cost, have high spatial and temporal resolution and provide flexibility in the types of sensors used (GOFC-GOLD 2011).

2.2 Utilization in Forest Monitoring

The use of remote sensing in forest resource assessment provides different types of information such as the spatial extent of forest cover and its change over time, forest types and biophysical and biochemical properties of forests (Boyd and Danson 2005). Assessments using remote sensing commonly involve visual interpretation, e.g., to generate training or validation data sets, particularly when analyzing the spatial extent of forests and their change over time. For example, a global forest cover change product at a 30 m resolution was calibrated and validated based on reference data created via visual interpretation of HR and VHR satellite data (Hansen et al. 2013). Similarly, an assessment of the suitability of this global forest change product for forest area estimation at the national level in Gabon was made using an independent validation data set, visually interpreted using available satellite imagery (Sannier et al. 2016).

A hybrid approach that combines automated processing of satellite data and visual interpretation is commonly used to monitor forest extent and its change over time. For example, Duveiller et al. (2008) assessed deforestation rates in Central Africa using sample-based estimation, which was created from automated image segmentation and visual interpretation of both HR and VHR satellite imagery. Sy et al. (2015) assessed the subsequent land use types after deforestation in South America using sample-based deforestation data from the UN-FAO Forest Resources Assessment (FRA), created from automated segmentation, and visual interpretation of VHR satellite imagery. In a study to create reference data for forest and tree cover fraction, Pengra et al. (2015) automatically classified 500 VHR images distributed globally and then manually edited them to obtain reference maps at a 2 m resolution. Bey et al. (2016) applied “augmented visual interpretation” by combining visual interpretation of VHR images with vegetation indices computed for all dates available back to the year 2000 from MODIS, Landsat and Sentinel.

Visual interpretation of VHR images is also used in other forest assessments such as forest structure and biomass estimation. Examples include studies on selective forest logging and its impacts (Read et al. 2003; Furusawa et al. 2004; Pithon et al. 2013), forest species classification (Clark et al. 2005; Valérie and Marie-Pierre 2006; Kim et al. 2009; Bilous et al. 2017), tree crown identification (Garzon-Lopez et al. 2013; Karlson et al. 2014) and biomass estimation (Hussin et al. 2014). In these studies, visual interpretation of VHR images was often used to create training data in the absence of in situ data.

Visual interpretation can also be used to produce statistics directly from sampling point approaches. In contrast to a map-based area estimation approach, this method has several advantages. First, when collecting information with photointerpretation, the land cover data collected are independent from other environmental layers. Secondly, there is no loss in information when upscaling to the map resolution (i.e., to Landsat, MODIS or any other product) because the raw information is still available. This is particularly relevant for regions with sparse tree cover. Moreover, this approach of directly obtaining statistics through visual interpretation is based on a transparent data assessment where the only uncertainty is potential photointerpretation errors, e.g., as debated in Schepaschenko et al. (2017) and the subsequent response by Bastin et al. (2017b). If such a sampling approach is applied at a larger scale, then independent statistical information can even be obtained globally, e.g., in the global dryland assessment by Bastin et al. (2017a), in which global statistics on forest cover from a visual interpreted sample were derived.

Visual interpretation of HR and VHR data has been recommended as one of the approaches for national forest monitoring and reporting by international committees such as the GOFC-GOLD (2011). As a result, several tools and software applications have emerged for collecting reliable reference data for map calibration and validation for forest and forest change monitoring purposes. Some of these tools are described in more detail in the next section.

3 Tools for Visual Interpretation

This section provides an overview of the main tools available for visual interpretation of satellite imagery. The final part of this section covers tools that exist but which are not openly available or they have been developed in the context of awareness raising and the monitoring of deforestation.

3.1 Geo-Wiki

Geo-Wiki was originally developed as an online tool to collect information on the accuracy of class descriptions from three global land cover maps (GLC-2000, MODIS and Globcover 2005) when viewed on top of VHR satellite imagery from Google Earth. Users could choose any location on Earth, and validation was on pixel-by-pixel basis (Fritz et al. 2009). Users could also view the main land cover products in a single application, along with layers highlighting the areas of disagreement in cropland and forest classes. Hence, Geo-Wiki served as an application for exploring the accuracy of existing land cover products using satellite imagery, which was not previously possible via an open application.

Since 2009, the Geo-Wiki tool has evolved in a number of ways. First, the Geo-Wiki interface was expanded into multiple branches, each related to a specific land cover or land use theme. For example, the biomass branch of Geo-Wiki (https://biomass.geo-wiki.org) was designed to allow users to visualize all the major data sets of above ground and forest woody biomass on VHR imagery in order to compare the layers with one another but also to provide feedback on the biomass estimates at any location on Earth (Schepaschenko et al. 2015a).

A second key adaptation to Geo-Wiki was the use of crowdsourcing campaigns to collect data related to a specific research question. The first six campaigns are documented in See et al. (2015a) and resulted in the visual interpretation of around 250 K locations around the world (Fritz et al. 2017). The campaigns were used to validate a map of the land availability for biofuels (Fritz et al. 2013), to develop maps of cropland and agricultural field size (Fritz et al. 2015; Lesiv et al. 2018a), wilderness (See et al. 2016) and global land cover (See et al. 2015b). The campaigns employed gamification and incentives such as Amazon vouchers and coauthorship.

New campaigns have since been run to collect a global validation data set for cropland (Laso Bayas et al. 2017a). At the same time, Geo-Wiki was also set up to run internal data

Fig. 1 An example of a Geo-Wiki interface used to collect data on forest cover (Schepaschenko et al. 2015b)

collection campaigns to produce high-quality training and validation data sets for different tasks related to the development of forest maps. For example, data on forest cover were collected by various experts using an interface like that shown in Fig. 1. Combined data from the Geo-Wiki database, in combination with several existing forest maps, have been used to develop a hybrid map of global forest cover (Schepaschenko et al. 2015b). More details are provided in Sect. 5.1. Other products include a high-resolution hybrid forest cover map of Ukraine along with a detailed tree species map (Lesiv et al. 2018c).

The "Human impact on forest" branch of Geo-Wiki focuses on the collection of the following forest and land use features: mature forest; plantations; fruit plantations; planted forest; clearcut, thinning; unpaved forest roads; paved roads; mosaic tree cover/cropland; and mosaic tree cover/urban. The "Drivers of forest cover change" Geo-Wiki branch allows users to record the following classes of land use change: no changes: stable tree cover, stable non-tree cover; tree cover loss: expansion of agriculture, shifting cultivation, urban/infrastructure expansion, mining, wildfire/windfall/dieback, timber harvest (without land use change); tree cover gain: reforestation, afforestation, tree crops.

3.2 LACO-Wiki

LACO-Wiki is a free platform for undertaking land cover accuracy assessment (https://laco-wiki.net). It contains the complete workflow from uploading a map for validation, creating a sample and interpreting the sample using VHR satellite imagery from Google Maps or Microsoft Bing Maps (or other imagery provided by a Web Map Service). An accuracy report is then produced, which contains the confusion matrix and the indicators of accuracy chosen by the user. Figure 2 provides an example of the visual interpretation process, illustrating validation of the GlobeLand30 land cover map in Kenya. The online

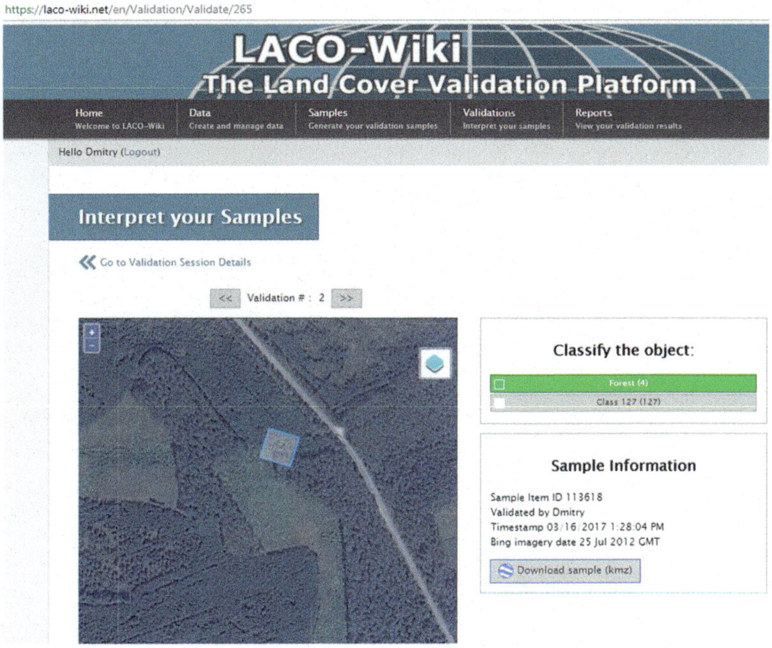

Fig. 2 An example of the visual interpretation of a sample unit using the LACO-Wiki tool

tool, the workflow and the Kenyan validation example are described in more detail in See et al. (2017).

Since the legend for visual interpretation is determined by the land cover map uploaded to the system, LACO-Wiki can be used for collecting data on forest cover, expressed as either forest cover classes or percentage forest cover, or any other features defined by the users, which are visible from VHR satellite imagery. LACO-Wiki also allows users to upload predefined sample locations, e.g., produced using a GIS, and undertake visual interpretation.

The vision of LACO-Wiki goes beyond that of just providing an online tool for accuracy assessment. It is also intended to be a repository for sharing land cover maps and the reference data sets generated as part of the accuracy assessment process (See et al. 2017). In this way, hybrid land cover or forest cover maps could be created from a mosaic of existing layers from the repository using a potentially much larger training data set than could be collected by individuals or organizations on their own.

3.3 Picture Pile

Picture Pile is a mobile and online application designed for the rapid classification of VHR satellite imagery and geotagged photographs. It is a generalized version of the Cropland Capture game (Sturn et al. 2015; Baklanov et al. 2016; Salk et al. 2016), which focused on visually classifying imagery for cropland using a simple question: Is there evidence of cropland? Users had the choice of yes, no or maybe, swiping the image in the direction of the answer on mobile devices or using the cursor keys on the browser version. In this way, it was possible to collect millions of interpretations of the presence/absence of cropland over a 6-month period.

Fig. 3 An example of the rapid interpretation of deforestation from a pair of images in picture pile

In Picture Pile, the concept is that piles of pictures corresponding to any subject area (not just cropland) could be visually interpreted in a rapid manner using the same simple question. The first campaign run provided users with pairs of satellite images from different time periods; the question they were asked was: "Do you see tree loss over time?" as shown in Fig. 3.

The idea behind this campaign was to validate the Hansen et al. (2013) tree loss and gain products for Tanzania and Indonesia. Preliminary results show that there are examples where Hansen indicates deforestation, but visual interpretation shows no tree loss and vice versa, i.e., no deforestation is indicated by Hansen et al. (2013), but visual interpretation shows clear patterns of forest loss. A new tool called Picture Paint is currently being developed that will take areas with deforestation as inputs from Picture Pile and allow users to shade those locations where deforestation has occurred. In this way, more accurate locations of where deforestation occurred can be recorded, which can be used to improve maps of deforestation and forest cover for specific time periods. Hence, Picture Pile and Picture Paint are two visual interpretation tools that can be used together to gather rapid and then more detailed forest observations.

3.4 Collect Earth

Collect Earth is a free and open-access software tool developed by the UN-FAO for the monitoring of land cover and land use (Bey et al. 2016). Collect Earth allows for the collection of plot-level information (e.g., a square of 0.5 ha) of current and historic land properties of a given location using various sources of remote sensing data. Combining the information collected for numerous plots through a systematic or randomized sampling design, Collect Earth outputs can then be used (1) for the development of statistics at national, regional or even global scale or (2) as a set of training and validation points for wall-to-wall mapping. To develop statistics with a low level of error requires considerable photointerpretation efforts, which rely on human capacity. For example, to characterize tree cover in drylands with less than 1% error, Bastin et al. (2017a) collected information at more than 200 000 points, which required the coordination of 236 operators in the photointerpretation effort. The list of recorded parameters includes: land cover type, land use category, land use change (2000–2015), year of changes, tree or shrub count, vegetation type, length of linear objects (vegetation, paved or unpaved roads), type and impact of disturbances and the accuracy of the assessment.

Collect Earth is built on Google Earth technologies and, in particular, two different tools: Google Earth Desktop and Google Earth Engine Code Editor. The interface of Collect Earth is completely included in the Google Earth Desktop. It provides access to all available satellite archives of VHR images, which are used for the photointerpretation of predefined plots. In parallel, for each plot, Collect Earth provides access to preprocessed satellite information at a high-temporal resolution computed using the Google Earth Engine Code Editor. In particular, it provides information on the normalized difference

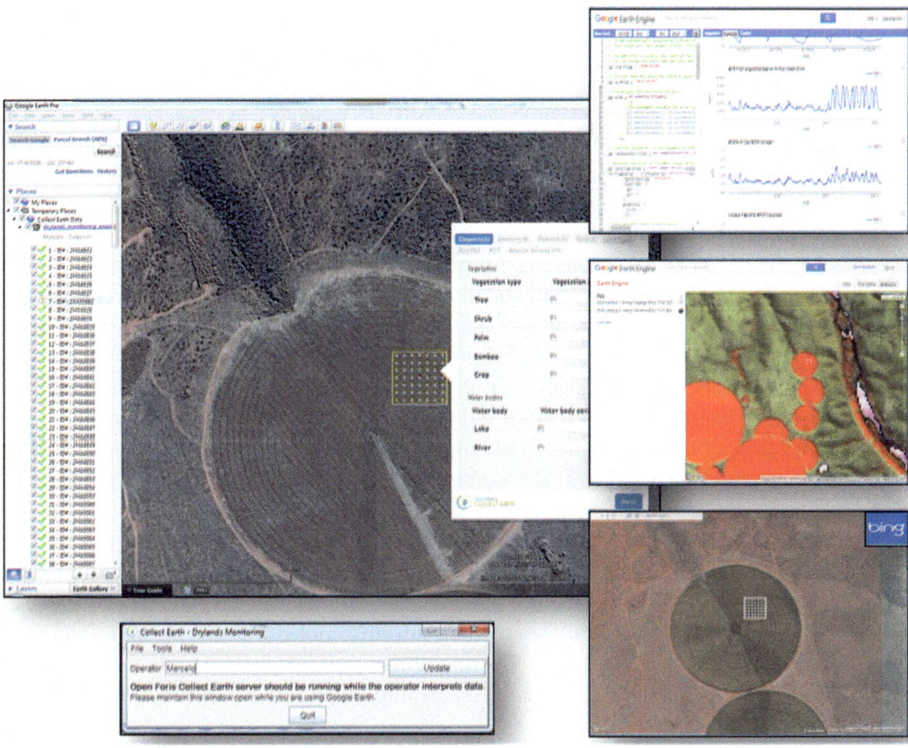

Fig. 4 An example of the rapid interpretation of land use and land use change using Collect Earth, combining VHR images available in Google Earth and Bing Maps with NDVI computed since the 2000s from Google Earth Engine Code Editor

vegetation index (NDVI), computed since the year 2000 from MODIS, and for Landsat and Sentinel 2, where available. Combining both high spatial and temporal resolution for land monitoring, Collect Earth allows the operator to collect a considerable amount of information about the land properties through augmented visual interpretation (Fig. 4).

3.5 Other Tools

Some tools for visual interpretation are not openly available, but some examples have been documented in the literature. For example, the VIEW-IT (Virtual Interpretation of Earth Web-Interface Tool) project was developed to collect land cover reference data including classes for woody and mixed woody vegetation in order to support ongoing research in land cover change (Clark and Aide 2011). The system is similar to Geo-Wiki in that users estimate the percentage of land cover in a 250 m MODIS pixel using VHR imagery from Google Earth. The tool was used to collect more than 46,000 reference samples across Latin America and the Caribbean with a team of 30 staff and students. The authors admit that the first version was meant to be a proof of concept with future developments to include expansion to global coverage, opening up to crowdsourcing, customized samples, grid sizes and data collection (similar to what is available in LACO-Wiki) and multiple interpretations of the same sample unit. Since the publication of that paper, the tool does not appear to have been opened up and the reference data have not been made openly available.

Focusing more on the forest domain, the UN-FAO complemented the Forest Resources Assessment (FRA) exercise in 2010 with a remote sensing survey (RSS) using an interface developed for interpretation of Landsat imagery (Lindquist et al. 2012). There were 13,066 sample sites in the RSS, each of which is a 20 km^2 area located at each 1 degree intersection of latitude and longitude, except in Canada and the Russian Federation where a different approach was used. In the framework of the TREES-3 project, the Joint Research Centre of the European Commission (JRC) took the responsibility for producing and verifying a large component of this global data set, particularly focusing on tropical forests. Using a segment-based classification and the commercial eCognition software, Landsat imagery from 1990, 2000 and 2005 was automatically classified to a legend that allowed the detection, at segment level, of tree cover loss and change between epochs. The individual segments were visually verified against Google Earth and other VHR imagery by both JRC staff and national experts, using a standalone tool developed in IDL (Simonetti et al. 2011). The results of the exercise (Achard et al. 2014) are available for free public download.[1]

A Web version of the verification/image interpretation tool described above was implemented as a research prototype in the context of the EuroGEOSS project, supported by open standards and application programming interfaces (APIs) (Bastin et al. 2012). This demonstrated the feasibility of Web-enabling a well-established validation process with significant policy impact, and offered the advantage that no imagery needed to be downloaded to the desktop. However, the standalone nature of the original IDL toolkit has proved to be particularly useful in areas where internet connectivity is limited or sporadic, and it is still available for free public download and use.[2] The JRC tools for mapping of forest change and loss have subsequently been improved and made more accessible—in

[1] http://forobs.jrc.ec.europa.eu/trees3/.

[2] http://forobs.jrc.ec.europa.eu/products/software/other.php.

particular, the classification step now uses entirely open-source tools and libraries (Simonetti et al. 2015). The entire workflow from image selection to validation is now embedded in a free and open-source toolkit named IMPACT.[3] An example of the analysis possible with IMPACT is documented in Szantoi et al. (2016), and the labeled data from that analysis is available through open OGC-standard Web Map and Feature Services hosted at JRC.

The Web-based version of the JRC land cover validation tool was adapted into a system that allowed the labeling of point samples using a combination of visual interpretation with inspection of NDVI signatures (Bastin et al. 2013). This approach was used by the Royal Society for the Protection of Birds (RSPB) and Birdlife in an exercise that labeled user-specified sample points in and around Important Bird Area (IBAs) in order to test the hypothesis that legal site protection reduces the loss of natural land cover (Beresford et al. 2013). This interface also delivered NDVI information, by overlaying the value for the sample location being interpreted at the time of acquisition on a graph of historic mean/standard deviation for the same area across the year. The point of this was for a user to see (a) whether the particular image they were looking at was anomalous or representative of the land cover at that location, and (b) the characteristic vegetation profile for that location over time. The NDVI information was derived from a web service based on JRC's eStation, which compiles and delivers environmental information from remote sensing across Africa. An administrative interface allowed sample locations to be uploaded and allocated to authorized labelers, and the resulting information could then be downloaded. The resulting ground-truth data set was reused to validate an experimental water classification by Pekel et al. (2014), and is available on request from the original researchers. This web tool and underlying database was taken offline in 2015 because of a lack of resources, but the codebase is archived in an open repository.

A system for object delineation and visual interpretation of VHR imagery from Google Earth has been developed by the Universite Catholique de Louvain (UCL). This system, which is targeted at experts, was developed to collect reference data for the development of land cover products, e.g., GlobCover (Bontemps et al. 2011) and the ESA-CCI (European Space Agency Climate Change Initiative) land cover time series (ESA CCI LC 2017), which have been produced by researchers at UCL in collaboration with other partners. Medium-resolution time series or high-resolution snapshot images are also at hand to facilitate the labeling process. The interface was recently simplified to allow experts in remote sensing and agronomy from the EU-funded SIGMA project[4] to collect binary cropland data at 4147 locations (Waldner et al. 2018). Unfortunately, the reference data that have been collected through this bespoke interface over several years have not been made openly available.

Earth Watchers is a software tool developed by Geodan Inc. in the Netherlands for involving students in change detection in the context of deforestation monitoring.[5] Students are assigned an area, which is a hexagon of around 1.6 km^2. Each week, a new processed radar satellite image is provided (although Landsat has been used in the past) and students raise an alert if they see evidence of change. The ten most likely sites of change are surveyed by a team on the ground, which documents their findings using photographs and video or they deploy UAVs for safety reasons. Any illegal activities are reported to the

[3] http://forobs.jrc.ec.europa.eu/products/software/impact.php.

[4] http://www.geoglam-sigma.info/.

[5] https://dfa.tigweb.org/about/?section=earthwatchers.

local authorities who have pledged their support to the project and to halt deforestation. The project is currently operating in Indonesian Borneo.

3.6 Comparison of Geo-Wiki, LACO-Wiki and Collect Earth

This section compares and contrasts the three main tools for visual interpretation of VHR imagery to which anyone can openly contribute: Geo-Wiki, LACO-Wiki and Collect Earth, based on a range of different features as outlined in Table 1. There are a number of similarities between the tools, e.g., the research teaching communities are a main target group for all of the tools. However, Collect Earth also targets organizations involved in official statistical reporting, while LACO-Wiki has been recommended as a validation tool for European Environment Agency member states. Geo-Wiki, through its crowdsourcing campaigns, targets a broader audience that includes citizens. Each system also has a mobile app although the Geo-Wiki Pictures app is now being replaced by ongoing developments in LACO-Wiki Mobile.

Overall, there are greater differences between the tools, in particular between Collect Earth and the other two. This is because users work directly in Google Earth Engine in Collect Earth so this means historical HR imagery is directly available along with other ancillary layers and scripts for processing the data, e.g., generation of NDVI profiles. Different types of information are collected with each tool as outlined in Table 1. Both Collect Earth and LACO-Wiki are customizable, while Geo-Wiki must be set up internally in response to research needs while both Geo-Wiki (to a limited extent) and LACO-Wiki allow users to design their sample online while samples must be generated outside of Collect Earth using a GIS package or scripts and then uploaded. At this stage, only Collect Earth is fully open, with code available in Github; this is not yet the case for Geo-Wiki and LACO-Wiki although the new mobile version of LACO-Wiki will be open source. Conversely, the data are not necessarily open from Collect Earth as this depends on the group or persons who use the tool for a particular purpose while the data from Geo-Wiki can either be downloaded from the website or can be found in PANGAEA (Fritz et al. 2017; Laso Bayas et al. 2017a). The validation data in LACO-Wiki are theoretically open through the licensing agreement that users agree to when registering in the system although individual users can currently keep the data private or share it more broadly. Finally, each of the systems has various support mechanisms including a user forum (Collect Earth), newsletters (Collect Earth and Geo-Wiki), a facebook group (Geo-Wiki) and contact via email (all three systems). Hence, it is clear from this comparison that each of these packages has different strengths and weaknesses, which are partly a function of how they were designed in response to user requirements.

4 Issues Related to the Quality of VHR Imagery and Visual Interpretation

The quality of visual interpretation of VHR imagery is a function of the quality of the satellite imagery and the interpretation procedure applied to the imagery. The interpretation procedure can involve experts or crowdsourcing, both of which can be problematic (See et al. 2013; Coillie et al. 2014). In this section, we consider issues that influence the quality of both of these components.

Table 1 A comparison of the three main open tools for visual interpretation of VHR imagery

Feature	Collect Earth	Geo-Wiki	LACO-Wiki
Main users	Research and teaching community, organizations involved in the verification or preparation of national statistics on land cover, land use and land use change for IPCC, UNFCCC and FRA reporting	Research and teaching community, citizens interested in science that want to participate in data collection campaigns that involve visual interpretation	Research and teaching community, any organization requiring land cover map validation, e.g., it is recommended to European Environment Agency member states for validation of VHR components as part of Copernicus land monitoring activities
Information that can be collected using the tool	Land cover type, land use category, land use change, year of changes, tree or shrub count, vegetation type, length of linear vegetation, type and impact of disturbances, accuracy of the assessment	Land cover, land use type; forest fragmentation, plantations, drivers of forest cover change (roads, agriculture expansion, shifting cultivation, etc.)	Land cover type, land use category and accuracy of the assessment can be customized so that any features represented as attributes in the map could be collected or validated
Role of GEE	Users work directly in GEE and have access to all data and scripts in GEE. Or they can write new/modify existing scripts, which requires some technical expertise	There is no direct access to GEE. Instead, there are tools embedded in the interface (or will be shortly in the case of LACO-Wiki) that ingest results generated by GEE; see example of NDVI tool below	
High-resolution imagery	Direct access to current and historical imagery such as Landsat is available in GEE	Current and historical HR imagery is available through Sentinel Hub as a WMS	
Imagery dates	The dates of the Google VHR images are recorded manually. Microsoft Bing Map dates are read automatically via an API		
Ancillary layers	Only those layers that are available in GEE, but users can upload their own layers	Any ancillary layer can be displayed as a WMS.	
		Users must request the layer to be displayed	Users can specify the layers to be displayed from a WMS
NDVI time series	NDVI automatically generated for each plot photointerpreted. Generated for 2000 to present day for MODIS data, and for all available dates with no cloud cover for Landsat and Sentinel data	An NDVI tool is embedded in the interface. Users click on a location and can see NDVI values over time and mean values calculated from GEE	This feature will be implemented in 2019
Sample design	Users cannot generate a sample in Collect Earth, but scripts are provided on the Collect Earth website to generate a country-level sampling grid through the GEE code editor	A simple random sample can be created, but otherwise the sample is created using GIS and then hardcoded into the Geo-Wiki branch by developers	Users can generate different sample types (random, stratified, systematic) or upload a sample generated with a GIS

Table 1 (continued)

Feature	Collect Earth	Geo-Wiki	LACO-Wiki
Level of customization	Users can customize the interface for any specific data collection needs, including different data types and levels of spatial detail	Customization is currently not possible. Branches are created ad hoc to meet user needs	Users can customize a validation session to meet their needs (sampling design, data types)
Feedback and community support	There is an open forum for user queries and newsletters for subscribed users	Users can ask questions via email or discuss open issues via the Geo-Wiki facebook group. Newsletters are sent to subscribed users	
Code availability	Open access on GitHub	Not available	Not available but currently working on an open-source version of LACO-Wiki Mobile
Data access	The data collected using Google Earth belong to the group or persons who initiated the data collection campaign. Only they can decide if they share these data with others	The data can be downloaded from the Geo-Wiki website or from the PANGAEA repository	As part of the licensing, users agree to share their validation data. However, during individual sessions, users can keep the data private, share with individuals or share with everyone
Mobile app	Collect Mobile for Android	Geo-Wiki Pictures (Android and iOS) but now deprecated and will be replaced with LACO-Wiki Mobile	LACO-Wiki Mobile is currently being developed for Android and iOS as an open-source application

GEE Google Earth Engine; NDVI Normalized Difference Vegetation Index; API Application Programming Interface; WMS Web Map Service

4.1 Problems Associated with the Quality of VHR Imagery

Although the spatial coverage of VHR imagery in Google Earth and Microsoft Bing Maps continues to improve, visual interpretation may still be hampered by the actual quality of the imagery, which may lead to confusion in the resulting interpretation made by volunteers and experts. The most common reasons are image blurriness and the presence of clouds covering a scene. However, ubiquitous digital cameras have made the development of blur detection algorithms an active area of research (Tong et al. 2004; Li et al. 2016). These algorithms provide a quality score for an image, which allows a threshold to be specified that can be used to remove highly blurred images from an analysis. For example, the application of blur detection algorithms helped to detect around 2300 blurry images that were impossible to label correctly, even for experts (Baklanov et al. 2016, 2017). These images distorted the accuracy of the results in a crowdsourcing campaign due to high inconsistencies, i.e., volunteers and experts changed opinions when presented with the same blurry image, which affected the overall results when they were combined through a voting procedure.

VHR images from sources such as Google Earth and Microsoft Bing Maps come without a cloud mask, have very limited spectral information, and usually lack thermal bands that can be used for robust cloud detection (Zhu and Woodcock 2012). Thus, when working with VHR imagery from these sources, we only have a true-color RGB image. To handle these images, numerous approaches (Başeski and Cenaras 2015; Bai et al. 2016; Fan et al. 2017) of different complexity have been developed, which have an overall accuracy close to that when applying one of the algorithms that uses thermal bands.

The generalization of automated approaches from optical data is very dependent on sun-scene-sensor geometries (Barbier et al. 2011; Morton et al. 2014) and from specific interaction with object properties (e.g., the forest type) (Bastin et al. 2014), where this problem increases with spatial resolution. Hence this is a very important issue for the automated processing of VHR images. However, procedures for the correction of such geometries (Barbier and Couteron 2015) and the fusion of remote sensing metrics (Ploton et al. 2017) help to address some of these issues.

Finally, there are problems related to the comparison of interpretations of detailed information from VHR resolution imagery (e.g., tree count, tree cover) with ground observations. As raised previously and although this point has been debated in the literature (Schepaschenko et al. 2017; Bastin et al. 2017b), Bastin et al. (2017a) demonstrated that photointerpretation can lead to an error that is less than 10% in estimating the extent of forest at a global scale.

4.2 Approaches to Quality Assurance of Visual Interpretation of VHR Imagery

There are different approaches to quality assurance that have been applied to visual interpretation of VHR imagery, which includes comparison with in situ data/field measurements or other high-quality products such as a regional or national map, when available, comparison with an expert or "control" data set of visual interpretations and approaches for combining multiple observations at a single location.

In situ data from the Land Use/Cover Area frame Survey (LUCAS), which are collected every 3 years by Eurostat, have also been used as ground-truth data for the 2015 FotoQuest Austria campaign (Laso Bayas et al. 2016). Citizens were asked to go to specific locations across the Austrian landscape and collect information on land cover and land use with a mobile app. The results showed an agreement of up to 80% for high-level land cover and land use types in homogeneous areas.

Existing land cover maps have also been validated using a version of the Geo-Wiki platform that can be run offline and hence used more easily in countries with less reliable internet connections. Over 25,000 VHR Digital Globe images from Tanzania were classified by local volunteers for the degree of cropland and forests present. The land cover products included the ESA-CCI, GlobeLand30, FROM-GC and a regional product (Tanzania Land Cover 2010 Scheme II). Although the results reported in Laso Bayas et al. (2017b) were specifically focused on cropland, the results showed an overestimation of cropland by ESA-CCI and a large underestimation by the FROM-GC product. To ensure quality of the crowdsourced data, each location was interpreted by at least three different local people and combined using majority voting. A similar approach could be applied to forest cover data.

Geo-Wiki campaigns have made extensive use of an expert or control data set, i.e., a subset of the sample that is classified by a group of experts; this control data set is then used to calculate the performance of the crowd (See et al. 2013; Fritz et al. 2017; Laso Bayas et al. 2017a; Lesiv et al. 2018a). The information about performance can then be used to weight the data when used in subsequent applications, e.g., giving less weight to those interpreters who performed less well (e.g., Lesiv et al. 2018a). The control data set has also been used during campaigns to calculate a score based on both quality and quantity of interpretations, which was then used to determine the award of prizes or other incentives such as coauthorship.

The quality control of the data gathered through Collect Earth is not performed automatically; it should be implemented by the operator. In the study of Bastin et al. (2017a) on the extent of forest in dryland biomes, two main steps were implemented: a quality control and an uncertainty analysis. The quality control implied the reassessment of plots identified as potentially problematic, which were identified through the definition of several logical cross-control rules. For example, all plots classified as grassland reported to have $\geq 10\%$ tree canopy were reassessed by a team of experts and corrected as necessary. The uncertainty analysis was performed accounting for (1) sampling error and (2) measurement error (i.e., the mismatch between ground truth and photointerpretation). In the study, the two were combined and propagated in the original data 100 times, meaning a random error was applied 100 times to the tree cover of each plot, following a normal distribution centered on the original value and with a variance equal to the sum of the two types of errors. The forest extent of dryland biomes was then recalculated 100 times from which the final error was calculated with the standard deviation.

Multiple interpretations at a single location have also been used to ensure quality. For this, a simple majority vote rule can be a good way to aggregate votes if volunteers are correct most of the time (Baklanov et al. 2016). For example, Foody et al. (2018) used a simple weighted average to combine interpretations from multiple participants to improve the overall accuracy, while in the study by Laso Bayas et al. (2017b), quality was estimated based on the consistency between the interpreters. However, there are numerous state-of-the-art vote aggregation methods (e.g., see the surveys by Hung et al. 2013; Chittilappilly et al. 2016), which could be employed to deal with incorrect visual interpretations but have not yet been applied to this specific field. Salk et al. (2017) also showed that majority voting can have its limitations in visual interpretation of cropland and that the use of a control data set is recommended.

5 Success Stories of Using VHR Data for Forest Observation and Assessment

5.1 Global Forest Mask

A number of global and regional maps of forest extent are available, but when compared spatially, there are large areas of disagreement between them. Moreover, until recently, there has been no global forest map available that is consistent with the national statistics from the UN-FAO FRA. Geo-Wiki has been used to validate existing forest extent maps and to collect training data for combining diverse data sources into a single forest cover product (Schepaschenko et al. 2015b; Lesiv et al. 2016). From these data, it was possible to produce a global forest map that is more accurate (at the target 1 km resolution) than the individual input layers and to produce a map that is consistent with UN-FAO FRA statistics. Geographically weighted regression (GWR) was employed to integrate eight different forest products into three global hybrid forest cover maps for the reference year 2000. Input products included global land cover (GLC2000, GLC-NMO, MODIS LC, GlobCover) and forest maps at varying resolutions from 30 m (Sexton et al. 2013; Hansen et al. 2013) to 1 km, mosaics of regional land use/land cover products where available, and the MODIS Vegetation Continuous Fields product. The GWR algorithm was trained using crowdsourced data of visual interpretations collected via the Geo-Wiki platform, and the hybrid maps were then validated using an independent data set collected via the same system by experts. Three different hybrid maps were produced: two consistent with UN-FAO statistics, one at the country and one at the continental level, and a "best-guess" forest cover map that is independent of UN-FAO. Independent validation showed that the "best-guess" hybrid product had the best overall accuracy of 93% when compared with the individual input data sets. The global hybrid forest cover maps are available in a Geo-Wiki branch dedicated to biomass: http://biomass.geo-wiki.org.

5.2 Dryland Forest Assessment

Classical approaches employed to map tree cover at the global scale are generally based on medium-to-high-resolution satellite image processing, showing a typical spatial resolution of 10 to 250 meters. Such products are, however, known to produce very uncertain results in drylands (Sexton et al. 2016). Consequently, the UN-FAO has developed a new assessment of dryland forest resources based on a global photointerpretation effort coordinated with 236 operators, including researchers and country officials. Analyzing more than 210,000 0.5-ha sample plots through Collect Earth, this new assessment revealed that in 2015, 1327 million hectares of drylands had more than 10% tree-cover, and 1079 million hectares comprised forest. This estimate was 40–47% higher than previous estimates, corresponding to 467 million hectares of forest that have never been reported before. This means that previous efforts assessing the global forest extent were underestimating forest cover by at least 9% (Bastin et al. 2017a). This exercise has shown that a global photointerpretation effort coordinated between 236 operators can overcome, with some debated limitations (Schepaschenko et al. 2017; Bastin et al. 2017b), current flaws in state-of-the-art-automated mapping methods.

5.3 Landsat-Based Forest Loss and Gain

Landsat imagery provides the longest HR time series of land observations and has been used to estimate forest cover loss and gain (Hansen et al. 2013; Feng et al. 2016). To train their forest cover change algorithm, Feng et al. (2016) selected a stratified sample of points, which were visually classified as forest or non-forest cover by experienced image analysts. The decision was made based on Landsat time series images presented as multiple three-band combinations, e.g., near infrared, green, blue, and shortwave infrared, as well as auxiliary information including NDVI phenology from MODIS, HR satellite imagery and maps from Google Maps, and geotagged ground photos. Hansen et al. (2013) derived training data to relate to the Landsat metrics from image interpretation, including mapping of crown/no crown categories using VHR spatial resolution imagery such as Quickbird, and existing percentage tree cover layers derived from time series of Landsat data.

5.4 National Forest Reference Level: The Case of Papua New Guinea

Papua New Guinea submitted their national Forest Reference Level (FRL) to the UNFCCC in 2017 (Climate Change and Development Authority of Papau New Guinea 2017). The FRL is one of the elements to be developed by participating countries in the framework of REDD+ activity reporting, in agreement with decisions taken at several recent Conference of the Parties (COP) meetings. In their report, the Papua New Guinea governmental operators estimated their FRL for the year 2014 to 2018 in comparison with the reference period of 2001 to 2013. Carbon emissions for these periods were assessed from land use and land use change assessments using Collect Earth, which included deforestation, degradation and carbon stock enhancement, among others. The conversion to carbon emissions was obtained from national scientific literature and using the 2006 IPCC Guidelines for National Greenhouse Gas Inventories (IPCC 2006). This represents a good example of the nonscientific use of Collect Earth, which helped Papua New Guinea to report on LULUCF to the UNFCCC in a transparent fashion, demonstrating that developing countries can monitor their own carbon emissions.

5.5 Drivers of Forest Change

Global maps of forest loss show the scale and magnitude of forest disturbance, but they do not distinguish land use change (i.e., deforestation) from temporary loss of tree cover due to forestry or wildfires. Using visual interpretation of VHR satellite imagery, Curtis et al. (2018) developed a forest loss classification model to determine the spatial attribution of forest disturbance to the dominant drivers of land cover and land use change over the period 2001 to 2015. Their results indicate that 27% of global forest loss can be attributed to deforestation through permanent land use change. The remaining areas maintained the same land use over 15 years, and tree cover loss was attributed to forestry (26%), shifting cultivation (24%) and wildfire (23%).

6 Conclusions and the Future Perspectives

This review has shown that new sources of open satellite imagery have emerged over the last fifteen years, in particular HR imagery in the form of Landsat time series and Sentinel 2, as well as VHR imagery from Google Earth and Microsoft Bing Maps. This development has led to new tools for visual interpretation of VHR imagery such as Geo-Wiki, Collect Earth and more recently LACO-Wiki, which are collectively opening up visual interpretation of satellite imagery to crowdsourcing and nonscientific use, as well as providing new possibilities for research in the field of forest observation. Yet, the coverage of the Earth's surface by VHR imagery from freely open sources such as Google and Microsoft Bing Maps is not complete, limiting visual interpretation at these locations and potentially biasing the results from any probability-based sample. In fact, this number may be even higher in the northern latitude boreal forests where coverage by VHR imagery is even lower. However, the coverage is expected to continue to improve and there are instances where the coverage of VHR imagery from Google Maps can be complemented by that of Microsoft Bing Maps (Lesiv et al. 2018b). The incorporation of data from other new, emerging sources such as Planet may also help to fill in missing gaps in the future.

There should also be more joint efforts to bring together data sets based on visual interpretation from multiple sources, i.e., from the confidential data collected for individual projects that should be shared to the openly accessible data currently growing in size, which could contribute to a library of multipurpose reference data collections. Such an approach is already embedded in the philosophy of the LACO-Wiki system (See et al. 2017) but requires more coordinated efforts to achieve this at a global scale.

The future will also likely see more examples of online systems that allow for visualization of VHR imagery, particularly given the ease with which the imagery can now be accessed through Web Map Services, the growing archive of images available in Google Earth, which allows for some historical or change validation, and the trend in crowdsourcing and citizen science applications. We will also probably see a convergence in the functionality of the systems, particularly with regard to features that help with visual interpretation. More use of geotagged photographs and social media such as Twitter may also provide alternative sources of information that can aid visual interpretation efforts in the future.

Finally, it is anticipated that more VHR imagery will be opened up to the research community, which will allow for larger scale efforts in terms of automatic classification of VHR using the spectral information instead of just the RGB images from Google Earth or Microsoft Bing Maps, in particular deep learning approaches such as the use of Fully Convolutional Neural Networks (FCNs) applied to remote sensing (Maggiori et al. 2016; Fu et al. 2017; Baklanov et al. 2018a, b). These developments will only continue to improve the use of remote sensing data for forest observation in the future.

Acknowledgements Open access funding provided by International Institute for Applied Systems Analysis (IIASA). The study has been partly supported by the CCI Biomass (4000123662/18/I-NB) project funded by ESA, the FP7 ERC project CrowdLand (No. 617754) and the Horizon2020 LandSense project (No. 689812).

References

Achard F, Beuchle R, Mayaux P et al (2014) Determination of tropical deforestation rates and related carbon losses from 1990 to 2010. Glob Change Biol 20:2540–2554. https://doi.org/10.1111/gcb.12605

Airbus (2018) Satellite imagery—the constellation. https://www.intelligence-airbusds.com/. Accessed 8 May 2018
Baccini A, Laporte N, Goetz SJ et al (2008) A first map of tropical Africa's above-ground biomass derived from satellite imagery. Environ Res Lett 3(4):045011. https://doi.org/10.1088/1748-9326/3/4/045011
Bai T, Li D, Sun K et al (2016) Cloud Detection for high-resolution satellite imagery using machine learning and multi-feature fusion. Remote Sens 8:715. https://doi.org/10.3390/rs8090715
Baklanov A, Fritz S, Khachay M et al (2016) The cropland capture game: good annotators versus vote aggregation methods. Adv Comput Methods Knowl Eng AISC 453:167–180. https://doi.org/10.1007/978-3-319-38884-7_13
Baklanov A, Fritz S, Khachay M et al (2017) Vote aggregation techniques in the Geo-Wiki crowdsourcing game: a case study. Analysis of Images. Social Networks and Texts. Springer, Cham, pp 41–50
Baklanov A, Khachay M, Pasynkov M (2018a) Application of fully convolutional neural networks to mapping industrial oil palm plantations. Anal Images Soc Netw Texts LNCS 11179:155–167
Baklanov A, Khachay M, Pasynkov M (2018b) Fully convolutional neural networks for mapping oil palm plantations in Kalimantan. Learn Intell Optim LNCS 11353:427–432
Barbier N, Couteron P (2015) Attenuating the bidirectional texture variation of satellite images of tropical forest canopies. Remote Sens Environ 171:245–260. https://doi.org/10.1016/j.rse.2015.10.007
Barbier N, Proisy C, Véga C et al (2011) Bidirectional texture function of high resolution optical images of tropical forest: an approach using LiDAR hillshade simulations. Remote Sens Environ 115:167–179. https://doi.org/10.1016/j.rse.2010.08.015
Başeski E, Cenaras Ç (2015) Texture and color based cloud detection. In: 2015 7th international conference on recent advances in space technologies (RAST), pp 311–315
Bastin L, McInerney D, Revez G et al (2012) Web services for forest data, analysis and monitoring: developments from EuroGEOSS. In: Earthzine. https://earthzine.org/2012/07/25/web-services-for-forest-data-analysis-and-monitoring-developments-from-eurogeoss/. Accessed 5 Jun 2018
Bastin L, Buchanan G, Beresford A et al (2013) Open-source mapping and services for Web-based land-cover validation. Ecol Inform 14:9–16. https://doi.org/10.1016/j.ecoinf.2012.11.013
Bastin J-F, Barbier N, Couteron P, Adams B, Shapiro A, Bogaert J, De Cannière C (2014) Aboveground biomass mapping of African forest mosaics using canopy texture analysis: toward a regional approach. Ecol Appl 24(8):1984–2001. https://doi.org/10.1890/13-1574.1
Bastin J-F, Berrahmouni N, Grainger A et al (2017a) The extent of forest in dryland biomes. Science 356:635–638. https://doi.org/10.1126/science.aam6527
Bastin J-F, Mollicone D, Grainger A et al (2017b) Response to comment on "The extent of forest in dryland biomes". Science 358:eaao2070. https://doi.org/10.1126/science.aao2070
Beresford AE, Eshiamwata GW, Donald PF et al (2013) Protection reduces loss of natural land-cover at sites of conservation importance across Africa. PLoS ONE 8:e65370. https://doi.org/10.1371/journal.pone.0065370
Bey A, Sánchez-Paus Díaz A, Maniatis D et al (2016) Collect earth: land use and land cover assessment through augmented visual interpretation. Remote Sens 8:807. https://doi.org/10.3390/rs8100807
Bilous A, Myroniuk V, Holiaka D et al (2017) Mapping growing stock volume and forest live biomass: a case study of the Polissya region of Ukraine. Environ Res Lett 12:105001. https://doi.org/10.1088/1748-9326/aa8352
Bontemps S, Defourny P, van Bogaert E et al (2011) GLOBCOVER 2009: products description and validation report
Boyd DS, Danson FM (2005) Satellite remote sensing of forest resources: three decades of research development. Prog Phys Geogr Earth Environ 29:1–26. https://doi.org/10.1191/0309133305pp432ra
Butler D (2006) The web-wide world. Nature 439:776–778
Chittilappilly AI, Chen L, Amer-Yahia S (2016) A survey of general-purpose crowdsourcing techniques. IEEE Trans Knowl Data Eng 28:2246–2266. https://doi.org/10.1109/TKDE.2016.2555805
Clark ML, Aide TM (2011) Virtual interpretation of earth web-interface tool (VIEW-IT) for collecting land-use/land-cover reference data. Remote Sens 3:601–620. https://doi.org/10.3390/rs3030601
Clark ML, Roberts DA, Clark DB (2005) Hyperspectral discrimination of tropical rain forest tree species at leaf to crown scales. Remote Sens Environ 96:375–398. https://doi.org/10.1016/j.rse.2005.03.009
Climate Change and Development Authority of Papau New Guinea (2017) Papua New Guinea's national REDD+ forest reference level. Submission for UNFCCC technical assessment in 2017. http://redd.unfccc.int/files/png_frl__submission-15.01.2017.pdf. Accessed 1 June 2018
Coillie FMBV, Gardin S, Anseel F et al (2014) Variability of operator performance in remote-sensing image interpretation: the importance of human and external factors. Int J Remote Sens 35:754–778. https://doi.org/10.1080/01431161.2013.873152
Copernicus (2018) Copernicus open access hub. https://scihub.copernicus.eu/dhus/#/home

Curtis PG, Slay CM, Harris NL et al (2018) Classifying drivers of global forest loss. Science 361:1108–1111. https://doi.org/10.1126/science.aau3445
Defourny P, Vancustem C, Bicheron P et al (2006) GLOBCOVER: a 300 m global land cover product for 2005 using ENVISAT MERIS time series. In: Proceedings of the ISPRS commission VII mid-term symposium: remote sensing: from pixels to processes. Enscede NL
DiMiceli CM, Carroll ML, Sohlberg RA et al (2011) Annual global automated MODIS vegetation continuous fields (MOD44B) at 250 m spatial resolution for data years beginning Day 65, 2000–2010, collection 5 percent tree cover
Drusch M, Del Bello U, Carlier S et al (2012) Sentinel-2: ESA's optical high-resolution mission for GMES operational services. Remote Sens Environ 120:25–36. https://doi.org/10.1016/j.rse.2011.11.026
Duveiller G, Defourny P, Desclée B, Mayaux P (2008) Deforestation in central Africa: estimates at regional, national and landscape levels by advanced processing of systematically-distributed Landsat extracts. Remote Sens Environ 112:1969–1981. https://doi.org/10.1016/j.rse.2007.07.026
ESA (2018) The home of ESA earth online data. https://earth.esa.int/web/guest/data-access/
ESA CCI LC (2017) 300 m annual global land cover time series from 1992 to 2015 | ESA CCI land cover website. https://www.esa-landcover-cci.org/?q=node/175. Accessed 1 Jun 2017
Fan Z, Zhang W, Zhang D, Meng L (2017) An automatic accurate high-resolution satellite image retrieval method. Remote Sens 9:1092. https://doi.org/10.3390/rs9111092
Feng M, Sexton JO, Huang C et al (2016) Earth science data records of global forest cover and change: assessment of accuracy in 1990, 2000, and 2005 epochs. Remote Sens Environ 184:73–85. https://doi.org/10.1016/j.rse.2016.06.012
Foody G, See L, Fritz S et al (2018) Increasing the accuracy of crowdsourced information on land cover via a voting procedure weighted by information inferred from the contributed data. ISPRS Int J Geo-Inf 7:80. https://doi.org/10.3390/ijgi7030080
Friedl MA, Sulla-Menashe D, Tan B et al (2010) MODIS collection 5 global land cover: algorithm refinements and characterization of new datasets. Remote Sens Environ 114:168–182. https://doi.org/10.1016/j.rse.2009.08.016
Fritz S, McCallum I, Schill C et al (2009) Geo-Wiki.Org: the use of crowdsourcing to improve global land cover. Remote Sens 1(3):345–354
Fritz S, See L, van der Velde M et al (2013) Downgrading recent estimates of land available for biofuel production. Environ Sci Technol 47:1688–1694. https://doi.org/10.1021/es303141h
Fritz S, See L, McCallum I et al (2015) Mapping global cropland and field size. Glob Change Biol 21:1980–1992. https://doi.org/10.1111/gcb.12838
Fritz S, See L, Perger C et al (2017) A global dataset of crowdsourced land cover and land use reference data. Sci Data 4:170075. https://doi.org/10.1038/sdata.2017.75
Fu G, Liu C, Zhou R et al (2017) Classification for high resolution remote sensing imagery using a fully convolutional network. Remote Sens 9:498. https://doi.org/10.3390/rs9050498
Furusawa T, Pahari K, Umezaki M, Ohtsuka R (2004) Impacts of selective logging on New Georgia Island, Solomon Islands evaluated using very-high-resolution satellite (IKONOS) data. Environ Conserv 31:349–355. https://doi.org/10.1017/S0376892904001638
Garzon-Lopez CX, Bohlman SA, Olff H, Jansen PA (2013) Mapping tropical forest trees using high-resolution aerial digital photographs. Biotropica 45:308–316. https://doi.org/10.1111/btp.12009
GOFC-GOLD (2011) A sourcebook of methods and procedures for monitoring and reporting anthropogenic greenhouse gas emissions and removals caused by deforestation, gains and losses of carbon stocks in forests remaining forests, and forestation. Accessed 1 June 2018
Google (2018) Google earth downloaded more than one billion times. In: Google lat long. https://maps.googleblog.com/2011/10/google-earth-downloaded-more-than-one.html. Accessed 11 May 2018
Gorelick N, Hancher M, Dixon M et al (2017) Google earth engine: planetary-scale geospatial analysis for everyone. Remote Sens Environ 202:18–27. https://doi.org/10.1016/j.rse.2017.06.031
Hansen MC, Potapov PV, Moore R et al (2013) High-resolution global maps of 21st-century forest cover change. Science 342:850–853. https://doi.org/10.1126/science.1244693
Harris (2018) Satellite imagery—geospatial data & imagery | Harris Geospatial. http://www.harrisgeospatial.com/DataImagery/SatelliteImagery.aspx. Accessed 11 May 2018
Hudson-Smith A, Batty M, Crooks A, Milton R (2009) Mapping for the masses: accessing web 2.0 through crowdsourcing. Soc Sci Comput Rev 27:524–538. https://doi.org/10.1177/0894439309332299
Hung NQV, Tam NT, Tran LN, Aberer K (2013) An evaluation of aggregation techniques in crowdsourcing. In: Lin X, Manolopoulos Y, Srivastava D, Huang G (eds) Web information systems engineering—WISE 2013. Springer, Berlin, pp 1–15

Hussin YA, Gilani H, van Leeuwen L et al (2014) Evaluation of object-based image analysis techniques on very high-resolution satellite image for biomass estimation in a watershed of hilly forest of Nepal. Appl Geomat 6:59–68. https://doi.org/10.1007/s12518-014-0126-z

IPCC (2006) 2006 IPCC guidelines for national greenhouse gas inventories. IGES, Hayama

Justice CO, Giglio L, Korontzi S et al (2002) The MODIS fire products. Remote Sens Environ 83:244–262. https://doi.org/10.1016/S0034-4257(02)00076-7

Karlson M, Reese H, Ostwald M (2014) Tree crown mapping in managed woodlands (Parklands) of semi-arid west Africa using WorldView-2 imagery and geographic object based image analysis. Sensors 14:22643–22669. https://doi.org/10.3390/s141222643

Kim MH, Madden M, Warner TA (2009) Forest type mapping using object-specific texture measures from multispectral Ikonos imagery: segmentation quality and image classification issues. PEampRS Photogramm Eng Amp Remote Sens 75:819–829

Laso Bayas JC, See L, Fritz S et al (2016) Crowdsourcing in-situ data on land cover and land use using gamification and mobile technology. Remote Sens 8:905. https://doi.org/10.3390/rs8110905

Laso Bayas JC, Lesiv M, Waldner F et al (2017a) A global reference database of crowdsourced cropland data collected using the Geo-Wiki platform. Sci Data 4:170136. https://doi.org/10.1038/sdata.2017.136

Laso Bayas JC, See L, Perger C et al (2017b) Validation of automatically generated global and regional cropland data sets: the case of Tanzania. Remote Sens 9:815

Lesiv M, Moltchanova E, Schepaschenko D et al (2016) Comparison of data fusion methods using crowdsourced data in creating a hybrid forest cover map. Remote Sens 8:261. https://doi.org/10.3390/rs8030261

Lesiv M, Laso Bayas JC, See L et al (2018a) Estimating the global distribution of field size using crowdsourcing. Glob Change Biol. https://doi.org/10.1111/gcb.14492

Lesiv M, See L, Laso Bayas JC et al (2018b) Characterizing the spatial and temporal availability of very high resolution satellite imagery in google earth and microsoft bing maps as a source of reference data. Land 7:118. https://doi.org/10.3390/land7040118

Lesiv M, Shvidenko A, Schepaschenko D et al (2018c) A spatial assessment of the forest carbon budget for Ukraine. Mitig Adapt Strateg Glob Change. https://doi.org/10.1007/s11027-018-9795-y

Li L, Lin W, Wang X et al (2016) No-reference image blur assessment based on discrete orthogonal moments. IEEE Trans Cybern 46:39–50. https://doi.org/10.1109/TCYB.2015.2392129

Lindquist EJ, D'Annunzio R, Gerrand A et al (2012) Global forest land-use change from 1990–2005. Food and Agriculture Organization of the United Nations, Rome

Maggiori E, Tarabalka Y, Charpiat G, Alliez P (2016) Fully convolutional neural networks for remote sensing image classification. In: 2016 IEEE international geoscience and remote sensing symposium (IGARSS), pp 5071–5074

Morton DC, Nagol J, Carabajal CC et al (2014) Amazon forests maintain consistent canopy structure and greenness during the dry season. Nature 506:221–224. https://doi.org/10.1038/nature13006

NASA (2018) Earth data. https://earthdata.nasa.gov/. Accessed 11 May 2018

Ontikov PV, Schepaschenko DG, Karminov VN et al (2016) Dynamics of the area of tree cover in the Moscow region for the years 2000-2013. For Herold 1:184–188

Paz P (2017) Drone monitoring of land cover changes detected by Terra-I in Yurimaguas, Peru. In: Terra-Iorg. http://www.terra-i.org/news/news/Drone-monitoring-of-land-cover-changes-detected-by-Terra-I-in-Yurimaguas–Peru.html. Accessed 11 May 2018

Pekel J-F, Vancutsem C, Bastin L et al (2014) A near real-time water surface detection method based on HSV transformation of MODIS multi-spectral time series data. Remote Sens Environ 140:704–716. https://doi.org/10.1016/j.rse.2013.10.008

Pengra B, Long J, Dahal D et al (2015) A global reference database from very high resolution commercial satellite data and methodology for application to Landsat derived 30 m continuous field tree cover data. Remote Sens Environ 165:234–248. https://doi.org/10.1016/j.rse.2015.01.018

Pithon S, Jubelin G, Guitet S, Gond V (2013) A statistical method for detecting logging-related canopy gaps using high-resolution optical remote sensing. Int J Remote Sens 34:700–711. https://doi.org/10.1080/01431161.2012.706719

Planet (2018) High resolution monitoring. In: Planet. https://planet.com/products/hi-res-monitoring/. Accessed 11 May 2018

Ploton P, Barbier N, Couteron P, Antin CM, Ayyappan N, Balachandran N, Barathan N, Bastin J-F, Chuyong G, Dauby G, Droissart V, Gastellu-Etchegorry J-P, Kamdem NG, Kenfack D, Libalah M, Mofack G, Momo ST, Pargal S, Petronelli P, Proisy C, Réjou-Méchain M, Sonké B, Texier N, Thomas D, Verley P, Zebaze Dongmo D, Berger U, Pélissier R (2017) Toward a general tropical forest biomass prediction model from very high resolution optical satellite images. Remote Sens Environ 200:140–153. https://doi.org/10.1016/j.rse.2017.08.001

Read JM, Clark DB, Venticinque EM, Moreira MP (2003) Application of merged 1-m and 4-m resolution satellite data to research and management in tropical forests. J Appl Ecol 40:592–600. https://doi.org/10.1046/j.1365-2664.2003.00814.x

Running SW, Nemani RR, Heinsch FA et al (2004) A continuous satellite-derived measure of global terrestrial primary production. Bioscience 54:547–560

Saatchi SS, Harris NL, Brown S et al (2011) Benchmark map of forest carbon stocks in tropical regions across three continents. PNAS 108:9899–9904. https://doi.org/10.1073/pnas.1019576108

Salk CF, Sturn T, See L et al (2016) Assessing quality of volunteer crowdsourcing contributions: lessons from the Cropland Capture game. Int J Digit Earth 9:410–426. https://doi.org/10.1080/17538947.2015.1039609

Salk CF, Sturn T, See L, Fritz S (2017) Limitations of majority agreement in crowdsourced image interpretation. Trans GIS 21:207–223. https://doi.org/10.1111/tgis.12194

Sannier C, McRoberts RE, Fichet L-V (2016) Suitability of global forest change data to report forest cover estimates at national level in Gabon. Remote Sens Environ 173:326–338. https://doi.org/10.1016/j.rse.2015.10.032

Santoro M, Beaudoin A, Beer C et al (2015) Forest growing stock volume of the northern hemisphere: spatially explicit estimates for 2010 derived from Envisat ASAR. Remote Sens Environ 168:316–334. https://doi.org/10.1016/j.rse.2015.07.005

Schepaschenko D, Kraxner F, See L et al (2015a) Global biomass information: from data generation to application. In: Yan J (ed) Handbook of clean energy systems. Wiley, New York, pp 11–33

Schepaschenko D, See L, Lesiv M et al (2015b) Development of a global hybrid forest mask through the synergy of remote sensing, crowdsourcing and FAO statistics. Remote Sens Environ 162:208–220. https://doi.org/10.1016/j.rse.2015.02.011

Schepaschenko D, Fritz S, See L et al (2017) Comment on "The extent of forest in dryland biomes". Science 358:eaao0166. https://doi.org/10.1126/science.aao0166

See L, Comber A, Salk C et al (2013) Comparing the quality of crowdsourced data contributed by expert and non-experts. PLoS ONE 8:e69958. https://doi.org/10.1371/journal.pone.0069958

See L, Fritz S, Perger C et al (2015a) Harnessing the power of volunteers, the internet and Google Earth to collect and validate global spatial information using Geo-Wiki. Technol Forecast Soc Change 98:324–335. https://doi.org/10.1016/j.techfore.2015.03.002

See L, Schepaschenko D, Lesiv M et al (2015b) Building a hybrid land cover map with crowdsourcing and geographically weighted regression. ISPRS J Photogramm Remote Sens 103:48–56. https://doi.org/10.1016/j.isprsjprs.2014.06.016

See L, Fritz S, Perger C et al (2016) Mapping human impact using crowdsourcing. Mapping Wilderness. Springer, Dordrecht, pp 89–101

See L, Laso Bayas JC, Schepaschenko D et al (2017) LACO-Wiki: a new online land cover validation tool demonstrated using GlobeLand30 for Kenya. Remote Sens 9:754. https://doi.org/10.3390/rs9070754

Sexton JO, Song X-P, Feng M et al (2013) Global, 30-m resolution continuous fields of tree cover: landsat-based rescaling of MODIS vegetation continuous fields with lidar-based estimates of error. Int J Digit Earth 6:427–448. https://doi.org/10.1080/17538947.2013.786146

Sexton JO, Noojipady P, Song X-P et al (2016) Conservation policy and the measurement of forests. Nat Clim Change 6:192–196. https://doi.org/10.1038/nclimate2816

Sheppard SRJ, Cizek P (2009) The ethics of Google Earth: crossing thresholds from spatial data to landscape visualisation. J Environ Manage 90:2102–2117. https://doi.org/10.1016/j.jenvman.2007.09.012

Simard M, Pinto N, Fisher JB, Baccini A (2011) Mapping forest canopy height globally with spaceborne lidar. J Geophys Res Biogeosciences 116:45. https://doi.org/10.1029/2011jg001708

Simonetti D, Beuchle R, Eva H (2011) User manual for the JRC land cover/use change validation tool. Publications Office of the European Union, Ispra

Simonetti D, Marelli A, Eva H (2015) IMPACT: portable GIS toolbox for image processing and land cover mapping—EU Science Hub—European Commission. Publications Of fice of the European Union, Luxembourg

Solano-Correa YT, Bovolo F, Bruzzone L (2018) An approach for unsupervised change detection in multitemporal VHR Images acquired by different multispectral sensors. Remote Sens 10:533. https://doi.org/10.3390/rs10040533

Sturn T, Wimmer M, Salk C et al (2015) Cropland capture—a game for improving global cropland maps. In: Foundation of digital games conference (FDG 2015). Pacific Grove, CA

Sy VD, Herold M, Achard F et al (2015) Land use patterns and related carbon losses following deforestation in South America. Environ Res Lett 10:124004. https://doi.org/10.1088/1748-9326/10/12/124004

Szantoi Z, Brink A, Buchanan G et al (2016) A simple remote sensing based information system for monitoring sites of conservation importance. Remote Sens Ecol Conserv 2:16–24. https://doi.org/10.1002/rse2.14

Thurner M, Beer C, Santoro M et al (2014) Carbon stock and density of northern boreal and temperate forests. Glob Ecol Biogeogr 23:297–310. https://doi.org/10.1111/geb.12125

Tong H, Li M, Zhang H, Zhang C (2004) Blur detection for digital images using wavelet transform. In: 2004 IEEE international conference on multimedia and expo (ICME) (IEEE Cat. No.04TH8763), vol 1, pp 17–20

USGS (2018) EarthExplorer. https://earthexplorer.usgs.gov/. Accessed 11 May 2018

Valérie T, Marie-Pierre J (2006) Tree species identification on large-scale aerial photographs in a tropical rain forest, French Guiana—application for management and conservation. For Ecol Manag 225:51–61. https://doi.org/10.1016/j.foreco.2005.12.046

Waldner F, Schucknecht A, Lesiv M et al (2018) Toward seamless integration of expert and non-expert reference data to validate global binary thematic maps. Submitt Remote Sens Environ 221:235–246

Wulder MA, Masek JG, Cohen WB et al (2012) Opening the archive: how free data has enabled the science and monitoring promise of Landsat. Remote Sens Environ 122:2–10. https://doi.org/10.1016/j.rse.2012.01.010

Zahawi RA, Dandois JP, Holl KD et al (2015) Using lightweight unmanned aerial vehicles to monitor tropical forest recovery. Biol Conserv 186:287–295. https://doi.org/10.1016/j.biocon.2015.03.031

Zhu Z, Woodcock CE (2012) Object-based cloud and cloud shadow detection in Landsat imagery. Remote Sens Environ 118:83–94. https://doi.org/10.1016/j.rse.2011.10.028

Publisher's Note Springer Nature remains neutral with regard to jurisdictional claims in published maps and institutional affiliations.

Affiliations

Dmitry Schepaschenko[1] · Linda See[1] · Myroslava Lesiv[1] · Jean-François Bastin[2,3] · Danilo Mollicone[2] · Nandin-Erdene Tsendbazar[4] · Lucy Bastin[5,6] · Ian McCallum[1] · Juan Carlos Laso Bayas[1] · Artem Baklanov[1,7] · Christoph Perger[1] · Martina Dürauer[1] · Steffen Fritz[1]

1 Ecosystems Services and Management Program, International Institute for Applied Systems Analysis (IIASA), Schlossplatz 1, 2361 Laxenburg, Austria

2 Food and Agriculture Organization of the United Nations, Vialle delle Terme di Caracalla, 00153 Rome, Italy

3 Department of Environmental Systems Science, Institute of Integrative Biology, ETH Zürich, 8092 Zurich, Switzerland

4 Laboratory of Geo-information Science and Remote Sensing, Wageningen University, Droevendaalsesteeg 3, 6708 PB Wageningen, The Netherlands

5 Knowledge Management Unit, Joint Research Centre of the European Commission, Via Enrico Fermi 2749, 21027 Ispra, Italy

6 School of Engineering and Applied Science, Aston University, Birmingham B4 7ET, UK

7 National Research University Higher School of Economics, Soyuza Pechatnikov Street 16, St. Petersburg, Russian Federation 190121

Surveys in Geophysics (2019) 40:863–880
https://doi.org/10.1007/s10712-019-09528-w

Ground Data are Essential for Biomass Remote Sensing Missions

Jérôme Chave[1] · Stuart J. Davies[2] · Oliver L. Phillips[3] · Simon L. Lewis[3,8] · Plinio Sist[4] · Dmitry Schepaschenko[5] · John Armston[6] · Tim R. Baker[3] · David Coomes[7] · Mathias Disney[8,9] · Laura Duncanson[6] · Bruno Hérault[4,10] · Nicolas Labrière[1] · Victoria Meyer[11] · Maxime Réjou-Méchain[12] · Klaus Scipal[13] · Sassan Saatchi[11]

Received: 15 October 2018 / Accepted: 30 March 2019 / Published online: 27 April 2019

Abstract
Several remote sensing missions will soon produce detailed carbon maps over all terrestrial ecosystems. These missions are dependent on accurate and representative in situ datasets for the training of their algorithms and product validation. However, long-term ground-based forest-monitoring systems are limited, especially in the tropics, and to be useful for validation, such ground-based observation systems need to be regularly revisited and maintained at least over the lifetime of the planned missions. Here we propose a strategy for a coordinated and global network of in situ data that would benefit biomass remote sensing missions. We propose to build upon existing networks of long-term tropical forest monitoring. To produce accurate ground-based biomass estimates, strict data quality must be guaranteed to users. It is more rewarding to invest ground resources at sites where there currently is assurance of a long-term commitment locally and where a core set of data is already available. We call these 'supersites'. Long-term funding for such an inter-agency endeavour remains an important challenge, and we here provide costing estimates to facilitate dialogue among stakeholders. One critical requirement is to ensure in situ data availability over the lifetime of remote sensing missions. To this end, consistent guidelines for supersite selection and management are proposed within the Forest Observation System, long-term funding should be assured, and principal investigators of the sites should be actively involved.

Keywords Biomass · Calibration · Forest · In situ data · Validation

1 Introduction

The global carbon cycle is being altered by anthropogenic activities: carbon dioxide and other economy-related greenhouse gas emissions have steadily increased since the 1960s (Le Quéré et al. 2018). This has already had detectable consequences on the mean

✉ Jérôme Chave
jerome.chave@univ-tlse3.fr

Extended author information available on the last page of the article

temperature of our planet (IPCC 2018). Land ecosystems hold a large potential for carbon storage. For example, it has been estimated that allowing Neotropical secondary forests to regenerate, without further human intervention, may enable Latin America and the Caribbean to be carbon–neutral for decades (Chazdon et al. 2016). Also, protecting intact forests is essential to ensuring carbon storage and many other ecosystem services (Pan et al. 2011). Thus, conserving existing intact forests, in combination with restoring and managing sustainably degraded forests, is almost certain to be a key action to help meet the Paris Accord targets. The idea of financially incentivizing local and national initiatives to spare forest land and favour reforestation has thus received further attention, as evidenced by the United Nations' Reduced Emissions from Deforestation and forest Degradation (REDD+) program.

The REDD+ framework is predicated on the ability to measure the differential amount of carbon stored in land ecosystems as a result of a change in policy compared to a defined business-as-usual scenario. This presupposes that instruments and methods are in place for monitoring, reporting and verification of land carbon budgets, yet there remain great challenges in this area (Grassi et al. 2017). In many temperate countries, which have largely built their political system around wood as a key commodity, elaborate systems of forest resource assessment and management were established early on, and they have been operated by national forest services. Thus, nationally determined carbon contributions are relatively reliable in the temperate zone, where forest biomass stocks are based on well-established sample-based forest inventories (Fridman et al. 2014). However, the political history of many tropical or subtropical countries has been such that national forest inventory systems are either young or absent, in spite of efforts by the UN Food and Agriculture Organization (FAO) to set up such systems in several countries since the 1990s (Schimel et al. 2015). This situation is now changing, with national forest inventories being developed in Brazil and the Democratic Republic of Congo (Xu et al. 2017).

Remotely sensed approaches to estimate carbon stocks have emerged as a solution to this problem, and several missions are planned in the 2018–2022 period, including BIOMASS (P-band radar satellite by ESA), NISAR (L-band radar by NASA and ISRO), and GEDI (lidar onboard the ISS by NASA). These missions will not measure carbon stocks directly, but they will use proxies of forest structure, volume, and biomass components that correlate with the aboveground carbon stocks. Canopy height is measured by lidar and polarimetric interferometry, and tall forests tend to hold more carbon than shorter ones. The second physical quantity related to forest carbon store is the wood volume and water content which influence the backscattered electromagnetic energy measured at P-band (~70 cm) or L-band (~25 cm) wavelengths (Le Toan et al. 2011; Saatchi et al. 2011a, b; Shugart et al. 2010). Thus, these missions will collect data that can be empirically related to forest carbon content.

Because forest carbon stores are indirectly inferred from satellite sensors, with questionable assumptions about their dependence on forest structure and water content, it is essential that the planned missions make use of accurate ground estimates of carbon stocks to train their inversion algorithms and validate their products. However, estimating biomass on the ground is a challenge in itself and ecologists and foresters have struggled with this problem for a long time. Inevitably, providing inaccurate carbon stock estimates to the Earth Observation (EO) community will result in uncertain (and potentially biased) carbon maps, and this would have serious downstream effects on the usefulness of these maps in policy. For instance, even though pantropical biomass maps inferred from remote sensing have been available for some time now (Saatchi et al. 2011a, b; Baccini et al. 2012), the UN Intergovernmental Panel on Climate Change (IPCC) has been reluctant to recommend their widespread use over national inventories because of possible calibration issues. Here, we offer a perspective from the ground up and propose a strategy for gathering reliable ground-based measurements and biomass estimates that will be useful to the various Earth Observation missions aimed at quantifying forest structure and carbon stock at a global scale.

Overarching principles are summarized here, and echo meetings jointly held on ground data and upcoming land Earth Observation missions (NASA-ESA-Smithsonian Workshop 2016; ISSI ESA Meeting, Bern, 2017). First, the focus of all of these missions is primarily tropical. Many forested extratropical countries already have a forest inventory assessment in operation. In contrast, ground-based monitoring systems are sorely lacking in the tropics (Schimel et al. 2015). Forest extent in the tropics is still very substantial, by far the most living biomass is located in the tropics (63% of carbon in intact tropical forests, against 15% in boreal forests and 13% in temperate forests, according to a comprehensive estimate, Pan et al. 2011). Second, in order to map change in forest ecosystems, ground sites need to be regularly revisited (Frolking et al. 2009). It is impossible financially and logistically to maintain thousands of sites without long-lasting governmental or international support. These observation systems need to be maintained at least over the lifetime of the planned missions, but it is likely that they will find even greater value if made permanent through binding agreements—here we provide costing estimates to facilitate the discussion of this question, while acknowledging that informed recommendations for the calibration and validation of the missions are dependent on the nature of the algorithm, on the resolution of the data, and on the mission duration, and are therefore beyond the scope of the present study. Third, estimating biomass correctly in situ remains a delicate business, and strict data quality control must be guaranteed to users.

2 Principles of Ground-Based Biomass Estimation

Aboveground biomass (AGB) is the total amount of dry matter of live trees held above-ground in a plot. It is a key parameter for a range of applications, including greenhouse gas accounting, forest fire assessment, management of the timber industry, monitoring of land-use change, and ecosystem science. Currently, accurate AGB estimates can be obtained only by labour-intensive fieldwork (plot inventories) conducted by trained operators. The AGB of each tree is estimated from measured variables using an allometric model AGB = f(ρ, D, H), where ρ is the stem wood density of the focal tree, D its trunk diameter, and H its total height. Trunk diameter is usually measured at breast height (130 cm aboveground) in forests, but in non-forest habitats, trees tend to branch low, and standard measurements are lower, e.g. at 10 cm aboveground in Australia (Paul et al. 2016). For the largest trees, diameter needs to be measured above buttresses (Sillett et al. 2019). Such allometric models are constructed from destructively harvested trees in which AGB and the other variables are all measured directly. Oven-dry biomass is approximately 47% carbon, the conversion from AGB to aboveground carbon stores is easy, and the two notions often used interchangeably. Importantly, a thorough and recent analysis of wood carbon content showed that carbon content varies predictably among plant functional types and ranges from 41 to 51%, and this source of variation should be carefully considered (Martin et al. 2018). Since measuring wood density for each tree would be too labour-intensive, recent studies have used species-mean wood density values, taken from publicly available lookup tables. It follows that the reliable taxonomic identification of each tree is essential for an accurate estimate of its AGB.

The tree-level AGB estimates are then summed across trees and over the plot, to produce an aboveground biomass value at the plot level, usually expressed in Mg/ha of dry matter. AGB is not measured in forests, but estimated (Clark and Kellner 2012), and errors due to the estimation of height, wood density, and choice of the allometric model usually result in a plot-based AGB uncertainty that is non-negligible. AGB is estimated to within 50% absolute error for a single tree (Chave et al. 2014). This absolute error, when propagated at stand level, is around 10% at the 1-ha scale (Réjou-Méchain et al. 2014).

In most current applications, live below ground biomass is usually inferred from AGB using standard root-shoot ratios (Mokany et al. 2006; Paul et al. 2019). Although this poses important and specific challenges, the issue of estimating biomass components other than AGB (belowground biomass, soil carbon, coarse woody debris) is not covered in this contribution.

In the past few decades, estimation of carbon stocks and AGB in the tropics has been based on permanent forest plots with measurements of tree trunk diameters, and often tree mapping and species identification of the trees. AGB can then be estimated per tree and then summed over all trees in a stand (Brown and Lugo 1982). Permanent sampling forest plots are thus the basic unit of biomass measurement. Their size ranges from 0.01 ha (10×10 m) to almost 100 ha (1000×1000 m). Within these plots, all trees above a given trunk diameter threshold (usually 10 cm) are counted in the census. These trees are mapped (usually to the nearest m or ideally better—but unfortunately quite often much less accurately), they are marked permanently with a tag, and their trunk diameter is measured at a standard position on the trunk (usually 130 cm aboveground, or 50 cm above buttresses or irregularities, if present). The point of measurement is noted by a permanent paint mark. A sample of trees may also have their height dimension measured, and as far as possible, all trees are identified taxonomically. The numbers of trees ≥ 10 cm in one hectare of mature moist forest varies from 300 to 1000. The number of tree species per hectare is up to 30 in temperate forests and up to 300 in the tropics (Gentry 1988). Thus, species identification can be a considerable challenge.

Temperate and boreal forests present different challenges than tropical forests, and several temperate countries can rely on National Forest Inventories (NFIs) constructed using statistical sampling theory. These are designed to provide inferences of biomass and other commodity values at regional or jurisdictional levels using a large number of small plots (Smith 2002). These plots can be used to validate the Earth Observation biomass products, but their use for algorithm development and training is more limited due to plot size and potentially large plot-level uncertainty of ground biomass. The challenge of mobilizing temperate-country ground data is an important one but, because most of the world's high-biomass forests are in the tropics, we here emphasize the tropical zone. We also note that vast regions like extratropical Asia are missing NFIs, and it would be important to better account for in situ forest information in these regions.

Aerial lidar scanning (ALS) has been intensively used for estimating tropical forest biomass (Drake et al. 2002; Asner et al. 2010), and the literature suggests that, if ALS data can be calibrated locally with permanent sampling plots, the resulting biomass maps are unbiased and reliable. (They have a relative uncertainty of $<20\%$ at the 1-ha scale.) Establishing high-resolution biomass maps at 1000-ha (10 km^2) scale would result in a 100-fold increase over plot data, and the 1000-ha scale is typically the area surveyed around permanent field stations in the tropics: this means that the sites are within walking distance and the ALS-derived biomass map can be thoroughly ground-truthed. However, since forests are constantly changing, algorithm training and validation of ALS data is impossible without near-contemporaneous fieldwork.

3 High-Quality Carbon Estimates Require Long-Term Study Sites

3.1 Forest Dynamics and Tree Inventories

Collections of small plots can offer a representative sample of the landscape-scale variability of biomass, but lack the temporal dimension that is also critically important for understanding the system. Forest changes include (1) secular changes in mature forests driven externally by climate (e.g., increasing/declining growth rate), (2) sudden, stochastic changes (e.g., drought-, flood-, wind-, pest-, fire-induced mortality), (3) successional development (e.g., savanna thickening into forest, floodplain forests). Anthropogenic impacts on forests are equally important, complex, and ubiquitous. As a result, it is not enough to use forest biomass estimates at a given point in time, but forests should be measured repeatedly in situ. Most existing permanent census plots are recensused by trained teams of foresters and botanists every 4–5 years, funds permitting, a monitoring revisit frequency that is sufficient if tree turnover rates are around 1–3%/year.

Another reason for measuring forest stands repeatedly is that some measurement errors may affect biomass estimates far more than others. A small number of large trees holds a large fraction of the biomass in a stand, and these are the most difficult to measure in the field. Assume a 100×100 m stand of tropical forest contains around 500 trees of trunk diameter > 10 cm, and the oven-dry aboveground biomass is 300 tons or more, a typical situation in moist tropical forests. Thus, on average, a tree weighs 0.6 ton. However, the distribution of tree weights is hugely skewed, since according to one study conducted in intact tropical forests, 41% of the aboveground biomass was held in trees above 60 cm in trunk diameter (Lutz et al. 2018). In tropical forests, historical permanent plots were often established by botanists to explore plant diversity. In these cases, initially, little attention may have been paid to carefully measuring the largest trees, and plots were often located based on convenience more than based on a sampling protocol (plots installed by foresters tended to have the opposite focus: large trees measured well, but little attention paid to the botany of non-commercial species). Clark and Clark (2000) provided the first comprehensive study comparing different carbon sampling strategies in the tropical forest of La Selva, Costa Rica. They showed that measuring trunk diameter above buttresses was key to a proper estimate of AGB (Fig. 1; see also Condit 1998).

It is essential to realize that, for many sites, the history of the plots is complex, and data quality may have changed over time. Therefore, the issue is not only to process pre-existing data, but also to critically appraise the field collection protocols to ensure that legacy data are made available and associated with an uncertainty assessment that accounts for the history of data acquisition, that varies greatly from site to site and among groups of data collectors.

In the tropics, a major contribution is required from developing country scientists and technicians. The ground data they produce are hard-won and need to be repeated at regular intervals. The participants in our plot networks span hundreds of tropical forested localities across up to 50 different nations. Such work involves obvious logistical complexities of organizing ground data collection, institutional collaborations, intellectual property, permits, and health and safety protocols to allow remote fieldwork and plant collection across

Fig. 1 Permanent plot in intact lowland tropical forest at the Nouragues station, French Guiana. Left: Large, buttressed, trees encompass the majority of the aboveground biomass stock. Right: A pioneer tree (*Cecropia sciadophylla* Mart.) has grown from the top of a palm, causing issues of trunk diameter measurement. Situations like this one are resolved only with proper field protocols

so many countries while complying with protected area regulations. In addition, a key challenge is to harmonize datasets and differing existing ground biomass protocols. Consequently, because of this major effort and the heavy dependence on often specialized human labour, local researchers who lead collection campaigns and own ground-based measurements must be involved as scientific collaborators, and field teams adequately trained, equipped, insured, and paid.

3.2 Calibration and Validation Strategies for Earth Observation Missions

The aboveground biomass data from in situ plot networks are an essential requirement of current EO missions. Several biomass EO missions are currently in the process of developing their algorithms or preparing their validation plans utilising this type of ground data. This includes the Global Ecosystem Dynamics Investigation (GEDI) mission, launched in December 2018 (NASA), the NISAR mission (NASA–ISRO synthetic aperture radar, launch in 2021), and the BIOMASS mission (ESA, launch in 2022). The ground data already collected as part of these efforts are remarkably similar, even if the requirements differ slightly.

The major requirement is that ground biomass values be available, based on intensive tree inventories, and reliable biomass estimation methods. EO missions have included requirements about quality assessment of these plot-based biomass estimates, because improperly estimated ground biomass values are not rare, and failure to account for unreliable data will result in serious problems in the calibration and validation plan.

The GEDI, NISAR and BIOMASS missions have each complied and collected data from plot networks, and have recently shared the metadata of their efforts. The GEDI science team, the most advanced to the date of the submission of this contribution, has assembled a dataset of 105 sites. These data were contributed by a variety of projects and are thus in-kind contributions. They span the major biomes and represent almost 1400 ha of surveyed plots, of which 40% are in the Neotropics (557 ha), 12% in Africa (173 ha), and 7% in tropical Asia (108 ha). The NISAR mission cal/val team has assembled data for 77 sites, with quite some overlap with that of the GEDI science team. BIOMASS is the least advanced, including 6 sites, two in French Guiana (Neotropics) and four in Gabon, Africa (Labrière et al. 2018), and a total sampled area of 227 ha. In addition to permanent plot data, all three missions include airborne lidar scanning (ALS) in their ground dataset. ALS has been shown to be a critically component to EO missions, because it provides invaluable information on forest structure, in a carefully georeferenced format, and this can be used to upscale plot-based biomass estimates to landscape-scale biomass maps.

In comparison with these datasets, forest plot networks include far more information (Table 1). For instance, the Smithsonian Institute's ForestGEO coordinates 245 ha of forest across 4 sites in Africa, and over 250 ha in tropical Asia (Anderson-Teixeira et al. 2015). The Rainfor network in South America, AfriTRON in tropical Africa and T-Forces in Asia, managed by University of Leeds, coordinates no less than 400 ha of plots in Amazonia alone (Mitchard et al. 2014) and 315 ha in tropical Africa (Lewis et al. 2013). These latter three networks are all stored in ForestPlots.net allowing quality control and data processing chains to be applied across all three networks. Among these various networks, there is almost no overlap, and they do not include independent large projects such as forest management experiments. These have recently been coordinated by the Tropical managed Forest Observatory (TmFO), with almost 1200 ha of forests permanently monitored (Sist et al. 2015). Also, a network of secondary forest plots has been established in the Neotropics and coordinates effort on forest regeneration (Chazdon et al. 2016). Our estimate is that the area of tropical forests that are currently monitored globally is in excess of 2500 ha by these four networks and not accounting for many more projects. This, however, remains a minuscule fraction of the total area covered by forest worldwide, and the biomass estimation challenge is therefore one of upscaling.

3.3 Supersites

Based on our knowledge of available data within the partners, it would be more cost-effective to prioritize a limited number of ecologically representative sites around the world. We call this the 'supersites' concept. Such sites combine intensive and long-term fieldwork data, airborne vegetation monitoring, and ancillary information, such that reliable landscape-scale biomass estimation is possible (Fig. 2). This idea of establishing long-term sampling sites with EO applications in mind is fundamentally the same as that of the US long-term ecological research sites (LTER) in place since the 1970s, and the International Biological Program (Golley 1993). The Committee on Earth Observing Satellites (CEOS) Working Group on Land Product Validation has officially endorsed a supersite concept not only for biomass, but to identify and promote the collection of validation data for the wide range of Essential Climate Variables products that are currently available or expected in the coming years.

Meeting high data quality in the tropics is possible at or near research stations with existing infrastructure and resources, and with the potential to upgrade datasets, funds permitting. Based on previous experience with the TropiSAR campaign (French Guiana; Dubois-Fernandez et al. 2012), and AfriSAR mission (Gabon), we propose that these supersites be selected based on the following specific requirements: (1) availability of at least 10 already established 1-ha permanent sampling plots, ideally well-distributed across the landscape, capturing local gradients of biomass. The plots should be established according the best tropical forestry standards (see RAINFOR or Center for Tropical Forest Science protocols, e.g., Condit 1998); (2) availability of tree height measurements at each of these plots (for

Table 1 List of characteristics for some of the international tropical forest-monitoring networks currently in operation

Network	Countries	Plots	Plot sample area (ha)	Trees	Species	Measurements	Forest types	Regional focus
ForestGEO	26	65	16–120	6.5 million	12,000	20 million	Primary	Global
RAINFOR	9	400	0.2–9, mostly 1	280,000	5500	2 million	Primary	South America
AfriTRON	11	320	0.2–10, mostly 1	170,000	1800	600,000	Primary	Africa
TmFO	10	517	0.25–27, mostly 1	300,000–400,000	–	~6 million	Logged	Pantropical

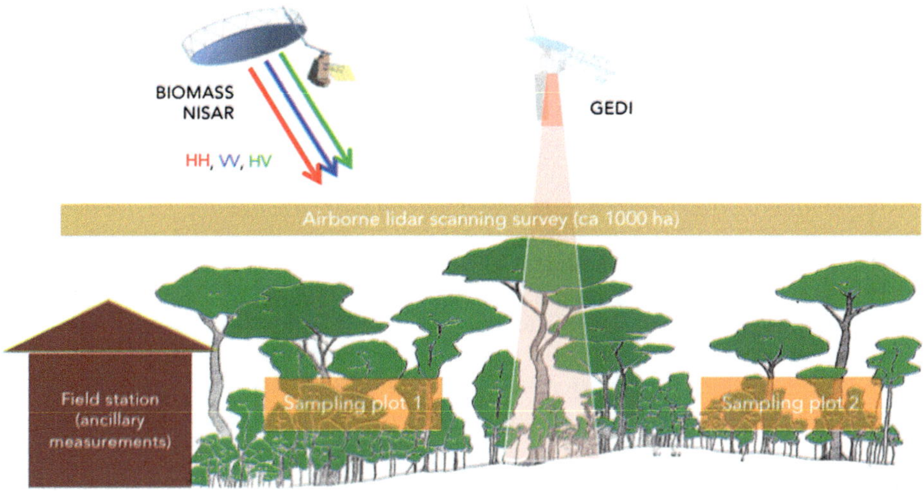

Fig. 2 The supersite concept. Relatively few sites with long-term investment by plot principal investigators, and the potential to upgrade the sites. The background is taken from Ashton (1964, Kuala Belalong, Brunei)

all trees or at least a representative sample of trees); (3) availability or potential future collection of ALS coverage over at least 1000 ha, flown over the permanent plots, with minimal quality requirements (i.e. such that 1-m canopy elevation models can be constructed); (4) availability of a weather station and, optionally, automated soil moisture monitoring (ideally encompassing the landscape-scale variation of soil moisture). Inasmuch as possible, these data should be acquired at the same time.

We also propose to implement terrestrial lidar scanning (TLS) surveys of the permanent plots. TLS surveys are no substitute for forest tree inventories, but they have the potential to complement them usefully: they provide an accurate measure of individual tree volume, a measure of total tree height, and an accurate correction of stem geolocation (relative, at stand scale). They also give access to the details of forest structure that may be important in modelling canopy reflectance at these sites (Calders et al. 2018). This considerably increases the quantity and quality of much of the key parameters on which all of the other estimates rely. The resulting tree volume data can also be used to augment existing allometric relationships used to generate tree biomass estimates, particularly across a much larger range of tree size and including many more large trees (Disney et al. 2018). Tree volume and tree mass may differ significantly in the case of hollow trunks, and large trees tend to be more often hollow than small ones (Nogueira et al. 2006). Recently, drone-based alternatives for terrestrial lidar scanning have been proposed and they present the additional advantages of scanning the canopy tops and of producing already-stitched point clouds, over large areas typically several hectares (Brede et al. 2017). While this technology still requires development, it would be important to explore its applicability to the establishment and monitoring of forest supersites.

3.4 Drivers of Biomass Stocks and Geographical Coverage

Forest structure varies at all spatial scales. Determining the optimal sampling strategy for Earth's forests requires research into the drivers of biomass stocks, which in turn depends on spatially explicit maps of forest structure that will not be available until the new mission products come online. It is difficult to segment tropical forests worldwide into forest types that would both make ecological sense and would be optimal for the training of biomass retrieval algorithms. For instance, two forests may have a similar structure, yet display species with different wood densities, resulting in very different biomass estimates. Also, the forest lower canopy may play a significant role in the radar backscattering properties, and, like wood density, this is not readily assessed remotely.

We therefore provide ecologically informed guiding principles for the selection of sites. Tropical forests vary in their structure and floristic composition, and this in turn impacts their biomass storage capacity (Malhi et al. 2004; Stegen et al. 2011). The four main driving factors of this variation are soil fertility, moisture supply, elevation, and disturbance regime. Therefore, forests often hold less biomass on very infertile or very fertile soils (Castilho et al. 2006). Also, dry tropical forests have less biomass, but there is also potentially a hump-shaped distribution of biomass with respect to annual precipitation whereby ever-wet forests tend to have lower biomass stocks than moist forests (Brown and Lugo 1982). Elevation is another important factor, and biomass usually declines with increasing altitude, although some exceptions exist, for example when trees of the oak family are present (Phillips et al. 2016). Finally, disturbed forests have a lower biomass than undisturbed ones. The foremost cause of disturbance in the tropics is anthropic, but other causes exist including wildfires, wind storms, insect predation, or diseases, and the frequency and intensity of natural disturbance exerts a critical control on intact forest wood density and biomass (Keeling and Phillips 2007; Johnson et al. 2016).

In addition, tropical forests have almost zero floristic overlap between the Neotropics (South America), Africa, Asia, and Oceania (including Papua New Guinea and Australia) with each biogeographic region having thousands of tree species whose architecture and unique identity helps to determine forest structure and biomass in that region. Assuming that three conditions (low, medium, high) are selected for each of the four major gradients in each of the four continents, the number of possibilities is $4 \times 3^4 = 324$.

Practically, when selecting sites for algorithm training or product validation, it is essential to include the full range of variability in biomass, i.e. high-biomass forests, typically moist tropical forests with biomass stocks more than 300–400 tons/ha, and up to 600 tons/ha, but also low-biomass forests, typically $<$100–200 tons/ha (Sullivan et al. 2017). For instance, a relatively young secondary forest of ca. 20 years regrowing from clear-cutting holds about 100 tons/ha in tropical areas (assuming an accumulation rate of 5 tons/ha/year). Also, woodlands store 30–150 tons/ha. It would be important to include both secondary vegetation in the study landscapes, and to select dry vegetation types. These vegetation types are particularly important for the NISAR mission, which aims at estimating biomass up to 100 Mg/ha, above which L-band backscatter signals saturate with respect to biomass (Chapman and Kasischke 2018).

4 Building on Long-Term Forest Plots

4.1 The Forest Observation System

Permanent plots provide the most accurate method for forest biomass estimation, which not only depends on biometric variables, but also on species-dependent estimates of wood density. Many sites across the forested tropics have on the order of ten 1-ha plots, because this sampling intensity is manageable. Much larger sampling intensities do exist, but they are rare. Further, while plots are often established randomly with respect to small-scale disturbance events, they are rarely formally randomised within strata across the landscape. Plots are often located to monitor specific environments (e.g. old-growth forest), or test hypotheses, and were not established to estimate biomass stocks at the landscape scale, as required by the various EO missions.

The European Space Agency has funded the Forest Observation System (FOS) as an effort to coordinate in situ activities in relation with the BIOMASS mission. The FOS includes several large international consortia who are addressing the issues of ground data sharing and standardization: RAINFOR, AfriTRON, and T-FORCES (led from the University of Leeds), ForestGEO (including CTFS; Smithsonian Institution). These consortia have a solid record in tackling key scientific questions, in engaging a community of collaborators and in standardizing forestry data. For up to 40 years now, they have been devoted to coordinating long-term research with permanent sampling forest plots. They have (1) established permanent sampling plots in tropical and temperate forests, (2) encouraged and carried out extensive plant collection and identification, (3) proposed robust protocols for accurate tree mapping, and measurement, (4) monitored existing plots repeatedly, and (5) established databases with a special emphasis on data quality control at the tree level, and have successfully incorporated historical databases.

Two additional networks of permanent forest plots have now been invited to join the Forest Observation System: TmFO (Tropical managed Forest Observatory; Sist et al. 2015) and AusCover (CSIRO). We are aware that many more groups of scientists and networks of plots have been established, but when examining inclusion of new sites into the FOS, it is essential to consider upstream quality assessment. It is preferable to build upon projects that have already established a data sharing policy, quality assessment procedures, and instruments for communication with principal investigators at each of the sites.

NASA and ESA are also in the process of establishing a Multi-mission Analysis and Algorithm platform (MAAP), which will house field plot, airborne lidar, and spaceborne datasets, including data from NISAR, GEDI, and BIOMASS. This will be a virtual open and collaborative environment, bringing together data, cloud-based computing resources, and collaborative tools. It will establish a collaboration framework between ESA and NASA to share data, science algorithms and computing resources in order to foster and accelerate scientific research conducted by NASA and ESA scientists. We intend that the Forest Observation System (FOS) will become an integral part of this multi-mission analysis platform, facilitating provision of field plot data such as from existing plots and new supersite data acquisitions.

Gathering calibration and validation data relevant to biomass for the Earth Observation community faces a number of challenges, and the FOS aims to address the most important ones. We here list the priorities: (1) ensuring the respect of intellectual property rights, (2) providing site principal investigators with a knowledge of the scientific challenges undertaken with their data, and (3) ensuring that datasets included in FOS are of the highest possible quality and are representative of all forest ecosystems.

A key aspect of the collaboration is that the intellectual property of the primary data remains with the site's principal investigator. This principle is upheld in the FOS data sharing policy (under preparation). Official data sharing policies are found, for instance for the Smithsonian Institute (white paper 'Sharing Smithsonian Digital Scientific Research Data from Biology', March 2011), for the RAINFOR project (white paper 'Ethical Code, Data Sharing & Publication Policy for RAINFOR Participants', June 2009), and in TmFO's Memorandum of Understanding. Within the FOS, plot consortia are acting on behalf of the site principal investigators. Importantly, data providers are not asked to provide their primary (tree-by-tree) data. The data shared in the FOS are stand-level descriptors, including aboveground biomass estimates, that are obtained from a standardized procedure.

One of the most frequent complaints voiced by site principal investigators is that the data they are providing serve projects downstream of which they are not aware. This is to a large extent a communication problem, and one that can be solved through constant interaction with site principal investigators through a mailing list. Allied to this, sometimes data is used by scientists who make implausible scientific interpretations, without knowing the site, its history and other details. Some interaction with data providers often improves papers and quickens the pace of scientific discovery.

4.2 Plot Data Requirements

Minimal data requirements are here discussed. These data should be produced by the partners and provided to the Forest Observation System database.

A minimum set of site descriptors are included in the metadata. These include: (a) the name and contact (email) of the plot principal investigator(s); they should agree to be mentioned in the database (for privacy protection, this information is made available online in the password-protected part of the database); (b) the name of the partner institution(s) and individual in charge of data management; (c) the names of the funding bodies; and (d) some characteristic photographs of the forest.

The following plot information is important: (a) plot coordinates, which should be checked for the geodetic system and be provided in World Geodetic Survey WGS84; GPS coordinates should be of high accuracy, typically to within 10 m (but ideally with surveying GPS to within cm), so as to facilitate co-registration with other datasets (ALS, TLS, and EO); plot coordinates should ideally refer to the centre and the four corners of the plot; (b) collection date and periodicity; number and date of censuses carried out should also be known; the census number for which AGB data are provided should be given; (c) the total sampled area, i.e. the horizontal projection of the on-ground sampled area (i.e. topography effects are ignored), and plot geometry; most plots are squares or rectangles; (d) the dataset should also document the relief (slope, exposition); in situations where aerial LiDAR is available, this usually provides accurate measurements of ground relief; (e) forest type (i.e. wet, moist, dry forests) and successional status should be documented. Note that networks have already faced the issue of post-field data standardization/filtering. However, it is not established that they all have settled to a common practice in all instances.

We also report on metadata for the tree inventory itself: (a) the number of trees $\geq$ 10 cm in trunk diameter; note that trees < 10 cm and other life forms are usually excluded in AGB estimates in case their contribution to AGB is < 5%; (b) a quality assessment index, reporting on whether points of measurement have been properly recorded for each tree; (c) an index reporting on the quality of taxonomic identification will also be needed; as a rough measure, the proportion of trees identified to species level, genus level, and family level is reported. In tropical forests, identification of < 50% of the trees to species level is far from unusual. Careful botanical identification by botany experts typically results in identification rates of > 90% of the trees, but may entail climbing trees to collect and significant downstream identification effort with botanists and herbaria; (d) plot-averaged wood density is the basal-area-weighted wood density of the trees in a plot. For plots with reliable taxonomic identifications, this may be deduced from census data and species-average wood density values; (e) mean canopy height of the plot, as inferred from direct tree height measurements or from airborne LiDAR measurements; if necessary, several canopy height metrics should be provided; quality-control metric: height of the largest measured tree, trunk diameter of the largest measured tree.

Finally, aboveground biomass and confidence intervals are computed and provided at the plot scale, following an agreed single methodology across partners; the methodology will be made accessible for each database release, and data users should be prepared to changes in the methodology. An efficient strategy is to jointly develop a statistical routine such that several database formats can be accommodated and that perform the tasks of calculating biomass and canopy height at each site. The R statistical software is recommended because it is free, already widely used in the ecological research community, and networks such as ForestGEO or ForestPlots already have developed R routines for parsing the datasets and performing quality checks. We have established a package called BIOMASS that calculates biomass values and propagates uncertainty from tree measurement to stand-level estimates (Réjou-Méchain et al. 2017). This package is flexible and makes it possible to use user-supplied conditions or allometric equations.

4.3 Candidate Supersites and Their Coverage of Environmental Gradients

There are around 50–100 supersites already potentially available worldwide, and here, we discuss a list of 78 sites included as priority sites by the ESA-NASA cross-mission working group. All sites share a number of basic features including a long-term presence of scientists, existing forest monitoring programs, and willingness to collaborate in international scientific projects on the part of the principal investigators.

Taken together, these sites encompass much of the variability in forest types, and within each 1000-ha region of interest, these sites display a large spectrum in biomass ranges and disturbance histories. Figure 3 illustrates the location of sites that could be prioritized as supersites. A majority of the supersites are located in the tropics, reflecting the more pressing need for data in tropical forest environments. However, several sites were also selected outside of the tropical belt.

We also illustrate the coverage of these sites in terms of biomes and bioclimatic conditions (Fig. 4). The 78 sites currently being considered for the network of supersites span broad bioclimatic conditions, and although they are mostly located below 1000 m in

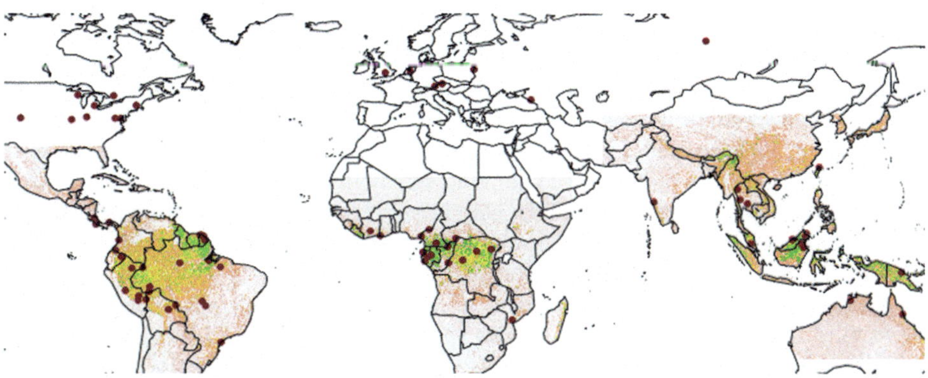

Fig. 3 Potential location of 78 candidate supersites. Proposed sites were selected to maximize geographical coverage, environmental and forest structure conditions, and logistical constraints of maintaining long-term sites. The background is Avitabile et al. (2016) carbon stock map

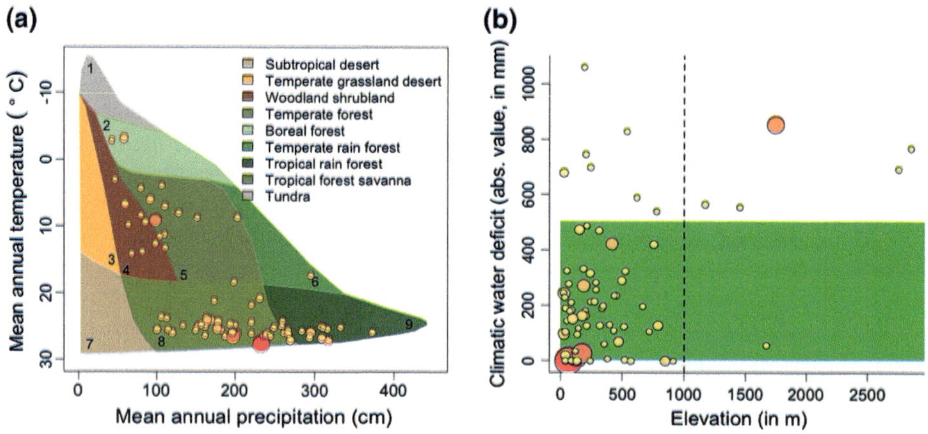

Fig. 4 **a** Environmental coverage of the 78 candidate supersites in bioclimatic space (Whittaker diagram). **b** Distribution of the candidate supersites across the range of elevation (in m above sea level); drought stress, as measured by the climate water deficit: larger values represent more stressed environments (sites above 500 are usually ascribed to xeric habitats). In both panels, disturbance intensity is displayed with large red dots representing highly disturbed sites, while small yellow dots represent undisturbed habitats. Disturbance intensity was measured by the proportion of forest pixels lost between 2000 and 2017 in a buffer of 5 km radius around the supersites, using the global 30-m resolution Landsat (Hansen et al. 2013)

elevation, a few sites ($n=6$) are above this limit. As seen in Fig. 4, the current list of supersites does not include many dry forests, semi-deciduous tropical forests, or boreal forests. Also, warm temperate forests are currently under-sampled in this dataset. Finally, a large proportion of the sites are currently located in areas with less than 1% disturbance from 2000 to 2017 (28 out of 78) but some are in highly disturbed landscapes. One example is the STREK site in Indonesia (TmFO), in which over 60% of the surrounding landscape has been deforested since 2000, another example being the Pasoh plot (ForestGEO) with over 40% of deforestation since 2000.

5 Building a Ground-Based Earth Observation Mission

There has been a shift over the past few decades toward freely available Earth Observation data, and NASA and ESA have adopted open data policies with the aim of accelerating science and applications (Turner et al., 2015). Earth Observation data are typically delivered free at the point of use, using technology that has cost space agencies and their funding governments hundreds of millions of dollars or euros to develop and launch. Once the sensors are installed in orbit, they continue to supply data at, relatively, limited recurrent cost. Because nations have provided the core investment, they can rightly insist that Earth Observation data are provided for free to the entire scientific community (although conditions of use may vary among space agencies).

In situ ground-based datasets stand in stark contrast with this situation, because the more reliable data are obtained by human specialists, who are paid for gathering them, verifying them, and maintaining databases over decades. In addition to data collection costs, data curation and coordination is also costly, and these costs do not come down with time. Most of the ground forest stand data available as of 2018, and summarized above, were collected and processed through long-term collaborations and with funds mobilized, typically, by individual scientists within the scientific community of many countries, rather than national efforts, and for a multitude of purposes. Few were collected with the express purpose of calibrating or validating remote imagery. It should not be assumed that the level of funding provided to these science projects will persist with the same intensity from 2019 to 2029. The majority of the principal investigators reside in countries with limited support from national science funding, hence relying on international collaboration to sustain their activities. It is reasonable to suppose that if the substantial future ground effort proposed in this paper is to be effectively used to support remote sensing missions, then it needs to be funded to do so.

Costs of collecting ground data are difficult to summarise given vastly differing national circumstances of operating, for example, in Costa Rica compared to the Democratic Republic of Congo. It was estimated by FOS partners that the full cost of recensusing a single 1-ha plot in high-diversity tropical forests is on the order of 15 k€ (2016 economic conditions). This reflects the entire cost from concept to delivery of the highest possible quality data with accurate tree dimensions and identification. For instance, to fully recensus 600 1-ha plots across all four tropical continents included in the ForestPlots database, the full cost would be 9 M€ per remeasurement cycle. A similar figure is to be expected for the ForestGEO network. These costs are indicative, but result from decades of experience in establishing and maintaining tropical forest plots across the world. If satellite estimates of biomass are to be of high quality and serve the widest use, these costs should be factored into the calibration and validation strategies of EO missions.

A coordinated and global ground-based monitoring of forests would benefit several sectors of science and the society and would be of direct use to biomass-related spaceborne missions. It would allow the collection and maintenance of ground-based databases for the lifetime of the currently planned missions and potentially for longer periods. In addition, this 'ground mission' would help consolidate the remote sensing/ecology nexus, helping to bridge the gap between these two scientific communities and increase the value of both ground and remotely sensed data. Finally, the study sites could become valuable beyond the currently planned biomass missions. For instance, several missions are committed to measuring photosynthetic activity through solar-induced fluorescence or aim to monitor biodiversity using hyperspectral imagery. In these situations, it is also essential to validate the concept of these missions at a set of reference sites that can be appropriately accessed, equipped, and maintained.

Acknowledgements We thank the organizers of the ISSI ESA Meeting held in Bern, Switzerland, in November 2017 for stimulating discussions and for their invitation to submit this paper. We gratefully acknowledge funding by 'Investissement d'Avenir' programs managed by Agence Nationale de la Recherche (CEBA, ref. ANR-10-LABX-25-01; TULIP, ref. ANR-10-LABX-41), from CNES and from ESA (IFBN Project 4000114425/15/NL/FF/gp, as part of the BIOMASS mission program).

References

Anderson-Teixeira KJ, Davies SJ, Bennett AC, Gonzalez-Akre EB, Muller-Landau HC, Wright SJ et al (2015) CTFS-Forest GEO: a worldwide network monitoring forests in an era of global change. Glob Change Biol 21:528–549

Ashton PS (1964) Ecological studies in the mixed dipterocarp forests of Brunei State. Clarendon Press, Oxford

Asner GP, Powell GV, Mascaro J, Knapp DE, Clark JK, Jacobson J et al (2010) High-resolution forest carbon stocks and emissions in the Amazon. Proc Natl Acad Sci 107:16738–16742

Avitabile V, Herold M, Heuvelink GB, Lewis SL, Phillips OL, Asner GP et al (2016) An integrated pan-tropical biomass map using multiple reference datasets. Glob Change Biol 22:1406–1420
Baccini A, Goetz SJ, Walker WS, Laporte NT, Sun M, Sulla-Menashe D, Hackler J et al (2012) Estimated carbon dioxide emissions from tropical deforestation improved by carbon-density maps. Nat Clim Change 2:182–185
Brede B, Lau A, Bartholomeus HM, Kooistra L (2017) Comparing RIEGL RiCOPTER UAV LiDAR derived canopy height and DBH with terrestrial LiDAR. Sensors 17:2371
Brown S, Lugo AE (1982) The storage and production of organic matter in tropical forests and their role in the global carbon cycle. Biotropica 14(3):161–187
Calders K, Origo N, Burt A, Disney MI, Nightingale J, Raumonen P, Åkerblom M, Malhi Y, Lewis P (2018) Realistic forest stand reconstruction from terrestrial LiDAR for radiative transfer modelling. Remote Sens 10:933
Castilho CV, Magnusson WE, de Araújo RN, Luizao RC, Luizao FJ, Lima AP, Higuchi N (2006) Variation in aboveground tree live biomass in a central Amazonian Forest: effects of soil and topography. For Ecol Manag 234:85–96
Chapman B, Kasischke ES (2018) Evaluation of above study region sites for future calibration and validation of Nisar Science requirements. In IGARSS 2018-2018 IEEE International Geoscience and Remote Sensing Symposium, pp 8279–8281
Chave J, Réjou-Méchain M, Búrquez A, Chidumayo E, Colgan MS, Delitti WB et al (2014) Improved allometric models to estimate the aboveground biomass of tropical trees. Glob Change Biol 20:3177–3190
Chazdon RL, Broadbent EN, Rozendaal DM, Bongers F, Zambrano AMA, Aide TM et al (2016) Carbon sequestration potential of second-growth forest regeneration in the Latin American tropics. Sci Adv 2(5):e1501639
Clark DB, Clark DA (2000) Landscape-scale variation in forest structure and biomass in a tropical rain forest. For Ecol Manag 137:185–198
Clark DB, Kellner JR (2012) Tropical forest biomass estimation and the fallacy of misplaced concreteness. J Veg Sci 23:1191–1196
Condit R (1998) Tropical forest census plots: methods and results from Barro Colorado Island, Panama and a comparison with other plots. Springer, Berlin
Disney MI, Boni Vicari M, Calders K, Burt A, Lewis S, Raumonen P, Wilkes P (2018) Weighing trees with lasers: advances, challenges and opportunities. R Soc Interface Focus 8(2):20170048
Drake JB, Dubayah RO, Clark DB, Knox RG, Blair JB, Hofton MA et al (2002) Estimation of tropical forest structural characteristics using large-footprint lidar. Remote Sens Environ 79:305–319
Dubois-Fernandez PC, Le Toan T, Daniel S, Oriot H, Chave J, Blanc L et al (2012) The TropiSAR airborne campaign in French Guiana: objectives, description, and observed temporal behavior of the backscatter signal. IEEE Trans Geosci Remote Sens 50:3228–3241
Fridman J, Holm S, Nilsson M, Nilsson P, Ringvall AH, Ståhl G (2014) Adapting National Forest Inventories to changing requirements—the case of the Swedish National Forest Inventory at the turn of the 20th century. Silva Fennica 48(3):29
Frolking S, Palace MW, Clark DB, Chambers JQ, Shugart HH, Hurtt GC (2009) Forest disturbance and recovery: a general review in the context of spaceborne remote sensing of impacts on aboveground biomass and canopy structure. J Geophys Res Biogeosci 114(G2):222
Gentry AH (1988) Changes in plant community diversity and floristic composition on environmental and geographical gradients. Ann Mo Bot Gard 75(1):1–34
Golley FB (1993) A history of the ecosystem concept in ecology: more than the sum of the parts. Yale University Press, New Haven
Grassi G, House J, Dentener F, Federici S, den Elzen M, Penman J (2017) The key role of forests in meeting climate targets requires science for credible mitigation. Nat Clim Chang 7:220
Hansen MC, Potapov PV, Moore R, Hancher M, Turubanova SAA, Tyukavina A et al (2013) High-resolution global maps of 21st-century forest cover change. Science 342:850–853
IPCC (2018) Special report on global warming of 1.5 °C. https://www.ipcc.ch/sr15/
Johnson MO, Galbraith D, Gloor M, De Deurwaerder H, Guimberteau M, Rammig A et al (2016) Variation in stem mortality rates determines patterns of above-ground biomass in Amazonian forests: implications for dynamic global vegetation models. Glob Change Biol 22:3996–4013
Keeling HC, Phillips OL (2007) The global relationship between forest productivity and biomass. Glob Ecol Biogeogr 16:618–631
Labrière N, Tao S, Chave J, Scipal K, Le Toan T, Abernethy K et al (2018) In Situ reference datasets from the TropiSAR and AfriSAR campaigns in support of upcoming spaceborne biomass missions. IEEE J Sel Top Appl Earth Obs Remote Sens 99:1–11

Lutz JA, Furniss TJ, Johnson DJ, Davies SJ, Allen D, Alonso A et al (2018) Global importance of large-diameter trees. Glob Ecol Biogeogr 27(7):849–864
Le Quéré C, Andrew RM, Friedlingstein P, Sitch S, Pongratz J, Manning AC et al (2018) Global carbon budget 2017. Earth Syst Sci Data 10:405–448
Le Toan T, Quegan S, Davidson MWJ, Balzter H, Paillou P, Papathanassiou K et al (2011) The BIOMASS mission: mapping global forest biomass to better understand the terrestrial carbon cycle. Remote Sens Environ 115:2850–2860
Lewis SL, Sonké B, Sunderland T, Begne SK, Lopez-Gonzalez G, Van Der Heijden GM et al (2013) Above-ground biomass and structure of 260 African tropical forests. Philos Trans R Soc B 368:20120295
Malhi Y, Wood D, Baker TR, Wright J, Phillips OL, Cochrane T, Meir P et al (2004) The regional variation of aboveground live biomass in old-growth Amazonian forests. Glob Change Biol 12:1107–1138
Martin AR, Doraisami M, Thomas SC (2018) Global patterns in wood carbon concentration across the world's trees and forests. Nat Geosci 11(12):915
Mitchard ET, Feldpausch TR, Brienen RJ, Lopez-Gonzalez G, Monteagudo A, Baker TR et al (2014) Markedly divergent estimates of Amazon forest carbon density from ground plots and satellites. Glob Ecol Biogeogr 23:935–946
Mokany K, Raison RJ, Prokushkin AS (2006) Critical analysis of root:shoot ratios in terrestrial biomes. Glob Change Biol 12:84–96
NASA-ESA-Smithsonian Workshop on Calibration and Validation of Upcoming Satellite Missions on Forest Structure and Biomass, Washington DC, 2016. (https://nisar.jpl.nasa.gov/files/nisar/NISAR_Vegetation_Biomass_Workshop_Report.pdf)
Nogueira EM, Nelson BW, Fearnside PM (2006) Volume and biomass of trees in central Amazonia: influence of irregularly shaped and hollow trunks. For Ecol Manag 227:14–21
Pan Y, Birdsey RA, Fang J, Houghton R, Kauppi PE, Kurz WA et al (2011) A large and persistent carbon sink in the world's forests. Science 1(2):1–3. https://doi.org/10.1126/science.1201609
Paul KI, Roxburgh SH, Chave J, England JR, Zerihun A, Specht A et al (2016) Testing the generality of above-ground biomass allometry across plant functional types at the continent scale. Glob Change Biol 22:2106–2124
Paul KI, Larmour J, Specht A, Zerihun A, Ritson P, Roxburgh SH et al (2019) Testing the generality of below-ground biomass allometry across plant functional types. For Ecol Manag 432:102–114
Phillips J, Duque A, Scott C, Wayson C, Galindo G, Cabrera E et al (2016) Live aboveground carbon stocks in natural forests of Colombia. For Ecol Manag 374:119–128
Réjou-Méchain M, Muller-Landau HC, Detto M, Thomas SC, Toan TL, Saatchi SS et al (2014) Local spatial structure of forest biomass and its consequences for remote sensing of carbon stocks. Biogeosciences 11:5711
Réjou-Méchain M, Tanguy A, Piponiot C, Chave J, Hérault B (2017) Biomass: an r package for estimating above-ground biomass and its uncertainty in tropical forests. Methods Ecol Evol 8:1163–1167
Saatchi SS, Harris NL, Brown S, Lefsky M, Mitchard ET, Salas W, Zutta BR et al (2011a) Benchmark map of forest carbon stocks in tropical regions across three continents. Proc Natl Acad Sci USA 108:9899–9904
Saatchi S, Marlier M, Chazdon RL, Clark DB, Russell AE (2011b) Impact of spatial variability of tropical forest structure on radar estimation of aboveground biomass. Remote Sens Environ 115:2836–2849
Schimel D, Pavlick R, Fisher JB, Asner GP, Saatchi S, Townsend P et al (2015) Observing terrestrial ecosystems and the carbon cycle from space. Glob Change Biol 21:1762–1776
Shugart HH, Saatchi S, Hall FG (2010) Importance of structure and its measurement in quantifying function of forest ecosystems. J Geophys Res Biogeosci. https://doi.org/10.1029/2009JG000993
Sillett SC, Van Pelt R, Carroll AL, Campbell-Spickler J, Coonen EJ, Iberle B (2019) Allometric equations for Sequoia sempervirens in forests of different ages. For Ecol Manag 433:349–363
Sist P, Rutishauser E, Peña-Claros M, Shenkin A, Hérault B, Blanc L et al (2015) The Tropical managed Forests Observatory: a research network addressing the future of tropical logged forests. Appl Veg Sci 18:171–174
Smith WB (2002) Forest inventory and analysis: a national inventory and monitoring program. Environ Pollut 116:S233–S242
Stegen JC, Swenson NG, Enquist BJ, White EP, Phillips OL, Jørgensen PM, Weiser MD, Mendoza AM, Vargas PN (2011) Variation in above-ground forest biomass across broad climatic gradients. Glob Ecol Biogeogr 20:744–754

Sullivan MJ, Talbot J, Lewis SL, Phillips OL, Qie L, Begne SK et al (2017) Diversity and carbon storage across the tropical forest biome. Scientific Rep 7:39102

Turner W, Rondinini C, Pettorelli N, Mora B, Leidner AK, Szantoi Z et al (2015) Free and open-access satellite data are key to biodiversity conservation. Biol Cons 182:173–176

Xu L, Saatchi SS, Shapiro A, Meyer V, Ferraz A, Yang Y et al (2017) Spatial distribution of carbon stored in forests of the Democratic Republic of Congo. Sci Rep 7:15030

Affiliations

Jérôme Chave[1] · Stuart J. Davies[2] · Oliver L. Phillips[3] · Simon L. Lewis[3,8] · Plinio Sist[4] · Dmitry Schepaschenko[5] · John Armston[6] · Tim R. Baker[3] · David Coomes[7] · Mathias Disney[8,9] · Laura Duncanson[6] · Bruno Hérault[4,10] · Nicolas Labrière[1] · Victoria Meyer[11] · Maxime Réjou-Méchain[12] · Klaus Scipal[13] · Sassan Saatchi[11]

1 Université Toulouse 3 Paul Sabatier, CNRS, ENFA, UMR 5174 Evolution et Diversité Biologique (EDB), 31062 Toulouse, France

2 Center for Tropical Forest Science-Forest Global Earth Observatory, Smithsonian Tropical Research Institute, Washington, DC 20560, USA

3 School of Geography, University of Leeds, Leeds LS2 9JT, UK

4 Cirad, Univ Montpellier, UR Forests & Societies, 34398 Montpellier Cedex 5, France

5 International Institute for Applied Systems Analysis, Schlossplatz 1, 2361 Laxenburg, Austria

6 Department of Geographical Sciences, University of Maryland, College Park, MD 20742, USA

7 Department of Plant Sciences, Forest Ecology and Conservation group, University of Cambridge, Cambridge CB2 3EA, UK

8 Department of Geography, University College London, London WC1E 6BT, UK

9 UKNERC National Centre for Earth Observation (NCEO), Swindon SN2 1EU, UK

10 Institut National Polytechnique Félix Houphouët-Boigny, INP-HB, Yamoussoukro, Ivory Coast

11 Jet Propulsion Laboratory, California Institute of Technology, Pasadena, CA 91109, USA

12 AMAP, IRD, CNRS, CIRAD, INRA, Univ Montpellier, 34090 Montpellier, France

13 ESA-ESTEC, 2201 AZ Noordwijk, The Netherlands

Surveys in Geophysics (2019) 40:881–911
https://doi.org/10.1007/s10712-019-09532-0

Upscaling Forest Biomass from Field to Satellite Measurements: Sources of Errors and Ways to Reduce Them

Maxime Réjou-Méchain[1] · Nicolas Barbier[1] · Pierre Couteron[1] · Pierre Ploton[1] · Grégoire Vincent[1] · Martin Herold[2] · Stéphane Mermoz[3,8] · Sassan Saatchi[4] · Jérôme Chave[5] · Florian de Boissieu[6] · Jean-Baptiste Féret[6] · Stéphane Momo Takoudjou[1,7] · Raphaël Pélissier[1]

Received: 16 October 2018 / Accepted: 12 April 2019 / Published online: 30 May 2019

Abstract
Forest biomass monitoring is at the core of the research agenda due to the critical importance of forest dynamics in the carbon cycle. However, forest biomass is never directly measured; thus, upscaling it from trees to stand or larger scales (e.g., countries, regions) relies on a series of statistical models that may propagate large errors. Here, we review the main steps usually adopted in forest aboveground biomass mapping, highlighting the major challenges and perspectives. We show that there is room for improvement along the scaling-up chain from field data collection to satellite-based large-scale mapping, which should lead to the adoption of effective practices to better control the propagation of errors. We specifically illustrate how the increasing use of emerging technologies to collect massive amounts of high-quality data may significantly improve the accuracy of forest carbon maps. Furthermore, we discuss how sources of spatially structured biases that directly propagate into remote sensing models need to be better identified and accounted for when extrapolating forest carbon estimates, e.g., through a stratification design. We finally discuss the increasing realism of 3D simulated stands, which, through radiative transfer modelling, may contribute to a better understanding of remote sensing signals and open avenues for the direct calibration of large-scale products, thereby circumventing several current difficulties.

Keywords Biomass · Calibration · Carbon · Error propagation · Field data · Modelling

1 Introduction

The large uncertainty associated with the global spatio-temporal dynamics of forest carbon (C) is a major obstacle to the projection of future atmospheric CO_2 concentrations and the implementation of mitigation strategies. Satellite observations of land-use changes are currently reliable enough to provide robust information on deforestation dynamics (Hansen

✉ Maxime Réjou-Méchain
maxime.rejou@ird.fr

Extended author information available on the last page of the article

et al. 2013). However, the associated C release remains uncertain because broad-scale C mapping applications from satellite data still convey large uncertainty (Mitchard et al. 2013; Huang et al. 2015; Rodriguez-Veiga et al. 2017). Even higher uncertainty is associated with C fluxes related to forest degradation and regrowth because small changes in the C stocks in closed-canopy forests are challenging to detect remotely (Bustamante et al. 2016), resulting in an underestimation of forest degradation globally (Pearson et al. 2017).

Forest aboveground biomass (AGB), the main proxy for forest C stock, is rarely directly measured, either in the field or via remote sensing (RS) (Clark and Kellner 2012). AGB estimations are thus derived from statistical models with their own underlying assumptions that can generate random and/or systematic errors when violated. The way the errors that are associated with these models propagate up to the final AGB estimate is, however, generally poorly understood and accounted for. A simple illustration of how RS–AGB models may produce inaccurate AGB density maps is shown in Fig. 1. First, the propagation of a uniform bias over calibration data (e.g., field plot AGB estimates) conveys a systematic over- or underestimation, which may be straightforward to correct in the final density map provided that the bias can be quantified (Fig. 1b). A more overlooked but common effect is the propagation of a non-uniform bias, leading the RS model to overestimate small AGB values and underestimate large AGB values, resulting in density maps flattened around the mean (Fig. 1c; Avitabile et al. 2016; Xu et al. 2016). Such bias may originate from three sources: (1) a non-uniform bias exists in the calibration dataset, such as an AGB over- or underestimation in young versus old successional forests, which can be corrected if the non-uniform pattern of errors is well identified a priori; (2) the low sensitivity of RS signals to high AGB values generally results in an AGB underestimation (i.e. signal saturation), while in low AGB open areas, non-forest land surfaces (i.e. surface roughness, non-woody vegetation) also contribute to the remotely sensed signal, leading to an overestimation of forest AGB; (3) in regression models, violation of the assumption of low/non-existent error in the independent variable (the "observed" AGB) results in a systematic underestimation of the model slope (Fuller 1987; Réjou-Méchain et al. 2014). Finally, the major source of

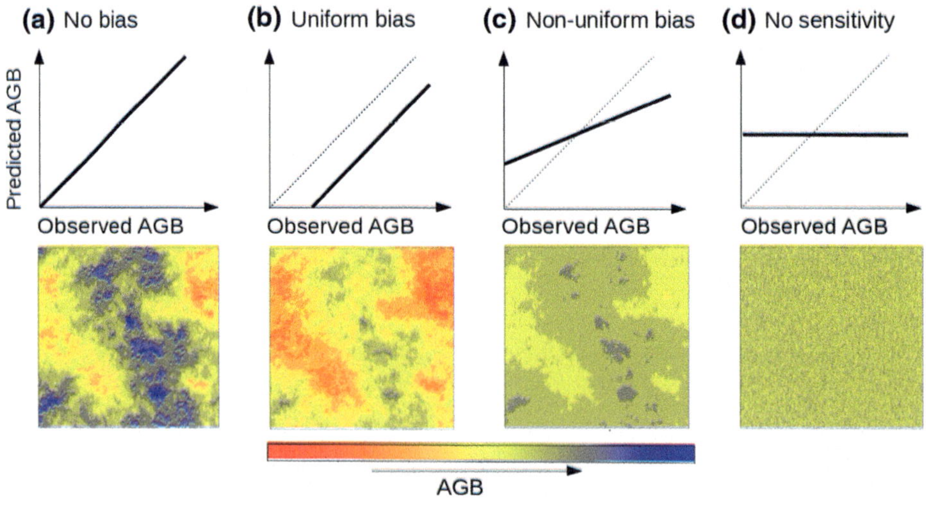

Fig. 1 Conceptual representation of common bias patterns found in remote sensing AGB models

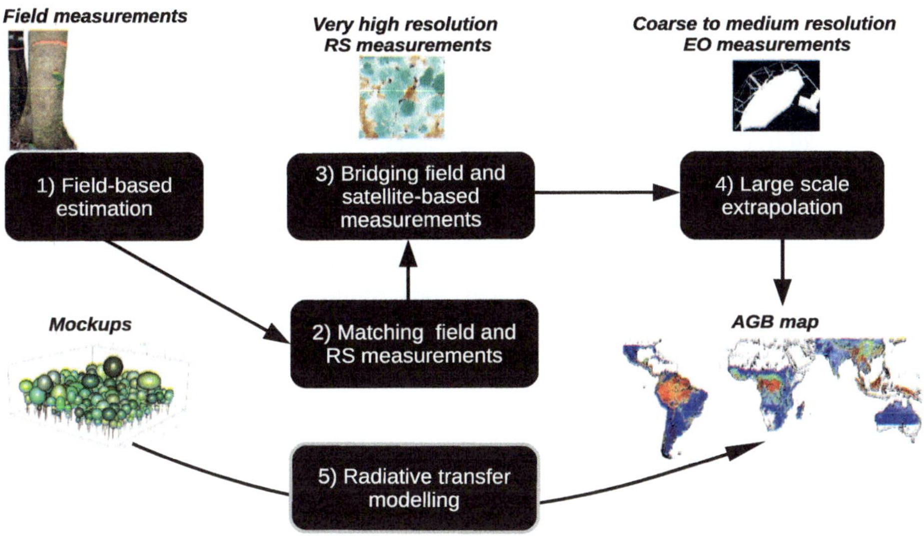

Fig. 2 Flowchart illustrating the main sections of this paper

uncertainty in RS-based models is obviously the poor signal sensitivity to AGB variations, which makes the post-correction of density maps impossible (Fig. 1d).

In this paper, we review the sources of errors that may occur during the process of upscaling AGB estimates from ground to RS data and provide suggestions for reducing these errors (Fig. 2). Section 2 illustrates that field-based estimates—often improperly seen as "ground truth"—most likely convey more important errors than generally assumed. Section 3 discusses the issues associated with mismatching field-based AGB estimates to RS data. Section 4 presents a multi-step calibration strategy aimed at bridging field and satellite measurements through airborne or spaceborne very high-resolution RS products. This section also highlights the pitfalls and promises of such a strategy. Section 5 discusses the common problems associated with broad-scale AGB extrapolations for regional mapping through inexpensive satellite imagery. Finally, Sect. 6 reviews physical modelling approaches and discusses how they may improve our ability to map AGB. These sections mostly focus on tropical forest examples, where uncertainty in C dynamics is the highest globally (Mitchard 2018), although most of the issues that are discussed are also concerns in other forest biomes.

2 Improving Field-Based AGB Estimation

Uncertainty in field-based AGB estimation has recently become a matter of concern for the RS community (e.g., Mermoz et al. 2014, 2015; Longo et al. 2016; Xu et al. 2017; Bouvet et al. 2018; Jucker et al. 2018a). For instance, Chen et al. (2015) developed an analytical framework to track the sources of errors from field measurements in airborne LiDAR-AGB predictions. The authors estimated that field-based uncertainty contributed only 10% of the total pixel-level uncertainty at a 0.16-ha resolution (see also Longo et al. 2016). However, the statistical framework developed by these authors rested upon the assumption that tree-level AGB estimation errors are independent within and between field plots; thus, they average out in RS models. While this assumption is common (e.g., Chave et al. 2004; McRoberts and Westfall 2013; Longo et al. 2016), it is often violated, generating overlooked biases in AGB maps.

Field-based AGB estimates at the plot level are obtained by summing individual AGB values for all trees within the plot. At the tree level, AGB is generally estimated using an allometric model that combines key tree features, typically the stem diameter (*D*), total tree height (*H*) and wood density (WD; Chave et al. 2005, 2014). While *D* may suffer from important measurement errors (Clark 2002), significant efforts have been made towards protocol standardization (Phillips et al. 2009; Condit et al. 2014) to warrant negligible errors, at least in scientific plots (Chave et al. 2004; Molto et al. 2013). However, as shown below, the two other predictors, as well as the AGB model itself, may convey systematic error components within and between plots that may directly propagate into RS models.

2.1 Uncertainty in Wood Density

Individual tree AGB is directly proportional to WD, which varies by as much as tenfold among tree species (Chave et al. 2009). However, WD is rarely measured on individual trees in forest inventories, and in most cases, a mean specific value is assigned from independent databases to all trees of a given taxonomic category, often higher than the species level. This approximation is very crude beyond the genus level (Flores and Coomes 2011); however, it is common in the tropics because of inaccurate botanical identifications (Gomes et al. 2013) or the limited availability of WD data (Chave et al. 2009). For example, in the well-known pantropical forest inventory dataset generated by Al Gentry, which includes 43,000 trees (> 8000 taxa), more than one-third of the trees could not be assigned a WD value at the species or genus level from the Global Wood Density Database (Chave et al. 2009; Zanne et al. 2009). Given that most tropical tree taxa are spatially aggregated at various scales (Condit et al. 2000), uncertainty in WD estimates is expected to cause spatially structured errors, potentially generating biases in plot-level AGB estimates (Baker et al. 2004). One way to mitigate this error source is to rely on permanent scientific sites for model calibration. For instance, in the 50-ha Center for Tropical Forest Science (CTFS) permanent plot at Barro Colorado Island, Panama (Condit 1998), less than 2% of the trees have WD values assigned at a coarser taxonomic resolution than the genus level. In addition, repeated censuses also strongly minimize field measurement or encoding errors and reduce the number of missing trees (Lopez-Gonzalez et al. 2011), providing high-quality calibration data for RS applications (Chave et al. this issue).

The assignment of a WD value to an individual tree at the species level does not, however, ensure bias-free estimates. WD is known to vary widely between individuals of the same species and even within individual trees (Tarelkin et al. 2019; Swenson and Enquist 2008; Bastin et al. 2015a; Wassenberg et al. 2015). For example, species having at least 10 observations in the Global Wood Density Database ($n = 109$) exhibit a mean WD variation of 9% around the mean. Given that two-thirds of the taxa in that dataset are documented with a single measure, the effects of intra-specific variations in WD on AGB estimates are currently difficult to quantify. Intra-individual variations may also have a strong impact on field AGB estimations, e.g., when converting volume estimations (e.g., from terrestrial LiDAR, see below) into biomass. A recent study conducted in Cameroon showed a significant decrease in WD with height for most species; thus, the use of WD at the trunk base to convert volume into biomass led to an AGB overestimation of approximately 10% at

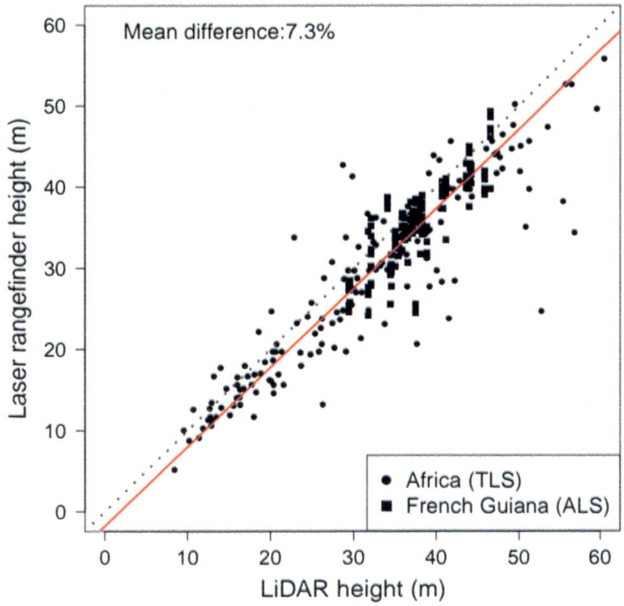

Fig. 3 Comparison between height measurements inferred from airborne or terrestrial LiDAR acquisitions and measured in the field using a laser hypsometer (TruPulse 360R for the 151 trees from Cameroon and Haglöf Laser Vertex for the 306 French Guiana trees). The relationship suggests that tree height measurements from laser rangefinders underestimate the tree height by 7.3%. The red line illustrates the output of a standard major axis regression (i.e. minimizing errors in both X and Y), and the dotted line represents the 1:1 line

the plot level (Sagang et al. 2018). The authors, however, demonstrated that vertical variations could be predicted from basal WD; thus, these variations could be controlled for in biomass estimations, which was recently confirmed in a large dataset from 6 different countries in central Africa (Momo Takoujdou et al. in preparation).

A lesson learned from this short review is that more efforts are needed to collect WD data for tropical taxa using standardized protocols (Williamson and Wiemann 2010) while accounting for all major sources of variation. Given the time and cost of wood core analyses, non-destructive and rapid WD measurement techniques are appealing, such as torsiometers, Pilodyns, nail withdrawal tools or emerging electronic devices to measure drilling resistance (reviewed in Gao et al. 2017). For instance, the use of an empirical model developed from Pilodyn measurements led to a predicted WD error of 15% for 1427 trees from four continents.

2.2 Uncertainty in Tree Height

As for WD, tree height is not systematically available from field plot inventories, and *H–D* models are often used to estimate the height (*H*) of individual trees from stem diameter (*D*) measurements. These models are built from *H* measurements that are subjected to errors, especially in dense forest canopies. Indeed, in addition to a strong operator effect, Larjavaara and Muller-Landau (2013) showed that the two most common *H* measurement methods (the so-called tangent method, which combines horizontal distance to the trunk and the angle to the treetop, and the "sine" method, which combines the angle and distance measurements to the treetop) led to significantly contrasting results, with an underestimation of ca. 20% with the "sine" method. Based on an original dataset of *H* measurements of 457 trees using both a laser hypsometer in the field and LiDAR systems (airborne or terrestrial), we confirm that the "sine" method results in a systematic bias of −7% in the *H* estimations (Fig. 3). The way this bias propagates into AGB estimates through direct measurements or *H–D* models depends on whether a similar bias occurred in the destructive reference dataset used to build the AGB allometric model.

Beyond measurement errors, *H–D* allometric models represent a major source of uncertainty in field-based AGB estimates because this allometry exhibits large spatially structured variations at local, regional and continental scales (Ketterings et al. 2001; Feldpausch et al. 2011; Vieilledent et al. 2012; Vincent et al. 2012b; Réjou-Méchain et al. 2015). To account for broad-scale variations, both regionally averaged *H–D* relationships (Feldpausch et al. 2012) and bioclimatic proxies (Eq. 6 in Chave et al. 2014) have been proposed. These approaches significantly reduce AGB model errors in tropical regions, but certain local deviations may remain quite large. For instance, in the central Congo Basin, forest AGB was overestimated by 24% when the Feldpausch et al. (2012) regional *H–D* relationship was used instead of a local relationship (Kearsley et al. 2013). In the Khao Yai Forest Reserve in Thailand, we found that the bioclimatic proxy proposed by Chave et al. (2014) underestimated *H* by 26% (Jha, Chanthorn, Réjou-Méchain, unpublished). Finally, both approaches resulted in an AGB underestimation $>20\%$ in a Brazilian forest (Hunter et al. 2013). Considering such uncertainty in only *H* estimations, any RS–AGB model calibrated in these regions would result in a systematic bias $> \pm 20\%$ in AGB estimations (case B in Fig. 1), i.e. an error margin larger than the commonly admitted accuracy.

H–D relationships may also strongly vary between plots within landscapes. Average differences in *H* of approximately 10% for a given *D* were found between plots within areas < 10 km^2 in French Guiana (Vincent et al. 2012a, 2014; Réjou-Méchain et al. 2015). More specifically, the *H–D* relationship was found to strongly vary along a successional gradient in Thailand (Chanthorn et al. 2017); hence, this relationship varied concomitantly with forest AGB, potentially leading to a non-uniform bias in RS models (case of Fig. 1c). Various strategies may be adopted to better account for these inter-plot variations. The most straightforward strategy is to acquire *H* data for at least a subsample of trees to calibrate plot-specific *H–D* allometries (Vieilledent et al. 2012; Hunter et al. 2013), even though care must be taken with respect to the *H* measurement method and sampling strategy (see above and Sullivan et al. 2018). Another way to better account for systematic variations in *H–D* allometry between plots would consist of finding local covariates. Forest structural metrics that reflect stand competition intensity or soil fertility were found to explain some variation, but to a relatively small extent (Feldpausch et al. 2011; Banin et al. 2012). In contrast, mean canopy height has been found to correlate well with tree slenderness (a higher *H* for a given *D*), both within species and at the community level (Vincent et al. 2012a). Hence, when a high-resolution canopy height map is available, the accuracy of individual *H* estimations from *D* measurements can be significantly improved by adding local canopy height metrics to the *H–D* model (e.g., median of the canopy height model in a 50-m window centred on each tree). Using this approach, the individual *H* error was reduced by 14%, leading to a threefold decrease in the plot-level AGB bias at the Paracou site, French Guiana ($n = 2141$ trees; Vincent et al. unpublished). This result adequately illustrates that if field-based estimates generally feed into RS models, they may also themselves be improved by RS measurements. With the recent advances in our ability to extract individual trees from dense LiDAR point clouds (Ferraz et al. 2016), *H* measurements may be directly assigned to individual canopy trees that account for a large share of the stand AGB (Bastin et al. 2015b). In conclusion, better integration of LiDAR technologies and field measurements has the potential to substantially reduce the uncertainties associated with tree height variations.

2.3 Uncertainty in AGB Allometric Equations

The allometric conversion of individual tree attributes into AGB is a major source of uncertainty in tree and plot AGB estimates (Chave et al. 2004; Molto et al. 2013). At the tree level, the predictive error associated with the pantropical allometric model of Chave et al. (2014) reaches almost 40% (the total error is slightly above 50%), but if considered random, this error rapidly averages out as the number of trees increases in the predicted population (Réjou-Méchain et al. 2017). A cross-validation assessment of this model indeed revealed a mean bias across sites of only 5% but exceeded 30% in 8 out of 58 sites (Chave et al. 2014). Models that assume universal allometry may thus be liable to strong and uncontrolled local bias, which is generally ignored in error propagation schemes. This site effect may reflect differences in species composition or biotic (e.g., competition intensity) or abiotic (e.g., soil fertility) conditions. For instance, the use of biome-specific rather than species-specific allometries resulted in an AGB prediction bias between − 18% and + 14% for a Brazilian mangrove forest (de Souza Pereira et al. 2018).

Crown allometry variations among sites have also been identified as a major source of uncertainty in pantropical allometric models (Goodman et al. 2014). The proportion of the crown to the total tree AGB is indeed highly variable among trees (3–88%) and tends to increase with the increase in tree AGB beyond 10 Mg, reaching 50% on average for trees > 45 Mg and 34% for trees < 10 Mg (Ploton et al. 2016). This systematic change in tree shape with total tree AGB explained 20% of the underestimation of the AGB of large trees reported by Chave et al. (2014). Using simulations, Ploton et al. (2016) showed that, for 1-ha plots, this bias led to systematic errors ranging from − 26% to + 16% that depended on the frequency of large trees, thus introducing a non-uniform bias in AGB maps (Fig. 1c). Integrating crown dimensions in allometric models strongly reduces this allometric uncertainty. Jucker et al. (2017b) even proposed an AGB allometry using crown size and tree height, i.e. tree dimensions potentially directly measurable by means of RS (e.g., Ferraz et al. 2016).

Another potential source of uncertainty in the allometric conversion of tree attributes into AGB is the existence of hollow parts in the trunks or branches. Hollow parts are implicitly accounted for in AGB estimates when allometric equations are built on directly weighed trees. However, direct weighing is rarely possible for large trees, which are also more prone to hollows (Fig. 4). As a consequence, volume estimation is usually preferred for large trees (Henry et al. 2010; Fayolle et al. 2013; Chave et al. 2014), which results in potential estimation bias (Moundounga Mavouroulou et al. 2014). For instance, Nogueira et al. (2006) found that hollows occurred in ca. 10% of the trees > 5 cm in diameter in Brazil, but hollows occurred in up to 50% of trees > 80 cm in diameter (but $n = 4$ in that study). This condition, however, resulted in a bias ≤ 1% in the AGB in a 1-ha plot, which is consistent with the results of Clark and Clark (2000) in Costa Rica (hollow parts accounted for only 1.7% of the outer volume). Conversely, other studies reported much higher errors in other sites (Rodrigues and Valle 1964 cited in Nogueira et al. 2006; Dickinson and Tanner 1978), suggesting that the frequency of hollow trees may vary between sites and may also be a source of uncertainty in AGB maps. For instance, in over 523 forested sites in south-eastern Australia, the number of hollow trees strongly varied between sites with a skewed distribution, i.e. many study sites contained few or no hollow trees, but some sites contained up to 13 hollow trees per ha (Lindenmayer et al. 1991). Interestingly, the occurrence of hollow trees was well predicted by topography, stand age, region, logging history and the dominant species. Similarly, in a subtropical forest in China, the density of hollow

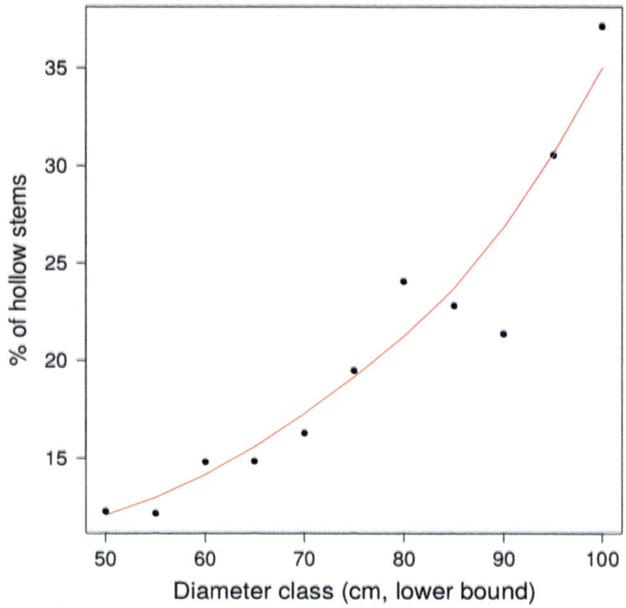

Fig. 4 Increase in hollow occurrence with tree diameter in a French Guiana dataset. In total, 17% of a sample of 3746 stems with DBH > 50 cm in French Guiana were found to be hollow. This proportion appeared to be strongly positively related to tree size. Courtesy of L. Descroix, unpublished

trees was approximately 90 trees per ha, which is much higher than that in temperate forests, and hollow occurrence varied significantly with species, crown position and topographical context (Liu et al. 2018). However, more studies are needed to better quantify the impact of hollows or simply wood decay on stand AGB estimates, using, for instance, non-destructive wood imaging techniques (Arciniegas et al. 2014). Marra et al. (2018) used sonic and electrical resistance tomography to identify internal wood decay without apparent cavities that resulted in C loss in the trunk ranging from 0.1 to 24% (5–37% with an apparent cavity).

2.4 Perspective on Field-Based AGB Estimation

The use of terrestrial laser scanners (TLS) has recently emerged as a credible alternative to destructive approaches for the estimation of tree AGB (Calders et al. 2014; Disney 2018; Gonzalez de Tanago et al. 2018; Lau et al. 2018; Momo Takoudjou et al. 2018). This technology and the associated processing methods are rapidly progressing but still face a number of challenges in tropical moist forests, such as the high degree of occlusion, the difficulty in segmenting individual trees in intricate vegetation (Trochta et al. 2017; Calders et al. 2018) and in filtering out leaves (Béland et al. 2014; Ma et al. 2016; Calders et al. 2018), which is a required step prior to tree structure reconstruction from the point cloud (Momo Takoudjou et al. 2018). Overcoming these issues would take us one step closer to an automated routine for reconstructing entire forest stands, thereby producing large amounts of data to feed allometric models (Raumonen et al. 2013). However, the above-mentioned problems associated with poor WD estimates and the presence of hollow trees remain a serious difficulty for TLS to overcome, which inherently measures the outer volume.

Another perspective from the advent of LiDAR technology is the possibility of directly estimating stand-level AGB instead of summing the AGBs of individual trees. For instance, Vincent et al. (2014) analytically derived a stand-level AGB equation based on stand volume-weighted WD, tree density, mean quadratic diameter and mean canopy height, which, as discussed above, controls the local *H–D* allometry. Stand density and mean quadratic diameter can both be extracted from TLS data (e.g., Bauwens et al. 2016) more accurately and more easily than individual tree volumes, while a canopy height model within a scanned plot can be derived from the same TLS data. Once such a relationship is calibrated, airborne laser scanners (ALS, including those on unmanned aerial vehicles (UAVs)) suffice to derive AGB estimates from standard stem diameter measurements and WD estimates. However, the calibration step is challenging because it would ideally rely on destructive stand-level measurements. Such an approach would probably limit the propagation of individual errors, especially when neither tree height measurements nor a reliable local *H–D* allometry is available.

3 Matching Field and RS Measurements

When an RS signal is related to a field-based estimate of forest AGB, the basic assumption is that they both measure the same area and objects. This assumption is often violated in practice, leading to large uncertainty in the final AGB density maps (Gobakken and Naesset 2009; Frazer et al. 2011; Mascaro et al. 2011; Réjou-Méchain et al. 2014; Saatchi et al. 2015). Here, we review important sources of mismatch between field and RS measurements and identify ways to overcome them.

3.1 Geolocation Uncertainty

Geolocation mismatch between RS and field measurements is an obvious source of uncertainty (Frazer et al. 2011). Ground measurements are generally geolocated using a global navigation satellite system (GNSS) receiver, whose accuracy is known to vary with receiver quality or topographical and vegetation conditions by up to 2 orders of magnitude (Johnson and Barton 2004). In particular, GNSS accuracy decreases exponentially with the increase in canopy cover (Sigrist et al. 1999). From simultaneous acquisitions, Johnson and Barton (2004) found that 20% of the measurements had geolocation errors > 10 m under forest cover (due to multipathing effects), but only 2% in a nearby open area. Under unfavourable satellite conditions, the error even exceeded 200 m under forest cover. Under a dense forest cover in Gabon, a recent high-grade GNSS resulted in a mean measurement error of 5 m with 2.5% of extremes greater than 70 m (Fig. 5; Réjou-Méchain and Barbier, unpublished). Differential GNSS corrections did not reduce this uncertainty because vegetation cover prevents the use of phase-shift information. However, geolocation errors rapidly averaged out as the number of measurement points increased in either space (over a few tens of metres) or time (over several hours or days), stabilizing below 5 m with 20 measurement points and below 3 m with 50 measurement points. This result confirms the recommendation of Segrist et al. (1999) to collect at least 20 GNSS measurement points at different locations or times to accurately geolocate a field plot in dense forest conditions.

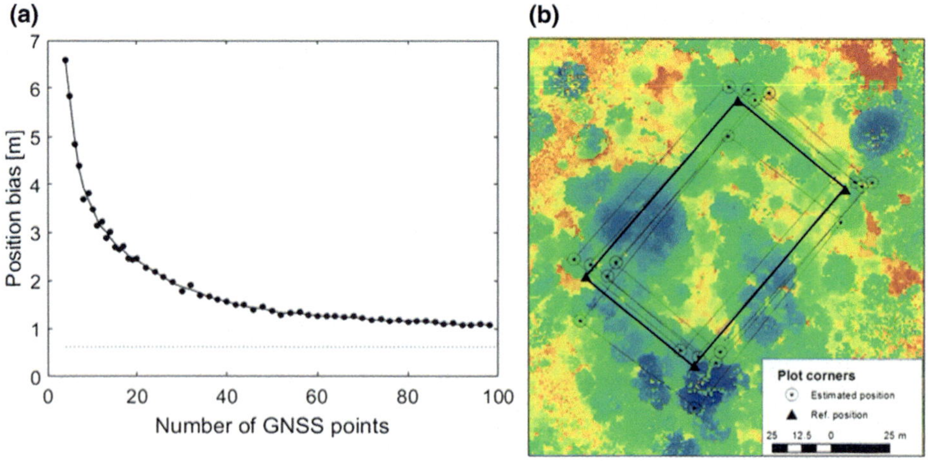

Fig. 5 Plot positioning error in a dense forest from Gabon. **a** The 75th percentile of mean position bias as a function of the number of GNSS points taken along plot limits from random point sets taken among 207 GNSS points (using a Trimble L1/L2 GEOXT 7000 rover GNSS device). The dotted line represents the background error level (e.g., due to imperfections in the relative positions of field marks). **b** An example of plot position bias for a 60×100 m plot at the 75th percentile of the error distribution using random sets of 4 GNSS points to estimate the transformation from local to UTM coordinates. Reference positions were provided by a professional topography consultant using a total station. Background colours show maximum canopy height (from brown to blue) from terrestrial laser scanning data

3.2 Acquisition Angle

Satellite-borne instruments do not always measure forests vertically (i.e. in nadir mode). When an active penetrating signal (e.g., RADAR or LiDAR) is sent with a large (pluri-metric) footprint size and a significant incidence angle (typically > 30° for RADAR sensors; Saatchi et al. 2011a; Robinson et al. 2013), the forest volume measured in field plots and intercepted by the satellite sensor does not correspond to the same physical objects, especially for small plots or in topographically complex areas (Villard and Le Toan 2015). Using theoretical simulations, Réjou-Méchain and Barbier (unpublished) found that an incidence angle of 30° may lead to volume differences of ca. 55, 20 and 10% for plot and pixel sizes of 20, 50 and 100 m, respectively, even if existing surrounding trees partly compensate for this mismatch. Thus, any RS signal acquired with a significant incidence angle should not be calibrated and validated with small field plots or in small patches of heterogeneous vegetation.

3.3 Mismatch in Tree Representation

The mismatch between the RS and field measurements may also be due to differences in tree representation. Conventional field measurements for AGB estimations are "trunk-based", i.e. a tree is considered to belong to a given field plot when half its trunk base section is within the plot limits, irrespective of the proportion of its vertical crown projection falling outside the plot. With RS approaches, AGB is measured from an area-based perspective, i.e. only the plant material with a ground projection within the area of interest is considered. Using LiDAR data, Mascaro et al. (2011) showed that accounting for this edge effect reduces the LiDAR-AGB model error by 55, 21 and 4% at 20-, 50- and 100-m resolution, respectively, i.e. with a decrease in the perimeter-to-area ratio, the associated edge effects logically vanish because the proportion of trees with crowns crossing the plot edge decreases with the increase in plot size.

Differences in tree representation may also impact the estimates of forest C dynamics. Réjou-Méchain et al. (2015) compared the 4-year AGB dynamics inferred from repeated field plot censuses and LiDAR acquisitions and found relatively poor agreement between the two estimates; these authors concluded that different components of forest dynamics are measured by the two approaches. Natural canopy dynamics is dominated by many small-scale events that are captured by LiDAR data but not by field measurements (Kellner and Asner 2009; Leitold et al. 2018). Similarly, leaf density, which influences LiDAR estimates but not field estimates, regenerates faster than woody biomass after a disturbance (Asner et al. 2006). Thus, any dynamics inferred from RS measurements should be interpreted differently from the dynamics inferred from the ground.

3.4 Temporal Mismatch

The temporal difference between RS and field measurements is a common problem in RS studies. The impact of such temporal mismatches is difficult to predict in natural forests due to the high stochasticity of tree mortality. This stochasticity, however, decreases with the increase in plot size (Chambers et al. 2013); hence, the error associated with the temporal difference between measurements is likely to decrease with increasing plot size in the absence of major climate anomalies (e.g., El Niño events) or major disturbances, such as hurricanes or fires, which may be detected with Landsat (Kennedy et al. 2010) or MODIS (Justice et al. 2002) time series data. (MODIS is the Moderate Resolution Imaging Spectroradiometer aboard the NASA Terra and Aqua satellites.) In this case, estimated growth rates may be used to compensate for the time lag in AGB estimates from field and RS data (Avitabile and Camia 2018).

3.5 Scale Mismatch

The issue of scale mismatch between calibration field plots and RS data pixels is challenging, particularly when coarse-resolution RS products are calibrated with numerous small field plots such as national forest inventory data collected in sampling units less than 0.1 ha in size. For instance, the pantropical 1-ha resolution AGB map of Saatchi et al. (2011b) was calibrated using field plots that were typically ≤ 1 ha in size, hence representing a local sampling rate of 1% at best. For instance, Réjou-Méchain et al. (2014) showed that the use of calibration plots smaller than the RS pixels generates large sampling errors, producing significantly biased AGB maps (case Fig. 1c). For a given pixel-to-plot size ratio, the error due to local AGB variability is larger for small plots and pixels than for large plots and pixels (Réjou-Méchain et al. 2014). Thus, studies aiming at building coarse-resolution maps face the challenge of minimizing this sampling error. A common practice is to spatially aggregate field estimates to better match the RS pixel resolution; however, this approach results in large uncertainty in the associated sampling error (Avitabile and Camia 2018). A more reliable approach consists of using intermediate-resolution RS data, such as very high-resolution images, to bridge the gap.

4 Bridging Field and Satellite-Based Measurements

Field plots are generally too costly and time-consuming to establish to be densely and evenly distributed within landscapes. Consequently, RS models are often calibrated with field plots that are not representative of the area of interest (Marvin et al. 2014). Furthermore, field plot size rarely approaches the size of coarse-resolution RS pixels (e.g., MODIS, ca. 250 m), which are the current basis of the wall-to-wall biome or global mapping. This problem generates the mismatch issues discussed in the previous section. To overcome these problems, a multi-step upscaling approach involving RS data of very high spatial resolution (VHSR) can provide intermediate-scale biophysical maps to be then used as reference to calibrate wall-to-wall RS data of coarse resolution and broad swaths (Asner et al. 2013; Baccini and Asner 2013; Xu et al. 2017). In this section, we first review the VHSR RS products that may be used in such an approach and then discuss the caveats and perspectives associated with such a multi-step upscaling strategy.

4.1 Airborne LiDAR

Airborne LiDAR (ALS), with its ability to penetrate the canopy, provides a fine 3D description of the forest structure and represents an excellent option for linking field plots to broad-scale RS data. Over the last decade, a large number of studies have aimed at retrieving forest AGB from LiDAR-derived metrics, usually with good accuracy at the 1-ha scale (i.e. with a model prediction error typically below 15%; Zolkos et al. 2013). ALS allows for the computation of a variety of metrics related to the vertical or horizontal forest profile (Lefsky et al. 2002). Many studies have relied on statistical fittings (e.g., stepwise linear models) to identify the most predictive metrics in a given context, resulting in site-specific ALS–AGB model forms with limited transferability across sites (Vincent et al. 2012b; Zolkos et al. 2013). For this reason, some studies aimed at designing generic ALS–AGB models that were expected to perform consistently across sites. For instance, Bouvier et al. (2015) identified predefined metrics of strong complementarity and interpretability to build a generic model transferable across temperate forest types. In the tropics, another attempt used a single ALS metric, the mean top-of-canopy height (TCH), combined with minimal field data to predict forest AGB. Asner and Mascaro (2014) used data from 14 tropical areas to calibrate a model of the form $\mathrm{AGB} = a\mathrm{TCH}^{b1}\mathrm{BA}^{b2}\ \mathrm{WD}^{b3}$, where the basal area (BA) was locally predicted from TCH, and where WD relied on regional estimates. However, the fitted equation, which was presented as universal, led to underestimations of 7% (Jucker et al. 2017a) and 16% (Réjou-Méchain et al. 2015) in two independent sites compared with locally adjusted models. A modification of the model form was suggested by Vincent et al. (2014), where the scaling of AGB estimates from tree to plot used the stem number (N) and average stem cross-sectional area (or quadratic diameter) instead of BA to avoid unwarranted errors in the scaling process. A universal predictive equation that relates ALS metrics and AGB should continue to be sought to maximize the benefits from the increasing availability of ALS data (Labriere et al. 2018). However, as long as *H–D* variability is not accounted for, a single predictive equation would remain elusive. Furthermore, airborne data acquisition remains costly (ca. 200–500€ km^{-2}) and is subjected to flight authorization, which hampers ALS acquisitions in certain countries.

4.2 Unmanned Aerial Vehicle (UAV) Systems

UAVs may reduce the cost of airborne acquisitions and are thus increasingly utilized. UAV systems can generate 3D point clouds through the acquisition of very high-resolution passive imagery. The overlap between UAV acquisitions enables stereoscopic image processing with potentially very high spatial resolution and low cost compared to LiDAR systems (Puliti et al. 2015). Under the generic appellation of structure-from-motion approaches, the automation of traditional photogrammetry by the detection of invariant features in 2D images allows for the derivation of dense point clouds (Fig. 6). However, such systems mostly capture variation in the top of the canopy and have a rather low ability to penetrate dense forests, limiting the computation of forest height metrics (Roşca et al. 2018). UAV-borne LiDAR systems are rapidly developing but are still relatively new (Brede et al. 2017). With an optimized sampling strategy, these systems are able to generate a large number of points (thousands of pts m^{-2}), opening the door for individual tree volume reconstruction using toolboxes developed for TLS (Morsdorf et al. 2017), even if UAV-specific methods would be preferable due to an inverted vertical distribution of point density. However, flight authorization remains an issue in some countries, and even where authorized, maximum legal UAV piloting distances limit the acquisitions to rather small areas (e.g., ca. 300 ha in France).

4.3 VHSR Spaceborne Systems

Spaceborne data may be mobilized to complement ALS or UAV data and move beyond the relatively small extents of ALS and UAV campaigns, notably because these data can be acquired at a low cost and with few issues regarding authorizations. For instance, VHSR satellite optical images (≤ 2 m resolution) are able to characterize canopy texture, which informs the size distribution of canopy crowns and inter-crown gaps (Couteron et al. 2005; Frazer et al. 2005), thus indirectly informing stand structure and AGB. Texture has historically been widely used in forest science for visual interpretation of aerial photographs. Similar interpretations from automated processing can be carried out from VHSR images via canopy texture analysis whenever images have sufficient effective resolution (the maximum acceptable seems to be approximately 2 m; Proisy et al. 2007). For instance, the Fourier-based textural ordination (FOTO) method aims to ordinate canopy image windows along texture gradients based on Fourier spectra (Couteron et al. 2005), and this method successfully retrieved the AGB gradients in several case studies across the tropics with an accuracy only slightly lower than that of ALS approaches (relative prediction error $\leq 20\%$; Proisy et al. 2007; Ploton et al. 2012; Bastin et al. 2014; Singh et al. 2014; Pargal et al. 2017) but at 50–100 times lower cost (2–10€ km^{-2} for optical images). Hence, the trade-off between data interpretability versus affordability should suggest designing scaling-up strategies based on nested sampling of field, airborne and VHSR satellite data. Satellite-borne optical sensors offering VHSR may thus be part of an effective chain for mapping AGB at landscape-regional scales. However, there are also situations in which texture analyses fail to properly retrieve certain AGB gradients (Ploton et al. 2013; Blanchard et al. 2015) because the way in which texture features relate to AGB strongly varies across forest types (Ploton et al. 2017). Due to the limited number of field plots available for calibration and validation, biophysical drivers of canopy texture have not been fully identified in previous studies. The increasing availability of ALS data will help to better evaluate such texture-based approaches and ultimately allow for the calibration and validation of texture indices in sufficiently diverse forest types. However, canopy texture,

Fig. 6 Illustration of the structure-from-motion approach allowing for the derivation of an orthomosaic (lower left) and a surface model (lower right) by automated stereophotogrammetry from drone imagery (upper panel) in a forest-savanna mosaic from central Cameroon

such as reflectance, is strongly impacted by the Sun–scene–sensor geometry, although correction techniques have been proposed (Barbier et al. 2011; Barbier and Couteron 2015). The number of textural descriptors, therefore, needs to be kept to a minimum to control these effects in large collections of images acquired in diverse geometrical configurations.

Apart from texture indices, stereophotogrammetric approaches from VHSR satellite images allow for the retrieval of canopy surface models that may then be used to predict

AGB. This retrieval implies more cloud-free images and a higher budget for data acquisition and, contrary to LiDAR, stereophotogrammetric approaches do not provide information on fine-scale topographical variation, which is detrimental for building canopy height models. Thus, such approaches remain very scarce in the literature (St-Onge et al. 2008; Lagomasino et al. 2015). Finally, non-penetrating X-band RADAR satellite images used in stereo mode (Tandem-X) have also shown to have the potential to predict AGB gradients linked to logging and forest degradation and to assess AGB losses through diachronic comparisons (Schlund et al. 2015; Solberg et al. 2018).

4.4 Limits and Perspectives on Bridging Field and Satellites

Regardless of the VHSR product or model structure used for bridging field plots and large-scale RS data, there are several caveats. First, both the number of calibration field plots and their size remain important for restricting VHSR-based AGB prediction errors to acceptable limits. For instance, in a meta-analysis of LiDAR-based AGB estimations, Zolkos et al. (2013) showed an asymptotic decrease in the prediction error with the increase in the size of the calibration plots, with a slow decay on average above a plot size of 0.2 ha, corresponding to a prediction error of ca. 20%. However, highly contrasted results can be found across case studies, suggesting that prediction errors vary with forest structure at a given plot size. For instance, in tropical forests of Ghana, Chen et al. (2015) found a LiDAR-based prediction error on AGB of ca. 42% at a spatial resolution of 0.16 ha. Thus, AGB estimates derived from VHSR data and used for calibrating broad-scale products should be performed at a rather coarse resolution, with 1 ha being a conservative resolution to ensure an acceptable accuracy.

Another major problem that may arise during the plot to VHSR step is that significant biases may occur during the extrapolation of plot AGB estimates through the VHSR products. For instance, the prediction of AGB from local stand height or related RS metrics relies on knowledge of local WD and height–diameter allometry. As discussed in Sect. 2, both measures are known to vary, sometimes abruptly, within landscapes, and there is thus no guarantee against biases when extrapolating AGB estimates over areas where these variations are unknown. Stratifying the landscape by forest types prior to AGB model calibration thus appears essential to guarantee model robustness, although there are not many accounts of thorough pre-stratification processes in the literature (Gregoire et al. 2016). However, sources of complementary data are developing. Notably, hyperspectral sensors now appear to able be to provide information on canopy species composition and help infer WD (Jucker et al. 2018b). Moreover, some LiDAR metrics (e.g., canopy gap fractions) can help assess the disturbance or successional status of the forest with relevance for the explicit estimation of WD (Guitet et al. 2018).

5 Global Wall-to-Wall Extrapolation

Earth system models and international policy initiatives both require wall-to-wall data (Sitch et al. 2008; Romijn et al. 2018; Herold et al. 2019). Thus, there is a strong need for going beyond local to regional AGB estimates to produce unified AGB density maps at the biome or global scale. Pantropical, temperate, boreal and global AGB maps have been proposed in the last decade, most often by extrapolating field-based and/or LiDAR-based AGB estimations through the use of global optical, RADAR and environmental datasets. In this section, we discuss the issues associated with large-scale AGB mapping and the promise of new methods and upcoming spaceborne missions.

5.1 The Problem of Interpolation

None of the RS systems available to date that display fair to good correlation with AGB (with no signal saturation) offers wall-to-wall coverage at very broad scales (i.e. regions, countries, continents). At best, these systems provide a fairly systematic sampling of the earth surface under the form of belt transects (e.g., ICESat/GLAS or the upcoming GEDI and MOLI missions). This condition implies that an interpolation step is necessary to produce a continuous AGB map from this discrete sampling. To that end, all published approaches have used or combined statistical interpolation (e.g., kriging or co-kriging with environmental drivers) and predictions from high- to medium-resolution spaceborne RS data to produce wall-to-wall coverage (e.g., Landsat, MODIS, QuickScat, ALOS PALSAR). For instance, Saatchi et al. (2011b) and Baccini et al. (2012) mapped AGB over the tropics by spatially interpolating discrete AGB estimates using Quick Scatterometer and/or MODIS data. Avitabile et al. (2016) then combined these two maps and used additional reference data (field plots and locally calibrated maps) in an attempt to generate an improved AGB map. Finally, the map from Santoro et al. (2018) was derived at 100-m resolution using a combination of maps based on ALOS PALSAR, ASAR data, Landsat data and ICESat GLAS transects, without any calibration from in situ data.

The use of these RS data for estimating AGB at a large scale raises two problems, which remain largely unsolved. First, most current satellite-based data relate to signals for which relationships to AGB are saturating and/or highly context-dependent (Steininger 2000), due to either varying relationships between AGB and RS signals or signal artefacts. Optical images are affected by geometrical and atmospheric effects that may produce significant spatial and/or seasonal artefacts in surface reflectance (Morton et al. 2014). For instance, the high degree of cloudiness in western Gabon and Cameroon is known to strongly impact MODIS reflectance data and is likely the reason why Baccini et al. (2012) underestimated the AGB values in this region (Fig. 7). Similarly, RADAR data are affected by factors related to forest structure and the environment (e.g., soil and vegetation moisture, topography) and the uncertainties associated with RADAR data acquisition (Villard and Le Toan 2015). Second, in natural tropical forests, stand structure variables often strongly vary over short distances. Variograms computed on field plot networks reported spatial correlation in AGB or BA that barely reached 5 km (Guitet et al. 2015; Hajj et al. 2017). This result means that locations that are more than 4–5 km apart are virtually independent; thus, any reliable AGB estimate conveys no useful information for interpolation beyond this distance.

Thus, it is not surprising that strong inconsistencies have been reported among broad-scale AGB maps (Mitchard et al. 2013; Huang et al. 2015; Rodriguez-Veiga et al. 2017). For example, Mitchard et al. (2013) compared two widely used pantropical maps (Saatchi et al. 2011b, Baccini et al. 2012) and found fairly consistent estimates at the continental scale but large regional differences (> 100%) within continents. In Amazonia, a comparison with field estimates suggested that both pantropical maps failed to capture the AGB gradient across regions with some regions over- or underestimated by > 25% (Mitchard et al. 2014). However, without a systematic field sampling of AGB at a large scale, as done

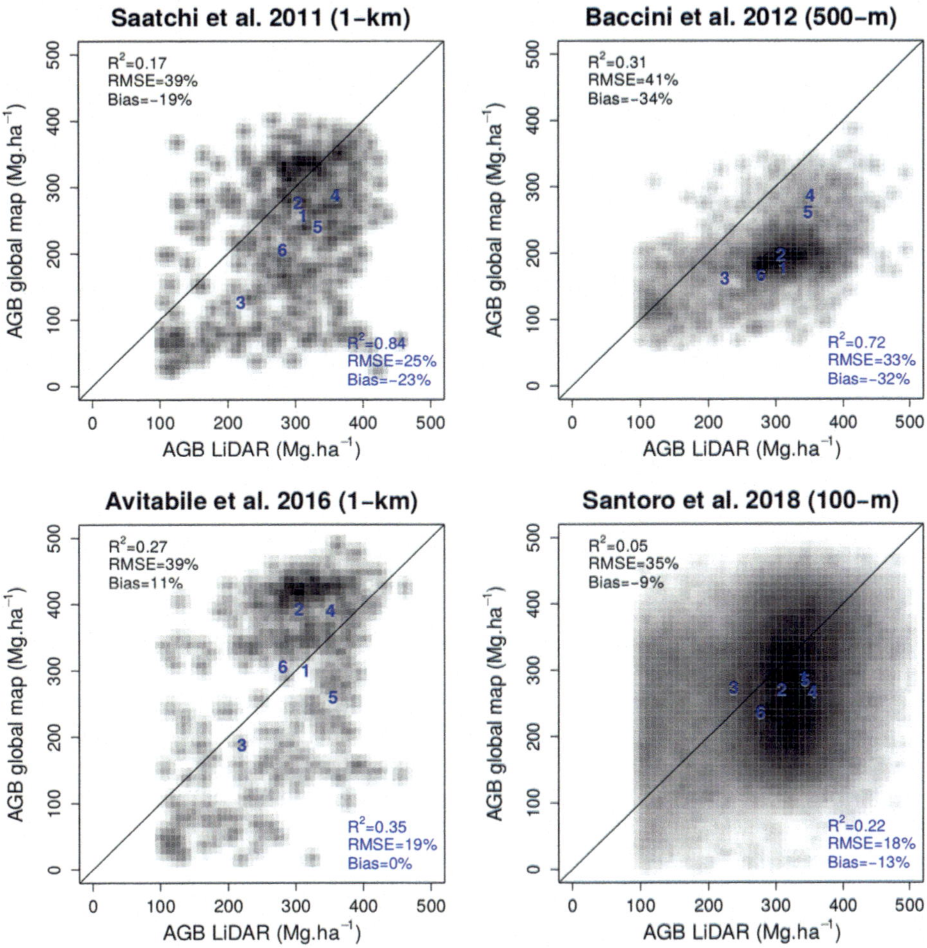

Fig. 7 Validation of global and pantropical maps using independent AGB estimates from six airborne LiDAR (ALS) campaigns in French Guiana and Gabon in forests > 100 Mg ha^{-1}. The results are reported at both the pixel (black) and mean AGB site (blue) level (1: Lope; 2: Mabounie; 3: Mondah; 4: Nouragues; 5: Paracou; 6: Rabi). ALS–AGB density maps at 50 m resolution were first resampled to the resolution of the pantropical and global maps using the Geospatial Data Abstraction Library (GDAL) *gdalwarp* routine in average mode. ALS- and global AGB density maps were then superimposed using bicubic interpolation (Inglada and Vadon 2005) through the *otbcli_Superimpose* routine of the Orfeo ToolBox (https://www.orfeo-toolbox.org/). Spatial correlations using fast Fourier transforms in moving windows confirmed that the spatial match between both AGB estimates was high (uncertainty was below the pixel resolution)

in national forest inventories (Kleinn 2017), it is difficult to validate whether a map captures the regional AGB gradients.

In Fig. 7, we report the result of a comparison of five existing pantropical/global AGB maps with six LiDAR-based AGB maps from Africa and French Guiana (Labriere et al. 2018). The results confirm that in dense forests (> 100 Mg ha^{-1}), broad-scale maps display relatively weak ($R^2 \leq 0.3$) if any correlation at the pixel level with independent AGB data of improved quality. However, some maps exhibit much higher correlations than others considering the very coarse-resolution gradients (here the mean AGB per LiDAR site), with the R^2 value for the map of Saatchi et al. (2011b) exceeding 0.8. Although this result is based on six observations and should be confirmed with more sites, it could suggest that broad-scale AGB variations may be successfully captured with such maps. These comparisons shed some light on the uncertainty associated with these AGB map estimates, but they also convey their own errors for validation or accuracy assessments for at least two reasons. First, the maps are all produced for different time periods, e.g., the Saatchi et al. (2011b) map is for ca. 2000, the Baccini et al. (2012) map is for ca. 2008 and Avitabile et al. (2016) combined these two maps using ground- and LiDAR-estimated AGB at different spatial scales and time periods. Second, the definition of forest and AGB may be different for each map. In Saatchi et al. (2011b), the biomass estimate at 1 km referred to the biomass of remaining forest in each 1 km grid cell, whereas in all other maps, the estimates referred to the mean biomass at the grid cell of the map (average of forest and non-forest biomass).

5.2 The Use of Environmental Proxies

One possible way to mitigate the aforementioned difficulties that hinder interpolation is to rely on spatialized environmental proxies to enhance the mapping. In the central Amazon, Saatchi et al. (2007) used a multivariate decision tree approach featuring both environmental variables (e.g., elevation, climate) and saturating RS metrics (e.g., Qscat_SV, NDVI) and found that AGB classes referring to values above 250 Mg ha^{-1} were mainly located in areas below 190 m elevation and of limited ruggedness (both assessed from SRTM). AGB is indeed known to covary with topography globally (Réjou-Méchain et al. 2014), but the strength and direction of the relationship strongly vary across sites (de Castilho et al. 2006; McEwan et al. 2011; Detto et al. 2013; Réjou-Méchain et al. 2015). Some studies successfully relied on spatialized climatic variables (Simard et al. 2011; Fayad et al. 2016) or geomorphological units (Guitet et al. 2015) to predict AGB or forest height in particular regional contexts. Thus, the ongoing development and improvement of global databases of environmental variables open avenues for geographically enlarging such approaches bearing in mind that relationships between AGB and the environment are intrinsically context-dependent. Indeed, AGB is largely driven by complex edaphic processes, such as soil structure (Gourlet-Fleury et al. 2011; Jucker et al. 2018b) or water table and bedrock depth (Emilio et al. 2013), which often vary at a relatively fine scale and for which continuous information is lacking. Furthermore, past and present anthropogenic activities, which are not randomly distributed according to abiotic conditions, may largely blur such environmental determinism at large scales. Therefore, calibration and prediction should be designed according to a relevant pre-stratification of the territory, accounting for anthropogenic activities, with sufficient reference data in each stratum.

5.3 Perspectives on Extrapolation

Upcoming RADAR satellite missions, such as NISAR in 2021 (Rosen et al. 2017) and BIOMASS in 2022 (Le Toan et al. 2011), should be pivotal in global-scale AGB mapping by providing wall-to-wall active measurements. While NISAR will be relevant for low AGB areas due to the use of an L-band saturating signal, BIOMASS will operate with a 50-m resolution P-band signal that is sensitive to large AGB values. (AGB products will, however, be delivered at 200-m resolution.) Preliminary analyses from P-band airborne missions revealed that polarimetric intensities (PolSAR) and polarimetric RADAR interferometry (PolInSAR) correlated well with AGB values up to 300 Mg ha^{-1} in temperate and tropical areas (ESA 2012). Above 300 Mg ha^{-1}, such approaches failed to detect AGB variations, but tomographic approaches may significantly improve the retrieval of AGB up to 500 Mg ha^{-1} with no obvious saturation and a prediction error of only ca. 10% at a 1.5-ha resolution (Minh et al. 2014). A cross-validation performed between two sites of tropical forests in French Guiana suggested that tomographic models are transferable with an error within 20% at 1-ha resolution (Minh et al. 2016). However, these approaches rely on multiple acquisitions with different geometries over the same site, constituting a challenge for a spaceborne sensor and limiting the areas that may be covered in such a way. As a consequence, tomography will be implemented in only the first year of the mission; the three subsequent years will instead correspond to an interferometry phase. Beyond AGB mapping, the upcoming missions, including GEDI and MOLI, also constitute a unique opportunity to better understand the relationship between forest AGB, productivity and spatialized environmental proxies in tropical dense forests.

Finally, if future missions will considerably improve our ability to monitor AGB in space and time, the full potential of current RS products has not yet been completely assessed. For instance, using multi-angle MODIS observations to characterize the anisotropy of forests, de Moura et al. (2016) were able to generate metrics that displayed a good correlation ($r^2 > 0.5$) with ALS-derived metrics in high AGB forests in the Brazilian Amazon, with no apparent saturation. This result adequately illustrates that there is also room for progress through the development of new methodologies to be used with the current RS products.

6 Radiative Transfer Modelling for Understanding Error Sources

A prerequisite to controlling the sources of errors in satellite-based assessments of forest AGB is to understand how the electromagnetic signal of interest interacts with forest components. Contrary to sea surface temperature, for instance, there is not yet any sensor that directly measures woody AGB per se. We thus use the variations in proximal measurements or indices to infer the AGB variations; however, these measures may be prone to interferences and artefacts that sometimes blur the true response of the forest AGB. Thus, radiative transfer (RT) modelling has become a standard approach in RS to analyse sensor sensitivity, and inverse modelling is used to measure the extent at which specific forest information is recoverable from the recorded signals. Some recent prominent examples can be found in studies that investigated the seasonal variation in leaf area in tropical forests (Morton et al. 2014; Wu et al. 2018) or texture retrieval of forest AGB from VHSR optical images (Ploton et al. 2017). Even crude modelling approaches may prove useful to untangle complex interactions in the sensor–atmosphere–scene–light source system (Barbier et al. 2011; Barbier and Couteron 2015), but a key point in such simulation studies is the way in which the complex 3D organization of tropical forest components can be represented. We will focus here on examples involving passive and active optical signals, but similar progress can be observed in RADAR studies (Villard 2009), which were historically pioneering (Ishimaru 1978).

6.1 An Increase in Model Realism

In the early days, geometric-optic models were used to understand the variation in directional reflectance within large pixels comprising a number of objects (e.g., trees) (Egbert 1977; Li and Strahler 1992; Roujean et al. 1992; Ni et al. 1999). Scene reflectance was computed as the sum of nominal reflectance of the components (e.g., shaded and lit surfaces) of a projection of simple, generally opaque, geometrical shapes within a scene. These efforts allowed for the derivation of analytical functions (kernels) that proved useful to fit actual directional reflectance variations with a small number of parameters (e.g., Lyapustin et al. 2011).

As computing power increased, it became possible to increase scene complexity, as well as to improve the level of detail in the descriptions of light–matter interactions. Discrete ray-tracing models based on Monte Carlo simulations of photon behaviour and (multiple) scattering within the scene have now become standard. This phenomenon is, for instance, evidenced by the periodic RAMI (RAdiation transfer Model Intercomparison) experiment, which aims at benchmarking and testing a number of existing models (Widlowski et al. 2013). Initially, scenes represented homogeneous parallel layers (Verhoef 1985; Myneni et al. 1989) of a turbid medium, in which the macroscopic transfer equation was statistically deduced from the microscopic distribution of elements, such as leaves or aerosols, within the cell. These elements were characterized by a density and angle distribution, as well as optical properties, which imply a probability of intercepting light rays depending on the incoming and outgoing directions. In addition to turbid elements, solid surfaces can be added in the form of triangle meshes to represent scene elements other than leaves (woody parts, roads, ground, building or water bodies). Triangles can be translucid and display Lambertian, specular or other optical properties. Progressively, spatial heterogeneity was introduced in the turbid medium distribution (Myneni 1991; Gastellu-Etchegorry et al. 1996; North 1996). Indeed, as opposed to most annual crops, crown shading is fundamental to understanding scene reflectance in forests. This fundamental need is even more pronounced when the spatial resolution of a sensor increases to a point where the pixels are smaller than the tree crowns, as biophysical descriptors can then be found in the texture features, and describing the spatial arrangement of reflectance values is crucial (Couteron et al. 2005; Barbier et al. 2011; Barbier and Couteron 2015). The degree of realism of a 3D description has evolved from sets of floating spheres or ellipsoids to more complex envelopes describing tree crown contours (Cescatti 1997; Widlowski et al. 2013). The dimensions and spatial distributions of these envelopes can generally be inferred from standard forestry measurements or allometric equations (Barbier et al. 2011).

To further improve the realism of the forest scenes, LiDAR data constitute a giant leap forward, as they provide both access to a detailed description of the geometry and topology of solid surfaces (trunks and branches) and a statistical sampling of foliage directly compatible with the turbid representation via voxelization procedures accounting for spatial sampling variations (Grau et al. 2017; Tymen et al. 2017; Vincent et al. 2017). In a temperate montane forest, a voxel-based approach performed better at simulating hyperspectral data than an approximation of individual tree envelopes (Schneider et al. 2014). Given the importance of LiDAR technology in detailed measurements of vegetation from space, air or ground stations, simulating the transfer of this active signal in vegetation is now itself an active field of study (Sun and Ranson 2000; Gastellu-Etchegorry et al. 2015). Figure 8 illustrates the current possibilities and challenges for the realistic simulation of dense tropical forests. On the one hand, the integration of the structure derived from airborne LiDAR

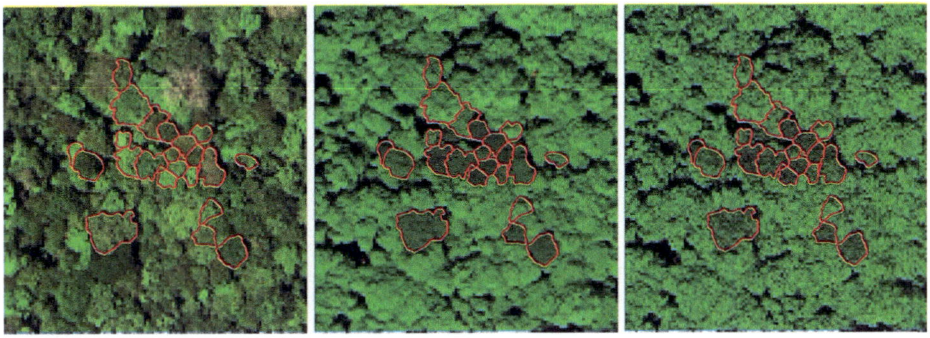

Fig. 8 Simulation of airborne optical imaging in a tropical forest from French Guiana (Paracou). The simulation is based on the integration of airborne LiDAR and field spectroscopy in the DART 3D radiative transfer model (Gastellu-Etchegorry et al. 2015). The 3D mock-ups were computed from airborne LiDAR point clouds using AMAPvox (Vincent et al. 2017), and leaf optical properties corresponding to sampled trees (delineated in red) were measured with a field spectroradiometer and assigned to leaf elements from all voxels on the vertical column of the mock-up. A generic set of leaf optical properties was applied to all trees with undocumented leaf optical properties using the PROSPECT leaf model (Féret et al. 2017). Left: original image (red = 640 nm; green = 549 nm; blue = 458 nm); centre: simulation using a turbid representation for leaf elements; right: simulation using triangle approach for leaf elements

acquisitions allows accurate simulation of canopy grain and gaps with metric resolution. On the other hand, the lack of information about the proportion of leaf versus woody elements (Malenovskỳ et al. 2008) and the difficulty in accurately characterizing the diversity of leaf optical properties at the canopy scale contribute to the relatively low radiometric agreement. Better accounting for the differences in radiometric properties among species or individuals and possibly within individuals would require a better understanding of the spatial heterogeneity of spectral properties (Féret and Asner 2014; Rocchini et al. 2018).

6.2 Limits and Perspectives on RT Modelling

The current difficulties that need to be overcome to effectively derive forest mock-ups from TLS data are the same as those mentioned above for deriving AGB estimates from TLS. For instance, individual tree segmentation is required to allocate species-specific radiometric properties to individual trees. In addition, accurate segmentation of leaves from wood and proper estimation of leaf angle distribution are required to accurately simulate light–vegetation interactions (Antin et al. 2015). These steps become more difficult in the upper canopy as the density of the TLS point cloud decreases dramatically due to occlusion and increasing distance from the laser source. This difficulty may, however, be minimized through the fusion of TLS and UAV data. Finally, a better understanding of leaf demography and its integration into advanced modelling of leaf optical properties, combined with the improved description of non-photosynthetic tissues in LiDAR-derived 3D mockups, will increase the realism of simulations derived from physical modelling.

An alternative to producing a forest scene with a turbid description of leaves is to use an architectural model to simulate the tree branching structure and individual leaf distribution (De Reffye et al. 1995; Dauzat et al. 2001; Côté et al. 2011; Widlowski et al. 2013; Calders et al. 2018). Independently of the biological realism of such detailed scene descriptions, these models provide a means to assess the degradation of the transfer modelling outcome

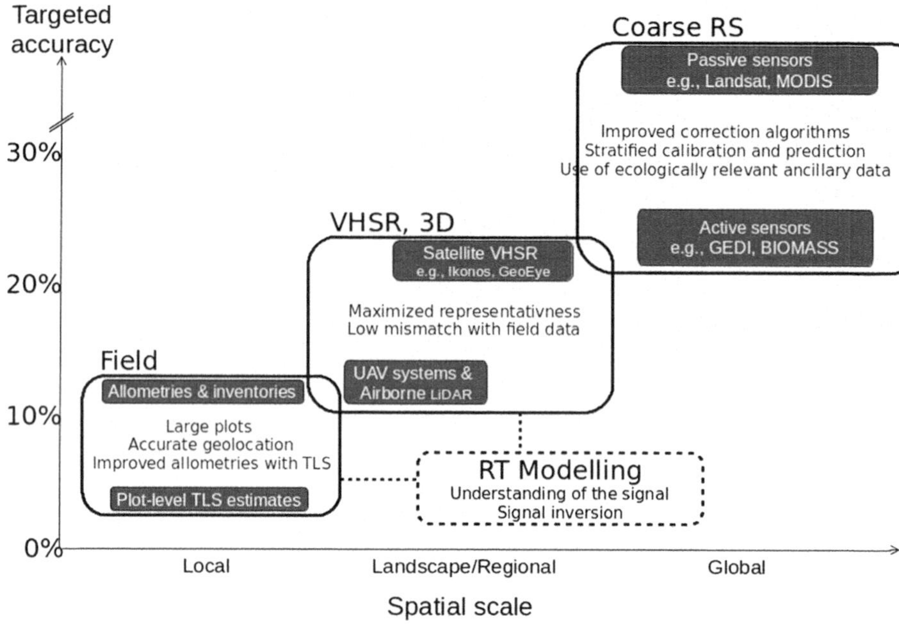

Fig. 9 Best-practice guidelines to upscale AGB estimates from local to global scales. The targeted accuracy (*Y* axis) refers to a 1-ha scale

as the scene is progressively simplified in terms of representation (Jonckheere et al. 2006; Widlowski et al. 2013; Antin et al. 2015).

7 Conclusions and Perspectives

By reviewing the advances and limitations in biomass estimation at each level of the processing chain, we propose a logical workflow that combines what could be considered as best practices given the current state of knowledge (Fig. 9).

Because almost all steps of the processing chain rely on statistical models, the quality of calibration/validation data is crucial, yet often neglected. In this review, we first discussed common-sense procedures to drastically reduce the error levels at the field plot scale, such as improving georeferencing, increasing field plot size (1 ha or more) or developing allometric equations that better account for local ecological variations in plant composition and/or architecture. Terrestrial LiDAR scanning appears to be a promising avenue for individual-level and plot-level measurements of aboveground volumes. Combined with representative WD estimates that account for the intra- and inter-individual patterns of WD variations, robust predictions of AGB may be achievable. Furthermore, the combination of TLS and airborne LiDAR measurements in the upscaling scheme may significantly reduce the source of mismatches described in Sect. 3, allowing for the easy-to-perform control of coregistration and ensuring a continuum in the measured objects (Kükenbrink et al. 2017).

The extrapolation of local estimates to global scales should ideally be performed in successive steps, relying on intermediate, high-resolution data, such as VHSR spaceborne optical imagery, airborne or UAV-based LiDAR and (UAV-based) stereophotogrammetry. These data can be used to estimate AGB with an accuracy close to the accuracy of ground estimates, but they dramatically increase the representativeness of calibration data when earth observation data of decreased sensitivity (but also decreased cost) are to be used for regional to global predictions. The recent and rapid development of UAV systems, which currently fill the gap between terrestrial and airborne laser scanning systems, is particularly promising because these systems have the potential to bridge different spatial scales, from within-individual-tree to Earth observation pixels of medium resolution (e.g., MODIS).

In parallel, efforts should be expanded to better understand the interactions between different types of electromagnetic signals in varying contexts, e.g., RADAR, LiDAR and passive optical systems, to fully assess the potential of multisensor data fusion. These interactions may be largely blurred by numerous sources of noise and bias—instrumental, atmospheric, geometric, etc. Radiative Transfer (RT) modelling and the prospect to attain a high degree of realism in 3D stand modelling have made this approach a central tool for performing sensitivity analysis and designing signal correction algorithms. Assuming that 3D simulated stands will soon closely reflect reality, they may even be used to directly calibrate inversion models, bypassing many issues such as plot geolocation problems.

The last promise comes from the launch of new space missions (BIOMASS, GEDI, Icesat-2, MOLI, NISAR), which will improve our extrapolation capacities by providing both global coverage and signals of increased relevance. However, whether used individually or in fusion, the calibration of these products should rely on a pre-stratification of forested lands, with enough reference data in each stratum, as it is unlikely that a single universal model can be transferable across forest types and regions without biases.

To conclude, even high-quality Earth observation data currently do not meet the requirements of international environmental policies (e.g., IPCC GPG, REDD+, SDGs, see Herold et al. 2019). Understanding the sources of error and pathways to improve biomass estimation at all levels (as explained in this paper) is fundamental to implementing the best possible practices at each step, resulting in high-quality, consistent biomass estimates with uncertainties quantified to serve the various applications and users.

Acknowledgements We gratefully thank the organizers of the Workshop held at ISSI Bern in November 2017 that was at the origin of this Special Issue. This review has been conducted under the project 3DForMod funded by ERA-GAS (ANR-17-EGAS-0002-01, NWO-3DForMod-5160957540) and has also benefited from the “Investissement d’Avenir” programs managed by Agence Nationale de la Recherche (CEBA, ref. ANR-10-LABX-25-01).

References

Antin C, Grau E, Vincent G et al (2015) From leave scale to tree scale: which structural parameters influence a simulated full-waveform large-footprint LiDAR signal? SilviLaser 2015:110–112

Arciniegas A, Prieto F, Brancheriau L, Lasaygues P (2014) Literature review of acoustic and ultrasonic tomography in standing trees. Trees 28:1559–1567

Asner GP, Mascaro J (2014) Mapping tropical forest carbon: calibrating plot estimates to a simple LiDAR metric. Remote Sens Environ 140:614–624

Asner GP, Broadbent EN, Oliveira PJC et al (2006) Condition and fate of logged forests in the Brazilian Amazon. Proc Natl Acad Sci 103:12947–12950. https://doi.org/10.1073/pnas.0604093103

Asner GP, Mascaro J, Anderson C et al (2013) High-fidelity national carbon mapping for resource management and REDD+. Carbon Balance Manag 8:1–14. https://doi.org/10.1186/1750-0680-8-7

Avitabile V, Camia A (2018) An assessment of forest biomass maps in Europe using harmonized national statistics and inventory plots. For Ecol Manag 409:489–498. https://doi.org/10.1016/j.foreco.2017.11.047
Avitabile V, Herold M, Heuvelink GBM et al (2016) An integrated pan-tropical biomass map using multiple reference datasets. Glob Change Biol 22:1406–1420. https://doi.org/10.1111/gcb.13139
Baccini A, Asner GP (2013) Improving pantropical forest carbon maps with airborne LiDAR sampling. Carbon Manag 4:591–600
Baccini A, Goetz SJ, Walker WS et al (2012) Estimated carbon dioxide emissions from tropical deforestation improved by carbon-density maps. Nat Clim Change 2:182–185. https://doi.org/10.1038/nclimate1354
Baker TR, Phillips OL, Malhi Y et al (2004) Variation in wood density determines spatial patterns in Amazonian forest biomass. Glob Change Biol 10:545–562
Banin L, Feldpausch TR, Phillips OL et al (2012) What controls tropical forest architecture? Testing environmental, structural and floristic drivers. Glob Ecol Biogeogr 21:1179–1190. https://doi.org/10.1111/j.1466-8238.2012.00778.x
Barbier N, Couteron P (2015) Attenuating the bidirectional texture variation of satellite images of tropical forest canopies. Remote Sens Environ 171:245–260
Barbier N, Proisy C, Véga C et al (2011) Bidirectional texture function of high resolution optical images of tropical forest: an approach using LiDAR hillshade simulations. Remote Sens Environ 115:167–179
Bastin J-F, Barbier N, Couteron P et al (2014) Aboveground biomass mapping of African forest mosaics using canopy texture analysis: toward a regional approach. Ecol Appl 24:1984–2001. https://doi.org/10.1890/13-1574.1
Bastin J-F, Fayolle A, Tarelkin Y et al (2015a) Wood specific gravity variations and biomass of central african tree species: the simple choice of the outer wood. PLoS ONE 10:e0142146
Bastin J-F, Barbier N, Réjou-Méchain M et al (2015b) Seeing Central African forests through their largest trees. Sci Rep 5:13156
Bauwens S, Bartholomeus H, Calders K, Lejeune P (2016) Forest inventory with terrestrial LiDAR: a comparison of static and hand-held mobile laser scanning. Forests 7:127
Béland M, Baldocchi DD, Widlowski J-L et al (2014) On seeing the wood from the leaves and the role of voxel size in determining leaf area distribution of forests with terrestrial LiDAR. Agric For Meteorol 184:82–97
Blanchard E, Birnbaum P, Proisy C et al (2015) Prédire la structure des forêts tropicales humides calédoniennes: analyse texturale de la canopée sur des images Pléiades. Rev Fr Photogrammétrie Télédétection 209:141–147
Bouvet A, Mermoz S, Le Toan T et al (2018) An above-ground biomass map of African savannahs and woodlands at 25 m resolution derived from ALOS PALSAR. Remote Sens Environ 206:156–173
Bouvier M, Durrieu S, Fournier RA, Renaud J-P (2015) Generalizing predictive models of forest inventory attributes using an area-based approach with airborne LiDAR data. Remote Sens Environ 156:322–334
Brede B, Lau A, Bartholomeus HM, Kooistra L (2017) Comparing RIEGL RiCOPTER UAV LiDAR derived canopy height and DBH with terrestrial LiDAR. Sensors 17:2371
Bustamante MMC, Roitman I, Aide TM et al (2016) Toward an integrated monitoring framework to assess the effects of tropical forest degradation and recovery on carbon stocks and biodiversity. Glob Change Biol 22:92–109. https://doi.org/10.1111/gcb.13087
Calders K, Newnham G, Burt A et al (2014) Nondestructive estimates of above-ground biomass using terrestrial laser scanning. Methods Ecol Evol. https://doi.org/10.1111/2041-210x.12301
Calders K, Origo N, Burt A et al (2018) Realistic forest stand reconstruction from terrestrial LiDAR for radiative transfer modelling. Remote Sens 10:933
Cescatti A (1997) Modelling the radiative transfer in discontinuous canopies of asymmetric crowns. I. Model structure and algorithms. Ecol Model 101:263–274
Chambers JQ, Negron-Juarez RI, Marra DM et al (2013) The steady-state mosaic of disturbance and succession across an old-growth Central Amazon forest landscape. Proc Natl Acad Sci 110:3949–3954. https://doi.org/10.1073/pnas.1202894110
Chanthorn W, Hartig F, Brockelman WY (2017) Structure and community composition in a tropical forest suggest a change of ecological processes during stand development. For Ecol Manag 404:100–107. https://doi.org/10.1016/j.foreco.2017.08.001
Chave J, Condit R, Aguilar S et al (2004) Error propagation and scaling for tropical forest biomass estimates. Philos Trans R Soc Lond Ser B-Biol Sci 359:409–420
Chave J, Andalo C, Brown S et al (2005) Tree allometry and improved estimation of carbon stocks and balance in tropical forests. Oecologia 145:87–99. https://doi.org/10.1007/s00442-005-0100-x

Chave J, Coomes D, Jansen S et al (2009) Towards a worldwide wood economics spectrum. Ecol Lett 12:351–366
Chave J, Réjou-Méchain M, Búrquez A et al (2014) Improved allometric models to estimate the aboveground biomass of tropical trees. Glob Change Biol 20:3177–3190. https://doi.org/10.1111/gcb.12629
Chen Q, Laurin GV, Valentini R (2015) Uncertainty of remotely sensed aboveground biomass over an African tropical forest: propagating errors from trees to plots to pixels. Remote Sens Environ 160:134–143
Clark DA (2002) Are tropical forests an important carbon sink? Reanalysis of the long-term plot data. Ecol Appl 12:3–7. https://doi.org/10.1890/1051-0761(2002)012%5b0003:atfaic%5d2.0.co;2
Clark D, Clark D (2000) Landscape-scale variation in forest structure and biomass in a tropical rain forest. For Ecol Manag 137:185–198. https://doi.org/10.1016/s0378-1127(99)00327-8
Clark DB, Kellner JR (2012) Tropical forest biomass estimation and the fallacy of misplaced concreteness. J Veg Sci 23:1191–1196. https://doi.org/10.1111/j.1654-1103.2012.01471.x
Condit R (1998) Tropical forest census plots: methods and results from Barro Colorado Island, Panama and a comparison with other plots. Springer, Berlin
Condit R, Ashton PS, Baker P et al (2000) Spatial patterns in the distribution of tropical tree species. Science 288(5470):1414–1418
Condit R, Lao S, Singh A et al (2014) Data and database standards for permanent forest plots in a global network. For Ecol Manag 316:21–31
Côté J-F, Fournier RA, Egli R (2011) An architectural model of trees to estimate forest structural attributes using terrestrial LiDAR. Environ Model Softw 26:761–777. https://doi.org/10.1016/j.envsoft.2010.12.008
Couteron P, Pelissier R, Nicolini EA, Paget D (2005) Predicting tropical forest stand structure parameters from Fourier transform of very high-resolution remotely sensed canopy images. J Appl Ecol 42:1121–1128
Dauzat J, Rapidel B, Berger A (2001) Simulation of leaf transpiration and sap flow in virtual plants: model description and application to a coffee plantation in Costa Rica. Agric For Meteorol 109:143–160
de Castilho CV, Magnusson WE, de Araújo RNO et al (2006) Variation in aboveground tree live biomass in a central Amazonian forest: effects of soil and topography. For Ecol Manag 234:85–96. https://doi.org/10.1016/j.foreco.2006.06.024
de Moura YM, Hilker T, Goncalves FG et al (2016) Scaling estimates of vegetation structure in Amazonian tropical forests using multi-angle MODIS observations. Int J Appl Earth Obs Geoinf 52:580–590
De Reffye P, Houllier F, Blaise F et al (1995) A model simulating above-and below-ground tree architecture with agroforestry applications. Agrofor Syst 30:175–197
de Souza Pereira FR, Kampel M, Gomes Soares ML et al (2018) Reducing uncertainty in mapping of mangrove aboveground biomass using airborne discrete return LiDAR data. Remote Sens 10:637
Detto M, Muller-Landau HC, Mascaro J, Asner GP (2013) Hydrological networks and associated topographic variation as templates for the spatial organization of tropical forest vegetation. PLoS ONE 8:e76296. https://doi.org/10.1371/journal.pone.0076296
Dickinson TA, Tanner EVJ (1978) Exploitation of hollow trunks by tropical trees. Biotropica 10:231–233. https://doi.org/10.2307/2387908
Disney M (2018) Terrestrial LiDAR: a 3D revolution in how we look at trees. New Phytol. https://doi.org/10.1111/nph.15517
Egbert DD (1977) A practical method for correcting bidirectional reflectance variations. In: LARS symposia, p 203
Emilio T, Quesada CA, Costa FRC et al (2013) Soil physical conditions limit palm and tree basal area in Amazonian forests. Plant Ecol Divers 10:1. https://doi.org/10.1080/17550874.2013.772257
ESA (2012) Report for mission selection: biomass, ESA SP-1324/1 (3 volume series). European Space Agency Noordwijk, The Netherlands
Fayad I, Baghdadi N, Guitet S et al (2016) Aboveground biomass mapping in French Guiana by combining remote sensing, forest inventories and environmental data. Int J Appl Earth Obs Geoinf 52:502–514
Fayolle A, Doucet J-L, Gillet J-F et al (2013) Tree allometry in Central Africa: testing the validity of pantropical multi-species allometric equations for estimating biomass and carbon stocks. For Ecol Manag 305:29–37. https://doi.org/10.1016/j.foreco.2013.05.036
Feldpausch TR, Banin L, Phillips OL et al (2011) Height–diameter allometry of tropical forest trees. Biogeosciences 8:1081–1106
Feldpausch TR, Lloyd J, Lewis SL et al (2012) Tree height integrated into pantropical forest biomass estimates. Biogeosciences 9:3381–3403. https://doi.org/10.5194/bg-9-3381-2012
Féret J-B, Asner GP (2014) Mapping tropical forest canopy diversity using high-fidelity imaging spectroscopy. Ecol Appl 24:1289–1296

Ferraz A, Saatchi S, Mallet C, Meyer V (2016) LiDAR detection of individual tree size in tropical forests. Remote Sens Environ 183:318–333

Féret JB, Gitelson AA, Noble SD, Jacquemoud S (2017) PROSPECT-D: towards modeling leaf optical properties through a complete lifecycle. Remote Sens Environ 193:204–215

Flores O, Coomes DA (2011) Estimating the wood density of species for carbon stock assessments. Methods Ecol Evol 2:214–220. https://doi.org/10.1111/j.2041-210x.2010.00068.x

Frazer GW, Wulder MA, Niemann KO (2005) Simulation and quantification of the fine-scale spatial pattern and heterogeneity of forest canopy structure: a lacunarity-based method designed for analysis of continuous canopy heights. For Ecol Manag 214:65–90

Frazer GW, Magnussen S, Wulder MA, Niemann KO (2011) Simulated impact of sample plot size and co-registration error on the accuracy and uncertainty of LiDAR-derived estimates of forest stand biomass. Remote Sens Environ 115:636–649

Fuller WA (1987) Measurement error models. Wiley, New York

Gao S, Wang X, Wiemann MC et al (2017) A critical analysis of methods for rapid and nondestructive determination of wood density in standing trees. Ann For Sci 74:27

Gastellu-Etchegorry J-P, Demarez V, Pinel V, Zagolski F (1996) Modeling radiative transfer in heterogeneous 3-D vegetation canopies. Remote Sens Environ 58:131–156

Gastellu-Etchegorry J-P, Yin T, Lauret N et al (2015) Discrete anisotropic radiative transfer (DART 5) for modeling airborne and satellite spectroradiometer and LiDAR acquisitions of natural and urban landscapes. Remote Sens 7:1667–1701

Gobakken T, Naesset E (2009) Assessing effects of positioning errors and sample plot size on biophysical stand properties derived from airborne laser scanner data. Can J For Res 39:1036–1052

Gomes ACS, Andrade A, Barreto-Silva JS et al (2013) Local plant species delimitation in a highly diverse Amazonian forest: do we all see the same species? J Veg Sci 24:70–79. https://doi.org/10.1111/j.1654-1103.2012.01441.x

Gonzalez de Tanago J, Lau A, Bartholomeus H et al (2018) Estimation of above-ground biomass of large tropical trees with terrestrial LiDAR. Methods Ecol Evol 9:223–234

Goodman RC, Phillips OL, Baker TR (2014) The importance of crown dimensions to improve tropical tree biomass estimates. Ecol Appl 24:680–698. https://doi.org/10.1890/13-0070.1

Gourlet-Fleury S, Rossi V, Réjou-Méchain M et al (2011) Environmental filtering of dense-wooded species controls above-ground biomass stored in African moist forests. J Ecol 99:981–990. https://doi.org/10.1111/j.1365-2745.2011.01829.x

Grau E, Durrieu S, Fournier R et al (2017) Estimation of 3D vegetation density with terrestrial laser scanning data using voxels. A sensitivity analysis of influencing parameters. Remote Sens Environ 191:373–388

Gregoire TG, Næsset E, McRoberts RE et al (2016) Statistical rigor in LiDAR-assisted estimation of aboveground forest biomass. Remote Sens Environ 173:98–108

Guitet S, Pélissier R, Brunaux O et al (2015) Geomorphological landscape features explain floristic patterns in French Guiana rainforest. Biodivers Conserv 24:1215–1237

Guitet S, Sabatier D, Brunaux O et al (2018) Disturbance regimes drive the diversity of regional floristic pools across Guianan rainforest landscapes. Sci Rep 8:3872

Hajj ME, Baghdadi N, Fayad I et al (2017) Interest of integrating spaceborne LiDAR data to improve the estimation of biomass in high biomass forested areas. Remote Sens 9:213

Hansen MC, Potapov PV, Moore R et al (2013) High-resolution global maps of 21st-century forest cover change. Science 342:850–853

Henry M, Besnard A, Asante WA et al (2010) Wood density, phytomass variations within and among trees, and allometric equations in a tropical rainforest of Africa. For Ecol Manag 260:1375–1388

Herold M, Carter S, Avitabile V et al (2019) The role and need for space-based forest biomass-related measurements in environmental management and policy. Surv Geophys. https://doi.org/10.1007/s10712-019-09510-6

Huang W, Swatantran A, Johnson K et al (2015) Local discrepancies in continental scale biomass maps: a case study over forested and non-forested landscapes in Maryland, USA. Carbon Balance Manag 10:19

Hunter MO, Keller M, Victoria D, Morton DC (2013) Tree height and tropical forest biomass estimation. Biogeosciences 10:8385–8399

Inglada J, Vadon H (2005) Fine registration of SPOT5 and Envisat/ASAR images and ortho-image production: a fully automatic approach. In: Proceedings 2005 IEEE international geoscience and remote sensing symposium, 2005. IGARSS'05. IEEE, Vol. 5, pp 3510–3512

Ishimaru A (1978) Wave propagation and scattering in random media, vol 2. Academic press, New York, pp 336–393

Johnson CE, Barton CC (2004) Where in the world are my field plots? Using GPS effectively in environmental field studies. Front Ecol Environ 2:475–482. https://doi.org/10.1890/1540-9295(2004)002%5b0475:witwam%5d2.0.co;2
Jonckheere I, Nackaerts K, Muys B et al (2006) A fractal dimension-based modelling approach for studying the effect of leaf distribution on LAI retrieval in forest canopies. Ecol Model 197:179–195
Jucker T, Asner GP, Dalponte M et al (2017a) A regional model for estimating the aboveground carbon density of Borneo's tropical forests from airborne laser scanning. arXiv Prepr arXiv170509242
Jucker T, Caspersen J, Chave J et al (2017b) Allometric equations for integrating remote sensing imagery into forest monitoring programmes. Glob Change Biol 23:177–190
Jucker T, Asner GP, Dalponte M et al (2018a) Estimating aboveground carbon density and its uncertainty in Borneo's structurally complex tropical forests using airborne laser scanning. Biogeosciences 15:3811–3830
Jucker T, Bongalov B, Burslem DF et al (2018b) Topography shapes the structure, composition and function of tropical forest landscapes. Ecol Lett 21:989–1000
Justice CO, Giglio L, Korontzi S et al (2002) The MODIS fire products. Remote Sens Environ 83:244–262
Kearsley E, De Haulleville T, Hufkens K et al (2013) Conventional tree height–diameter relationships significantly overestimate aboveground carbon stocks in the Central Congo Basin. Nat Commun 4:2269
Kellner JR, Asner GP (2009) Convergent structural responses of tropical forests to diverse disturbance regimes. Ecol Lett 12:887–897
Kennedy RE, Yang Z, Cohen WB (2010) Detecting trends in forest disturbance and recovery using yearly Landsat time series: 1. LandTrendr—temporal segmentation algorithms. Remote Sens Environ 114:2897–2910
Ketterings QM, Coe R, van Noordwijk M et al (2001) Reducing uncertainty in the use of allometric biomass equations for predicting above-ground tree biomass in mixed secondary forests. For Ecol Manag 146:199–209. https://doi.org/10.1016/s0378-1127(00)00460-6
Kleinn C (2017) The renaissance of National Forest Inventories (NFIs) in the context of the international conventions—a discussion paper on context, background and justification of NFIs. Pesqui Florest Bras 37:369–379
Kükenbrink D, Schneider FD, Leiterer R et al (2017) Quantification of hidden canopy volume of airborne laser scanning data using a voxel traversal algorithm. Remote Sens Environ 194:424–436. https://doi.org/10.1016/j.rse.2016.10.023
Labriere N, Tao S, Chave J et al (2018) In situ reference datasets from the TropiSAR and AfriSAR campaigns in support of upcoming spaceborne biomass missions. IEEE J Sel Top Appl Earth Obs Remote Sens 99:1–11
Lagomasino D, Fatoyinbo T, Lee S-K, Simard M (2015) High-resolution forest canopy height estimation in an African blue carbon ecosystem. Remote Sens Ecol Conserv 1:51–60. https://doi.org/10.1002/rse2.3
Larjavaara M, Muller-Landau HC (2013) Measuring tree height: a quantitative comparison of two common field methods in a moist tropical forest. Methods Ecol Evol 4:793–801. https://doi.org/10.1111/2041-210x.12071
Lau A, Bentley LP, Martius C et al (2018) Quantifying branch architecture of tropical trees using terrestrial LiDAR and 3D modelling. Trees 32(5):1219–1231
Le Toan T, Quegan S, Davidson MWJ et al (2011) The BIOMASS mission: mapping global forest biomass to better understand the terrestrial carbon cycle. Remote Sens Environ 115:2850–2860. https://doi.org/10.1016/j.rse.2011.03.020
Lefsky MA, Cohen WB, Parker GG, Harding DJ (2002) LiDAR remote sensing for ecosystem studies LiDAR, an emerging remote sensing technology that directly measures the three-dimensional distribution of plant canopies, can accurately estimate vegetation structural attributes and should be of particular interest to forest, landscape, and global ecologists. Bioscience 52:19–30
Leitold V, Morton DC, Longo M et al (2018) El Niño drought increased canopy turnover in Amazon forests. New Phytol 219:959–971
Li X, Strahler AH (1992) Geometric-optical bidirectional reflectance modeling of the discrete crown vegetation canopy: effect of crown shape and mutual shadowing. IEEE Trans Geosci Remote Sens 30:276–292
Lindenmayer DB, Cunningham RB, Tanton MT et al (1991) Characteristics of hollow-bearing trees occupied by arboreal marsupials in the montane ash forests of the Central Highlands of Victoria, south-east Australia. For Ecol Manag 40:289–308
Liu J-Y, Zheng Z, Xu X et al (2018) Abundance and distribution of cavity trees and the effect of topography on cavity presence in a tropical rainforest, southwestern China. Can J For Res 48:1058–1066

Longo M, Keller M, dos-Santos MN et al (2016) Aboveground biomass variability across intact and degraded forests in the Brazilian Amazon. Glob Biogeochem Cycles 30:1639–1660
Lopez-Gonzalez G, Lewis SL, Burkitt M, Phillips OL (2011) ForestPlots.net: a web application and research tool to manage and analyse tropical forest plot data. J Veg Sci 22:610–613
Lyapustin A, Martonchik J, Wang Y et al (2011) Multiangle implementation of atmospheric correction (MAIAC): 1. Radiative transfer basis and look-up tables. J Geophys Res Atmos 116:1–9
Ma L, Zheng G, Eitel JU et al (2016) Improved salient feature-based approach for automatically separating photosynthetic and nonphotosynthetic components within terrestrial LiDAR point cloud data of forest canopies. IEEE Trans Geosci Remote Sens 54:679–696
Malenovský Z, Martin E, Homolová L et al (2008) Influence of woody elements of a Norway spruce canopy on nadir reflectance simulated by the DART model at very high spatial resolution. Remote Sens Environ 112:1–18
Marra RE, Brazee NJ, Fraver S (2018) Estimating carbon loss due to internal decay in living trees using tomography: implications for forest carbon budgets. Environ Res Lett. https://doi.org/10.1088/1748-9326/aae2bf
Marvin DC, Asner GP, Knapp DE et al (2014) Amazonian landscapes and the bias in field studies of forest structure and biomass. Proc Natl Acad Sci 111:E5224–E5232. https://doi.org/10.1073/pnas.1412999111
Mascaro J, Detto M, Asner GP, Muller-Landau HC (2011) Evaluating uncertainty in mapping forest carbon with airborne LiDAR. Remote Sens Environ 115:3770–3774. https://doi.org/10.1016/j.rse.2011.07.019
McEwan RW, Lin Y-C, Sun I-F et al (2011) Topographic and biotic regulation of aboveground carbon storage in subtropical broad-leaved forests of Taiwan. For Ecol Manag 262:1817–1825. https://doi.org/10.1016/j.foreco.2011.07.028
McRoberts RE, Westfall JA (2013) Effects of uncertainty in model predictions of individual tree volume on large area volume estimates. For Sci 60(1):34–42
Mermoz S, Le Toan T, Villard L et al (2014) Biomass assessment in the Cameroon savanna using ALOS PALSAR data. Remote Sens Environ 155:109–119. https://doi.org/10.1016/j.rse.2014.01.029
Mermoz S, Réjou-Méchain M, Villard L et al (2015) Decrease of L-band SAR backscatter with biomass of dense forests. Remote Sens Environ 15:307–317. https://doi.org/10.1016/j.rse.2014.12.019
Minh DHT, Le Toan T, Rocca F et al (2014) Relating P-band synthetic aperture radar tomography to tropical forest biomass. IEEE Trans Geosci Remote Sens 52:967–979
Minh DHT, Le Toan T, Rocca F et al (2016) SAR tomography for the retrieval of forest biomass and height: cross-validation at two tropical forest sites in French Guiana. Remote Sens Environ 175:138–147
Mitchard ETA (2018) The tropical forest carbon cycle and climate change. Nature 559:527–534. https://doi.org/10.1038/s41586-018-0300-2
Mitchard ET, Saatchi SS, Baccini A et al (2013) Uncertainty in the spatial distribution of tropical forest biomass: a comparison of pan-tropical maps. Carbon Balance Manage 8:10
Mitchard ETA, Feldpausch TR, Brienen RJW et al (2014) Markedly divergent estimates of Amazon forest carbon density from ground plots and satellites. Glob Ecol Biogeogr 23:935–946. https://doi.org/10.1111/geb.12168
Molto Q, Rossi V, Blanc L (2013) Error propagation in biomass estimation in tropical forests. Methods Ecol Evol 4:175–183. https://doi.org/10.1111/j.2041-210x.2012.00266.x
Momo Takoudjou S, Ploton P, Sonké B et al (2018) Using terrestrial laser scanning data to estimate large tropical trees biomass and calibrate allometric models: a comparison with traditional destructive approach. Methods Ecol Evol 9:905–916
Morsdorf F, Eck C, Zgraggen C et al (2017) UAV-based LiDAR acquisition for the derivation of high-resolution forest and ground information. Lead Edge 36:566–570
Morton DC, Nagol J, Carabajal CC et al (2014) Amazon forests maintain consistent canopy structure and greenness during the dry season. Nature 506:221–224. https://doi.org/10.1038/nature13006
Moundounga Mavouroulou Q, Ngomanda A, Engone Obiang NL et al (2014) How to improve allometric equations to estimate forest biomass stocks? Some hints from a central African forest. Can J For Res 44:685–691
Myneni RB (1991) Modeling radiative transfer and photosynthesis in three-dimensional vegetation canopies. Agric For Meteorol 55:323–344
Myneni RB, Ross J, Asrar G (1989) A review on the theory of photon transport in leaf canopies. Agric For Meteorol 45:1–153
Ni W, Li X, Woodcock CE et al (1999) An analytical hybrid GORT model for bidirectional reflectance over discontinuous plant canopies. IEEE Trans Geosci Remote Sens 37:987–999

Nogueira EM, Nelson BW, Fearnside PM (2006) Volume and biomass of trees in central Amazonia: influence of irregularly shaped and hollow trunks. For Ecol Manag 227:14–21. https://doi.org/10.1016/j.foreco.2006.02.004
North PR (1996) Three-dimensional forest light interaction model using a Monte Carlo method. IEEE Trans Geosci Remote Sens 34:946–956
Pargal S, Fararoda R, Rajashekar G et al (2017) Inverting aboveground biomass–canopy texture relationships in a landscape of Forest mosaic in the Western Ghats of India using very high resolution Cartosat imagery. Remote Sens 9:228
Pearson TR, Brown S, Murray L, Sidman G (2017) Greenhouse gas emissions from tropical forest degradation: an underestimated source. Carbon Balance Manag 12:3
Phillips OL, Baker TR, Brienen R, Feldpausch TR (2009) Field manual for plot establishment and remeasurement. https://www.forestplots.net/upload/ManualsEnglish/RAINFOR_field_manual_EN.pdf
Ploton P, Pélissier R, Proisy C et al (2012) Assessing aboveground tropical forest biomass using Google Earth canopy images. Ecol Appl 22:993–1003
Ploton P, Pélissier R, Barbier N et al (2013) Canopy texture analysis for large-scale assessments of tropical forest stand structure and biomass. In: Devy S, Ganesh T, Lowman MD (eds) Treetops at risk. Springer, Berlin, pp 237–245
Ploton P, Barbier N, Momo ST et al (2016) Closing a gap in tropical forest biomass estimation: taking crown mass variation into account in pantropical allometries. Biogeosciences 13:1571–1585
Ploton P, Barbier N, Couteron P et al (2017) Toward a general tropical forest biomass prediction model from very high resolution optical satellite images. Remote Sens Environ 200:140–153
Proisy C, Couteron P, Fromard F (2007) Predicting and mapping mangrove biomass from canopy grain analysis using Fourier-based textural ordination of IKONOS images. Remote Sens Environ 109:379–392. https://doi.org/10.1016/j.rse.2007.01.009
Puliti S, Ørka HO, Gobakken T, Næsset E (2015) Inventory of small forest areas using an unmanned aerial system. Remote Sens 7:9632–9654
Raumonen P, Kaasalainen M, Åkerblom M et al (2013) Fast automatic precision tree models from terrestrial laser scanner data. Remote Sens 5:491–520
Réjou-Méchain M, Muller-Landau HC, Detto M et al (2014) Local spatial structure of forest biomass and its consequences for remote sensing of carbon stocks. Biogeosciences 11:6827–6840
Réjou-Méchain M, Tymen B, Blanc L et al (2015) Using repeated small-footprint LiDAR maps to infer spatial variation and dynamics of a high-biomass neotropical forest. Remote Sens Environ 169:93–101
Réjou-Méchain M, Tanguy A, Piponiot C et al (2017) BIOMASS: an r package for estimating above-ground biomass and its uncertainty in tropical forests. Methods Ecol Evol 8:1163–1167
Robinson C, Saatchi S, Neumann M, Gillespie T (2013) Impacts of spatial variability on aboveground biomass estimation from L-band radar in a temperate forest. Remote Sens 5:1001–1023
Rocchini D, Luque S, Pettorelli N et al (2018) Measuring β-diversity by remote sensing: a challenge for biodiversity monitoring. Methods Ecol Evol 9:1787–1798
Rodrigues WA, Valle RC (1964) Ocorrência de troncos ocos em mata de baixio da regiao de Manaus, 16th edn. Publicacao. Botanica - Instituto Nacional de Pesquisa da Amazonia (Brazil), Manaus
Rodriguez-Veiga P, Wheeler J, Louis V et al (2017) Quantifying forest biomass carbon stocks from space. Curr For Rep 3:1–18
Romijn E, De Sy V, Herold M et al (2018) Independent data for transparent monitoring of greenhouse gas emissions from the land use sector—what do stakeholders think and need? Environ Sci Policy 85:101–112
Roşca S, Suomalainen J, Bartholomeus H, Herold M (2018) Comparing terrestrial laser scanning and unmanned aerial vehicle structure from motion to assess top of canopy structure in tropical forests. Interface Focus 8:20170038
Rosen P, Hensley S, Shaffer S et al (2017) The NASA-ISRO SAR (NISAR) mission dual-band radar instrument preliminary design. In: 2017 IEEE international geoscience and remote sensing symposium (IGARSS), pp 3832–3835
Roujean J-L, Leroy M, Deschamps P-Y (1992) A bidirectional reflectance model of the Earth's surface for the correction of remote sensing data. J Geophys Res Atmos 97:20455–20468
Saatchi SS, Houghton RA, Alvalá DS et al (2007) Distribution of aboveground live biomass in the Amazon basin. Glob Change Biol 13:816–837. https://doi.org/10.1111/j.1365-2486.2007.01323.x
Saatchi S, Marlier M, Chazdon RL et al (2011a) Impact of spatial variability of tropical forest structure on radar estimation of aboveground biomass. Remote Sens Environ 115:2836–2849. https://doi.org/10.1016/j.rse.2010.07.015

Saatchi SS, Harris NL, Brown S et al (2011b) Benchmark map of forest carbon stocks in tropical regions across three continents. Proc Natl Acad Sci 108:9899–9904. https://doi.org/10.1073/pnas.1019576108

Saatchi S, Mascaro J, Xu L et al (2015) Seeing the forest beyond the trees. Glob Ecol Biogeogr 24:606–610

Sagang LBT, Momo ST, Libalah MB et al (2018) Using volume-weighted average wood specific gravity of trees reduces bias in aboveground biomass predictions from forest volume data. For Ecol Manag 424:519–528. https://doi.org/10.1016/j.foreco.2018.04.054

Santoro M, Cartus O, Mermoz S et al (2018) A detailed portrait of the forest aboveground biomass pool for the year 2010 obtained from multiple remote sensing observations. In: EGU general assembly conference abstracts, p 18932

Schlund M, von Poncet F, Kuntz S et al (2015) TanDEM-X data for aboveground biomass retrieval in a tropical peat swamp forest. Remote Sens Environ 158:255–266

Schneider FD, Yin T, Gastellu-Etchegorry J et al (2014) At-sensor radiance simulation for airborne imaging spectroscopy. In: 2014 6th workshop on hyperspectral image and signal processing: evolution in remote sensing (WHISPERS), pp 1–4

Sigrist P, Coppin P, Hermy M (1999) Impact of forest canopy on quality and accuracy of GPS measurements. Int J Remote Sens 20:3595–3610. https://doi.org/10.1080/014311699211228

Simard M, Pinto N, Fisher JB, Baccini A (2011) Mapping forest canopy height globally with spaceborne LiDAR. J Geophys Res Biogeosci 116:1–12

Singh M, Malhi Y, Bhagwat S (2014) Biomass estimation of mixed forest landscape using a Fourier transform texture-based approach on very-high-resolution optical satellite imagery. Int J Remote Sens 35:3331–3349

Sitch S, Huntingford C, Gedney N et al (2008) Evaluation of the terrestrial carbon cycle, future plant geography and climate-carbon cycle feedbacks using five dynamic global vegetation models (DGVMs). Glob Change Biol 14:2015–2039

Solberg S, May J, Bogren W et al (2018) Interferometric SAR DEMs for forest change in Uganda 2000–2012. Remote Sens 10:228

Steininger MK (2000) Satellite estimation of tropical secondary forest above-ground biomass: data from Brazil and Bolivia. Int J Remote Sens 21:1139–1157

St-Onge B, Vega C, Fournier RA, Hu Y (2008) Mapping canopy height using a combination of digital stereo-photogrammetry and LiDAR. Int J Remote Sens 29:3343–3364

Sullivan MJ, Lewis SL, Hubau W et al (2018) Field methods for sampling tree height for tropical forest biomass estimation. Methods Ecol Evol 9:1179–1189

Sun G, Ranson KJ (2000) Modeling LiDAR returns from forest canopies. IEEE Trans Geosci Remote Sens 38:2617–2626

Swenson NG, Enquist BJ (2008) The relationship between stem and branch wood specific gravity and the ability of each measure to predict leaf area. Am J Bot 95:516–519

Tarelkin Y, Hufkens K, Hahn S et al (2019) Wood anatomy variability under contrasted environmental conditions of common deciduous and evergreen species from central African forests. Trees Struct Funct 33:893–909. https://doi.org/10.1007/s00468-019-01826-5

Trochta J, Krůček M, Vrška T, Král K (2017) 3D Forest: an application for descriptions of three-dimensional forest structures using terrestrial LiDAR. PLoS ONE 12:e0176871

Tymen B, Vincent G, Courtois EA et al (2017) Quantifying micro-environmental variation in tropical rainforest understory at landscape scale by combining airborne LiDAR scanning and a sensor network. Ann For Sci 74:32

Verhoef W (1985) Earth observation modeling based on layer scattering matrices. Remote Sens Environ 17:165–178

Vieilledent G, Vaudry R, Andriamanohisoa SF et al (2012) A universal approach to estimate biomass and carbon stock in tropical forests using generic allometric models. Ecol Appl 22:572–583

Villard L (2009) Forward and inverse modeling of synthetic aperture radar in the bistatic configuration: applications in forest remote sensing. Ph.D. thesis, ONERAISAE-Universite Paul Sabatier

Villard L, Le Toan T (2015) Relating P-band SAR intensity to biomass for tropical dense forests in hilly terrain: γ^0 or t^0? IEEE J Sel Top Appl Earth Obs Remote Sens 8:214–223

Vincent G, Caron F, Sabatier D, Blanc L (2012a) LiDAR shows that higher forests have more slender trees. Bois For Trop 314:51–56

Vincent G, Sabatier D, Blanc L et al (2012b) Accuracy of small footprint airborne LiDAR in its predictions of tropical moist forest stand structure. Remote Sens Environ 125:23–33. https://doi.org/10.1016/j.rse.2012.06.019

Vincent G, Sabatier D, Rutishauser E (2014) Revisiting a universal airborne light detection and ranging approach for tropical forest carbon mapping: scaling-up from tree to stand to landscape. Oecologia 175:439–443

Vincent G, Antin C, Laurans M et al (2017) Mapping plant area index of tropical evergreen forest by airborne laser scanning. A cross-validation study using LAI2200 optical sensor. Remote Sens Environ 198:254–266

Wassenberg M, Chiu H-S, Guo W, Spiecker H (2015) Analysis of wood density profiles of tree stems: incorporating vertical variations to optimize wood sampling strategies for density and biomass estimations. Trees 29:551–561

Widlowski J-L, Pinty B, Lopatka M et al (2013) The fourth radiation transfer model intercomparison (RAMI-IV): proficiency testing of canopy reflectance models with ISO-13528. J Geophys Res Atmos 118:6869–6890. https://doi.org/10.1002/jgrd.50497

Williamson GB, Wiemann MC (2010) Measuring wood specific gravity... correctly. Am J Bot 97:519–524

Wu X, Liu H, Li X et al (2018) Differentiating drought legacy effects on vegetation growth over the temperate Northern Hemisphere. Glob Change Biol 24:504–516

Xu L, Saatchi SS, Yang Y et al (2016) Performance of non-parametric algorithms for spatial mapping of tropical forest structure. Carbon Balance Manag 11:18

Xu L, Saatchi SS, Shapiro A et al (2017) Spatial distribution of carbon stored in forests of the Democratic Republic of Congo. Sci Rep 7:15030

Zanne AE, Lopez-Gonzalez G, Coomes DA, Ilic J, Jansen S, Lewis SL, Miller RB, Swenson NG, Wiemann MC, Chave J (2009) Data from: towards a worldwide wood economics spectrum. Dryad Digit Repos. https://doi.org/10.5061/dryad.234

Zolkos SG, Goetz SJ, Dubayah R (2013) A meta-analysis of terrestrial aboveground biomass estimation using LiDAR remote sensing. Remote Sens Environ 128:289–298. https://doi.org/10.1016/j.rse.2012.10.017

Publisher's Note Springer Nature remains neutral with regard to jurisdictional claims in published maps and institutional affiliations.

Affiliations

Maxime Réjou-Méchain[1] · Nicolas Barbier[1] · Pierre Couteron[1] · Pierre Ploton[1] · Grégoire Vincent[1] · Martin Herold[2] · Stéphane Mermoz[3,8] · Sassan Saatchi[4] · Jérôme Chave[5] · Florian de Boissieu[6] · Jean-Baptiste Féret[6] · Stéphane Momo Takoudjou[1,7] · Raphaël Pélissier[1]

[1] AMAP, IRD, CNRS, CIRAD, INRA, Univ Montpellier, Montpellier Cedex 05, France

[2] Laboratory of Geo-Information Science and Remote Sensing, Wageningen University and Research, Droevendaalsesteeg 3, 6708 PB Wageningen, The Netherlands

[3] CESBIO, CNES/CNRS, IRD/UPS, Université de Toulouse, Toulouse, France

[4] Jet Propulsion Laboratory, California Institute of Technology, Pasadena, CA 91109, USA

[5] CNRS, ENFA; UMR5174 EDB (Laboratoire Evolution et Diversité Biologique), Université Paul Sabatier, 118 Route de Narbonne, 31062 Toulouse, France

[6] TETIS, Irstea, AgroParisTech, CIRAD, CNRS, University of Montpellier, 34000 Montpellier, France

[7] Plant Systematic and Ecology Laboratory, Higher Teacher's Training College, University of Yaoundé I, P.O. Box 047, Yaoundé, Cameroon

[8] GlobEO, Avenue Saint-Exupery, Toulouse, France

Surveys in Geophysics (2019) 40:913–935
https://doi.org/10.1007/s10712-019-09540-0

Species Matter: Wood Density Influences Tropical Forest Biomass at Multiple Scales

Oliver L. Phillips[1] · Martin J. P. Sullivan[1] · Tim R. Baker[1] · Abel Monteagudo Mendoza[2] · Percy Núñez Vargas[3] · Rodolfo Vásquez[2]

Received: 9 October 2018 / Accepted: 6 May 2019 / Published online: 3 June 2019

Abstract

The mass of carbon contained in trees is governed by the volume and density of their wood. This represents a challenge to most remote sensing technologies, which typically detect surface structure and parameters related to wood volume but not to its density. Since wood density is largely determined by taxonomic identity this challenge is greatest in tropical forests where there are tens of thousands of tree species. Here, using pan-tropical literature and new analyses in Amazonia with plots with reliable identifications we assess the impact that species-related variation in wood density has on biomass estimates of mature tropical forests. We find impacts of species on forest biomass due to wood density at all scales from the individual tree up to the whole biome: variation in tree species composition regulates how much carbon forests can store. Even local differences in composition can cause variation in forest biomass and carbon density of 20% between subtly different local forest types, while additional large-scale floristic variation leads to variation in mean wood density of 10–30% across Amazonia and the tropics. Further, because species composition varies at all scales and even vertically within a stand, our analysis shows that bias and uncertainty always result if individual identity is ignored. Since sufficient inventory-based evidence based on botanical identification now exists to show that species composition matters biome-wide for biomass, we here assemble and provide mean basal-area-weighted wood density values for different forests across the lowand tropical biome. These range widely, from 0.467 to 0.728 g cm^{-3} with a pan-tropical mean of 0.619 g cm^{-3}. Our analysis shows that mapping tropical ecosystem carbon always benefits from locally validated measurement of tree-by-tree botanical identity combined with tree-by-tree measurement of dimensions. Therefore whenever possible, efforts to map and monitor tropical forest carbon using remote sensing techniques should be combined with tree-level measurement of species identity by botanists working in inventory plots.

Electronic supplementary material The online version of this article (https://doi.org/10.1007/s10712-019-09540-0) contains supplementary material, which is available to authorized users.

✉ Oliver L. Phillips
o.phillips@leeds.ac.uk

[1] School of Geography, University of Leeds, Leeds LS2 9JT, UK

[2] Jardín Botánico de Missouri, Jr. Bolognesi, 19230 Oxapampa, Peru

[3] Universidad de San Antonio Abad del Cusco, Av. de La Cultura 773, 08000 Cuzco, Peru

Keywords Amazon · Tropical forests · Species · Identity · Carbon · Biomass · Dynamics

1 Introduction

Tropical forests contain more species and biomass than any other biome on Earth. While they are being rapidly degraded and deforested, large areas of relatively intact tropical forest still exist, particularly in the Amazon and Congo basins. Wherever they persist, tropical forests contribute hugely to societies, economies, and human well-being, providing vital services that sustain people and nations (Watson et al. 2018). For example, dozens of the tree species in South American forests are also cultivated or domesticated, and hundreds more are close relatives (Levis et al. 2017). Meanwhile, the carbon sink into mature forests has mitigated deforestation and fossil fuel emissions in many Amazon nations for decades (Phillips and Brienen 2017; Phillips 2018; Vicuña Miñano et al. 2018), so slowing the rate of climate change. These services are all under threat, however, with climate change itself a leading concern. Tropical lands have been warming fast, and continued warming is projected to combine with stronger droughts and potentially lead to crossing ecological thresholds (e.g., Good et al. 2018), bringing increased risks to biomass storage, tree species, and human societies.

This unique nexus of values and threats in tropical forests means that measuring and mapping their biophysical properties—and then tracking changes—are central goals of global environmental science. Yet because of their extent and complexity, tropical forests are challenging to measure and monitor with precision. For the key property of biomass—from which we may derive carbon storage per unit area—space-borne and airborne sensor technologies are increasingly used to infer biomass [Zolkos et al. 2013; Minh et al. 2014; Coomes et al. 2017; Jucker et al. 2018a; Duncanson et al. (this volume)]. Laser scanning enables precise measurement of canopy height, and if done at sufficient intensity can reveal the three-dimensional structure of trees, while space-borne radar offers potentially global-scale assessment of forest structure. Optical sensing of canopies is widely used to infer vegetation state, such as distinguishing forest from non-forest. However, a key technical limitation is that no technology directly measures a critical determinant of every tree's biomass—its identity. This represents a fundamental challenge, especially given that the single most remarkable and celebrated feature of tropical forests is their extraordinary diversity of species and variation in tree composition (e.g., ter Steege et al. 2013). Indeed, tropical tree species composition varies at all scales from a few metres to across the whole biome due to factors that include climate, geomorphology, nutrient supply, evolutionary history, and anthropogenic impacts (e.g., Salo et al. 1986; Gentry 1988; Tuomisto et al. 1995; Condit et al. 2002; ter Steege et al. 2006; Honorio Coronado et al. 2009; Asner et al. 2017; Levis et al. 2017).

Accurate measurement of most tropical trees' biological identity requires direct observation from the ground, supported by collection and subsequent careful identification of the herbarium vouchers by trained botanists (Baker et al. 2017). Since biological composition determines the physical composition of forests in terms of leaf (e.g., Fyllas et al. 2009; Asner et al. 2017) and wood properties (Muller-Landau 2004; Patiño et al. 2009; Baraloto et al. 2011), then an inability to perceive biodiversity may significantly hinder estimation of biomass and carbon storage. Yet how much the variation in tropical forest species actually matters for biomass mapping remains controversial. The aims of this paper are to explore and quantify this issue for tropical forests and then suggest how the difficulties faced by current remote Earth Observation techniques in mapping tropical forest species compositional variation and biomass density might be mitigated. By combining literature and new analysis, we examine the issue from the scale of individual tropical tree up to whole continents and assess its impact on Amazon biomass estimates.

1.1 In Practice, Does Diversity Matter?

Forests are made mostly of trees, and in tropical forests these come in extraordinary variety. There can be 300 tree species in a 100-by-100-m patch of Amazon forest. Remarkably, these single-hectare tropical samples contain more woody plant species than are found in all of Earth's boreal forests—an area some nine orders of magnitude greater. Tropical Peru has almost 5000 tree species recorded, with new species being discovered every year (Vásquez et al. 2018), while the temperate UK has less than 50. In Amazonia, there are as many as 16,000 tree species (ter Steege et al. 2013). With huge floristic diversity, it is reasonable to expect a high degree of functional diversity too, including in the key attributes that affect tree biomass (Baker et al. 2009; Baker 2018). There is ongoing debate as to whether diversity helps support higher biomass, and if so how (e.g., Bunker et al. 2005; Sullivan et al. 2017), but here we are interested in the question of how *differences* in the composition of species from one diverse forest to another impact biomass. Thus, here it is the different taxonomic and evolutionary identities of the tree species present which are hypothesised to matter, not the number of species per se.

While the biomass contained by any individual tree is determined by many factors, these are reducible to just two: (1) its size—the volume of wood—and (2) the amount of matter per unit volume or its density. (Here we use the standard definition of 'basic specific gravity', defined as the ratio of the oven-dry mass of a wood sample divided by its green volume, e.g., Chave et al. 2006.) The genetic identity of a tree affects both how big it can become *and* how dense it is (Baker et al. 2004; Fauset et al. 2015; Coelho de Souza et al. 2016). Various studies have shown that these effects are largely independent (e.g., Turner 2001; Coelho de Souza et al. 2016; Hietz et al. 2017): across tropical tree species, maximum size (height, diameter, volume) and density of wood are largely uncorrelated. Since size is a poor predictor of wood density it follows we cannot use the dimensions of trees to infer their density. The fundamental disconnect between size and wood density means that measuring size alone can never capture all information needed to derive biomass.

Consequently, in species-diverse tropical forests tree biomass varies greatly even for a fixed tree size. In southern Peru, Goodman et al. (2012, 2014a) identified, harvested, and painstakingly weighed 51 individual trees as large as 169 cm diameter. We plot these data here to illustrate how volume and wood density combine to determine biomass (Fig. 1). Canopy trees with similar dimensions have very different biomass. For example, a *Cavanillesia umbellata* canopy tree with wood density measured at 0.132 g cm^{-3} had a dry aboveground mass of 2.3 Mg, while an *Apuleia leiocarpa* individual with wood density of 0.855 g cm^{-3} weighed 12.2 Mg, in spite of having slightly *less* wood volume. Further, in multivariate allometric models of tree biomass based on harvested tropical trees, wood density is the most important factor after stem diameter in explaining tree biomass—entering models *before* height (e.g., Chave et al. 2014; Goodman et al. 2014a)—with biomass scaling almost linearly with wood density (Chave et al. 2014). With the huge range in wood density of species present locally, it is essential to know identity in order to estimate tropical tree biomass with confidence.

Yet it does not necessarily follow that the impact of identity on biomass will persist at the larger scales of interest to most Earth Observation questions. Here, we seek to address

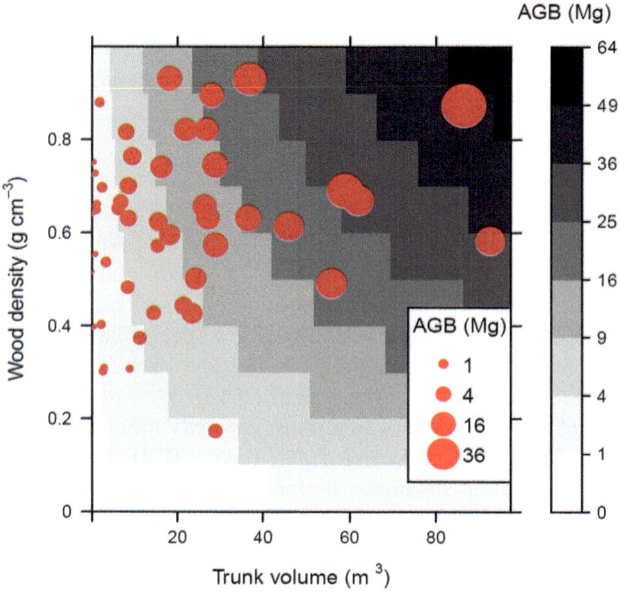

Fig. 1 Direct measurement of tropical trees shows that wood density and size each independently control biomass. Red points represent 51 forest trees destructively sampled and weighed by Goodman et al. (2014a, b) in Amazonian Peru. Point areas are proportional to the *actual, directly measured* aboveground biomass (AGB) of each tree, plotted against their trunk volume and directly measured wood density. Trunk volume was estimated as basal area multiplied by tree height. The greyscale background depicts a quasi-continuous allometric estimate of AGB for combinations of tree volume and wood density. To do this, the Chave et al. (2014) allometric equation was solved for each combination of diameter and wood density, with tree height estimated using a three-parameter Weibull model fitted to all trees in the Goodman et al. (2014a, b) dataset

the question of the extent to which species composition impacts on forest biomass at stand, landscape, and larger scales by reviewing current evidence and developing new analyses. Ultimately, we wish to shed light on the pervasiveness of species effects, focussing on South American tropical forests due to the relatively larger literature here.

At the very largest scales, some spatial variation in forest biomass driven by wood density is already recognised. For example in the neotropics, dry forest trees generally have greater wood density (Chave et al. 2006), and it has long been known that successional forests have lower wood density than mature forests (e.g., Brown and Lugo 1990). However, across the tropical moist forest biome we could not find any analysis at the pan-tropical level as to whether wood density varies systematically continent to continent.

At the sub-continental scale, concerted, species-driven differences in wood density prevail even within the same biome and same successional stage. In mature African moist forests, soil-related compositional differences cause significant differences in basal-area-weighted wood density, with forests on relatively fertile acrisols and cambisols having 10% lower values (0.609 and 0.617 g cm^{-3}) than on arenosols (0.660 g cm^{-3}) and 20% lower than swamp forests on histosols (0.728 g cm^{-3}) (Lewis et al. 2013). Basal-area-weighted wood density is also significantly higher for Central African forests than their West or East African counterparts (Lewis et al. 2013). Likewise, South American forests differ greatly when comparing central with western Amazonia, with 16% lower per-stem wood density in the west caused by differences in floristic composition (Baker et al. 2004). This is in spite of similar climate and instead is linked to differences in stem turnover rates, with the western forests much more dynamic (Phillips et al. 2004), often associated with more cation-rich and structurally poor soils (Quesada et al. 2012): trees grow and die faster here and this favours species which are adapted to exploiting gaps quickly. While different life-history strategies are found in *all* Amazon forests, the mean wood density in the slow turnover forests on the Guiana Shield is 50% greater than in the fast turnover forests in south-west Amazonia (ter Steege et al. 2006), helping drive much greater standing biomass in the north-east (Malhi et al. 2006; Johnson et al. 2016). This large-scale species-driven difference in biomass is invisible in space-borne LiDAR-derived biomass estimates (c.f. Mitchard et al. 2014), but is accounted for in hybrid biomass maps that attempt to combine plot-derived measures of species differences across space with LiDAR measurements (Avitabile et al. 2016).

While the broad difference between north-east and south-west Amazon forests is clear, uncertainties remain, including the exact nature of the relationships between AGB and wood density, and between mortality rates and wood density, and crucially whether these relationships also persist at smaller geographical units. Some evidence suggests that at finer scales a more nuanced situation prevails. Within western Amazonia, Landsat-based analyses have revealed great variation in spectral types of forest, starting with the seminal study of Salo et al. (1986). In combination with fieldwork, spectral variation has been linked to variation in species and subtle geomorphological, edaphic, and geological controlling factors have been revealed (e.g., Tuomisto et al. 1995; Higgins et al. 2011). Recently, using airborne hyperspectral sensing variation in canopy function has been explored here at high resolution (Asner et al. 2017; Draper et al. 2019). These analyses all confirm both discrete and continuous variations in canopy function across Peru's forests.

Recent investigations also show how soil and species differences affect tropical forest wood density in regional and landscape scales. In the Central African Republic, Gourlet-Fleury et al. (2011) found differences in wood density of 20% between forest types in one landscape, mediated by soil nutrients and drainage. Similarly, in north-west Amazonia and in French Guiana, Baraloto et al. (2011) found that forests with richer soils tended to support trees with lower wood density. In particular, they report highest wood density in white sand forests, with these having nearly 20% higher average wood density than terra firme and seasonally flooded forests. In Borneo, Jucker et al. (2018a, b) also found large, landscape-level differences in community-mean wood density. As in Amazonia and French Guiana, in Borneo it is white sand forests that have highest wood density. Jucker's (2018a) analysis also reveals smaller but significant variation in basal-area-weighted wood density between and within different sites in north-east Borneo, including additional substrate-related variation between forests growing on alluvial versus sandstone-derived soils. Here, forest-wide wood density is, respectively, 10% lower and 10% greater than the Borneo-wide forest mean (Qie et al. 2017). Therefore, in every case—across African, South American, and Asian landscapes—failing to account for habitat-related variation in wood density significantly biases AGB estimates. In Borneo, where extensive field sampling was combined with hyperspectral imaging, wood density estimates inferred from canopy leaf spectra vary with topography by as much as 40% at the 1-ha scale (Jucker et al. 2018b). In sum, complex variation in biodiversity across tropical forest landscapes is the rule, not the exception, and this matters for biomass mapping.

While quantifying how floristic variation impacts AGB is critical for mapping purposes, the question of *why* tropical forests' wood density varies spatially is equally important. In the white sand case, Borneo and South America are biogeographically isolated from one another so the consistent response implies independent convergence in function driven by selective pressures, possibly as a result of low nutrient availability favouring more conservative, slow-growing species. Topographic differences in wind disturbance (Fortunel et al. 2014) and drought stress (Cosme et al. 2017) may also control local-scale wood density variation. In Amazonia, the large regional differences in wood density are related to greater dominance by light-wooded families in the south-west (e.g., palms and mimosoid legumes) and dense-wooded families in the north-east (e.g., Sapotaceae and caesalpinoid legumes). This may be ultimately driven by the unique biogeographic history of the Guiana Shield and the edaphic differences between deep, weathered soils in east-central Amazonia and less developed soils in the Andean forelands affecting forest dynamics (Fyllas et al. 2009; Baraloto et al. 2011; Quesada et al. 2012; Johnson et al. 2016). Forest structure and dynamics are not only causally linked, but impact other carbon pools too. For example, wood density also affects carbon storage after death: light-wooded forests store less necromass than dense-wooded systems (Chao et al. 2009). The mechanistic links between environment, structure, composition and dynamics are relevant for the practical task of remote sensing of biomass as they point to forest properties measureable remotely which may be used to infer composition and hence wood density.

In sum, species variation impacts AGB at landscape, regional and continental scales. Yet given the scale of the biome we have only begun to evaluate how biodiversity affects mature forest biomass and wood density. Across the 6 million km^2 extent of Amazonia, we lack case study analysis of impacts on AGB, especially at fine scales and in larger, biogeographic units. Here we aim to address these gaps. We first analyse plot-based inventories from one of the best-sampled Amazon landscapes, backed up by some of the most comprehensive botanical work anywhere in the tropics. Secondly, using the latest published data from the RAINFOR plot network we revisit the question of how much wood density matters for large-scale Amazon-wide forest biomass and forest dynamics, and to what extent these relationships hold in geoecological units within Amazonia. Finally, we combine the new and recent ecological work reporting wood density to document for the first time *basal-area-weighted forest wood density* estimates at multiple scales across the tropics.

2 Methods

For our landscape-scale evaluation of species impacts on biomass, we focus on the lower Tambopata region, in south-eastern Peru. Thirty-five years of botanical collecting have generated a relatively complete knowledge of the flora of the region, and forest inventory and monitoring provide sample plots in intact and human-modified forests, including 1-ha permanent plots and 0.1-ha inventories using modified Gentry transects (Gentry 1988; Clinebell et al. 1995; Phillips and Miller 2002; Phillips et al. 2003; Pallqui et al. 2014). Variation in fluvial disturbance, soil chemistry, and land use all affect tree species and human livelihoods here (e.g., Phillips et al. 2003; Lawrence et al. 2005). The landscape is divisible into two major landforms reflecting areas with recent Holocene deposition ('depositional') and more weathered Pleistocene sediments now being eroded ('erosional') (Kalliola et al. 1992; Räsänen et al. 1992; Osher and Buol 1998). This reflects the folk categories recognised by local dwellers, 'Altura' and 'Bajio', who account for subtle differences in elevation and forest resources (Phillips et al. 2006). Classifications derived from larger-scale maps of the Peruvian Amazon (e.g., Asner et al. 2017; Peru Ministerio de Ambiente 2015) are consistent with local perceptions of the natural forest environment but don't fully coincide. We therefore use here the local terms 'Altura' and 'Bajio' and the locally equivalent 'Pleistocene' and 'Holocene' terminology. We centre our analysis on the mature-forest landscape in a roughly 600-km^2 region of the lower Tambopata (Fig. 2), an area with more than 1000 tree species (Phillips et al. 2003). We only use mature-forest plots that are botanically identified as our interest is to identify floristic variation that is geomorphologically associated, rather than due to land-use change. Sampling was conducted between 1983 and 2007 (median date 1998) and stratified collaboratively with local residents by geomorphology using Landsat imagery, with exact locations randomised within target habitats. While the landscape has a fluvial signature, to our knowledge the samples included here have not been subject to recent river flooding.

Following established procedures (e.g., Baker et al. 2004; Lewis et al. 2013; Malhi et al. 2014), we derive taxon-specific wood density (WD) from a large-scale Global Wood Density database (Zanne et al. 2009) and estimate AGB at the tree and plot levels (Table S1). We use standard methods to estimate biomass (Chave et al. 2014) accounting for stem diameter, tree species identity, and height derived from forest-type-specific height-diameter allometries (Sullivan et al. 2018). We also accounted for palm-specific allometry (Goodman et al. 2014b) and implemented these procedures in the BiomasaFP R package (Lopez-Gonzalez et al. 2015).

Because we wanted to assess the impact of using incomplete biological information ('identity-poor') on forest biomass estimates, we first allocated the best available WD per stem and then calculated alternative averages at different scales, using these mean values instead of the best available values to test questions about *how spatial scale of identity impacts WD*. Thus wood density is allocated to individual trees *optimally* by accounting for actual tree-by-tree identity (to generate *community wood density*), and then *instead* by applying to each tree just: (2) the *plot-mean WD*; (3) the *forest-type mean WD*; (4) the Tambopata *landscape-mean WD*; and (5) the *Amazon-wide mean* WD values. Thus, for example, (2) represents a situation where we have perfect knowledge of plot average wood density but no knowledge of individual tree identity, (3) represents a situation where we have knowledge only of forest-type mean wood density, and so on. In each case, we compute mean values using simple abundance-weighting and by weighting by basal area of each species. The different procedures to compute forest wood density are summarised in Table 1. We use our AGB estimates to quantitatively address three linked questions for our study landscape: (1) To what extent does wood density vary among Pleistocene and Holocene landscapes at the tree level and the plot level?; (2) Does accounting for wood density change the expected relationship between forest basal area and biomass?; and (3) What are the consequences in terms of bias and uncertainty of using different 'identity-free' estimators of aboveground biomass?

Then, to explore the links between composition, structure, and function at Amazon scale we use the latest published data from the RAINFOR long-term plot network (Malhi et al. 2002; Peacock et al. 2007). This includes plots monitored for as long as 30 years (Johnson et al. 2016), with standardised protocols applied to field data collection (Phillips et al. 2010) and data management (Lopez-Gonzalez et al. 2011). This enables us to address for the first time the relationships between forest functional *composition* (WD), forest *structure* (AGB), and forest *dynamics* (stem mortality, AGB mortality) in one analysis. We do this at pan-Amazon level, and also for each of the sub-regions of Amazonia defined by geography and substrate origin (Fittkau 1971; Feldpausch et al. 2012): Western Amazonia (Colombia, Ecuador, and Peru), where soils mostly derive from recent Andean deposits; the Brazilian Shield (Bolivia and Brazil); the Guiana Shield (Guyana, French Guiana, Venezuela); and eastern central Amazonia (Brazil), largely comprised of old sedimentary substrates derived from the other three regions (Quesada et al. 2012; Schargel 2011). We thus assess AGB

Fig. 2 Multiple perspectives on Amazon forest diversity. The figure depicts the study region and forest-type ▶ variation sensed with imagery acquired contemporaneously with the floristic and ecological inventories. **a** Top left. *South American forest cover* in the year 2000 and location of Peru. **b** Top right. *Western Amazon forest 'Functional Classes'* inferred from hyperspectral imagery by Asner et al. (2017) in Peru, with location of the lower Tambopata region in south-east Peru highlighted in red box. **c** Centre. *Our sample landscape* outlined as 15-by-40-km zone oriented along the lower Tambopata river. Young or disturbed vegetation regenerating after fluvial and anthropogenic clearing represents $\approx 10\%$ of the landscape and was not sampled. Black icons represent locations of floristic sample plots in 'Altura' forest (Pleistocene sediments); red icons sample plots in 'Bajio' forest (Holocene sediments). In this false-colour image, the purple-green hued vegetation closer to the river corresponds to 'Bajio'; the brighter green away from the river is 'Altura'. Landsat imagery from https://landsat.usgs.gov/landsatlook-images, level-1 data product using imagery from 1999 to 2001, centred on Landsat path 114 row 175 and treated with a three-standard-deviation stretch. **d** Below left. *The best-sampled forests* centred on Tambopata reserved zone. Note the fine-scale variation in canopy composition and structure driven by small elevational differences. The total elevational range within this IKONOS image is ≈ 30 m. **e** Below right: *Ground-truthed interpretation of IKONOS imagery based on direct observation of geomorphology, hydrology and vegetation species and structure*. Colours correspond to ten distinct local forest types (Gentry 1988, Conservation International and Foster 1994): among-habitat diversity in species composition and associated functional traits is greater than the basic Altura–Bajio dichotomy. 'Altura' forest is dark green here (ancient Pleistocene river terrace); 'Bajio' forest includes orange and pink (different levels of Holocene terraces) as well as swamp and fluvial successional systems. Images from Palmero (2004)

as a function of WD across more than 150 permanent plots distributed across Amazonia and compare these to long-term measured rates of stem turnover and carbon turnover for the same forests. We ask, (4) Does wood density of Amazon forests correlate with AGB at regional and Amazon-wide scales? (5) Is mean wood density predictable from the long-term dynamics of the same forests?

2.1 Data Analysis

2.1.1 Landscape-Level Analyses

To examine whether species wood densities vary with tree size, we calculated the correlation between the diameter of individual trees and their species wood density. This was performed separately for Pleistocene and Holocene forests, pooling data from plots in each landscape. We used nonparametric Kendall's tau as tree diameter was not normally distributed. We tested whether plot-level mean WD, total basal area and AGB differed between forest types using *t* tests, or Mann–Whitney tests when the response variable was not normally distributed (basal area, AGB). To assess whether landscape-level differences in wood density alter the relationship between basal area and biomass, we used linear models to relate AGB (log-transformed to meet linear model assumptions of normality and homogeneity of variances) to BA, forest type and their interaction, the latter indicating whether the relationship differed between forest types. To quantify the impact of using identity-free estimators of wood density instead of species wood density, we recalculated the AGB of each plot substituting species WD with the different levels of identity-poor WD metrics (Table 1).

2.1.2 Amazon-Wide Analyses

We assessed the relationship between biomass and wood density and basal area using bivariate linear regression, fitted both to the pan-Amazon data set and separately to each

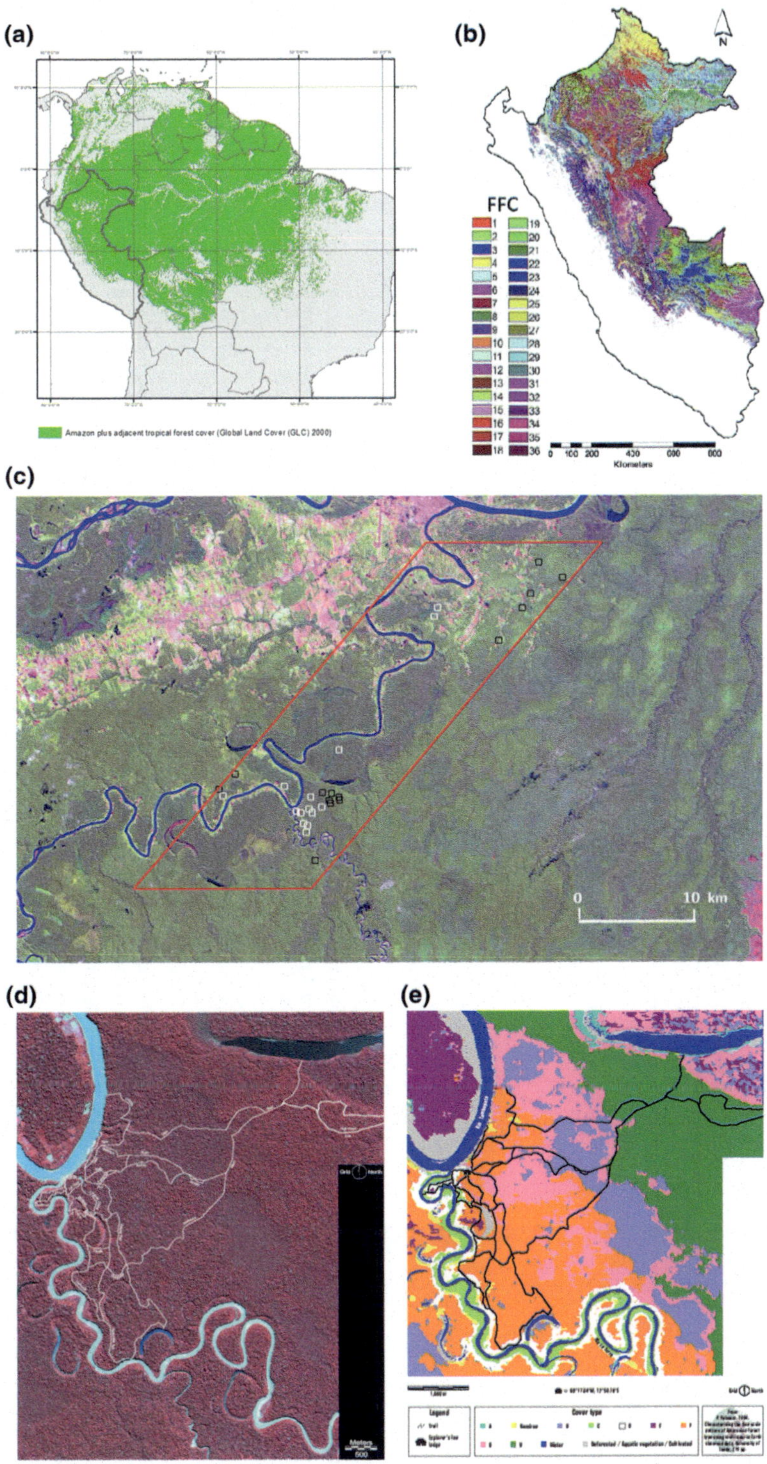
(a)
Amazon plus adjacent tropical forest cover (Global Land Cover (GLC) 2000)
(b)
N
FFC
1
2
3
4
5
6
7
8
9
10
11
12
13
14
15
16
17
18
19
20
21
22
23
24
25
26
27
28
29
30
31
32
33
34
35
36
Kilometers
(c)
0
10 km
(d)
Meters
500
(e)
Legend
Cover type
Water

Table 1 Terms used to report tropical forest wood density (g cm^{-3}) in this paper, together with their definitions and data requirements. *Identity-rich metrics* use species identity to derive the wood density of each and every tree. *Identity-poor metrics* simply apply aggregate mean values to all trees, plots, forest types, or landscapes. We include these latter approaches which implicitly assume that species identity does not matter to assess the impact of using incomplete biological identities on forest biomass estimates. Note that whether identity-poor or identity-rich, the community-mean, plot-mean, forest-type-mean, landscape-mean, Amazon-mean wood density metrics can all be either abundance-weighted or basal-area-weighted. We recommend use of *identity-rich basal-area-weighted wood density* whenever possible (highlighted here and Table 2)

Term	Definition	Data required to estimate
Identity-rich metrics		
Species wood density	Species 'basic specific gravity', the oven-dry mass of a wood sample divided by its green volume (cf. Chave et al. 2006)	Ideally based on multiple individuals and accounting for radial variation from core to pith. Either from compilations (Zanne et al. 2009) or local measurements (e.g., Goodman et al. 2014a, b) If no species wood density measurements available, allocate the genus-level mean, else the family-level mean (Baker et al. 2004)
Community-mean wood density	Community-mean wood density (WD), based on each tree's species wood density weighted by the abundance of each species	Additionally requires species-abundance data for the plot
Community-mean wood density: basal-area-weighted	Community-mean WD, based on each tree's species WD and weighted by the basal area of each species (e.g., Lewis et al. 2013)	Additionally requires accurate, above-buttress diameter measurement of every individual tree
Identity-poor metrics		
Plot-mean wood density	The mean WD of all trees in the plot, based on species WD with species' contributions weighted by their abundance or basal area	
Forest-type-mean wood density	The mean value of 'plot-mean wood density' averaged across contributing plots in the forest type	In the case of the Tambopata landscape, computed separately for Altura and Bajio forest types[a]
Landscape (Tambopata-wide) mean wood density	The mean value of 'forest-type-mean wood density', averaged across contributing forest types in the landscape	In the case of the Tambopata landscape, the mean of the mean values for Altura and Bajio forests[a]
Amazon-mean wood density	The mean value of 'plot-mean wood density', averaged across contributing plots in Amazonia	Published wood density values from plots across Amazonia (Mitchard et al. 2014)

[a] *Altura* and *Bajio* forest types represent the two major units within the Tambopata landscape. The folk nomenclature used here corresponds to geomorphical units (erosional, depositional) and chronological units (Pleistocene, Holocene). See text for details

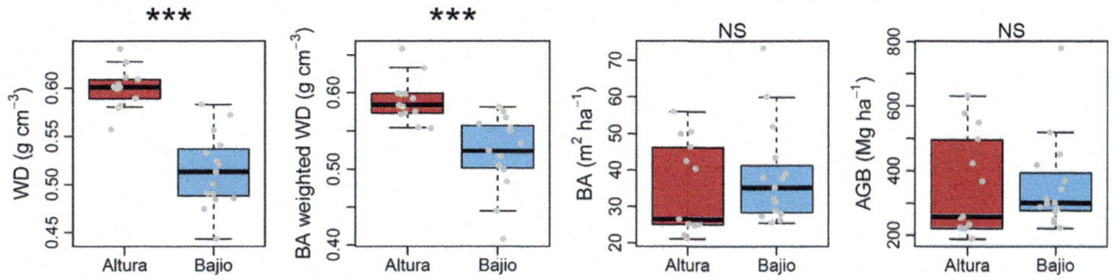

Fig. 3 Landscape variation in wood density, basal area and aboveground biomass. Boxplots show variation in each variable within Altura and Bajio forests, with grey points showing values from individual plots (jitter on x-axis for presentation purposes only). Differences between Altura and Bajio forests were tested using t tests (abundance-weighted wood density (WD), basal-area-weighted wood density) or Mann–Whitney tests (basal area, aboveground biomass), ***$P<0.001$; **$P<0.01$; *$P<0.05$, NS $P\geq0.05$

biogeographic region. We used variation partitioning to identify the independent contributions of each variable to explaining variation in AGB (Legendre and Legendre 2012); linear models were constructed with WD (M1), BA (M2) or both WD density and BA (M3) as explanatory variables, and R^2 values extracted. Shared variation due to both variables is calculated as $M1_{R2}+M2_{R2}-M3_{R2}$, which is subtracted from $M1_{R2}$ and $M2_{R2}$ to get the independent effect of each variable. Finally, we used linear regression to assess the bivariate relationships between WD and attributes of forest dynamics, and between AGB and measures of mortality.

3 Results

(1) *Wood density varies fivefold among species in Tambopata, with a similar range in both forest types*. Species wood density is only weakly associated with tree size, with correlations between species wood density and diameter slightly stronger in Holocene (Bajio) forests (Kendall's tau correlation, $\tau=-0.095$, $P<0.001$) than Pleistocene (Altura) forests ($\tau=-0.036$, $P<0.001$).

At the plot level (Table S1), there is a marked variation in wood density within landscapes. Abundance-weighted mean wood density was on average 16.6% higher in Altura forests than in Bajio forests ($t=7.37$, $df=22.5$, $P<0.001$, Fig. 3), and basal-area-weighted mean wood density was 13.4% higher ($t=4.66$, $df=23.4$, $P<0.001$, Fig. 3). In contrast, basal area was on average 9.0% higher in Bajio forest plots, but this difference was not statistically significant (Mann–Whitney test, $P=0.274$, Fig. 3). Aboveground biomass was similar in both forest types (Mann–Whitney test, $P=0.387$, Fig. 3).

(2) *Landscape-associated differences in wood density greatly alter the relationship between basal area and biomass*. In Altura forests, aboveground biomass increased by 3.6% per 1-m^2 increase in basal area (linear regression, ln (AGB)$=0.036\pm0.003$ BA, $t=12.8$, $P<0.001$, Fig. 4). Yet in Bajio forests, because average wood density was lower, AGB increased by just 2.3% per 1-m^2 increase in basal area (interaction between forest type and basal area, $\beta=-0.013\pm0.004$, $t=3.5$, $P=0.002$, Fig. 4). When analysis is repeated without any one of the three highest basal area outliers, the interaction term remains statistically significant.

Results were unaffected by plot size. Differences in wood density between Altura and Bajio forests remained when restricting analysis to plots of ≥1 ha ($t=5.15$, $df=7.0$,

Fig. 4 Relationship between stand basal area and aboveground biomass in Altura and Bajio forests. Note that aboveground biomass has been log-transformed to homogenise variances

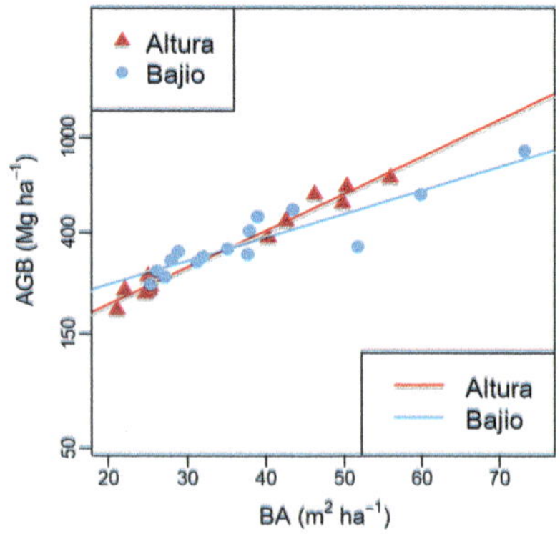

$P=0.001$), and the difference in the basal area–biomass relationship between forest types remained statistically significant when allowing it to be scale-dependent (no interaction between plot size and basal area, $t=0.05$, $P=0.638$; strong interaction between basal area and forest type, $t=3.49$, $P=0.002$).

(3) *Ignoring the actual species identity of each tree biases estimates of forest biomass* (Table S2). Using plot-mean wood density for all trees instead of their species-specific wood density (i.e. representing a situation where we have perfect knowledge of forest-wide spatial variation in wood density but no knowledge of individual identities) results in a mean error in estimating aboveground biomass of 15.0 ± 2.5 Mg ha^{-1} (~4.3% of AGB), with a maximum error of 39.8 Mg ha^{-1} (~11.9% of AGB) (Table S2). The maximum error was 77.1 Mg ha^{-1} (~17% of AGB). This bias itself varies between forest types, being negative for the Bajio forests but not for the Altura forests (Table S2).

Compared to estimates based on species' wood densities, when values were substituted with average wood density, all these 'identity-free' estimates of AGB had error and bias. Both the absolute error and bias increased with the spatial scale of the averaging process (Fig. 5). In particular, absolute bias increased markedly when moving from using a plot- or forest-type-mean wood density to a landscape or Amazon-wide mean wood density (e.g., absolute error was 26.6% higher when landscape-mean wood density was applied instead of the forest-type-mean). Yet even plot-level mean values introduce uncertainty and bias to the forest biomass estimates.

(4) *Across the Amazon basin, low forest biomass is strongly associated with both low wood density and low basal area* (Fig. 6). This is of course unsurprising, given that wood density and stem size are used to calculate tree biomass, but these Amazon-basin associations are worth noting especially given varying patterns at sub-regional scales.

(5) *The relationships between basal area, wood density and forest AGB vary, with different slopes and intercepts among regions*. In particular for western Amazonia, relationships are different to the other regions. There are also correlations between stand wood density and basal area, but these are weak and variable among regions (Fig. 6).

(6) Neither the rate of biomass production nor that of its loss is clearly associated with basal-area-weighted wood density. Thus, the *species traits of Amazon forests do not strongly control the rate at which carbon is being cycled by the forest* (Fig. 7, and see Fauset et al. 2019). Yet they are associated with the rates are which *individual trees* are cycled—stem mortality rates are clearly linked to the wood density of the forest, confirming that the lower the stand-level wood density is, the more rapidly the trees die (Fig. 7).

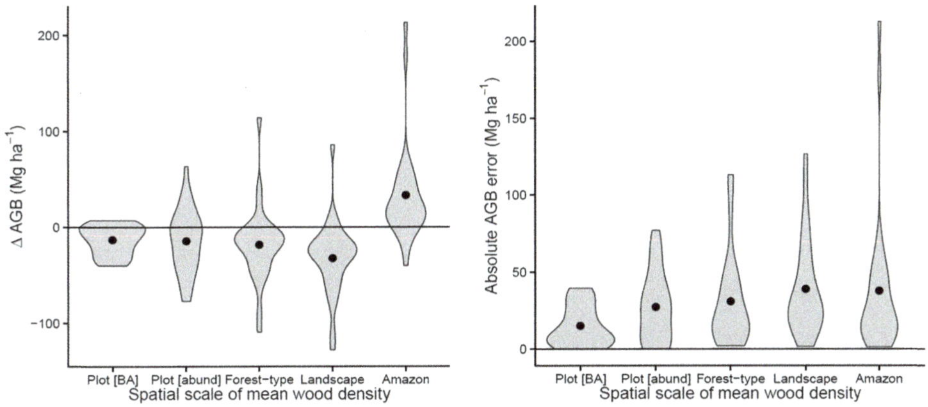

Fig. 5 Error in stand-level aboveground biomass estimates when using wood density means calculated as plot, forest type, landscape, and Amazon-wide scales, rather than the actual species values. Violin plots illustrate the distribution of values among plots, while points show the mean error across plots. Note the differences also between abundance-weighted WD and basal-area-weighted WD: the latter clearly entails less bias

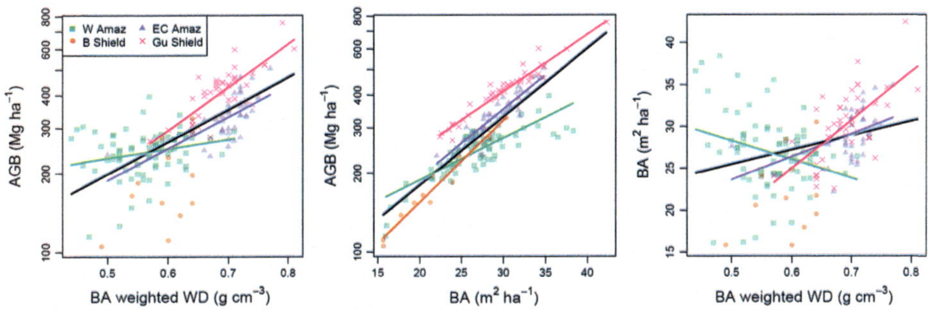

Fig. 6 Amazon and regional relationships between basal-area-weighted wood density, biomass, and basal area. For each variable pair, regression models were fitted across the whole data set and for each region. Regions are Western Amazon, Brazilian Shield, East-Central Amazon, Guyana Shield, following Feldpausch et al. 2012. Statistically significant relationships are plotted. Note that regression models with basal-area-weighted wood density predict Amazon biomass with much greater fidelity than simple relationships with basal area alone (Tables S3, S4). Model coefficients are given in Table S5

(7) *These relationships between wood density and forest dynamics propagate through to clear relationships between biomass and forest dynamics*. Thus, while standing biomass is not obviously associated with the rate at which wood biomass dies (Fig. 8 left), it is clearly related to the rate at which individual trees die (Fig. 8 right). Lower biomass forests typically have much lower wood density (Fig. 6) and much faster stem turnover (Fig. 8).

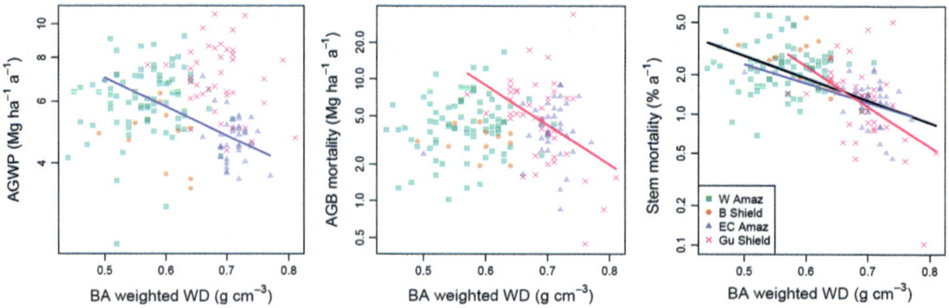

Fig. 7 Amazon and regional relationship between forest dynamic processes and wood density. Regression models were fitted across the whole data set and for each region. Statistically significant relationships are plotted. Model coefficients are given in Table S5

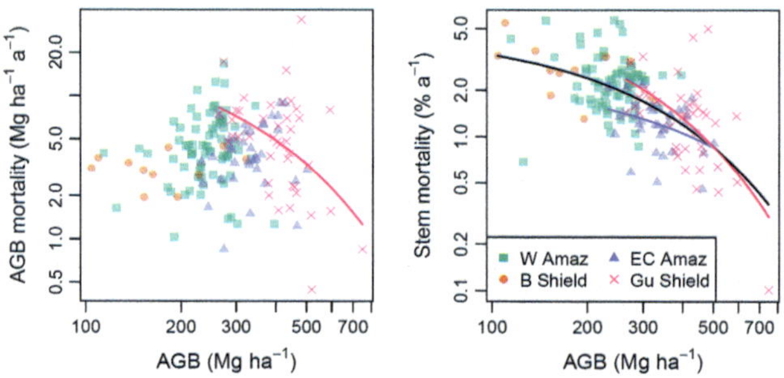

Fig. 8 Amazon and regional relationship between forest mortality and AGB. Regression models were fitted for the whole data set and for each region. Statistically significant relationships are plotted. Note the close similarity with the centre and right panels of Fig. 7: species wood density strongly determines biomass and is closely associated with the rate at which individual trees die (figures adapted from Fig. 8 in Johnson et al. 2016). Model coefficients are given in Table S5

4 Discussion

Our analysis shows that identity matters. Mapping tropical forest biomass and carbon always benefits from locally validated ground-based measurement of biological variation. In our well-studied Amazon landscape, forest wood density varies within and among landscape geomorphological units. Because relationships between wood density and forest size metrics such as basal area themselves vary, knowledge of forest dimensions and species composition is necessary for estimating tropical biomass. At larger Amazon-wide scales, we also find that lower wood density forests are closely associated with lower AGB and faster stem turnover. Yet within some Amazon regions and particularly western Amazonia, this relationship with forest dynamics breaks down. We explore the implications of these findings here.

In the specific case of Tambopata, the impacts of size and identity on mature-forest biomass are decoupled: while wood density is $\approx 15\%$ greater in the erosional Pleistocene landscape, basal area is equal or greater in depositional Holocene landscapes. As a result, while aboveground biomass is similar in both landscapes, *not* accounting for species effects (i.e. simply assuming a uniform wood density) would substantially underestimate AGB in erosional landscapes. Indeed, *all 'identity-free' approaches that ignore the floristic variation within these forests lead to biased estimates of forest biomass.* Notably, using a landscape-level Tambopata-wide mean wood density results in equally poor estimates of AGB as using an Amazon-wide mean wood density, although the sign of the errors is reversed. In this landscape, very good, *forest-type-specific* floristic knowledge is essential to estimate AGB with an error (bias) of less than 10%. While forest-type- and plot-mean wood density values give performance gains, the error (bias) on both is still 15–30 Mg AGB ha^{-1} compared with the baseline 'identity-rich' state in which *every tree in each plot is botanically identified and taxon-specific wood density values are used for every tree.* This bias itself varies between forest types, being negative for the Bajio forests but variable for the Altura forests, indicating that in the former canopy trees have greater wood density than understorey trees but not in the latter. This underlines the value of accounting for size and identifying all trees—large *and* small—to reduce bias and uncertainty in forest-level biomass estimation. Our results show that landscape-level mapping of tropical forest biomass requires both tree-level and landscape-level knowledge of biodiversity. Future analyses should account for this and also assess the extent to which accounting for within-species tree level and environmental variation in wood density (e.g., Patiño et al. 2009; Baraloto et al. 2011; Fortunel et al. 2014) might further improve the fidelity of AGB retrieval.

At larger scales, identity also matters: there are large differences in stand-level wood density across Amazonia. Basal-area-weighted wood density varies by 80% between the lowest and highest 1-ha plot values, and even within individual sub-regions it varies by as much as 40%. Low wood density Amazon forests have faster stem turnover, confirming that lower stand-level wood density is associated with more rapid rates of tree death. The relationship is especially marked for stem-level mortality, but much less so for biomass mortality. A key feature of stand-level wood density variation therefore is that it reflects forest dynamics and especially the rate of turnover in the tree population. However, the pan-Amazon associations between wood density and biomass, and wood density and mortality break down at some scales. In some regions (e.g., western Amazonia), variation in the size class distribution among forests may be more important than variation in mean wood density for determining variation in AGB (Fig. 6), but in most regions and across Amazonia, mean wood density is a major determinant of biomass.

4.1 Towards Integrating Species Effects into AGB Mapping

Currently, few attempts to map tropical forest biomass and carbon fully account for species effects—either because they are assumed to be unimportant, or else because ground-level data needed to parameterise and validate them is lacking. Yet foresters have long known that species impact on tropical biomass. An extensive compilation from the last century (Fearnside 1997) suggested that wood density in the Brazilian Amazon varies by 25% across forest types (from 0.60 to 0.75 g cm^{-3}), but as this is based on forestry surveys with imprecise identification it contains considerable uncertainty. A recent pan-tropical assessment of forest structure confirmed great variation of wood density within each continent (Sullivan et al. 2017, Fig. S16). This suggests that biodiversity-driven variation in wood density is a pervasive and multi-scalar feature of all tropical forests. Since sufficient inventory-based evidence based on botanical identification now exists to show that species composition matters biome-wide for biomass, we here compiled mean values where we could source *well-identified, well-measured plot data where basal-area weighting has been consistently applied* (Table 2).

Table 2 Multi-scale, basal-area-weighted mean community wood density for old-growth forests across the lowland tropical forest biome. Data assembled from the peer-reviewed ecological literature, from scales of 10^0 to 10^{10} hectares. *All values are basal-area-weighted and computed for each plot accounting for taxon-specific wood density*[a]. Thus, basal-area-weighted community-level WD of each plot was estimated as $\Sigma\, BA_{ij} \times WD_i$, where BA_{ij} is the relative basal area of species i in plot j, and WD_i is the mean wood density of species i. Values reported here represent the means of wood density from all available forest plots at the appropriate scale. The nested table structure illustrates how even these mean values vary at all scales, including among continents, among regions and nations, among landscapes within nations, and among forest types within landscapes within nations. Note how the scale at which WD is computed always matters. The best mean WD value to apply will depend on the spatial resolution of the remote sensing and mapping

Continent	Tropical forest climate	Region/nation	Landscape/forest type	Value	Source
Pan-tropical mean				0.619	Mean of Africa, Asia, S. America network mean values assembled here[b]
Africa	Moist			0.633 (CI = + 0.0080, n = 260 plots)	Lewis et al. (2013)
		West Africa		0.61	Lewis et al. (2013)
		Central Africa		0.64	*ibid.*
			Monodominant	0.696	*ibid.*
			Mixed	0.627	*ibid.*
		East Africa		0.61	*ibid.*
		West and Central Africa			
			Acrisols	0.609	Lewis et al. (2013)
			Cambisols	0.617	*ibid.*
			White Sand	0.660	*ibid.*
			Swamp	0.728	*ibid.*
		Central African Republic	Mbaiki: deep resource-rich soils	0.51[c]	Gourlet-Fleury et al. (2011)
			Mbaiki: deep resource-poor soils	0.59[c]	*ibid.*
			Mbaiki: physically constrained soils	0.525[c]	*ibid.*
Asia	Moist			0.594 (SD = 0.039, n = 71 plots)	Qie et al. (2017)
		Borneo		0.594	*ibid.*

Table 2 (continued)

Continent	Tropical forest climate	Region/nation	Landscape/forest type	Value	Source
			Old-growth, no edge effects	0.600 (SD = 0.038, n = 49 plots	*ibid.*
			Old-growth, edge effects	0.581 (SD = 0.039, n = 22 plots)	*ibid.*
		Borneo: Sabah	Sepilok: Alluvial	0.55	Jucker et al. (2018a, b)
			Sepilok: White Sand	0.64	*ibid.*
Central America				0.540 (SD = 0.063, n = 5 sites)	This paper, from literature sources
	Wet	Costa Rica	La Selva	0.47[d]	Muller-Landau (2004)
		Panama	Sherman	0.595	Stegen et al. (2009)
	Moist	Panama	Barro Colorado Island	0.51[d]	Muller-Landau (2004)
		Panama	Barro Colorado Island	0.545	Stegen et al. (2009)
	Dry	Panama	Cocoli	0.494	*ibid.*
		Costa Rica	San Emilio	0.614	*ibid.*
South America: Amazonia	Moist	All Amazon		0.629 (SD = 0.081, n = 165 plots)	This paper, from RAINFOR data
		Central Amazon		0.703 (SD = 0.041, n = 37 plots)	This paper; updating Baker et al. (2004), Mitchard et al. (2014)
		Brazilian Shield		0.591 (SD = 0.048, n = 11 plots)	*ibid.*
		Guyana Shield		0.688 (SD = 0.048, n = 41 plots)	*ibid.*
			Paracou: Terra Firme and Alluvial	0.67[e]	Baraloto et al. (2011)
			Paracou: White Sand	0.72	*ibid.*
		Western Amazon		0.566 (SD = 0.056, n = 76 plots)	This paper, updating Baker et al. (2004), Mitchard et al. (2014)
		Ecuador	Yasuni: Terra Firme	0.588	Stegen et al. (2009)
		Peru	Loreto: Terra Firme and Flooded	0.62[e]	Baraloto et al. (2011)
			Loreto: White Sand	0.64	*ibid.*

Table 2 (continued)

Continent	Tropical forest climate	Region/nation	Landscape/forest type	Value	Source
		Peru	Tambopata	0.554 (SD = 0.053, n = 28 plots)	This paper
			Tambopata: Holocene	0.521 (SD = 0.049, n = 15 plots)	*ibid.*
			Tambopata: Pleistocene	0.591 (SD = 0.029, n = 13 plots)	*ibid.*
			Tambopata: swamp	0.467 (SD = 0.034, n = 2 plots)	*ibid.*

[a]Multi-plot studies and compilations that present community-weighted wood density for tropical forests were only included if values were clearly basal-area-weighted and properly identified. Thus, (1) studies that apparently represent the average wood density of all species or stems in plots or other samples (e.g., ter Steege et al. 2006; Slik et al. 2010, Fortunel et al. 2014) were not included, because weighting by relative contribution to basal area is more likely to approximate the contribution of each species to carbon storage than weighting by its relative frequency or abundance (cf. the large differences in Amazon-dominant species as reported by Fauset et al. 2015 and ter Steege et al. 2013 when evaluated by basal area and when evaluated by stem abundance). Similarly, (2) studies based largely or entirely on vernacular name identifications are excluded, as in diverse tropical forests these are less reliable and precise than botanical identifications (cf. Fearnside 1997 for data and discussion of this). Sullivan et al. 2017 is not listed as a source in this table as data plotted in their Fig S16 are mostly available as continent-level mean values in other recent analyses (Lewis et al. 2013 for Africa, Qie et al. 2017 for Borneo, and the current paper for Amazonia)

[b]The simple unweighted mean of Amazon, Asian, and African moist forest values here from the plot networks across tropical forest Africa (AfriTRON), Asia (T-FORCES), and South America (RAINFOR), where trees ≥ 10 cm d.b.h. and a standard wood density data source (Zanne et al. 2009) were used. No pan-tropical value could be located in previous literature that was clearly based on plot measurements in which trees were identified to species and trees were all measured

[c]Trees ≥ 20 cm d.b.h.; from data plotted in Fig. 2 of Gourlet-Fleury et al. (2011)

[d]Trees > 30 cm d.b.h

[e]From data plotted in Fig. 5 of Baraloto et al. (2011)

Accounting for variation in such forest-wide means can improve biomass estimates. We find differences in intact forest basal-area-weighted wood density of as much as 20% in African, Bornean and Amazon landscapes of 10^1–10^3 km^2, of 20% within 10^5–10^6-km^2 geographic regions in north-west Amazonia and French Guiana, and 10–30% at the continental scale (between 10^7 km^2 units) between south-west and north-east Amazonia. The data compiled in Table 2 show that wood density impacts hugely on biomass even at continental levels. For example, applying a pan-tropical wood density mean to Central American forests could result in over-estimating aboveground carbon stocks there by 15%.

Our pan-Amazon results suggest a possible avenue for using technology like LiDAR to indirectly derive the key composition-based property of wood density. Thus, by quantifying tree mortality rates it may be possible to estimate wood density and so improve LiDAR's ability to estimate AGB. LiDAR is being increasingly used to sense tree biomass mortality (e.g., Espírito-Santo et al. 2014; Leitold et al. 2018). If these estimates can be produced over large-enough spatial and temporal scales to yield time-averaged tree mortality rates, it may be possible to derive proxies for wood density and validate them with plot species-level identifications. This would be a promising angle to explore, for example, with repeat-survey LiDAR data as they become available. Similarly, hyperspectral properties of forest canopies may correlate with wood density and underlying soil conditions, so these hold promise for deriving canopy wood density estimates that can be validated with full forest species-level identifications.

The multi-scale variability in forest wood density means that the next generation of tropical forest carbon maps and models needs to account better for species and functional variation. Mapping at all scales benefits from locally validated, ground-based identification of measured trees. Because most tropical forests are very diverse, this requires highly skilled professional botanists to collect and identify the trees, working in georeferenced plots, measured carefully and more-or-less synchronously with remote sensing measurements. There are currently just a handful of tropical forest landscapes where remote- and ground-based measurements exist with the requisite level of species identification (Chave et al. 2019). We need many more, distributed across key environmental and biodiversity gradients, if tropical forest nations are to realise the potential of remote sensing to help measure and validate their carbon stocks, fluxes, and nationally determined contributions to the Paris Climate Accord.

Acknowledgements We thank Klaus Scipal and the European Space Agency for organising the ISSI meeting in Berne in November 2017, Space-Based Measurement of Forest Properties for Carbon Cycle Research, and the invitation to OLP which stimulated this manuscript. Funding for the fieldwork reported here came from a Grant (ERP-196) from the U.K. Department for International Development, as well as grants from the U.K. Natural Environment Research Council (Grants NE/B503384/1, NE/D01025X/1, NE/D005590/1). Analysis was supported by NE/N012542/1 (BIORED) and an ERC Advanced Grant #291585 (T-FORCES). OLP was also supported by a Royal Society-Wolfson Research Merit Award. We thank Victor Chama, Camilo Diaz, Fernando Cornejo, Cesar Chacón, Alejandro Farfan, Washington Galeano, Euridice Honorio, Nestor Jaramillo, Kate Johnson, Mauricio Lopez, Gabriela Lopez-Gonzalez, Antonio Peña Cruz, Sam Rose, Wilmar Ramirez, Peggy Stern, Martín Timana, Miguel Alexiades, Joey Talbot and Adela Reategui for their field contributions, and the Peruvian national authorities (INRENA and SERNANP) for field and collection permissions. Paulo Palmero and Georgia Pickavance made graphical contributions. Elisban Armas, Terry Erwin, Gary Hartshorn, and in particular Alwyn Gentry played key roles in the first floristic inventories of Tambopata. We thank colleagues at Peruvian Safaris S.A., Explorer's Inn, Bahuaja Lodge, and the communities of La Torre, Infierno and Jorge Chavez for their hospitality and assistance. The manuscript has benefited from the insights and suggestions of two anonymous referees. Our paper is dedicated to the memory of the late Dr Max Gunther, whose remarkable vision and perseverance have helped shed much light on the extraordinary diversity of the Peruvian Amazon.

References

Asner GP, Martin RE, Knapp DE, Tupayachi R, Anderson CB, Sinca F, Vaughn NR, Llactayo W (2017) Airborne laser-guided imaging spectroscopy to map forest trait diversity and guide conservation. Science 355(6323):385–389

Avitabile V, Herold M, Heuvelink GB, Lewis SL, Phillips OL, Asner GP, Armston J, Ashton PS, Banin L, Bayol N et al (2016) An integrated pan-tropical biomass map using multiple reference datasets. Glob Change Biol 22(4):1406–1420

Baker TR (2018) biodiversity increases the resilience of tropical forests to climate change: implications for conservation policy. In: Rodríguez L, Anderson I (eds) Secretariat of the convention on biological diversity the Lima declaration on biodiversity and climate change: contributions from science to policy for sustainable development. Technical series no. 89, pp 24–31

Baker TR, Phillips OL, Malhi Y, Almeida S, Arroyo L, Di Fiore A, Erwin T, Killeen TJ, Laurance SG, Laurance WF et al (2004) Variation in wood density determines spatial patterns in Amazonian forest biomass. Glob Change Biol 10:545–562

Baker TR, Phillips OL, Laurance WF, Pitman NC, Almeida S, Arroyo L, DiFiore A, Erwin T et al (2009) Do species traits determine patterns of wood production in Amazonian forests? Biogeosciences 6:297–307

Baker TR, Pennington RT, Dexter KG, Fine PV, Fortune-Hopkins H, Honorio EN et al (2017) Maximising synergy among tropical plant systematists, ecologists, and evolutionary biologists. Trends Ecol Evol 32:258–267

Baraloto C, Rabaud S, Molto Q, Blanc L, Fortunel C, Herault B, Davila N, Mesones I, Rios M, Valderrama E, Fine PV (2011) Disentangling stand and environmental correlates of aboveground biomass in Amazonian forests. Glob Change Biol 17:2677–2688

Brown S, Lugo AE (1990) Tropical secondary forests. J Trop Ecol 6:1–32

Bunker DE, DeClerck F, Bradford JC, Colwell RK, Perfecto I, Phillips OL, Sankaran M, Naeem S (2005) Species loss and aboveground carbon storage in a tropical forest. Science 310:1029–1031

Chao KJ, Phillips OL, Baker TR, Peacock J, Lopez-Gonzalez G, Vásquez Martínez R, Monteagudo A, Torres-Lezama A (2009) After trees die: quantities and determinants of necromass across Amazonia. Biogeosciences 6(8):1615–1626

Chave J, Davies SJ, Phillips OL, Lewis SL, Sist P, Schepaschenko D, Armiston J, Baker TR, Coomes D et al (2019) Ground data are essential for biomass remote sensing missions. Surv Geophys. https://doi.org/10.1007/s10712-019-09528-w

Chave J, Muller-Landau HC, Baker TR, Easdale TA, ter Steege H, Webb CO (2006) Regional and phylogenetic variation of wood density across 2456 neotropical tree species. Ecol Appl 16:2356–2367

Chave J, Réjou-Méchain M, Búrquez A, Chidumayo E, Colgan MS, Delitti WBC, Duque A et al (2014) Improved allometric models to estimate the aboveground biomass of tropical trees. Glob Change Biol 20:3177–3190

Clinebell R, Phillips OL, Gentry AH, Stark N, Zuuring H (1995) Prediction of neotropical woody plant diversity from soil and climatic data. Biodivers Conserv 4:56–90

Coelho de Souza F, Dexter KG, Phillips OL, Brienen RJW, Chave J, Galbraith DR, Lopez Gonzalez G, Monteagudo Mendoza A et al (2016) Evolutionary heritage influences Amazon tree ecology. Proc R Soc B Biol Sci 283(1844):20161587. https://doi.org/10.1098/rspb.2016.1587

Condit R, Pitman N, Leigh EG, Chave J, Terborgh J, Foster RB, Núnez P, Aguilar S, Valencia R et al (2002) Beta-diversity in tropical forest trees. Science 295:666–669

Conservation International and Foster RB (1994) The Tambopata-Candamo reserved zone of southeastern Perú: a biological assessment. Conservation International, Washington, DC, p 184

Coomes DA, Dalponte M, Jucker T, Asner GP, Banin LF, Burslem DF, Lewis SL, Nilus R, Phillips OL et al (2017) Area-based vs tree-centric approaches to mapping forest carbon in Southeast Asian forests from airborne laser scanning data. Remote Sens Environ 194:77–88

Cosme LH, Schietti J, Costa FR, Oliveira RS (2017) The importance of hydraulic architecture to the distribution patterns of trees in a central Amazonian forest. New Phytol 215:113–125

Draper F, Baraloto C, Brodrick P, Phillips OLB, Vásquez R, Honorio Coronado E, Baker T et al. (2019) Imaging spectroscopy predicts variable distance decay across contrasting Amazonian tree communities. J Ecol (in press)

Duncanson L, Armston J, Disney M, Avitabile V, Barbier N, Calders K, Carter S, Chave J et al The importance of global land product validation: towards a standardized protocol for aboveground biomass. Surv Geophys (in press)

Espírito-Santo FD, Gloor M, Keller M, Malhi Y, Saatchi S, Nelson B, Junior RC, Pereira C, Lloyd J et al (2014) Size and frequency of natural forest disturbances and the Amazon forest carbon balance. Nat Commun 5:3434

Fauset S, Johnson MO, Gloor M, Baker TR, Monteagudo A, Brienen RJ, Feldpausch TR et al (2015) Hyperdominance in Amazonian forest carbon cycling. Nat Commun 6:6857

Fauset SF, Gloor M, Fyllas N, Phillips OL, Asner GP, Baker T, Bentley L, Brienen R et al (2019) Individual-based modelling of Amazon forests suggests that climate controls productivity while traits control demography. Front Earth Sci. https://doi.org/10.3389/feart.2019.00083

Fearnside PM (1997) Wood density for estimating forest biomass in Brazilian Amazonia. For Ecol Manag 90:59–87

Feldpausch TR, Lloyd J, Lewis SL, Brienen RJ, Gloor M, Monteagudo Mendoza A, Lopez-Gonzalez G, Banin L et al (2012) Tree height integrated into pantropical forest biomass estimates. Biogeosciences 27:3381–3403

Fittkau EJ (1971) Esboco de uma divisão ecologica da regiåo amazónica. Proc Symp Biol Trop Amaz Florencia y Leticia 1969:1363–1372

Fortunel C, Ruelle J, Beauchêne J, Fine PV, Baraloto C (2014) Wood specific gravity and anatomy of branches and roots in 113 Amazonian rainforest tree species across environmental gradients. New Phytol 202:79–94

Fyllas NM, Patiño S, Baker TR, Bielefeld Nardoto G, Martinelli LA, Quesada CA, Paiva R, Schwarz M et al (2009) Basin-wide variations in foliar properties of Amazonian forest: phylogeny, soils and climate. Biogeosciences 6:2677–2708

Gentry AH (1988) Changes in plant community diversity and floristic composition on environmental and geographical gradients. Ann Mo Bot Gard 75:1–34

Good P, Bamber J, Halladay K, Harper AB, Jackson LC, Kay G, Kruijt B, Lowe JA et al (2018) Recent progress in understanding climate thresholds. Prog Phys Geogr Earth Environ 42:24–60. https://doi.org/10.1177/0309133317751843

Goodman RC, Phillips OL, Baker TR (2012) Tropical forests: tightening up on tree carbon estimates. Nature 491:27

Goodman RC, Phillips OL, Baker TR (2014a) The importance of crown dimensions to improve tropical tree biomass estimates. Ecol Appl 24:680–698

Goodman RC, Phillips OL, del Castillo Torres D, Freitas L, Cortese ST, Monteagudo A, Baker TR (2014b) Amazon palm biomass and allometry. For Ecol Manag 310:994–1004

Gourlet-Fleury S, Rossi V, Rejou-Mechain M, Freycon V, Fayolle A, Saint-André L, Cornu G, Gérard J, Sarrailh JM, Flores O, Baya F (2011) Environmental filtering of dense-wooded species controls aboveground biomass stored in African moist forests. J Ecol 99:981–990

Hietz P, Rosner S, Hietz-Seifert U, Wright SJ (2017) Wood traits related to size and life history of trees in a Panamanian rainforest. New Phytol 213:170–180

Higgins MA, Ruokolainen K, Tuomisto H, Llerena N, Cardenas G, Phillips OL, Vásquez R, Räsänen M (2011) Geological control of floristic composition in Amazonian forests. J Biogeogr 38:2136–2149

Honorio Coronado EN, Baker TR, Phillips OL, Pitman NC, Pennington RT, Vasquez Martinez R, Monteagudo A et al (2009) Multi-scale comparisons of tree composition in Amazonian terra firme forests. Biogeosciences 6(11):2719–2731

Johnson MO, Galbraith D, Gloor M, De Deurwaerder H, Guimberteau M, Rammig A, Thonicke K et al (2016) Variation in stem mortality rates determines patterns of above-ground biomass in Amazonian forests: implications for dynamic global vegetation models. Glob Change Biol 22:3996–4013

Jucker T, Asner GP, Dalponte M, Brodrick PG, Philipson CD, Vaughn NR, Teh YA, Brelsford C, Burslem DF, Deere NJ, Ewers RM et al (2018a) Estimating aboveground carbon density and its uncertainty in Borneo's structurally complex tropical forests using airborne laser scanning. Biogeosciences 15(12):3811–3830

Jucker T, Bongalov B, Burslem DF, Nilus R, Dalponte M, Lewis SL, Phillips OL, Qie L, Coomes DA (2018b) Topography shapes the structure, composition and function of tropical forest landscapes. Ecol Lett 21(7):989–1000

Kalliola R, Salo J, Puhakka M, Rajasilta M, Häme T, Neller RJ, Räsänen ME, Arias WD (1992) Upper Amazon channel migration. Naturwissenschaften 79(2):75–79

Lawrence A, Phillips OL, Ismodes AR et al (2005) Local values for harvested forest plants in Madre de Dios, Peru: towards a more contextualised interpretation of quantitative ethnobotanical data. Biodivers Conserv 14:45–79
Legendre P, Legendre LF (2012) Numerical ecology. Elsevier, Amsterdam
Leitold V, Morton DC, Longo M, dos-Santos MN, Keller M, Scaranello M (2018) El Niño drought increased canopy turnover in Amazon forests. New Phytol 219(3):959–971
Levis C, Costa FR, Bongers F, Peña-Claros M, Clement CR, Junqueira AB, Neves EG et al (2017) Persistent effects of pre-Columbian plant domestication on Amazonian forest composition. Science 355:925–931
Lewis SL, Sonké B, Sunderland T, Begne SK, Lopez-Gonzalez G, van der Heijden GMF, Phillips OL et al (2013) Above-ground biomass and structure of 260 African tropical forests. Philos Trans R Soc B Biol Sci 368(1625):20120295
Lopez-Gonzalez G, Lewis SL, Burkitt M, Phillips OL (2011) ForestPlots.net: a web application and research tool to manage and analyse tropical forest plot data. J Veg Sci 22:610–613
Lopez-Gonzalez G, Sullivan MJP, Baker TR (2015) BiomasaFP: tools for analysing data downloaded from ForestPlots.net. R package version 1.1
Malhi Y, Phillips OL, Lloyd J, Baker T, Wright J, Almeida S, Arroyo L, Frederiksen T, Grace J, Higuchi N et al (2002) An international network to monitor the structure, composition and dynamics of Amazonian forests (RAINFOR). J Veg Sci 13:439–450
Malhi Y, Wood D, Baker TR, Wright J, Phillips OL, Cochrane T, Meir P, Chave J, Almeida S, Arroyo L et al (2006) The regional variation of aboveground live biomass in old-growth Amazonian forests. Glob Change Biol 12:1107–1138
Malhi Y, Farfán Amézquita F, Doughty CE, Silva-Espejo JE, Girardin CA, Metcalfe DB, Aragão LE et al (2014) The productivity, metabolism and carbon cycle of two lowland tropical forest plots in southwestern Amazonia, Peru. Plant Ecol Divers 7:85–105
Minh DH, Le Toan T, Rocca F, Tebaldini S, d'Alessandro MM, Villard L (2014) Relating P-band synthetic aperture radar tomography to tropical forest biomass. IEEE Trans Geosci Remote Sens 52:967–979
Mitchard ETA, Feldpausch TR, Brienen RJW, Lopez-Gonzalez G, Monteagudo A, Baker TR et al (2014) Markedly divergent estimates of Amazon forest carbon density from ground plots and satellites. Glob Ecol Biogeogr. https://doi.org/10.1111/geb.12168
Muller-Landau HC (2004) Interspecific and inter-site variation in wood specific gravity of tropical trees. Biotropica 36(1):20–32
Osher LJ, Buol SW (1998) Relationship of soil properties to parent material and landscape position in eastern Madre de Dios, Peru. Geoderma 83:143–166
Pallqui NC, Monteagudo A, Phillips OL, Lopez-Gonzalez G, Cruz L, Galiano W, Chavez W, Vasquez R (2014) Dinámica, biomasa aérea y composición florística en parcelas permanentes Reserva Nacional Tambopata, Madre de Dios, Perú. Rev Peru Biol 21:235–242
Palmero P (2004) Characterising the fine-scale pattern of Amazonian forest type using multi-source Earth observation data. Ph.D. thesis, University of Leeds, UK
Patiño S, Lloyd J, Paiva R, Baker TR, Quesada CA, Mercado LM et al (2009) Branch xylem density variations across the Amazon Basin. Biogeosciences 6(4):545–568. https://doi.org/10.5194/bg-6-545-9
Peacock J, Baker TR, Lewis SL, Lopez-Gonzalez G, Phillips OL (2007) The RAINFOR database: monitoring forest biomass and dynamics. J Veg Sci 18:535–542
Peru Ministerio de Ambiente (2015) Mapa del Patrimonio Forestal Nacional. https://sinia.minam.gob.pe/mapas/mapa-patrimonio-forestal-nacional
Phillips OL (2018) Recent changes in Amazon forest biomass and dynamics. In: Rodríguez L, Anderson I (eds) The Lima declaration on biodiversity and climate change: contributions from science to policy for sustainable development. Technical series no. 89. Secretariat of the Convention on Biological Diversity, pp 32–41
Phillips O, Miller JS (2002) Global patterns of plant diversity: Alwyn H. Gentry's forest transect data set. Monogr Syst Bot 89:1–319
Phillips OL, Núñez Vargas P, Monteagudo A, Peña Cruz A, Chuspe Zans ME, Galiano Sánchez W, Yli-Halla M, Rose S (2003) Habitat association among Amazonian tree species: a landscape-scale approach. J Ecol 91:757–775
Phillips OL, Baker TR, Arroyo L, Higuchi N, Killeen TJ, Laurance WF, Lewis SL, Lloyd J, Malhi Y, Monteagudo A et al (2004) Pattern and process in Amazon tree turnover, 1976–2001. Philos Trans R Soc Lond Ser B Biol Sci 359(1443):381–407. https://doi.org/10.1098/rstb.2003.1438
Phillips OL, Rose S, Monteagudo Mendoza A, Núñez Vargas P (2006) Resilience of southwestern Amazon forests to anthropogenic edge effects. Conserv Biol 20:1698–1710
Phillips OL, Baker TR, Brienen R, Feldpausch TR (2010) Field manual for plot establishment and remeasurement. http://www.geog.leeds.ac.uk/projects/rainfor

Phillips OL, Brienen RJW (2017) Carbon uptake by mature Amazon forests has mitigated Amazon nations' carbon emissions. Carbon Balance Manag. https://doi.org/10.1186/s13021-016-0069-2

Qie L, Lewis SL, Sullivan MJ, Lopez-Gonzalez G, Pickavance GC, Sunderland T, Ashton P, Hubau W et al (2017) Long-term carbon sink in Borneo's forests halted by drought and vulnerable to edge effects. Nat Commun 8:1966

Quesada CA, Phillips OL, Schwarz M, Czimczik CI, Baker TR, Patiño S, Fyllas NM, Hodnett MG, Herrera R et al (2012) Basin-wide variations in Amazon forest structure and function are mediated by both soils and climate. Biogeosciences 9:2203–2246. https://doi.org/10.5194/bg-9-2203-2012

Räsänen M, Neller R, Salo J, Jungner H (1992) Recent and ancient fluvial deposition systems in the Amazonian foreland basin, Peru. Geol Mag 129:293–306

Salo J, Kalliola R, Häkkinen I, Mäkinen Y, Niemelä P, Puhakka M, Coley PD (1986) River dynamics and the diversity of Amazon lowland forest. Nature 322:254

Schargel R (2011) Una resena de la geografıa fısica de Venezuela, con enfasis en los suelos. BioLlania Edicion Especial 10:11–26

Slik JW, Aiba SI, Brearley FQ, Cannon CH, Forshed O, Kitayama K, Nagamasu H, Nilus R et al (2010) Environmental correlates of tree biomass, basal area, wood specific gravity and stem density gradients in Borneo's tropical forests. Glob Ecol Biogeogr 19:50–60

Stegen JC, Swenson NG, Valencia R, Enquist BJ, Thompson J (2009) Above-ground forest biomass is not consistently related to wood density in tropical forests. Glob Ecol Biogeogr 18:617–625

Sullivan MJ, Talbot J, Lewis SL, Phillips OL, Qie L, Begne SK, Chave J et al (2017) Diversity and carbon storage across the tropical forest biome. Sci Rep 7:39102

Sullivan MJP, Lewis SL, Hubau W, Qie L, Baker TR, Banin LF, Chave J et al (2018) Field methods for sampling tree height for tropical forest biomass estimation. Methods Ecol Evol 9:1179–1189

ter Steege H, Pitman NC, Phillips OL, Chave J, Sabatier D, Duque A, Molino JF, Prévost MF, Spichiger R, Castellanos H, Von Hildebrand P (2006) Continental-scale patterns of canopy tree composition and function across Amazonia. Nature 443:444

ter Steege H, Pitman NC, Sabatier D, Baraloto C, Salomão RP, Guevara JE, Phillips OL, Castilho CV et al (2013) Hyperdominance in the Amazonian tree flora. Science 342:1243092

Tuomisto H, Ruokolainen K, Kalliola R, Linna A, Danjoy W, Rodriguez Z (1995) Dissecting Amazonian biodiversity. Science 269:63–66

Turner IM (2001) The ecology of trees in the tropical rain forest. Cambridge University Press, Cambridge

Vásquez R, Rojas R, Monteagudo AM, Valenzuela LG, Huamantupa I (2018) Catalogo de los Arboles del Perú. Q'ueña Revista de la Sociedad Botánica del Cusco 9(1), número especial

Vicuña Miñano E, Baker TB, Banda K, Honorio Coronado E, Monteagudo A et al (2018) El sumidero de carbono en los bosques primarios Amazónicos es una oportunidad para lograr la sostenibilidad de su conservación. Folia Amazónica 27:101–109

Watson JEM, Evans T, Venter O, Williams B, Tulloch A, Stewart C et al (2018) The exceptional value of intact forest ecosystems. Nat Ecol Evol 2:599–610

Zanne AE, Lopez-Gonzalez G, Coomes DA, Ilic J, Jansen S, Lewis SL, Miller RB, Swenson NG, Wiemann MC, Chave J (2009) Global wood density database. Dryad. http://hdl. handle.net/10255/dryad235

Zolkos SG, Goetz SJ, Dubayah R (2013) A meta-analysis of terrestrial aboveground biomass estimation using lidar remote sensing. Remote Sens Environ 128:289–298

Surveys in Geophysics (2019) 40:937–958
https://doi.org/10.1007/s10712-019-09527-x

Innovations in Ground and Airborne Technologies as Reference and for Training and Validation: Terrestrial Laser Scanning (TLS)

M. Disney[1,2] · A. Burt[1] · K. Calders[3] · C. Schaaf[4] · A. Stovall[5]

Received: 10 September 2018 / Accepted: 14 February 2019 / Published online: 29 March 2019

Abstract
The use of terrestrial laser scanning (TLS) to provide accurate estimates of 3D forest canopy structure and above-ground biomass (AGB) has developed rapidly. Here, we provide an overview of the state of the art in using TLS for estimating forest structure for AGB. We provide a general overview of TLS methods and then outline the advantages and limitations of TLS for estimating AGB. We discuss the specific type of measurements that TLS can provide, tools and methods that have been developed for turning TLS point clouds into quantifiable metrics of tree size and volume, as well as some of the challenges to improving these measurements. We discuss the role of TLS for enabling accurate calibration and validation (cal/val) of Earth observation (EO)-derived estimates of AGB from spaceborne lidar and RADAR missions. We give examples of the types of TLS equipment that are in use and how these might develop in future, and we show examples of where TLS has already been applied to measuring AGB in the tropics in particular. Comparing TLS with harvested AGB shows $r^2 > 0.95$ for all studies thus far, with absolute agreement to within 10% at the individual tree level for all trees and to within 2% in the majority of cases. Current limitations to the uptake of TLS include the capital cost of some TLS equipment, processing complexity and the relatively small coverage that is possible. We argue that combining TLS measurements with the existing ground-based survey approaches will allow improved allometric models and better cal/val, resulting in improved regional and global estimates of AGB from space, with better-characterised, lower uncertainties. The development of new, improved equipment and methods will accelerate this process and make TLS more accessible.

Keywords Terrestrial lidar · 3D forest structure · Forest structure · Above-ground biomass · Biomass

✉ M. Disney
mathias.disney@ucl.ac.uk

Extended author information available on the last page of the article

Fig. 1 Point clouds from three contrasting forests (all collected using a Riegl VZ-400 TLS) and detail ▶ of a single tree and branch. In the plot cases, 1 ha is shown either oblique or in elevation and the point clouds of individual trees are shown in different colours. Top row: from Calders et al. (2018), a 1 ha section of deciduous woodland, Wytham Woods, UK, containing 559 trees (232 with diameter-at-breast-height (DBH) > 0.2 m); second row: moist tropical forest at Caxiuanã, Brazil, showing 151 trees with DBH > 0.2 m; third row: moist tropical forest at Lopé, Gabon, with 107 trees of DBH > 0.2 m; bottom row, left panel: single tree from Caxiuanã (H = 30 m, DBH = 0.65 m); bottom row, right panel: ~ 1.5 m section of left-most branch, showing resolution of branches of < 4 cm diameter. These trees are the result of processing the raw plot TLS data to extract only the points for each tree, to remove all other points, i.e. understory and trees with DBH < 0.2 m and to remove any topographic variations

1 Introduction

Terrestrial laser scanning (TLS) is a relatively new technology, developed since the early 2000s for surveying applications, i.e. for measuring very precise distances and angles. TLS works by using laser pulses to measure distance and 3D (x, y, z) location to within mm potentially over hundreds of metres. TLS has largely been developed in its current form by commercial instrument developers such as RIEGL (http://www.riegl.com/), Leica Geosystems (https://leica-geosystems.com/en-gb), Faro (https://www.faro.com/), Zoller and Fröhlich (https://www.zf-laser.com/Home.91.0.html?&L=1) and others. The utility of TLS for measuring tree properties was explored in the early 1990s by the forestry community, particularly for providing new ways to measure properties already measured for forest surveying. These include tree height, diameter-at-breast-height (DBH), crown size and basal area. However, while TLS has proved useful for these applications (Hopkinson et al. 2004; Calders et al. 2015; Newnham et al. 2015), it is the potential to provide cm and even mm-accurate 3D structural detail (albeit at the individual branch, rather than plot scale), particularly for estimating 3D full-tree volume, and from this above-ground biomass (AGB), that has seen a rapid increase in interest in TLS for measurement of AGB. There is particular interest in how these measurements can be used in support of spaceborne Earth observation (EO) estimates of AGB, from new laser and RADAR missions, particularly via calibration and validation. The capability of TLS has also been extended to the spectral domain, via research-led development of multi- and even hyperspectral lidar instruments (Danson et al. 2014; Hakala et al. 2012; Douglas et al. 2015). Figure 1 shows an example of the 3D information provided by TLS, in this case TLS point clouds from deciduous broadleaf woodland in the UK (Calders et al. 2018) and tropical forests in Brazil and Gabon (Burt 2017).

The importance of AGB and forest structure more generally has driven the development of spaceborne EO missions targeting these properties (GCOS 2010). Estimating AGB accurately and consistently on regional to global scales is critical for reducing current uncertainties in carbon stocks and deforestation fluxes, which differ in magnitude but also location (Saatchi et al. 2011; Harris et al. 2012; Baccini et al. 2012;). Some of these uncertainties arise from observation and methodological differences (Avitabile et al. 2016) as well as in assumptions of wood density ρ (Mitchard et al. 2013, 2014), but there is also considerable uncertainty in the upscaling of spatial estimates and magnitude as well. New EO missions can reduce these uncertainties, but to fulfil their mission requirements in terms of both accuracy and precision will depend on accurate and independent ground-based estimates of AGB to calibrate EO retrieval methods and validate the resulting products. The difficulty of measuring AGB manually via forestry methods means that it is challenging to obtain validation data; TLS is one way to address this gap (Stovall and Shugart 2018). The Committee on Earth Observation Satellites (CEOS) has identified TLS as a key

approach to providing accurate calibration and validation data for EO-derived AGB (CEOS LPV WGCV Biomass 2019). Here, we summarise developments in TLS for AGB estimation in support of EO retrievals.

2 Above-Ground Biomass (AGB)

Estimating above- and belowground (root) biomass of a tree directly requires careful harvesting and weighing. This is destructive and difficult, particularly for anything other than 'small' trees, e.g., trees with diameter-at-breast-height DBH < 30 cm and height, $H < 30$ m. As such, destructive harvesting is done infrequently, tends to focus on smaller trees in more easily accessible areas, and of course precludes repeat measurements to capture dynamics. And yet spatial estimates of AGB are almost entirely based on extrapolating the masses of these harvested trees, using empirical size-to-mass allometric scaling equations (ASEs) (Baskerville 1972; Brown and Lugo 1984; Chave et al. 2005). The wide use of ASEs is a consequence of their reliance on only simple, rapid measurements, as well as being the only way we have to extrapolate estimates of AGB spatially. ASEs typically take the form $AGB = a\left(D^2 H\rho\right)^b$, where D and H are DBH and height, respectively, ρ is the wood density (g/cm^3), and a, b are empirically derived constants (Feldpausch et al. 2010; Chave et al. 2014). This ASE approach broadly represents a tree as a tapered cylindrical volume. Coefficients a, b can be derived via destructive sampling, and assuming a density ρ, AGB can then be estimated. However, the purely empirical nature of the ASE approach means a wide range of slightly different model forms and parameters have been developed, for site- and species-specific or more general regional versions, with often little to distinguish them in terms of suitability, particularly for upscaling (Ahmed et al. 2013). Further, the small sample of destructive harvests on which ASEs are based means that uncertainties in upscaled AGB is potentially large and hard to quantify (Ketterings et al. 2001; Sileshi 2014; Duncanson et al. 2015). As an example, the most comprehensive pan-tropical ASE yet developed is that of Chave et al. (2014) which contains around 4000 destructively harvested trees, only around 10% of which have DBH > 70 cm and/or height, $H > 30$ m. To estimate pan-tropical AGB, this sample is extrapolated across the hyperdiverse tropics (Ter Steege et al. 2013), containing an estimated 1.4 tn (1.4×10^{12}) trees of 46,500 species (Pan et al. 2011; Lewis et al. 2013; Crowther et al. 2015; Slik et al. 2015). Other ASEs contain far fewer samples, typically a few dozen in the case of most locally calibrated models (Stovall et al. 2018).

3 Uncertainty

Uncertainty is used here to mean the mismatch between an estimate of AGB and the true but unknown value, and can be described by a probability density function (PDF) characterising the range and likelihood of possible values (IPCC 2006). Uncertainty in AGB estimates arises from various sources (Chave et al. 2004; Réjou-Méchain et al. 2017). Firstly, the form and appropriateness of ASE will impact resulting AGB estimates (Molto et al. 2013; Burt et al. 2018). The fundamental ability (or otherwise) of a simple regression equation to represent tree volume and hence AGB, across regions and biomes, is limited. Estimating uncertainty due to model form is also hard in the face of small samples of destructively harvested and weighed trees, particularly as these are potentially biased to smaller trees (Ahmed et al. 2013; Sheil et al. 2016; Burt et al. 2018). Lutz et al. (2018) show that across a wide range of forests, the largest 1% of all trees with DBH > 1 cm comprise 50% of total live tree AGB overall, with trees of $D \geq 0.6$ m comprising 41% of total AGB. This disproportionate partitioning of AGB in large trees leads to potentially large uncertainties in AGB derived from ASEs with few large trees. This is further compounded by the fact that bigger trees have fundamentally greater variance in AGB (Mascaro et al. 2011; Clark and Kellner 2012). As a result, uncertainty in ASE-derived AGB grows rapidly with DBH.

Wood density, ρ, is another major source of uncertainty in estimating AGB using allometry (Niklas 1993; Chave et al. 2006; Swenson and Enquist 2007; Mitchard et al. 2013). The choice of ρ is what translates the ASE-derived volume to mass, but ρ is highly variable within individual trees radially and with height, let alone within and between species and across regions (Chave et al. 2006; Swenson and Enquist 2007). Databases of wood density do exist, but are still limited given the diversity observed in practice (Chave et al. 2009; Zanne et al. 2009; Momo et al. 2018).

A further fundamental issue is that all estimates other than destructive harvests are indirect, leading to the so-called fallacy of misplaced concreteness, i.e. indirect estimates of AGB are not measurements (Clark and Kellner 2012; Sileshi 2014). The problem is that if ground-based estimates of AGB, derived from plot-scale DBH measurements upscaled via ASEs, are used to 'validate' EO-derived estimates of AGB based on H to DBH ASEs, then the resulting spatial estimates (regional, pan-tropical, etc.) will have these uncertainties potentially exacerbated due to the upscaling process (Hill et al. 2013; Sileshi 2014). This leads to potentially unknown uncertainties as well as inconsistencies in comparing AGB estimates derived from different data and using different upscaling methods (Mitchard et al. 2013; Hill et al. 2013).

These issues of heteroscedasticity, i.e. growing variance with tree size, choice of ASE form, small sample size and variability within-species variability across locations and studies, are illustrated clearly in Fig. 2.

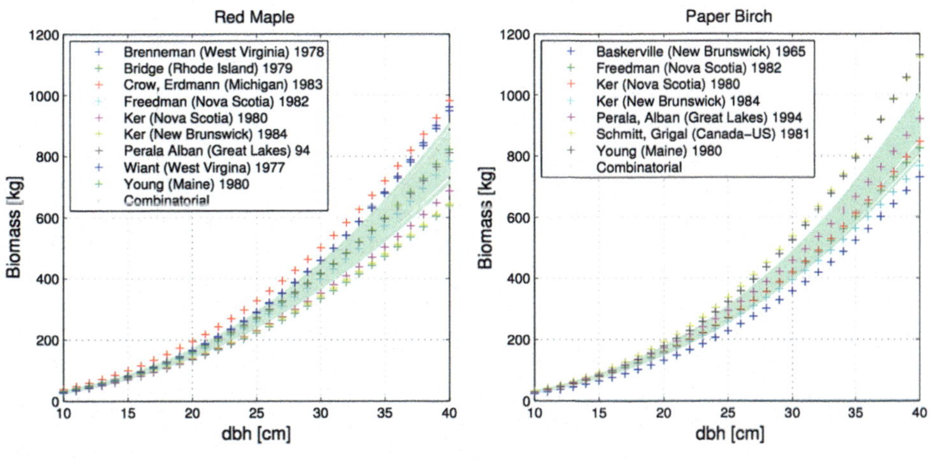

Fig. 2 From Ahmed et al. (2013). An illustration of the variation in estimated AGB values from different ASEs in Ter-Mikaelian and Korzukin (1997) for two deciduous broadleaf species: red maple (*Acer rubrum*) and paper birch (*Betula papyrifera*). The estimated biomass is plotted against dbh, the diameter-at-breast-height of the trees. The shaded region represents an estimate of what Ahmed et al. (2013) term σ_s, the site error in estimated AGB introduced by using ASE model coefficients developed at a different site (estimated via bootstrapping). The growth in uncertainty with D is clearly apparent, even for these relatively small trees

The most comprehensive attempts to assess the uncertainty of stand-scale AGB derived from ASEs suggest an uncertainty of around 10%, with potential variation between 5 and 23% (Chave et al. 2004; Réjou-Méchain et al. 2017) arising from plot-level structural variations. These estimates are based on analysis of the statistical properties of the underlying data and models but are potentially conservative in assumptions of measurement error distributions (normally distributed) and the ability of simple ASE models to capture large-scale ecological variations in tree form (Burt 2017; Burt et al. 2018). The uncertainty in AGB at the plot level is less than at the individual tree level, due to the impact of measurement errors reducing with increasing sample size. This plot-scale uncertainty, however, may be less important than the potential issues of bias and heteroscedasticity as outlined above, particularly for larger trees.

These issues mean that in developing accurate EO-derived estimates of AGB with realistic uncertainty, care must be taken to develop appropriate ASEs spanning the full range of biomass likely to be observed, with uncertainty due to ASE form, wood density and upscaling, quantified as far as possible (Ahmed et al. 2013; Hill et al. 2013). Terrestrial laser scanning (TLS) has the capability to address many of these problems, by providing tree- and plot-level AGB estimates which are independent of allometry, unbiased in terms of tree size distributions and with well-quantified uncertainty (Burt 2017; Stovall and Shugart 2018). TLS estimates collected widely and reliably can reduce current uncertainties in terrestrial C stocks, enable improved calibration and validation of AGB products, particularly from EO, and form the basis of improved allometric models. Here, we describe key developments in the use of TLS to estimate AGB, present analysis of uncertainties that should be addressed, and highlight challenges that remain.

4 TLS for AGB Applications

Terrestrial laser scanning has developed significantly over the last decade in terms of its ability to deliver highly detailed 3D structural measurements. This has been due in part to the development of robust new sensors and systems at prices that, while still high (approximately few \$10 s to ~ \$100 K), are becoming more affordable for researchers both in real terms, and as their benefits become more apparent. This in turn has led to the development of tools and methods to exploit the resulting point cloud data for ecological applications (Calders et al. 2015; Wilkes et al. 2017; Disney et al. 2018; Disney 2018). This has allowed the use of TLS to move from generalised structural metrics of tree size and shape such as height, DBH, crown size and volume (considered as an enclosing envelope), trunk size and shape, to full reconstruction of 3D tree architecture down to higher order branches, comprising the majority of tree woody biomass. This has been critical to the use of TLS for AGB applications: if full-tree volume can be reliably and near-automatically reconstructed, then estimating AGB is simply a case of multiplying by wood density, ρ. Perhaps even more importantly, given that ASEs represent the woody volume of trees, comparison between ASEs and TLS measurements can be made at the volume level, allowing the impact of uncertainty due to ρ to be considered separately from ASE model uncertainty. Reliable automation of 3D architecture reconstruction is important for comparisons with EO, where the benefits of TLS measurements of large numbers of tree measurements are key.

A range of methods have been proposed to extract full-tree 3D architecture automatically from TLS data (e.g., Yan et al. 2009; Preuksakarn 2012; Raumonen et al. 2013, 2015;

Hackenberg et al. 2014, 2015, etc.). Disney (2018) outlines some of the various assumptions and trade-offs underpinning them and challenges remaining to improve them. These include: mm-accurate co-registration of multiple point clouds in forest environments, and optimising data collection to achieve this (Wilkes et al. 2017; Paynter et al. 2018); reliable separation of individual trees from large point clouds, particularly in dense, tall forests (Burt et al. 2018); overcoming fundamental limitations of TLS data including occlusion and variable point density; separating wood and leaf points in leaf-on point clouds (Yun et al. 2016; Boni Vicari et al. 2019). Figure 3 shows an example of individual tree segmentation from a 1 ha TLS point cloud from a tropical forest canopy and highlights the challenges of extracting individual trees routinely from larger point clouds. As a general guide, taller and more dense canopies will require more dense TLS sampling to minimise the impacts of occlusion in the upper canopy. Wilkes et al. (2017) show that a 10-m grid sampling pattern allows for reconstruction of branching in dense tropical forest canopies

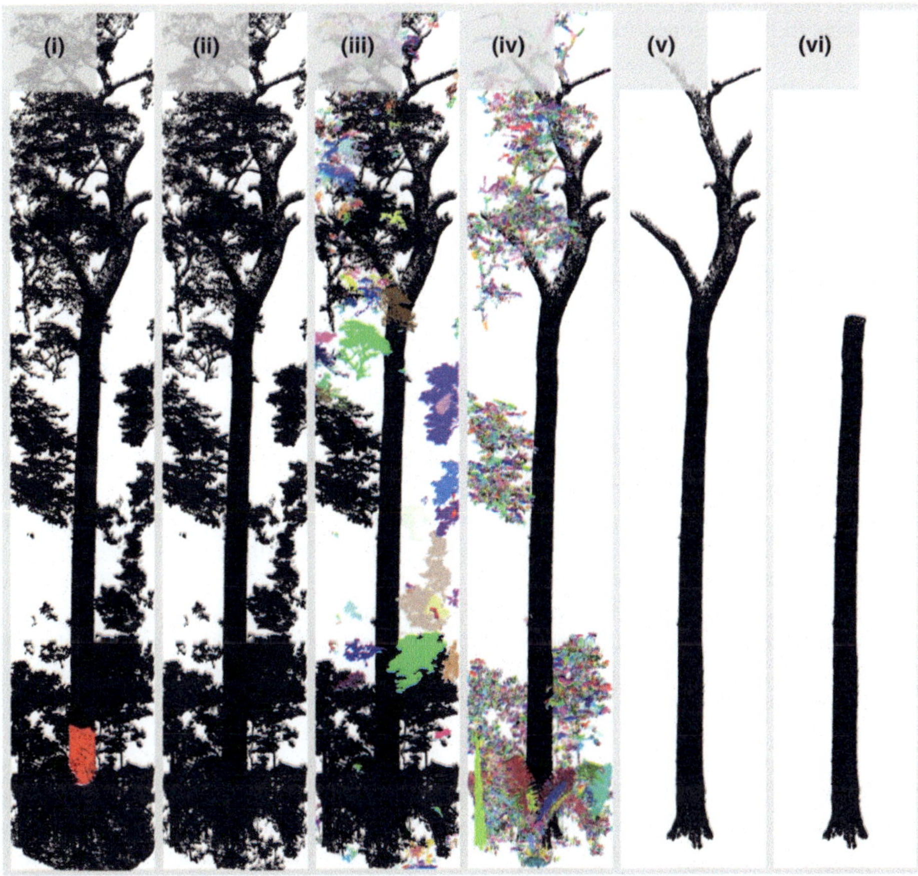

Fig. 3 (reproduced from Burt et al. 2018). Segmenting individual trees from large plot-scale TLS point clouds (collected with a Riegl VZ-400) using *Treeseg*. Panels show: (i) iterative pass-through filtering to fit enclosing cylinder (red); (ii) plane fitting to remove ground points; (iii), (iv) and (v) iterative clustering and region-based segmentation (multiple colours representing separate segments) to remove neighbouring vegetation; and finally (vi) cylinder fitting to determine the location of first branching and hence the height-to-crown

taller than 50 m, up to branches of ~ 5 cm diameter in the upper canopy. Increasing the scan density provides diminishing returns; the additional cost of data collection and processing more than offset the slight improvement in the very uppermost branches, which in any case contribute the least to AGB.

Various tools are now available to process and analyse TLS data for 3D reconstruction. Most commercial TLS instrument manufacturers provide commercial software tools to allow co-registration, including Leica Cyclone (https://hds.leica-geosystems.com/en/Leica-Cyclone_6515.htm), RIEGL RiScan Pro (http://www.RIEGL.com/products/software-packages/riscan-pro/), etc. Non-commercial software aimed at TLS processing include pylidar (http://www.pylidar.org/en/latest/) and 3D Forest (Trochta et al. 2017 and http://www.3dforest.eu/). Other non-commercial tools developed for 3D point clouds more generally (particularly ALS and now point clouds derived from structure-from-motion, e.g., from UAVs) are also useful for TLS, for example CloudCompare (https://www.danielgm.net/cc/). This has various features for downsampling, editing and manipulating point clouds as well as for extracting general metrics.

Specific tools developed for addressing 3D reconstruction include:

Treeseg (Burt et al. 2018): codes to segment and isolate individual tree point clouds from much larger point clouds, prior to reconstruction (https://github.com/apburt/treeseg);

TreeQSM (Raumonen et al. 2013): full 3D reconstruction based on fitting cylinder segments to TLS point cloud data. The resulting so-called quantitative structural model (QSM) contains the total and per-branch volume, as well as branch connectivity and insertion angles for each branch (https://github.com/InverseTampere/TreeQSM). Operates on single trees, or potentially at the stand-scale in less dense, more uniform canopies.

SimpleTree (Hackenberg et al. 2015): a full-tree reconstruction approach based on point cloud segmentation (https://github.com/SimpleTree).

3D Forest (http://www.3dforest.eu/): provides tree size and shape metrics based on manual segmentation and voxellisation, i.e. not explicit 3D and so differing slightly from the largely automated full 3D methods above.

Computree (Othmani et al. 2013): a more forestry-based software platform, designed to be extensible and to derive forestry-related metrics rather than 3D architecture per se (http://computree.onf.fr/?lang=en).

Forestr (Atkins et al. 2018): software tools (in the programming language R) to provide general structural metrics of canopy structural complexity (CSC) metrics, based on the x, z, or x, y, z position of LiDAR hit density expressed as vegetation area index (VAI) (https://github.com/atkinsjeff/forestr).

A key problem is how to assess the accuracy and uncertainty of the resulting 3D tree volumes. The only way to do this really is via destructive harvesting. The practical obstacles to this mean it is very difficult and so has been done in only a few cases. Figure 4, from Disney (2018) and reproduced here, compares examples from Calders et al. (2015 for three Eucalyptus species, Australia, Gonzalez de Tanago et al. (2018) in Guyana, Indonesia and Peru; and Momo Takoudjou et al. (2017) in eastern Cameroon. These efforts in validating TLS-derived AGB show that the method agrees with destructive harvest values with $r^2 > 0.95$ in all cases. This is robust across varying canopy types, including large tropical trees and trees with buttressed trunks. Absolute agreement between TLS-derived and harvested AGB agrees to within 10% at the individual tree level for all trees in these studies

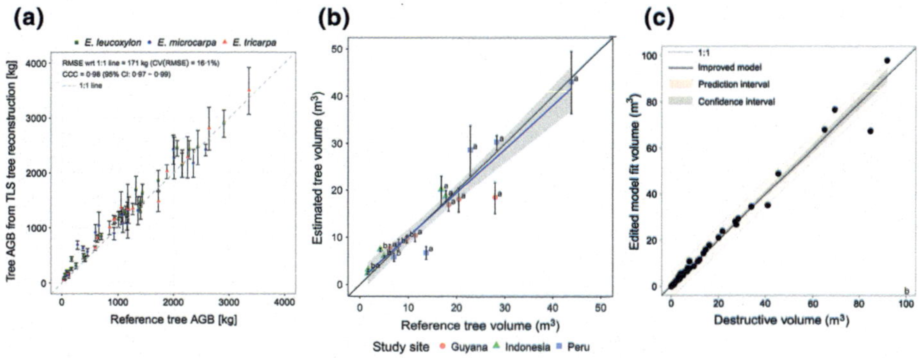

Fig. 4 From Disney (2018). Comparison of TLS-derived estimates of tree AGB and volume, with values from destructive harvests: **a** Calders et al. (2014) TLS-derived and weighed AGB for three Eucalyptus species, Australia; **b** Gonzalez de Tanago et al. (2018) TLS-derived and destructive volume across three tropical sites, with buttressed and non-buttressed trees marked a, b, respectively; **c** Momo Takoudjou et al. (2017) TLS-derived and destructive volume from eastern Cameroon. In the latter two cases, the shaded area is the 95% confidence interval of the regression fit

and to within 2% in the majority of cases. Buttressed trees are important for tropical forests as they can make up significant numbers of large trees in tropical plots. Disney et al. (2018) show that the buttressed trunks can introduce greater uncertainty in TLS-derived AGB at the single tree level (up to 10% over/under-estimation), due to the issue of fitting cylinders to non-cylindrical trunk forms. However, these fitting errors are randomly distributed and so tend to cancel out at the plot level. Buttressed trunks are known to be an issue for DBH-based allometry more generally, and TLS can help address this (Noelke et al. 2015). Less automated but more flexible volume fitting approaches such as mesh grids (*ibid.*) and outer hull models (OHMs; Stovall et al. 2017) can also be used to potentially address this issue. This is another area where TLS may help in cal/val of AGB derived from EO or survey methods, based on ASEs where buttressed trunk form is not well represented. Additionally, AGB derived from plot survey measurements of DBH have to contend with the issue of where and how DBH is measured for buttressed trees (http://www.rainfor.org/upload/ManualsEnglish/RAINFOR_field_manual_version_2016.pdf) and then how this is accounted for in the tapered cylinder assumption of ASE form.

These direct validation efforts illustrated in Fig. 4 demonstrate very clearly that TLS-derived AGB is both accurate and robust, across all tree size classes. This is the strongest and most direct evidence that TLS can be used to test the existing ASEs, as well as to augment allometric datasets, to include more, larger trees. This potentially addresses one of the fundamental limitations of ASEs highlighted above.

More generally, Disney et al. (2012) proposed using *simulated* TLS point clouds of 3D model trees where instrument, acquisition and tree architecture properties are known a priori. This allows for sensitivity and uncertainty analysis and has been used to test some of the approaches above (Raumonen et al. 2013; Hackenberg et al. 2014; Burt et al. 2018). The utility of this 3D simulation approach can also be tested by comparing simulations from the retrieved 3D architectural models with other, indirect estimates of canopy properties, including reflectance, leaf area index (LAI), etc. The advantage of this is that many of these EO-derived estimates of canopy properties have uncertainty which is much better-characterised than for estimates of AGB. Figure 5 illustrates this approach, showing a comparison of simulated and actual top-of-canopy

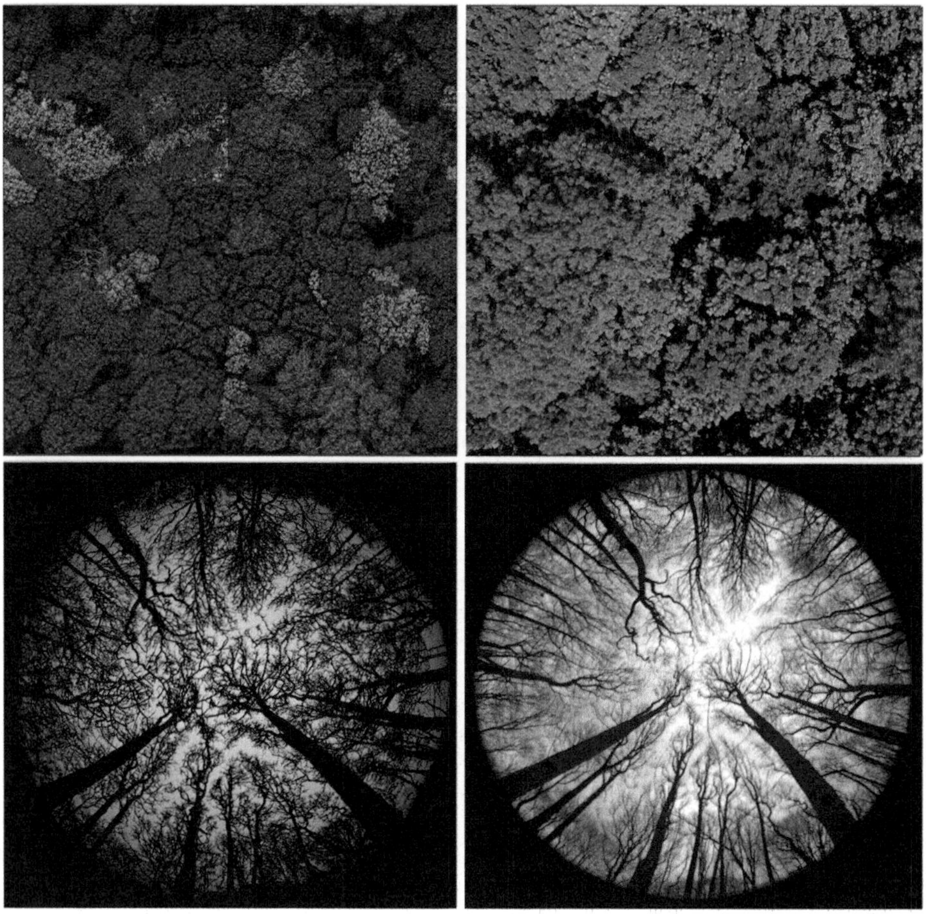

Fig. 5 Adapted from Calders et al. (2018). Upper row: simulated image of a deciduous woodland (left) and UAV RGB image over the canopy on which this model simulation is based (right). This is intended to be an indicative simulation rather than an exact representation, due to lack of leaf spectral and size information in the upper canopy. Lower row: simulated (left) and measured (right) hemiphotos in leaf-off conditions from the same location within the canopy. In this case, the model canopy structure is intended as a direct copy of the observed

red/green/blue (RGB) values obtained using an unmanned autonomous vehicle (UAV) and hemiphoto, where the simulations are derived from a 3D canopy architectural model derived from 1 ha TLS point cloud collected in Wytham Woods, UK (Calders et al. 2018).

An additional, very useful software tool in the context of AGB uncertainty, which can and should be applied once new ASEs are developed, from TLS or elsewhere, is the R package developed by Réjou-Méchain et al. (2017) to allow the uncertainty in new ASEs to be quantified (https://cran.r-project.org/web/packages/BIOMASS/index.html).

5 TLS to Underpin Satellite-Derived AGB: Developments and Challenges

As outlined above, TLS can provide several new advantages to AGB estimation which are likely to make it a significant development for satellite-derived estimates. All EO-derived estimates of AGB depend on indirect retrievals, typically via height allometry for lidar and RADAR estimates, but also via empirical backscatter–biomass relationships. The benefits of TLS can make vital contributions to calibration and validation (cal/val) efforts, including: improving the existing ASEs via the inclusion of many more trees across a much wider distribution of size and shape across different biomes; testing assumptions underlying, and form of, ASEs to better characterise uncertainty; to provide independent estimates of AGB to validate retrievals, particularly in tropical and other high biomass regions where ground-based estimates are rare or non-existent; to allow development of new highly detailed 3D radiative transfer (RT) models for simulating EO signals. This allows for new algorithm development and testing and benchmarking of simpler RT models, where canopy structural properties are known a priori (Widlowski et al. 2013; Calders et al. 2018; and Fig. 5 above).

The benefit of TLS for EO-derived AGB is exemplified by its inclusion by CEOS as a key method in its recommended protocol for cal/val of AGB. Stovall and Shugart (2018) present a framework for how TLS can be used to improve EO estimates of AGB, and this is reproduced in Fig. 6. They point out that TLS can figure prominently in both the measurement of plot-scale biomass and structure for improving EO retrievals, new allometric models, for pre-launch calibration and post-launch validation of EO estimates via plot measurements of biomass.

Some additional benefits of TLS for EO cal/val are shown in Fig. 7. This illustrates how volume estimates derived from TLS measurements can be used to understand the EO signal, as well as quantifying forest structural differences across biomes. Figure 7 illustrates differences between sites in how wood volume (and hence biomass) is distributed across branch size classes, with volume skewed towards smaller branches higher in the canopy in French Guiana; or across a wider range of branch sizes in Ghana. The two

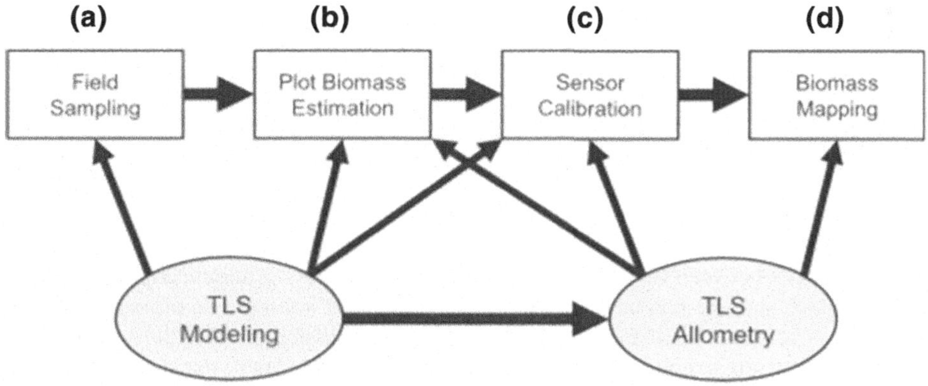

Fig. 6 From Stovall and Shugart (2018). A flow diagram outlining how TLS can be used to address the limitations of ASEs for EO-derived estimates of AGB, specifically via calibration and validation. They note that in **a** TLS can provide unbiased (by size class) estimates of standard forestry properties (DBH, height, etc.); **b** non-destructive estimates of AGB to test and develop improved ASEs; **c** and **d** plot-level volume/biomass estimates for calibration and validation of EO products

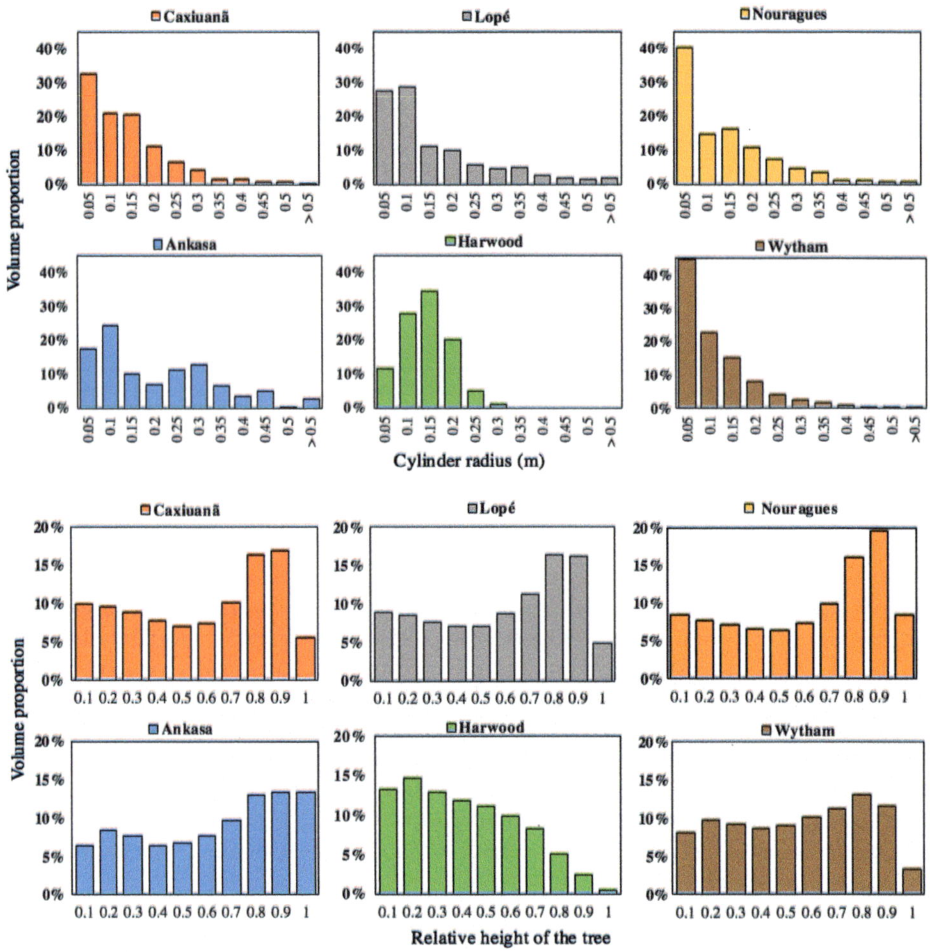

Fig. 7 From Tang (2017). Comparison of the TLS-derived volume as a function of tree size across four tropical sites (Caxiuanã, Brazil; Lopé, Gabon; Nouragues, French Guiana and Ankasa, Ghana), plantation conifer (Harwood, UK) and deciduous woodland (Wytham Woods, UK). Volume is derived from Quantitative Structural Model (QSMs) fitted to TLS data, separated into 11 × 5 cm size diameter classes from < 5 cm to ≥ 50 cm, with the proportion of volume in each size class shown (top panel). The lower panel shows the same information as a function of relative tree height

UK sites stand out as we would expect (and hope!), particularly the plantation conifer canopy (*Sitka spruce*). When expressed as a function of canopy height (lower panel), the differences between the tropical and UK forests are less pronounced, except for the conifer which shows a characteristic dominant trunk and small crown biomass. Information of this sort is critical for understanding the dependence of ASE-derived estimates of AGB based on height. This is particularly true for RADAR retrievals of biomass at longer wavelengths, e.g., for ESA BIOMASS at P-band (70 cm). These retrievals assume the dominant portion of the P-band backscatter signal over tropical forests arises from a peak of biomass in the upper part of the canopy (Le Toan et al. 2011; Garestier and Le Toan 2010). Smaller branches and leaves in the upper part of the canopy become more important for backscatter–AGB relationships from shorter wavelength RADAR, L-band for example (Mermoz et al. 2015). Measuring these components accurately from TLS is likely to be a greater challenge, as outlined below.

Figure 8, reproduced from Stovall and Shugart (2018), is an explicit example of how TLS can provide improved ASEs for biomass estimation from EO. They show that for species-specific ASEs, uncertainty in allometric estimates of AGB can be nearly halved through the use of TLS-derived ASEs compared to destructive harvest alone. Perhaps even more importantly, the TLS estimates have low or no bias (Calders et al. 2015; Gonzalez de

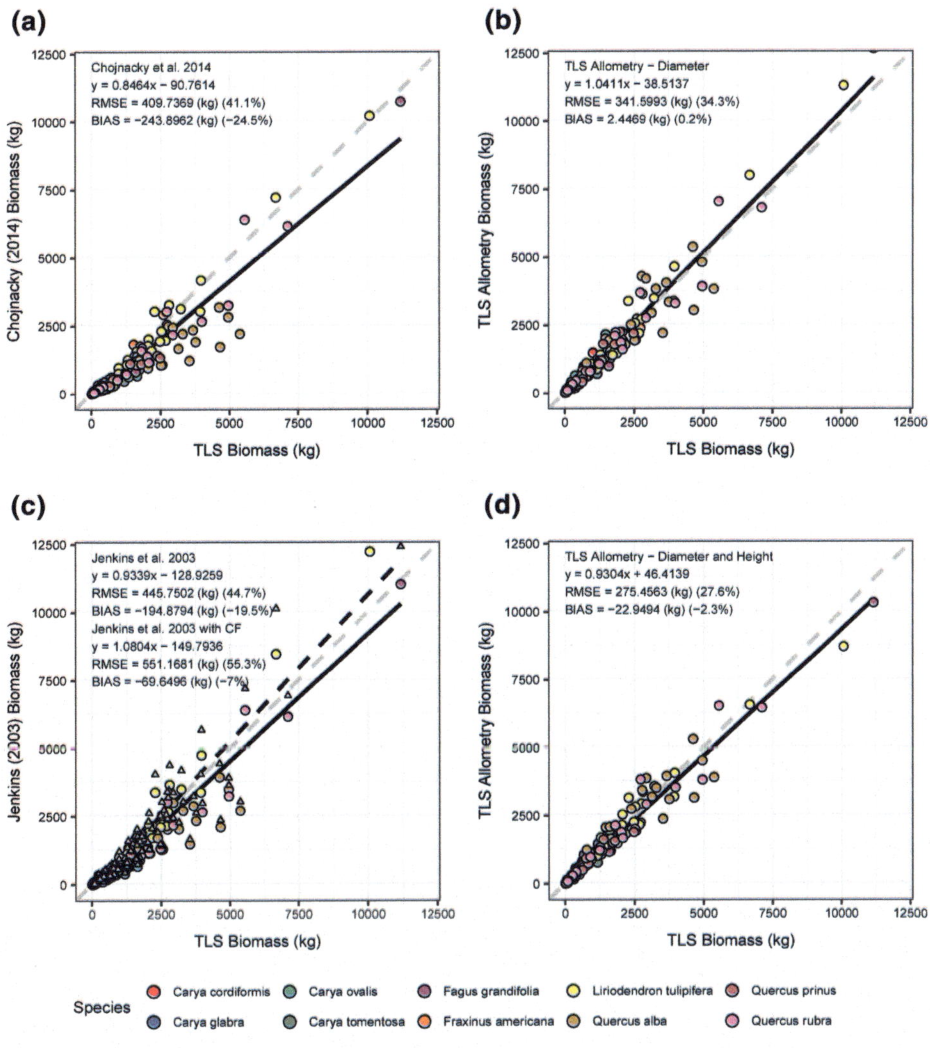

Fig. 8 From Stovall et al. (2018). Comparison of species-specific AGB derived from full volume QSM reconstruction of TLS measurements (x axis in each case), with previously published species-specific DBH and H-DBH ASEs (left column). The right hand panels compare AGB estimated from ASEs derived from TLS measurements of DBH (top right) and DBH and H (bottom right), with the full TLS-derived biomass

Tanago et al. 2018; Momo Takoudjou et al. 2017), compared to ~7 to 20% for the ASEs from destructive harvesting.

Stovall et al. (2018) conclude that TLS can produce unbiased AGB estimates, importantly by including larger diameter trees. They suggest that ASEs should be developed with sample sizes of at least 100 individuals to ensure equations have high precision and low bias. The issue of bias is discussed by Burt et al. (2018) who note that measurements of unbiased size distributions, and larger trees in particular, are vital to reduce uncertainty in AGB estimated via allometry. They note that the issue of heteroscedasticity means that uncertainty in ASE-derived AGB will always tend to be greater in higher biomass areas, but quantifying this uncertainty drastically improves the usefulness of resulting EO estimates.

One other area where TLS has the potential to improve EO-derived AGB is in the impact of uncertainty in wood density, ρ. Stovall et al. (2018) note that biomass estimates will vary as a function of intra-specific and site-level variation in wood density from published average values, as well as throughout individual trees. They conclude that quantifying wood density variability is essential for quantifying uncertainty in TLS-based biomass estimates. However, accurate estimates of wood density are critical to estimating AGB either from both allometry and TLS. As noted above, by comparing at the volume level wherever possible, as facilitated by TLS, the issue of wood density can be separated from the problem of uncertainty in the ASE or TLS-derived volume. Improved estimates of site and species level ρ can then be added where possible, to improve biomass estimates. 3D models from TLS are also opening the possibility for exploring the relationship between wood density and tree mechanical properties (Jackson et al. 2019).

The challenges to wider adoption of TLS for AGB estimation are broadly technical, cost and scaling and are all linked. The technical challenges lie in how to collect TLS data at plot scales and then retrieve structural properties, notably full 3D tree volume, with sufficient accuracy and quantified uncertainty. Resolving smaller branches in tall, dense leaf-on canopies is a challenge for TLS, which will tend to increase uncertainty in these components of AGB at the tree-scale. Comparisons between TLS-derived and harvested AGB at the individual tree level seem to show little or no bias, in part because the most poorly resolved branches are the smallest. This is potentially improved with higher density point clouds and with more effective methods to separate wood and leaf points (Boni Vicari et al. 2019). However, published comparisons tend not to look at these smaller components explicitly; more work is needed to quantify the impact of these possible sources of bias.

An additional challenge is how to do this for different groups using TLS instruments with different properties and across various environments. As outlined above, the development of standardised measurement protocols and software tools that aimed at processing TLS data for ecological applications specifically is helping to reduce these technical barriers. However, until acquisition and processing becomes 'routine', TLS for 3D volume estimates will still be the preserve of a relatively small number of groups and users. The cost of TLS is another potential barrier, with instruments costing upwards of \$100 K being out of reach for many/most researchers. However, if the value of the measurements provides demonstrably unique, important information for addressing key uncertainties in large-scale estimates of AGB, then funding is likely to become available. Wider adoption of TLS also provides impetus for development of new tools and instruments. Compact, lightweight and lower-cost TLS (e.g. ~\$20 K) has been developed which provides shorter-range, lower-precision measurements, but may provide some of the benefits of the higher-cost instruments for AGB estimates, e.g., by rapidly capturing tree volume for the larger parts of trees (Bauwens et al. 2016; Paynter et al. 2016). Figure 9 shows an example of Hemlock trees

Fig. 9 Left: TLS data from a Leica BLK360 of an Eastern Hemlock (*Tsuga canadensis*) measured in the Smithsonian Institute ForestGeo plot at Harvard Forest, MA, USA. The tree is suffering from defoliation due to infestation by the Hemlock Woolly Adelgid (HWA). From the TLS, the tree has DBH 81.5, H 24.1 m and volume of 10.5 m^3. Right: TLS data from the wide beam Canopy Biomass Lidar (CBL) of an Eastern Hemlock Tree unaffected by HWA, with TLS-derived DBH—12.3 cm, $H = 9.65$ m and volume 0.35 m^3. Insets in each case show close-up views of a branch from each tree, at the same scale

scanned in Harvard Forest, using lightweight TLS systems that are $10–20 K and have been developed for more general 3D measurements, both ecological or also for mobile mapping. The inset views in this case show where the TLS is unable to resolve separate branches and needles.

The rapid development of new low-cost TLS instruments has the potential to provide cal/val information with specific benefits. While it is desirable to make the most accurate ground measurements possible of tree volume and AGB for EO cal/val, CEOS LPV recognises the need for a hierarchical approach, to allow for collecting the best ground data within the constraints of resources (cost, time, personnel, access, etc.) and to allow for improvement over time (CEOS LPV 2019). Useful measurements can be taken using lower-cost, lower-precision TLS systems, potentially by more groups, over larger areas and/or more often, despite the resulting lower accuracy and detail. The accuracy requirement for cal/val for EO will depend on the nature of the EO estimates themselves and the region of interest. Duncanson et al. (2019 this volume) list the requirements for AGB accuracy for a range of space missions, all of which (where specified) set $\leq 20\%$ as the accuracy requirement. This suggests that ground measurements capable of doing better than this will provide useful cal/val data. Lower-cost TLS instruments will not produce accurate volume estimates of trees > 30 m in dense forests but are certainly capable of providing height and DBH for shorter, less dense canopies. The rapid development and improvement of these lower-cost instruments and associated processing software mean that their utility is only likely to increase. Such instruments will increasingly have a place in wider adoption of TLS methods for AGB estimation, particularly where cost is the primary barrier.

Lastly, the challenge of scale is how to integrate TLS, which excels at the individual tree to plot scale (~ha), with other measurements including manual survey measurements, manned and unmanned aerial lidar systems (ALS, UAS), and satellite observations. TLS has the potential to provide additional opportunities to exploit these measurements for AGB estimates, but the choice of scales at which to collect TLS and the type of plots to use are open questions. The application of TLS for AGB estimation and comparison to manual census, air- and spaceborne estimates is still new, and very much in development. Thus far, TLS survey plots have typically been selected to match the size and shape of the base unit of the other core measurements for a given application. For forest inventory applications, this is typically plots of 0.04–0.25 ha, circular or rectangular (Edwards 1998; USFS 2019); for tropical forest studies, plots are typically 1 ha square (Phillips et al. 2002). Wilkes et al. (2017) discuss the impact of varying density and arrangement of TLS scan locations for AGB estimation specifically. In particular, they show that the trade-off between more, smaller plots or fewer larger plots is not straightforward. The former may be tempting to enable sampling a wider range of forest type, climate, soil and other conditions, but the quality of the resulting TLS point cloud will be much lower than for larger plots. The larger the plot, the more TLS locations there will be that can contribute to the point cloud of a specific tree. This is particularly true for tall, dense canopies typical of tropical forests. For a 1 ha plot and with a 10-m scanning grid, there are 121 locations from which a tree might be 'seen', from 100 m away (or more from corner to corner). A 0.25 ha plot will only have 36 scan locations and the resulting extracted tree volume models will be of much lower quality, particularly in the upper canopy. Moreover, locations further from a tree often contribute more to the upper part of the tree than locations close by, due to the oblique angle through the canopy.

A related issue is how TLS measurements required for specific forest inventory applications may be used in the context of EO-derived AGB cal/val. Liang et al. (2018) present a comparison of forestry parameter extraction methods using a common TLS dataset that highlights some important issues of the impact of forest structure on the resulting point cloud. The comparison is carried out on smaller plots (0.1 ha) than those more suited for EO-derived AGB comparisons and using phase-shift TLS which has limitations of range and partial hits for tall forests (Newnham et al. 2012). These issues notwithstanding, an important conclusion is that the higher the density of smaller stems (DBH < 20 cm) in a plot, the greater the impacts of occlusion will be. TLS data collection for AGB estimation in this type of canopy, particularly young, dense or plantation forests, or forests dominated by lianas (Krishna Moorthy et al. 2018), will need to consider these possible limitations: increasing the scan density, plot size and by using time-of-flight longer range TLS will all help, but at increased cost.

An additional challenge comes for integration of TLS data with ALS or unmanned Aerial Lidar Systems (UAS). To match clouds at the crown-scale and finer requires accurate (cm-scale) co-location of the different lidar datasets, requiring very accurate RTK GPS or equivalent. Even then, the various point clouds may need manual co-registration, particularly for ALS point clouds where the absolute geolocation accuracy may be lower. The denser the airborne point cloud, the easier it will be to identify possible ground control points and leading to better manual matching, i.e. typically likely to be easier for UAS than ALS. Methods for merging these data are likely to develop and improve as they become more widely used. Specific issues of ALS and UAV lidar acquisition and characteristics for AGB estimation are discussed elsewhere.

The requirement for ASEs will remain, to enable wide-scale assessment of AGB, particularly from EO but also to provide continuity with the large body of the existing plot-scale census measurements. Consequently, we foresee the main role of TLS in EO-derived estimates of AGB will be initially to enable improved ASEs from a much larger sample of trees across a wider size distribution than is currently possible (calibration). TLS will also provide estimates of AGB at the same scale as those derived from EO, particularly from spaceborne measurements, allowing for improved validation of these products. In the longer term, TLS can potentially provide a better understanding of the impacts of structural variations (trunk form, crown size and shape) and within-tree wood density on the ASE approach. Lastly, 3D data derived from TLS enable development of more realistic 3D radiative transfer models to help quantify uncertainty in spaceborne observations (Calders et al. 2018). Structurally consistent models across optical and microwave domains will improve our ability to combine spaceborne data for AGB across these domains.

6 Conclusions

Terrestrial laser scanning offers considerable benefits to improving estimates of above-ground biomass from satellite observations. These benefits lie in a number of areas: improvement of the existing allometric scaling equations used to relate tree H and DBH to AGB, via the inclusion of many more samples of tree volume/biomass; the reduction of bias in ASEs by incorporating more large trees into ASEs to help address the fundamental issue of heteroscedasticity of AGB allometry; help develop new EO retrievals, through testing the assumptions of the existing retrieval models and by the use of full 3D radiative transfer models with which to simulate EO signals with known structural and radiometric uncertainties; by quantifying uncertainty in retrieved AGB estimates through pre-launch modelling and calibration, and post-launch validation at the plot scale; and lastly by providing a link between measurements made at the tree and plot scale, particularly manual forestry measurements and observations of varying spatial scales from airborne, unmanned aerial systems (UAS) and beyond. The interest in TLS has prompted many advances including new lower-cost TLS systems, methods to process TLS data more rapidly and consistently and protocols specifically aimed at collection of TLS data specifically for AGB applications. Further assessment of the uncertainty in the TLS model reconstructions is required, particularly in tall, dense canopies. Comparisons of TLS-derived and harvested AGB may agree very closely at the tree level, but this may overlook divergence within different size components. Further standardisation of data acquisition, automation of processing and more detailed testing of the resulting 3D reconstructions will bring TLS further into the mainstream for biomass estimation from space.

Acknowledgements MD acknowledges funding support from: NERC NCEO; NERC Standard Grants NE/N00373X/1 and NE/P011780/1; European Union's Horizon 2020 research and innovation programme under Grant Agreement No. 640176 for the EU H2020 BACI Project. MD gratefully acknowledges the support of the ISSI, Bern, for arranging and funding the Workshop Forest Biomass and Structure from Space, 6–9 November 2017, which provided the impetus for this article. CS was supported by NASA Grant

NNX16AP78A; all authors have benefited from the support of NSF RCN 1455636. We gratefully acknowledge the help of Peter Boucher at U Mass Boston for providing lidar examples for Fig. 9.

References

Ahmed R, Siqueira P, Hensley S, Bergen K (2013) Uncertainty of forest biomass estimates in north temperate forests due to allometry: implications for remote sensing. Remote Sens 5:3007–3036. https://doi.org/10.3390/rs5063007

Atkins JW, Bohrer G, Fahey RT, Hardiman BS, Morin TH, Stovall AE, Zimmerman N, Gough CM (2018) Quantifying vegetation and canopy structural complexity from terrestrial Li DAR data using the forestr r package. Methods Ecol Evol 9(10):2057–2066

Avitabile V et al (2016) An integrated pan-tropical biomass map using multiple reference datasets. Glob Change Biol 22(4):1406–1420. https://doi.org/10.1111/gcb.13139

Baccini A et al (2012) Estimated carbon dioxide emissions from tropical deforestation improved by carbon-density maps. Nat Clim Change 2:182–185

Baskerville GL (1972) Use of logarithmic regression in the estimation of plant biomass. Can J For Res 2:49–53. https://doi.org/10.1139/x72-009

Bauwens S, Bartholomeus H, Calders K, Lejeune P (2016) Forest inventory with terrestrial LiDAR: a comparison of static and hand-held mobile laser scanning. Forests 7(6):127

Boni Vicari M, Disney MI, Wilkes P, Burt A, Calders K, Woodgate W (2019) New framework for separating leaf and wood in terrestrial LiDAR point clouds. Methods Ecol Evol. https://doi.org/10.1111/2041-210X.13144

Brown S, Lugo E (1984) Biomass of tropical forests: a new estimate based on forest volumes. Science 223:1290

Burt A (2017) New 3D measurements of forest structure. Ph.D. thesis, University College London, London

Burt A, Disney MI, Calders K (2018) Extracting individual trees from lidar point clouds using treeseg. Methods Ecol Evol. https://doi.org/10.1111/2041-210X.13121

Calders K, Newnham G, Burt A, Murphy S, Raumonen P, Herold M, Culvenor D, Avitabile V, Disney MI, Armston J, Kaasalainen M (2014) Non-destructive estimates of above-ground biomass using terrestrial laser scanning. Methods Ecol Evol. https://doi.org/10.1111/2041-210X.12301

Calders K, Newnham G, Armston J, Disney MI, Schaaf C, Paynter I (2015) Terrestrial LIDAR for forest monitoring, Chapter 2.10 in GOFC-GOLD (2015) A sourcebook of methods and procedures for monitoring and reporting anthropogenic greenhouse gas emissions and removals associated with deforestation, gains and losses of carbon stocks in forests remaining forests, and forestation. GOFC-GOLD Report version COP21-1, (GOFC-GOLD Land Cover Project Office, Wageningen University, Thc Netherlands)

Calders K, Origo N, Burt A, Disney MI, Nightingale J, Raumonen P, Åkerblom M, Malhi Y, Lewis P (2018) Realistic forest stand reconstruction from terrestrial LiDAR for radiative transfer modelling. Remote Sensing 10(6):933. https://doi.org/10.3390/rs10060933

CEOS LPV (2019) https://lpvs.gsfc.nasa.gov. Accessed Feb 2019

CEOS LPV WGCV Biomass (2019): https://lpvs.gsfc.nasa.gov/Biomass/AGB_home.html. Accessed Feb 2019

Chave J, Condit R, Aguilar S, Hernandez A, Lao S, Perez R (2004) Error propagation and scaling for tropical forest biomass estimates. Phil Trans R Soc Lond B Biol Sci 359(1443):409–420

Chave J, Andalo C, Brown S, Cairns MA, Chambers JQ, Eamus D, Fölster H, Fromard F, Higuchi N, Kira T, Lescure JP (2005) Tree allometry and improved estimation of carbon stocks and balance in tropical forests. Oecologia 145(1):87–99

Chave J, Muller-Landau HC, Baker TR, Easdale TA, Steege HT, Webb CO (2006) Regional and phylogenetic variation of wood density across 2456 neotropical tree species. Ecol Appl 16(6):2356–2367

Chave J, Coomes D, Jansen S, Lewis SL, Swenson NG, Zanne AE (2009) Towards a worldwide wood economics spectrum. Ecol Lett 12(4):351–366

Chave J et al (2014) Improved allometric models to estimate the aboveground biomass of tropical trees. Glob Change Biol 20:3177–3190. https://doi.org/10.1111/gcb.12629
Clark DB, Kellner JR (2012) Tropical forest biomass estimation and the fallacy of misplaced concreteness. J Veg Sci 23:1191–1196
Crowther TW, Glick HB, Covey KR, Bettigole C, Maynard DS et al (2015) Mapping tree density at a global scale. Nature 525:201–205. https://doi.org/10.1038/nature14967
Danson FM, Gaulton R, Armitage RP, Disney MI, Gunawan O, Lewis P, Pearson G, Ramirez AF (2014) Developing a dual-wavelength full-waveform terrestrial laser scanner to characterize forest canopy structure. Agric For Meteorol 198:7–14
Disney MI (2018) Terrestrial lidar: a 3D revolution in how we look at trees. New Phytol. https://doi.org/10.1111/nph.15517
Disney MI, Lewis P, Raumonen P (2012) Testing a new vegetation structure retrieval algorithm from terrestrial lidar scanner data using 3D models, Silvilaser 2012, 16–19 September, Vancouver, BC, Canada
Disney MI, Boni Vicari M, Calders K, Burt A, Lewis S, Raumonen P, Wilkes P (2018) Weighing trees with lasers: advances, challenges and opportunities. Interface Focus 8(2):20170048. https://doi.org/10.1098/rsfs.2017.0048
Douglas ES, Martel J, Li Z, Howe G, Hewawasam K, Marshall RA, Schaaf CL, Cook TA, Newnham GJ, Strahler A, Chakrabarti S (2015) Finding leaves in the forest: the dual-wavelength Echidna lidar. IEEE Geosci Remote Sens Lett 12(4):776–780
Duncanson L, et al (2019) The importance of consistent global forest aboveground biomass product validation, chapter X in SSS ISSI book no. 71, eds. Scipal, Dubayah, Le Toan, Quegan, Cazenave and Lopez
Duncanson L, Rourke O, Dubayah R (2015) Small sample sizes yield biased allometric equations in temperate forests. Nature Sci. Rep. 5:17153. https://doi.org/10.1038/srep17153
Edwards PN (1998) Timber measurement: a field guide. Forestry Commission Booklet 49, ISBN 0 85538 403 4
Feldpausch TR, Banin L, Phillips OL, Baker TR, Lewis SL, Quesada CA, Affum-Baffoe K, Arets EJ, Berry NJ, Bird M, Brondizio ES (2010) Height-diameter allometry of tropical forest trees. Biogeosci Discuss 7:7727–7793
Garestier F, Le Toan T (2010) Estimation of the backscatter vertical profile of a pine forest using single baseline P-band (Pol-)InSAR data. IEEE Trans Geosci Remote Sens 48(9):3340–3348. https://doi.org/10.1109/TGRS.2010.2046669
GCOS (2010). Implementation plan for the global observing system for climate in support of the UNFCCC (2010 update). GCOS-138 (GOOS-184, GTOS-76, WMO-TD/No. 1523). World Meteorological Organization. url: http://wmo.int/pages/prog/gcos/Publications/gcos-138.pdf. visited on 21 July 2017
Gonzalez de Tanago J, Lau A, Bartholomeus H, Herold M, Avitabile V, Raumonen P, Martius C, Goodman RC, Manuri S, Burt A, Disney MI, Calders K (2018) Estimation of above-ground biomass of large tropical trees with terrestrial LiDAR. Methods Ecol Evol 9(2):223–234. https://doi.org/10.1111/2041-210X.12904
Hackenberg J, Morhart C, Sheppard J, Spiecker H, Disney MI (2014) Highly accurate tree models derived from terrestrial laser scan data—a method description. Forests 5:1069–1105. https://doi.org/10.3390/f5051069
Hackenberg J, Spiecker H, Calders K, Disney M, Raumonen P (2015) SimpleTree—an efficient open source tool to build tree models from TLS clouds. Forests 6(11):4245–4294
Hakala T, Suomalainen J, Kaasalainen S, Chen Y (2012) Full waveform hyperspectral LiDAR for terrestrial laser scanning. Opt Exp 20(7):7119–7127
Harris NL, Brown S, Hagen SC, Saatchi SS, Petrova S, Salas W, Hansen MC, Potapov PV, Lotsch A (2012) Baseline map of carbon emissions from deforestation in tropical regions. Science 336(6088):1573–1576
Hill TC, Williams M, Bloom AA, Mitchard ET, Ryan CM (2013) Are inventory based and remotely sensed above-ground biomass estimates consistent? PLoS ONE 8(9):e74170
Hopkinson C, Chasmer L, Young-Pow C, Treitz P (2004) Assessing forest metrics with a ground-based scanning lidar. Can J For Res 34:573–583. https://doi.org/10.1139/x03-225
IPCC (2006) Guidelines for national greenhouse gas inventories, Prepared by the national greenhouse gas inventories programme. In: Eggleston HS, Buendia L, Miwa K, Ngara T, Tanabe K (eds) IGES, Tokyo
Jackson T, Shenkin A, Wellpott A, Calders K, Origo N, Disney MI, Burt A, Raumonen P, Gardiner B, Herold M, Fourcaud T, Malhi Y (2019) Finite element analysis of trees in the wind based on terrestrial laser scanning data. Agric For Meteorol 265(15):137–144. https://doi.org/10.1016/j.agrformet.2018.11.014

Ketterings QM, Coe R, van Noordwijk M, Ambagau Y, Palm CA (2001) Reducing uncertainty in the use of allometric biomass equations for predicting above-ground biomass in mixed secondary forests. Forest Ecol Manag 146:199–209
Krishna Moorthy SM, Calders K, di Porcia e Brugnera M, Schnitzer SA, Verbeeck H (2018) Terrestrial laser scanning to detect liana impact on forest structure. Remote Sens 10(6):810
Le Toan T, Quegan S, Davidson M, Balzter H, Paillou P, Papathanassiou K, Plummer S, Rocca F, Saatchi S, Shugart H, Ulander L (2011) The BIOMASS Mission: mapping global forest biomass to better understand the terrestrial carbon cycle. Remote Sens Environ 115:2850–2860
Lewis SL, Sonké B, Sunderland T, Begne SK, Lopez-Gonzalez G et al (2013) Above- ground biomass and structure of 260 African tropical forests. Philos Trans R Soc Lond B Biol Sci 368:1625. https://doi.org/10.1098/rstb.2012.0295
Liang X, Hyyppä J, Kaartinen H, Lehtomäki M, Pyörälä J, Pfeifer N, Holopainen M, Brolly G, Francesco P, Hackenberg J, Huang H (2018) International benchmarking of terrestrial laser scanning approaches for forest inventories. ISPRS J Photogr Remote Sens 144:137–179
Lutz JA et al (2018) Global importance of large-diameter trees. Glob Ecol Biogeogr 27(7):849–864. https://doi.org/10.1111/geb.12747
Mascaro J, Litton CM, Hughes RF, Uowolo A, Schnitzer S (2011) Minimizing bias in biomass allometry: model selection and log-transformation of data. Biotropica 43(6):649–653. https://doi.org/10.1111/j.1744-7429.2011.00798.x
Mermoz S, Réjou-Méchain M, Villard L, Le Toan T, Rossi V, Gourlet-Fleury S (2015) Decrease of L-band SAR backscatter with biomass of dense forests. Remote Sens Environ 159:307–317
Mitchard ETA, Saatchi SS, Baccini A, Asner GP, Goetz SJ, Harris NL, Brown S (2013) Uncertainty in the spatial distribution of tropical forest biomass: a comparison of pan-tropical maps. Carbon Balance Manag. https://doi.org/10.1186/1750-0680-8-10
Mitchard ETA et al (2014) Markedly divergent estimates of Amazon forest carbon density from ground plots and satellites. Global Ecol Biogeogr 23(8):935–946
Molto Q, Rossi V, Blanc L (2013) Error propagation in biomass estimation in tropical forests. Methods Ecol Evol 4:175–183. https://doi.org/10.1111/j.2041-210x.2012.00266.x
Momo Takoudjou S, Ploton P, Sonké B, Hackenberg J, Griffon S, de Coligny F, Kamdem NG, Libalah M, Mofack GI, Le Moguédec G, Pelissiér R, Barbier N (2017) Using terrestrial laser scanning data to estimate large tropical trees biomass and calibrate allometric models: a comparison with traditional destructive approach. Methods Ecol Evol 9:905–916
Momo ST, Libalah MB, Rossi V, Fonton N, Mofack GI, Kamdem NG, Nguetsop VF, Sonké B, Ploton P, Barbier N (2018) Using volume-weighted average wood specific gravity of trees reduces bias in aboveground biomass predictions from forest volume data. For Ecol Manag 424:519–528
Newnham G, Armston J, Muir J, Goodwin N, Tindall D, Culvenor D, Püschel P, Nyström M, Johansen K (2012) Evaluation of terrestrial laser scanners for measuring vegetation structure. CSIRO, Canberra
Newnham GJ, Armston JD, Calders K, Disney MI, Lovell JL, Schaaf CB, Strahler AH, Danson FM (2015) Terrestrial laser scanning for plot scale forest measurement. Current For Rep 1(4):239–251. https://doi.org/10.1007/s40725-015-0025-5
Niklas KJ (1993) Influence of tissue density-specific mechanical properties on the scaling of plant height. Ann Bot 72:173–179. https://doi.org/10.1006/anbo.1993.1096
Noelke N, Fehrmann L, Nengah SJ, Tiryana T, Seidel D, Kleinn C (2015) On the geometry and allometry of big-buttressed trees - a challenge for forest monitoring: new insights from 3D-modeling with terrestrial laser scanning. For-Biogeosci For 8:574. https://doi.org/10.3832/ifor1449-007
Othmani A, Voon LFCLY, Stolz C, Piboule A (2013) Single tree species classification from Terrestrial Laser Scanning data for forest inventory. Pattern Recognit Lett 34:2144–2150
Pan Y, Birdsey RA, Fang J, Houghton R, Kauppi PE et al (2011) A large and persistent carbon sink in the world's forests. Science 333(6405):988–993. https://doi.org/10.1126/science.1201609
Paynter I, Saenz E, Genest D, Peri F, Erb A, Li Z, Wiggin K, Muir J, Raumonen P, Schaaf ES, Strahler A, Schaaf C (2016) Observing ecosystems with lightweight rapid-scanning terrestrial lidar scanners. Remote Sens Ecol Conserv 2(4):174–189. https://doi.org/10.1002/rse2.26
Paynter I, Genest D, Saenz E, Peri F, Li Z, Strahler A, Schaaf C (2018) Quality assessment of terrestrial laser scanner ecosystem observations using pulse trajectories. IEEE Trans Geosci Remote Sens. https://doi.org/10.1109/TGRS.2018.2836947
Phillips OL, Baker T, Feldspauch T Brienen RJW (2002) Field manual for plot establishment and remeasurement (RAINFOR). Amazon Forest Inventory Network, Sixth Framework Programme (2002–2006)
Preuksakarn C (2012) Reconstructing plant architecture from 3D laser scanner data (Doctoral dissertation, Université Montpellier II-Sciences et Techniques du Languedoc)

Raumonen P, Kaasalainen M, Åkerblom M, Kaasalainen S, Kaartinen H, Vastaranta M, Holopainen M, Disney MI, Lewis P (2013) Comprehensive quantitative tree models from terrestrial laser scanner data. Remote Sens 5(2):491–520. https://doi.org/10.3390/rs5020491

Raumonen P, Åkerblom M, Kaasalainen M, Casella E, Calders K, Murphy S (2015) Massive-scale tree modelling from TLS data. In: ISPRS annals of photogrammetry, remote sensing and spatial information sciences, p 2

Réjou-Méchain M, Tanguy A, Piponiot C, Châvé J, Hérault B (2017) biomass: an r package for estimating above-ground biomass and its uncertainty in tropical forests. Methods Ecol Evol 8(9):1163–1167. https://doi.org/10.1111/2041-210X.12753

Saatchi S et al (2011) Benchmark map of forest carbon stocks in tropical regions across three continents. Proc Natl Acad Sci 108(24):9899–9904

Sheil D, Eastaugh CS, Vlam M, Zuidema PA, Groenendijk P, der Sleen P, Jay A, Vanclay J (2016) Does biomass growth increase in the largest trees? Flaws, fallacies, and alternative analyses. Funct Ecol 31(3):568–581

Sileshi GW (2014) A critical review of forest biomass estimation models, common mistakes and corrective measures. For Ecol Manag 329:237–254

Slik JWF, Arroyo-Rodriguez V, Aiba S-I, Alvarez-Loayza P, Alves LF et al (2015) An estimate of the number of tropical tree species. Proc Natl Acad Sci 112(24):7472. https://doi.org/10.1073/pnas.1423147112

Stovall AE, Shugart HH (2018) Improved biomass calibration and validation with terrestrial LiDAR: implications for future LiDAR and SAR missions. IEEE J Sel Top Appl Earth Observ Remote Sens 99:1–11

Stovall AE, Vorster AG, Anderson RS, Evangelista PH, Shugart HH (2017) Non-destructive aboveground biomass estimation of coniferous trees using terrestrial LiDAR. Remote Sens Environ 200:31–42. https://doi.org/10.1016/j.rse.2017.08.013

Stovall AE, Anderson-Teixeira KJ, Shugart HH (2018) Assessing terrestrial laser scanning for developing non-destructive biomass allometry. For Ecol Manage 427:217–229

Swenson NG, Enquist BJ (2007) Ecological and evolutionary determinants of a key plant functional trait: wood density and its community-wide variation across latitude and elevation. Am J Bot 94:451–459

Tang S (2017) Quantifying Differences in Forest Structures with Quantitative Structure Models from TLS Data, UCL M.Sc. thesis (**unpublished**)

Ter Steege H, Pitman NCA, Sabatier D, Baraloto C, Salomaõ RP et al (2013) Hyperdominance in the amazonian tree flora. Science 342:6156. https://doi.org/10.1126/science.1243092

Ter-Mikaelian MT, Korzukin MD (1997) Biomass equations for sixty-five North American tree species. For Ecol Manag 97:1–24

Trochta J, Krůček M, Vrška T, Král K (2017) 3D Forest: an application for descriptions of three-dimensional forest structures using terrestrial LiDAR. PLoS ONE 12(5):0176871. https://doi.org/10.1371/journal.pone.0176871

USFS (2019) US Forestry Service Forest Inventory and Analysis Design, https://www.fia.fs.fed.us/library/fact-sheets/data-collections/Sampling%20and%20Plot%20Design.pdf. Accessed Feb 2019

Widlowski JL, Pinty B, Lopatka M, Atzberger C, Buzica D, Chelle M, Disney MI, Gastellu-Etchegorry JP, Gerboles M, Gobron N, Grau E, Huang H, Kallel A, Kobayashi H, Lewis P, Schlerf M, Stuckens J, Qin W, Xie D (2013) The 4th Radiation Transfer Model Intercomparison (RAMI-IV): using ISO standards to evaluate model proficiency. JGR Atmos 118:6869–6890. https://doi.org/10.1002/jgrd.50497

Wilkes P, Lau A, Disney MI, Calders K, Burt A, de Tanago Gonzalez, Menaca J, Bartholomeus H, Brede B, Herold M (2017) Data acquisition considerations for terrestrial laser scanning of forest plots. Remote Sens Environ 196:140–153. https://doi.org/10.1016/j.rse.2017.04.030

Yan D-M, Wintz J, Mourrain B, Wang W, Boudon F, Godin C (2009) Efficient and robust reconstruction of botanical branching structure from laser scanned points. In: 11th IEEE international conference on computer-aided design and computer graphics (CAD/Graphics 2009)

Yun T, An F, Li W, Sun Y, Cao L, Xue L (2016) A novel approach for retrieving tree leaf area from ground-based LiDAR. Remote Sens 8(11):942

Zanne AE et al (2009) Data from: towards a worldwide wood economics spectrum. Dryad Digit Repos. https://doi.org/10.5061/dryad.234

Affiliations

M. Disney[1,2] · A. Burt[1] · K. Calders[3] · C. Schaaf[4] · A. Stovall[5]

[1] Department of Geography, UCL, Gower Street, London WC1E 6BT, UK

[2] NERC National Centre for Earth Observation (NCEO), London WC1E 6BT, UK

[3] CAVElab - Computational and Applied Vegetation Ecology, Department of Environment, Ghent University, 9000 Ghent, Belgium

[4] School for the Environment, University of Massachusetts Boston, 100 Morrissey Blvd, Boston, MA 02125, USA

[5] Department of Environmental Sciences, University of Virginia, Charlottesville, VA 22903, USA

Surveys in Geophysics (2019) 40:959–977
https://doi.org/10.1007/s10712-019-09529-9

New Opportunities for Forest Remote Sensing Through Ultra-High-Density Drone Lidar

James R. Kellner[1,2] · John Armston[3] · Markus Birrer[4] · K. C. Cushman[1,2] · Laura Duncanson[3] · Christoph Eck[4] · Christoph Falleger[4] · Benedikt Imbach[4] · Kamil Král[5] · Martin Krůček[5] · Jan Trochta[5] · Tomáš Vrška[5] · Carlo Zgraggen[4]

Received: 14 October 2018 / Accepted: 2 April 2019 / Published online: 4 May 2019

Abstract
Current and planned space missions will produce aboveground biomass density data products at varying spatial resolution. Calibration and validation of these data products is critically dependent on the existence of field estimates of aboveground biomass and coincident remote sensing data from airborne or terrestrial lidar. There are few places that meet these requirements, and they are mostly in the northern hemisphere and temperate zone. Here we summarize the potential for low-altitude drones to produce new observations in support of mission science. We describe technical requirements for producing high-quality measurements from autonomous platforms and highlight differences among commercially available laser scanners and drone aircraft. We then describe a case study using a heavy-lift autonomous helicopter in a temperate mountain forest in the southern Czech Republic in support of calibration and validation activities for the NASA Global Ecosystem Dynamics Investigation. Low-altitude flight using drones enables the collection of ultra-high-density point clouds using wider laser scan angles than have been possible from traditional airborne platforms. These measurements can be precise and accurate and can achieve measurement densities of thousands of points · m^{-2}. Analysis of surface elevation measurements on a heterogeneous target observed 51 days apart indicates that the realized range accuracy is 2.4 cm. The single-date precision is 2.1–4.5 cm. These estimates are net of all processing artifacts and geolocation errors under fully autonomous flight. The 3D model produced by these data can clearly resolve branch and stem structure that is comparable to terrestrial laser scans and can be acquired rapidly over large landscapes at a fraction of the cost of traditional airborne laser scanning.

Keywords Drone · Global Ecosystem Dynamics Investigation (GEDI) · Lidar · Remote sensing · UAV

✉ James R. Kellner
james_r_kellner@brown.edu

Extended author information available on the last page of the article

1 Introduction

We are on the cusp of a golden age in satellite remote sensing. Current and forthcoming missions will produce measurements of the land surface with unprecedented spatial and temporal detail. The use of the International Space Station (ISS) as a platform for coordinated Earth observation will generate remote sensing data over diurnal timescales, including characterizations of canopy temperature, solar-induced fluorescence, and ecosystem structure (Stavros et al. 2017). The data products from these missions will enable new insight into the biological underpinnings of the Earth system and help to constrain uncertainties in the global distribution of aboveground stocks and fluxes of carbon and water.

Three missions will quantify aboveground biomass density (AGBD). The Global Ecosystem Dynamics Investigation (GEDI) is producing globally representative measurements of vertical height profiles (waveforms) and estimates of aboveground carbon stocks (Dubayah et al. 2014; Hancock et al. 2019). Two additional efforts are the NASA-ISRO Synthetic Aperture Radar (NISAR) and ESA P-band radar (BIOMASS), both of which are free-flying satellites that will produce AGBD data products (Scipal et al. 2010; NISAR Mission Science Handbook 2018).

Calibration and validation of these data products is critically dependent on the existence of globally representative field estimates of AGBD and coincident remote sensing measurements of vegetation structure (Duncanson et al. 2019). The best sources of remotely sensed vegetation structure are high-density airborne or terrestrial laser scanning (ALS and TLS, respectively), but there are relatively few places with field estimates of AGBD and coincident remote sensing data, and they are disproportionately in the northern hemisphere and temperate zone. In contrast, the tropics contain substantial aboveground forest carbon stocks (Pan et al. 2011), but are not as well represented in existing field inventories used for calibration and validation, especially those of Africa and Asia.

The use of traditional aircraft with lidar instruments to acquire remote sensing measurements of vegetation structure that is coincident with field estimates of AGBD is cost-prohibitive for a large number of remote locations, many of which are far from major centers. It would also be inefficient, because the critical areas necessary for calibration and validation activities are the plots and surrounding landscapes $< 10\ km^2$ (Held et al. 2015).

1.1 New Opportunities

An alternative to traditional airborne laser scanning is to obtain these data using drones. Lidar from low-altitude drones is fundamentally similar to traditional airborne laser scanning. Sensors record the return-time of emitted laser pulses and combine this information with location and attitude to project recorded laser pulses in a 3D coordinate system (Lefsky et al. 2002). But the methods of data collection from low-altitude flight create new opportunities to characterize the three-dimensional structure of forests in ways that have not been possible until now.

There are several important differences. Low-altitude flight at relatively slow speeds can produce point densities that are orders of magnitude greater than traditional airborne laser scanning. Coupled with wide scan angles, these large point densities can resolve individual tree and branch structure and are similar to TLS. Low-altitude flight reduces the impact of GPS and pointing uncertainties that accumulate with the distance between sensors and the terrain or vegetation surface, making wide-angle scans possible. Because drone flight is more firmly under the control of the investigator and about one order of magnitude less costly than traditional airborne laser scanning, flight plans can be developed to collect high-density measurements in novel ways that enable hypothesis testing, or evaluate the impact of collection scenarios on remote sensing measurements.

Below we summarize the potential for drones to produce new observations in support of mission science. In the first section, we describe technical considerations for producing high-quality measurements from autonomous platforms, and we highlight differences among commercially available laser scanners and drone aircraft. We next describe a case study using the Brown Platform for Autonomous Remote Sensing (BPAR, Fig. 1) in a temperate mountain forest in the southern Czech Republic in support of calibration and validation activities for GEDI. We demonstrate that drone measurements can facilitate large-area mapping and the high-quality measurements necessary for validation of mission data products. We conclude by addressing three areas where drone remote sensing is likely to make significant contributions to our understanding of the land surface and to meeting the calibration and validation needs of current and forthcoming space missions.

2 Technical Considerations for Drone Remote Sensing

Generating high-quality remote sensing data from an unmanned platform requires a stable aircraft with a robust flight control system, a survey-grade GPS and inertial motion unit (GPS-IMU), vibration isolation of sensor hardware, and post-processing using differential correction. In general, vibration isolation and stability are easier to control in heavier aircraft than lighter ones. Larger aircraft can resist wind and thermal activity, which allows them to operate under a wider range of conditions than smaller aircraft. Long flight durations are necessary to map areas larger than a few ha (Brede et al. 2017).

2.1 Light-Weight Airborne Laser Scanners

In the last few years, a number of commercially available, light-weight laser scanners have been developed that are suitable for drone-based flight operations (Table 1). Some of these sensors are developed by RIEGL Laser Measurement Systems, the Austrian manufacturer of survey-grade lidar instruments. Others were originally designed for automotive or industrial use, but have been repurposed for drone flight. These sensors differ in weight, range accuracy, beam divergence (and therefore footprint size and shape), wavelength, the maximum measurement rate, the number of returns recorded, the effective measurement range (EMR), and scan angle range (Table 1). The RIEGL sensors are heavier, more accurate, produce higher point densities, and are the only lightweight sensors capable of recording > 3 returns per emitted laser pulse. They also have the narrowest beam divergences and therefore result in the smallest footprints at a given flight altitude. For example, the RIEGL VUX-1 has a 0.5 mrad beam divergence, which produces a 5 cm circular footprint at a distance of 100 m. Contrast this with the Velodyne HDL-32, an industrial laser scanner repurposed for drone flight, which has a 3 × 1.5 mrad beam divergence (i.e., the beam divergence is different in the across-beam dimensions). This produces a rectangular footprint of 30 × 15 cm at a distance of 100 m.

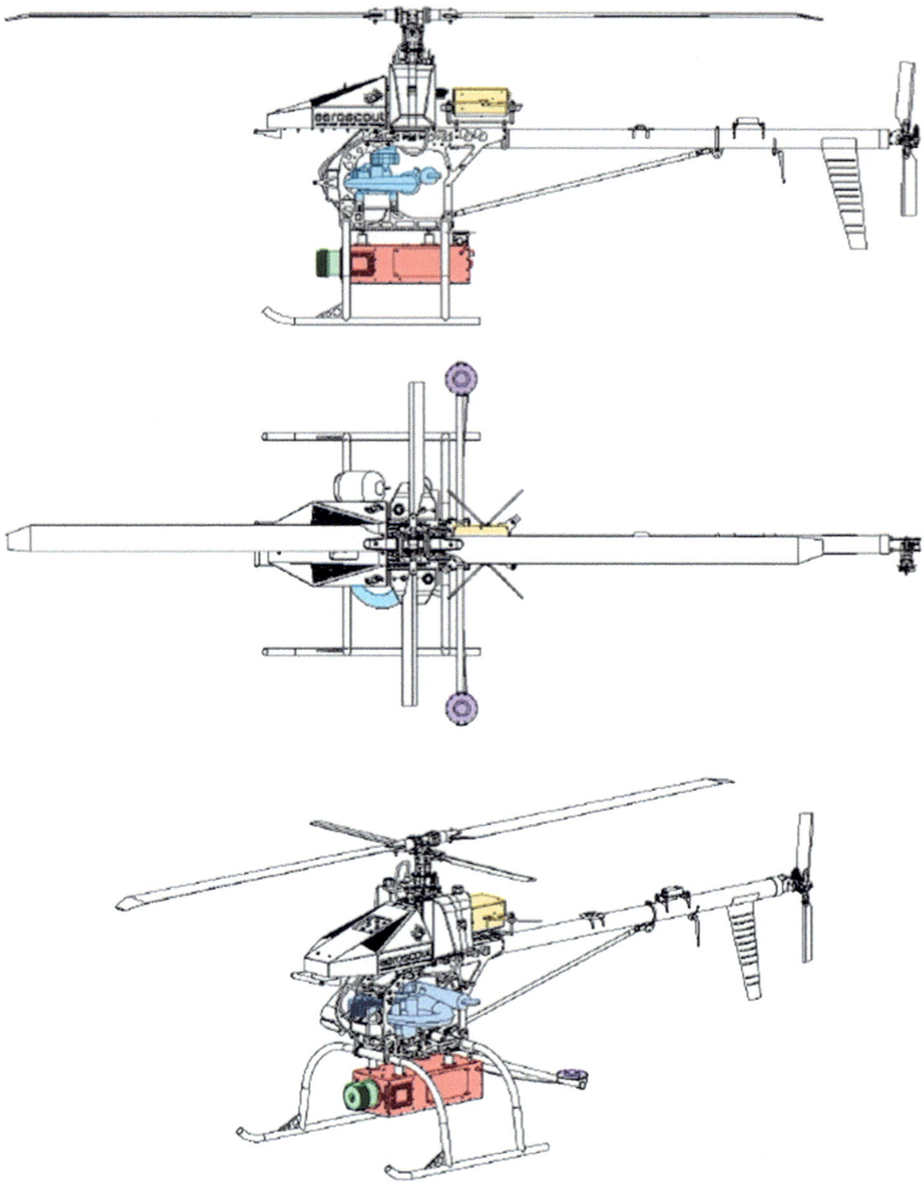

Fig. 1 The Aeroscout GmbH B100 heavy-lift autonomous helicopter. Components are distinguished in color: flight control system hardware (orange), engine (blue), dual GPS antennas (purple), RIEGL VUX-1 laser scanner (green), GPS-IMU and electrical supply (red). The main rotor is 3.2 m in length, and the aircraft weighs 77 kg with maximum payload

2.2 Precision and Accuracy

The precision and accuracy of all airborne laser scans is influenced by the stability of the sensor during flight, knowledge of the sensor attitude, typically from a

Table 1 Characteristics of lightweight commercial lidar instruments

Manufacturer	Model	Weight (kg)	Range accuracy (cm)	Beam divergence (mrad)	Laser Wavelength (nm)	Max. measurement rate (kHz)	No. returns	Max. measurement range (m)	FOV (deg.)
RIEGL	VUX-240	3.8	2.0	0.35×0.35	1550	1500	Multi	350	75
RIEGL	VUX-1	3.5	1.0	0.5×0.5	1550	500	Multi	170	330
RIEGL	VUX-1HA	5.0	0.5	0.5×0.5	1550	1000	Multi	120	360
RIEGL	VUX-1LR	3.5	1.5	0.5×0.5	1550	750	Multi	215	330
RIEGL	miniVUX-1	1.55	1.5	1.6×0.5	905	100	5	150	360
Velodyne	VLP-16	0.83	2.0	3.0×1.5	903	600	2	100	360
Velodyne	HDL-32	1.0	2.0	3.0×1.5	903	1390	2	100	360
Ibeo	LUX	1.0	4.0	14.0×1.4	905	38	3	100	110

Data provided by manufacturers unless otherwise indicated. Maximum measurement range depends on measurement rate and target reflectance. For the RIEGL instruments, the maximum measurement range is at the maximum measurement rate on a 20% reflectance target. For Velodyne and Ibeo instruments, the maximum measurement range is under unspecified conditions. Characteristics for the Ibeo LUX are from Lin et al. (2013)

survey-grade GPS-IMU, and flight altitude. Because GPS and pointing errors accumulate with distance from the target, low-altitude flight can partially compensate for GPS-IMU quality, although a poor GPS-IMU can result in data that are not useful (but see Sect. 3.2 on computer vision). Stated differently, the same GPS-IMU will result in better horizontal placement of projected laser pulses at a lower flight altitude than a higher one. For a sensor at altitude A, the horizontal position of a projected laser pulse along a single axis can be calculated as:

$$\text{Horizontal position} = A \times \tan(\text{scan angle} \pm \epsilon) \tag{1}$$

For the same value of A, a given ϵ, which denotes error in the scan angle measurement, has a larger impact on the horizontal position when the scan angle is larger. For a given scan angle and ϵ, error in the horizontal position is proportional to A. By operating at low altitude, drone flight minimizes error in horizontal position of projected laser pulses.

The best ranging accuracies reported for laser scanners are from terrestrial sensors, which are on the order of one to a few mm at distances of 100 m (Calders et al. 2017; Disney et al. 2018). No airborne sensors currently approach this degree of performance during flight operations (Table 1).

2.3 Footprint Size

The interpretation of ranging accuracies becomes challenging inside canopies due to differences in beam divergence, distance from the target, and consequently, footprint size. Beam divergence is the rate of increase in beam size with distance from the sensor. Just as a flashlight makes a larger circle on a wall from a greater distance, laser scans produce larger footprints when the sensor is farther away from a target reflective surface. The combination of footprint size, distance from the target, and target reflectivity fundamentally constrains the spatial resolving capacity of any laser scanner (Lichti and Jamtsho 2006; Milenković et al. 2018). Commercial TLS instruments produce footprint sizes in the 2–5 cm range at 100 m (Disney et al. 2018), allowing them to resolve small stems and branches. The RIEGL VUX-1 produces a footprint size of 5 cm at 100 m from the target. Other laser scanners that are being operated from low-altitude drones have variable beam divergences and a wide range of footprint sizes (Table 1). Traditional airborne laser scanners produce footprints in the 15–30 cm range from flight altitudes in the 500–1000 m range. Thus, airborne laser scanners that exhibit similar fundamental ranging accuracies may not be equally amenable to detection of branch and stem structure within canopies. This is because resolving fine structures requires high accuracy, high density, and small footprints (Disney et al. 2018). Because footprint size is a function of beam divergence and distance from the target, laser scanners operated from low-altitude produce footprints that are relatively more variable in size along the beam path. For example, at a flight altitude of 100 m over a 50 m canopy, footprint size for the VUX-1 is 2.5 cm on the canopy top and 5 cm on the ground. This is an absolute difference of 2.5 cm, but the ground footprints are twice as large. At 500 m altitude, the VUX-1 produces 22.5 cm footprints on a 50-m canopy and 25 cm footprints at ground level. The absolute difference is still 2.5 cm, but the ground footprints are now only 11% larger. This indicates that the potential for footprints to contain qualitatively different information along the beam path, such as individual leaf and small branch returns, is greater when the distance to the target is less.

2.4 Effective Measurement Range

EMR is the maximum distance over which a laser scanner can record a reflected laser pulse. The EMR depends on the size and reflectivity of the target and energy per emitted laser pulse (pulse repetition rate, PRR). For example, the EMR of the RIEGL VUX-1 at a PRR of 300 kHz is 230 m for a target of 20% reflectance, or 400 m for a target of 60% reflectance. The wavelength of the VUX-1 is 1550 nm. The reflectance of green vegetation is about 10% at 1550 nm, and woody vegetation has a reflectance of about 30% at 1550 nm (Asner and Heidebrecht 2002). The EMR of the VUX-1 increases to 400 and 660 m for targets of 20% and 60% reflectance, respectively, at a PRR of 100 kHz because there is more energy per emitted laser pulse. Velodyne does not report EMR for targets of varying reflectivity, but the reported EMR is 100 m. Lin et al. (2013) report an EMR for the Ibeo LUX lidar sensor of 100 m. These considerations are important, because the beam path must be $<$EMR to guarantee that returns can be recorded. At a scan angle of 60°, the beam path is two times the sensor altitude. A good rule of thumb is that flight altitude should not exceed $0.5\times$EMR when scan angles up to 60° are employed.

2.5 Drone Platforms

Just as the number of commercially available laser scanners has rapidly increased over the last few years, so too have the number of drone aircraft capable of carrying these instruments. These platforms range in size from battery-powered multirotor aircraft that weigh a few kg and can carry a small laser scanner for a few minutes, like those developed by the company Phoenix Lidar Systems, to heavy-lift gasoline powered helicopters like the Aeroscout GmbH B330 that can carry a 50-kg payload for three hours. Rotor-powered aircraft of intermediate flight duration and payload capacity include the electric powered Vapor 55 helicopter developed by Pulse Aerospace. This traditional helicopter can carry a 5-kg payload for 1 h. The RIEGL RiCOPTER is a multirotor developed by RIEGL Laser Measurement Systems to carry the VUX-line scanners. It has a 6.5-kg payload capacity and a maximum flight duration of 0.5 h. In general, fixed-wing aircraft have longer flight durations than helicopters or multirotor aircraft of the same size. For example, the Edinburgh-based company Carbomap is developing a long-endurance fixed-wing platform that will rival the large-area mapping ability of traditional airborne laser scanning from crewed aircraft. The long flight duration of fixed-wing aircraft comes at the cost of sacrificing control over flight speed.

Combinations of laser scanners, GPS-IMU, and aircraft dictate the capabilities and measurement quality that can be expected from drone-based laser scans. Smaller aircraft and sensors are nimble and can be deployed easily and within the scope of existing regulations. Larger aircraft may require explicit permission from national departments of civil aviation and a professional team of pilots and payload operators. Some aircraft and laser scanners can be subject to export control regulations. Larger aircraft can carry higher-quality laser scanners and GPS-IMU. The selection of an aircraft, laser scanner, and GPS-IMU hardware should therefore consider measurement requirements, including point density, accuracy, the required extent of areal coverage, and flight time.

3 Brown Platform for Autonomous Remote Sensing

The Brown Platform for Autonomous Remote Sensing is a suite of sensors carried by a heavy-lift helicopter developed by the Swiss company Aeroscout GmbH. The BPAR sensor package includes imaging spectrometers that cover the 400–2500 nm region of the solar spectrum at varying resolution and sampling intervals, a 24.3 MP digital camera, a RIEGL VUX-1 laser scanner (Table 1), and an Oxford Technical Solutions (OXTS) Survey +2 GPS-IMU. The VUX-1 is a 1550 nm laser capable of recording up to 500,000 measurements s^{-1}. The sensor records the return time of emitted laser pulses and processes the recorded waveform to identify discrete reflective surfaces (returns) within every emitted laser pulse. The OXTS GPS-IMU is a survey-grade instrument designed for airborne mapping applications. The nominal roll and pitch accuracy are 0.03°, and the nominal heading accuracy in dual-antenna mode is 0.05°. In our configuration, the instrument records at 250 Hz and has access to the GPS and GLONASS networks. During flight operations, we collect an independent global navigation satellite system (GNSS) data stream on the ground using a Novel FlexPak 6 Triple Frequency + L-band GNSS receiver. This data stream is used to differentially correct the OXTS GPS-IMU measurements in post-processing.

The aircraft and sensors are operated by three people. A certified pilot in command (PIC) is responsible for takeoff and landing under manual control. The PIC has access to a forward-facing video camera that is mounted on the front of the aircraft. The PIC can operate the aircraft using hand controls by line-of-sight or using a video-feed to the forward-facing camera. Upon reaching altitude, the PIC communicates with the ground-station pilot (GSP) by radio. The GSP interacts with the flight control system over an 868 MHz data link that provides information about location, altitude, speed, heading, engine performance, and exhaust characteristics. The GSP is responsible for engaging the flight control system to initiate autonomous flight, after which the aircraft will visit a set of locations in a specified order at a specified altitude and speed under autonomous control. During the flight, a payload operator (PO) communicates with onboard sensors over a 2.4 GHz data link to monitor the data collection in real time. The PO can communicate with the PIC and GSP to request changes to the flight plan when needed. When the mission is complete, the aircraft will return to a specified location and hover, awaiting further instructions from the PIC.

The aircraft used during campaigns in the Czech Republic was the Aeroscout B100, which is operated with rigorous failsafe mechanisms and multiple levels of redundancy. With the exception of the flight crew, no people are within 100 feet (30.5 m) of the aircraft during flight operations, and we design flight plans that minimize risk to people and property. When a road or trail crossing is unavoidable, we cross at a right angle and visually confirm that there are no people beneath the aircraft trajectory. The aircraft has a redundant flight control system and electrical supply. We program instructions into the flight control system so that the aircraft will safely land in the event that communication is severed. If this occurs, the aircraft will hover at a specified location for 10 min, or until remaining fuel falls below a critical level, then land autonomously. The engine is gasoline powered and can support an 18-kg payload for 1.5 h at sea level. The maximum takeoff weight is 77 kg, and the aircraft operates under certificate of the Swiss Civil Aviation Authority (BAZL).

3.1 Flight Campaign in the Czech Republic

We collected airborne laser scans using BPAR over a temperate mountain forest in the southern Czech Republic. The site contains the Zofin Forest Dynamics Plot, which is a 25 ha permanent-inventory plot in which all free-standing woody plants > 1 cm diameter at breast height (DBH) have been mapped and monitored since 2012 (Anderson-Teixeira et al. 2015; Janík et al. 2016). This forest is dominated by old-growth European beech (*Fagus sylvatica*) and Norway spruce (*Picea abies*), with occasional silver fir (*Abies alba*). The surrounding landscape has been actively managed for Norway spruce production and includes planted and clearcut areas.

The data were collected in two sets of orthogonal flight lines that we repeated 51 days apart. The first flights commenced on April 16, 2018. The second flights started on June 6, 2018. These dates were selected to produce observations under leaf-off and leaf-on conditions with little to no change in woody structure. The April campaign was completed in six flights over two consecutive days. The June campaign required six flights over three consecutive days. The total flight time for each campaign was about 5 h. For each campaign, there were 45 flight lines in the NW–SE direction and 45 flight lines in the NE-SW direction. The purpose of using two sets of flight lines over the same area was to ensure dense point coverage of stem and branch structure. The nominal flight altitude was 110 m aboveground, and the nominal flight speed was 6 m $\cdot$ s^{-1}. During the autonomous portion of the flight, the flight control system maintained stable control of the aircraft and sensors. For example, during a representative flight line the realized speed was 6 m $\cdot$ s^{-1} (standard deviation, SD = 0.06 m s^{-1}). The standard deviation in the pitch, roll, and heading axes was 0.3°, 0.6°, and 0.8°, respectively. The total areas covered were 1.72 and 1.60 km^2, respectively (Fig. 2).

Measurement density is large. The leaf-off point cloud contains 4,811,370,798 unique returns, corresponding to a mean density of 2801 $\cdot$ m^{-2}. The leaf-on point cloud, acquired over the same location 51 days later, contains 3,465,943,773 unique returns. This is a mean density of 2166 $\cdot$ m^{-2}. Within the primary area of interest, point density exceeds the landscape mean in forested areas, because the entire landscape includes edges with reduced overlap among flight lines, and open forest with fewer higher-order returns (Table 2). For example, a representative 2.25 ha area dominated by old-growth *F. sylvatica* contains 97,615,242 points under leaf-off conditions, corresponding to a density of 4438 $\cdot$ m^{-2}, and 67,769,379 points under leaf-on conditions. The leaf-on density is 3012 $\cdot$ m^{-2}.

3.2 Precision and Accuracy of Airborne Laser Scans Under Autonomous Flight

We quantified range precision and accuracy for the RIEGL VUX-1 during autonomous flight by extracting points along a 12 m profile on the stem of a fallen dead tree that was apparent in the high-density point cloud (Figs. 3, 4). Linear regressions under leaf-off and leaf-on conditions had very high coefficients of determination and similar slope and intercept terms (leaf off: $b0 = 745.0$ m, $b1 = -4.51$ cm, $r^2 = 0.930$, residual standard error (RSE) = 4.51 cm; leaf on: $b0 = 745.0$ m, $b1 = -4.65$ cm, $r^2 = 0.984$, RSE = 2.05 cm). The mean difference between the predicted values is 2.4 cm, and ranges from 3.2 to 1.7 cm. From this analysis, we conclude that the realized ranging accuracy in the post-processed point cloud has a standard deviation < 5 cm. We emphasize that this is the ranging accuracy net of all processing artifacts and geolocation errors under fully autonomous flight on a

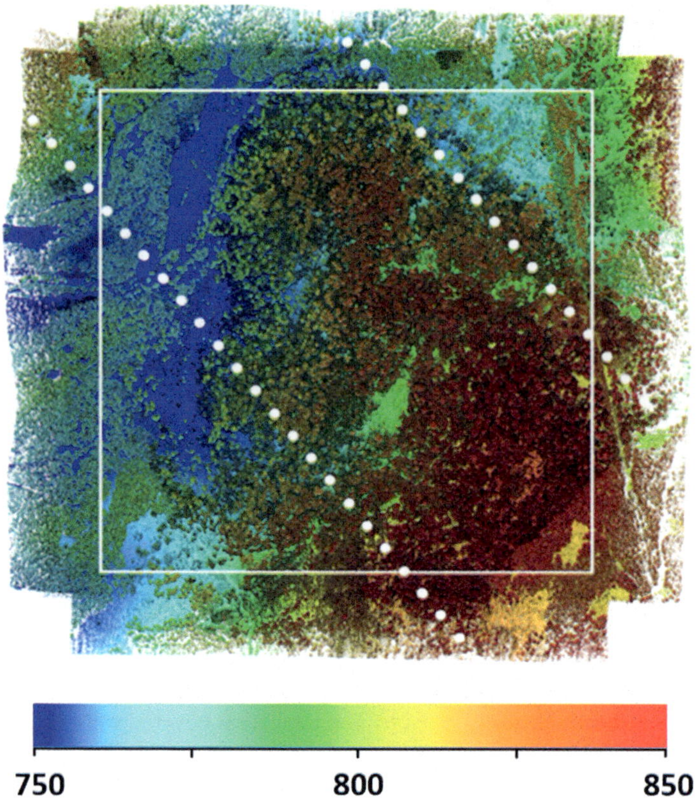

Fig. 2 Operational drone lidar in support of mission science. We collected these data using the Brown Platform for Autonomous Remote Sensing in a temperate beech forest in the southern Czech Republic in April, 2018. Colors indicate elevation (m), and the tallest trees are about 40 m. The white box is 1 by 1 km and equal in size to the GEDI L4B aboveground biomass density data product. White dots are simulated GEDI L4A ground tracks, drawn to scale (22 m footprint diameter, 60 m along-track spacing, 600 m across-track spacing, track inclination is arbitrary). These data were collected in five flight hours using 90 flight lines at 90° angles. Mean point density is 2801 points $\cdot$ m^{-2}. Collecting lidar data at 100 points $\cdot$ m^{-2} takes about 6 min $\cdot$ km^{-2} using this platform

Table 2 The distribution of return numbers (percentages) from airborne laser scans using a low-altitude drone under leaf-off and leaf-on conditions 51 days apart

	First	Second	Third	Fourth	≥Fifth
Closed forest leaf off	43.0	31.1	17.2	6.8	1.9
Closed forest leaf on	63.7	27.8	7.1	1.2	0.2
Open forest leaf off	83.2	13.6	2.8	0.4	0.0
Open forest leaf on	86.8	11.5	1.5	0.2	0.0

The measurements were collected using the Brown Platform for Autonomous Remote Sensing and contrast a closed-canopy old-growth forest dominated by European beech (*F. sylvatica*) with an open area dominated by Norway spruce (*P. abies*) and silver fir (*A. alba*). Higher-order returns occur more frequently in closed-canopy forest due to complex vertical structure. The presence of leaf area reduces return numbers > 3

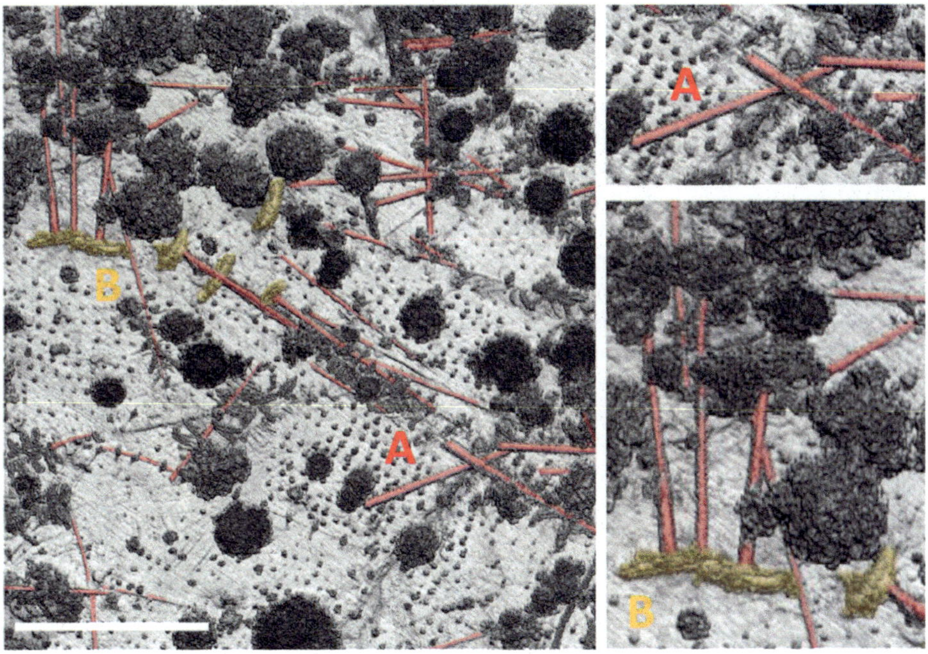

Fig. 3 A high-density digital surface model from drone lidar resolves recently fallen trees. The image is a 5 cm digital surface model and colors indicate sun-shaded intensity. Fallen trees are labeled in red. Tip-up mounds are orange. The white scale bar in the main image is 25 m

Fig. 4 Comparison of lidar surface elevation profiles from April 16, 2018 (green points and line) and June 6, 2018 (orange points and line). The profile is 2 cm wide and was extracted from the stem of a fallen tree (labeled A in Fig. 3). The mean difference in elevation between these lines is 2.4 cm. The residual standard error is 4.5 cm in April and 2.1 cm in June

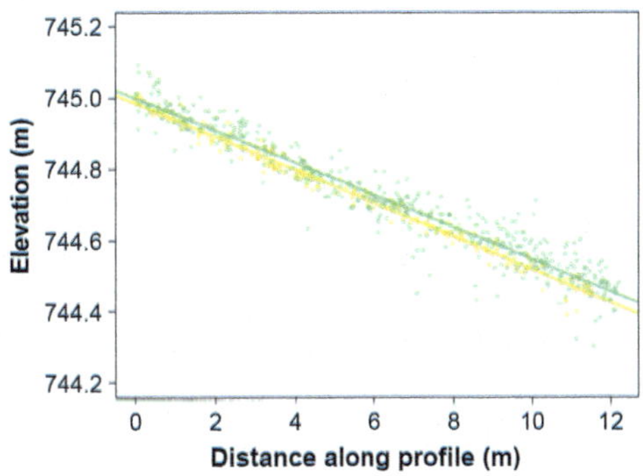

heterogeneous target (the fallen dead tree) and that this number is about one order of magnitude greater (worse) than the ranging accuracies reported for TLS (Calders et al. 2017; Disney et al. 2018). It is of a similar or smaller magnitude to ranging accuracies reported for traditional airborne laser scanning (Asner et al. 2007; Kellner et al. 2009).

Recent developments in computer vision may be able to improve the precision and accuracy of airborne laser scans, potentially compensating for low-quality GPS-IMU measurements deployed on aircraft with limited payload capacity and increasing the performance of survey-grade instruments. Simultaneous localization and mapping (SLAM) algorithms are used to facilitate autonomous navigation in unstructured environments lacking GPS

(Aouf et al. 2009; Qiang and Xin-sheng 2013). These methods work on 2D video streams from moving vehicles to solve the 3D geometry of the local environment. Although normally used for real-time navigation, this information provides a trajectory independent of the GPS-IMU that can constrain the position and pointing angles of the sensor. In a pioneering study, Wallace et al. (2012) collected airborne lidar from a multirotor aircraft using an Ibeo LUX laser scanner and low-quality GPS-IMU (Table 1). The horizontal RMSE was 0.61 m when data were processed using only the GPS-IMU. However, a high-definition video stream was collected at 30 frames per second during flight and used to produce an independent trajectory by applying the scale-invariant feature transform (SIFT) algorithm to the sequence of 2D images (Lowe 1999). The authors integrated the GPS-IMU and computer vision trajectories using a Kalman filter. This reduced the horizontal RMSE to 0.34 m. Thus, post-processing using computer vision approximately halved the horizontal uncertainty. No study has systematically evaluated the potential of this approach for improving the horizontal accuracy of airborne laser scans, but doing so is a high research priority.

4 Promising Applications of Drone Remote Sensing

4.1 Individual Tree Segmentation

One of the most promising applications of drone remote sensing is the segmentation of individual stems using methods originally developed for terrestrial laser scanning (Fig. 5). Applying these methods to ultra-high-density point clouds that contain millions of trees is a formidable computational problem (Raumonen et al. 2013). Overcoming this challenge would beat back a series of pervasive limitations to quantifying AGBD using remote sensing and create new opportunities for demographic analysis of tree populations using much larger samples than are currently available (Kellner and Hubbell 2017). The essential problem is that no remote sensing method or conventional forest inventory directly measures mass. The most commonly measured variables are stem diameter and height. These quantities are used to predict AGB using allometric scaling equations, which are themselves developed from a relatively small number of trees that have been measured, harvested, and weighed (Jenkins et al. 2003; Muukkonen 2007; Chave et al. 2014). The samples used to build these scaling equations are nonrandom and almost certainly biased in favor of trees with idealized forms (Clark and Kellner 2012).

The only way to evaluate the accuracy of these scaling equations is to compare predicted to measured AGB. The few studies that have undertaken this arduous task have shown that the equations are biased, especially among the largest trees (Calders et al. 2015; Gonzalez de Tanago et al. 2018). One way to lessen the impact of this problem is to compute wood volume using very high density laser scans (Raumonen et al. 2013; Calders et al. 2015; Gonzalez de Tanago et al. 2018). Multiplying wood volume by wood density results in an estimate of AGB that is independent of the allometries. Quantitative structure models (QSMs) segment individual trees within high-density point clouds and fit large numbers of cylinders to individual tree objects (Raumonen et al. 2013; Åkerblom et al. 2015). The accumulated volume of these cylinders is an estimate of tree-level wood volume. Three studies have compared these numbers to harvested trees. They found that wood volume and

Fig. 5 A single European beech (*F. sylvatica*) in the high-density point cloud acquired by the Brown Platform for Autonomous Remote Sensing. **a** All returns. **b** Only returns with 16-bit reflectance intensity > 45,000. The intensity filter removes leaf material and shows branch and stem structure necessary for automated segmentation

AGB quantified using QSMs are unbiased (Calders et al. 2015; Gonzalez de Tanago et al. 2018; Takoudjou et al. 2018).

Automated segmentation methods therefore hold great promise, because they will improve the quality of existing scaling equations and are likely to eventually replace them. The most important limitation to this approach is that it remains dependent on intensively collected field data from TLS and is therefore difficult to apply throughout large areas needed for the calibration and validation of data products from current and forthcoming space missions. For example, the GEDI L4B gridded AGBD data product has a resolution of 1 km^2 (Patterson et al. 2019). Low-altitude drone flight may be able to meet this challenge by producing high-density point clouds throughout areas of 1–10 km^2.

The high-density point clouds collected in the southern Czech Republic resolve individual stem and branch structure (Figs. 5, 6). It is clear that measurements from wide scan angles are disproportionately important sources of information about stem structure, a finding also reported by Brede et al. (2017). This is apparent by examining the

Fig. 6 Stem and branch structure from high-density lidar acquired by a low-altitude drone. The point density in this scene is $3981 \cdot m^{-2}$. Scale varies from this perspective. The length of the white bar is 30 m

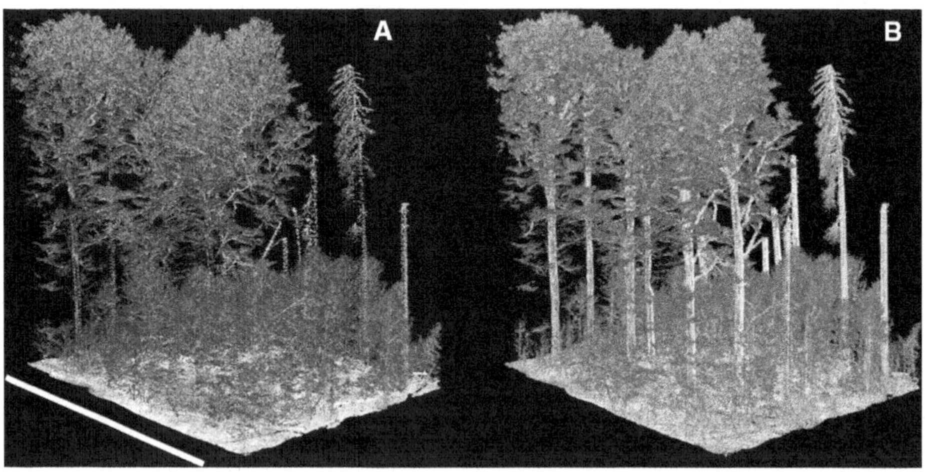

Fig. 7 The impact of scan angle on measurements of stem and branch structure. The area is the same as Fig. 6. **a** Only points acquired from absolute scan angles < 10°. **b** Only points acquired from absolute scan angles > 30°. Most returns from stem and branch positions are from wide scan angles. Scale varies from this perspective. The length of the white bar is 30 m

subset of old-growth forest in Fig. 7. The left panel contains only points from absolute scan angles < 10°, which is comparable to traditional airborne laser scanning, though much more densely sampled. The panel on the right shows the same location, but contains only points from absolute scan angles > 30°. Stem structure is more clearly resolved from measurements > 30° than from measurements < 10°. To determine the potential for automated stem segmentation using these data, we applied a low-intensity filter to remove non-wood returns. This indicates that branches are resolvable using measurements from low-altitude drone flight (Fig. 5).

4.2 High-Frequency Observation

Because drone flight is more firmly under the control of the investigator and much less costly than traditional airborne laser scanning, flight plans can be developed to collect high-density measurements in novel ways that enable hypothesis testing or evaluate the impact of collection scenarios on remote sensing data. There is considerable debate over the role of leaf phenology, age, sensor and atmospheric artifacts in Amazon greening apparent in broadband satellite data (Saleska et al. 2007; Samanta et al. 2010; Morton et al. 2014; Tang and Dubayah 2017). High-frequency observations from drones could serve as an important source of information in this debate, for example, by observing short-term changes in leaf area and condition. The data we collected in the southern Czech Republic were designed in part to determine the impact of leaf area on simulated GEDI waveforms in the absence of changes in woody structure. By collecting measurements 51 days apart, we are able to observe changes associated with leaf flushing and phenology that are independent of woody structure.

Comparing measurements from leaf-off and leaf-on conditions in the Czech Republic demonstrates that the difference in point density is due to a reduction in the number of higher-order returns in the presence of leaf area (Table 2). For example, the number of first returns under leaf-off conditions in the 2.25 ha forested area was 41,822,033, which is a mean density of $1859 \cdot m^{-2}$. Under leaf-on conditions, the number of first returns was 43,132,311 or $1917 \cdot m^{-2}$. The total point density and distribution of return numbers for the 2.25 ha forested area can be contrasted with an open, recently disturbed location of 0.75 ha, some of which is illustrated in Fig. 3. This location contains 14,620,517 first returns under leaf-off conditions, and 17,879,244 first returns under leaf-on conditions, corresponding to densities of $1949 \cdot m^{-2}$ and $2384 \cdot m^{-2}$, respectively. The distribution of return numbers is more strongly skewed toward first returns under leaf-on conditions (Table 2). Another opportunity is observing short-term changes in condition, for example due to leaf wilting. These observations could be made by conducing repeated drone flights over diurnal time scales.

4.3 Increasing the Representation of Ground-Based Field Inventories in Support of Mission Science

The Committee on Earth Observation Satellites (CEOS) is responsible for coordinating non-military Earth observations among national and international space agencies and other members (Duncanson et al. 2019). In concert with the Group on Earth Observations (GEO), the CEOS Working Group on Calibration and Validation (WGCV) Land Product Validation (LPV) subgroup has developed a forthcoming protocol focused on recommendations for validation of satellite AGBD data products. These recommendations include a call for increasing the number and quality of validation sites, highlighting the importance of permanent field plots (CEOS Strategy for Carbon Observations from Space 2014; CEOS Working Group on Calibration and Validation 2017). These recommendations recognize the critical role of ground-based field-inventory data in supporting the calibration and validation needs of current and forthcoming generations of land surface remote sensing.

Unfortunately, there are currently few locations in the world with large permanent-inventory plots suitable for calibration and validation of space mission data products. Even when field data exist, these inventories have usually not been professionally surveyed in an absolute coordinate system, making it difficult to align field data with airborne or satellite data products. Producing stem maps in large inventory plots is time-consuming and cannot be completed rapidly. These problems introduce a spatial and temporal mismatch between the field inventories and remotely sensed data (Réjou-Méchain et al. 2015). Other issues include changing points of measurement on individual tree stems over time due to the development of buttresses or other irregularities and inconsistent application of measurement protocols (Clark 2002; Metcalf et al. 2009).

All of these problems could be overcome by an operational framework to segment individual stems using ultra-high-density laser scanning from TLS and drones (Calders et al. 2015; Brede et al. 2017; Gonzalez de Tanago et al. 2018; Disney et al. 2018). Because the stem map can be derived directly from lidar data, spatial and temporal sources of uncertainty between field measurements and remote sensing are eliminated. Segmentation algorithms can be clearly described and placed in the public domain (Raumonen et al. 2013; Trochta et al. 2017). Uncertainties can be rigorously quantified and propagated, and algorithms can be applied consistently, ensuring that all data can be judged and evaluated by the same standard. This information could be rapidly generated using a coordinated campaign involving terrestrial and drone-based sensors over existing field inventories. For example, the Forest Global Earth Observatory plot network contains the largest permanent plots in the world, typically 25–50 ha in size. This network represents 66 sites in 27 countries, including 6 million individual trees and 10,000 species (Anderson-Teixeira et al. 2015).

Sampling these plots within regions of 1–10 km^2 would provide the measurements necessary for calibration and validation of space mission data products. Acquisition of these data is possible using a heavy-lift or long-duration drone at a fraction of the cost of traditional airborne surveys. What would it take to accomplish this task? Using our experiences in the southern Czech Republic as a guide, generating 5000 points $\cdot$ m^{-2} requires about 5 flight hours $\cdot$ km^{-2}. If lower point densities are acceptable, 50 points $\cdot$ m^{-2} can be achieved in about 6 min $\cdot$ km^{-2}.

5 Conclusions

Low-altitude flight using drones enables the collection of ultra-high-density point clouds using wider laser scan angles than have been possible from traditional airborne platforms. These measurements can be precise and accurate and can achieve measurement densities of thousands of points m^{-2}. The 3D model produced by these data can clearly resolve branch and stem structure that is comparable to TLS and can be acquired rapidly over large landscapes at a fraction of the cost of traditional airborne laser scanning. Drone remote sensing is not a replacement for traditional airborne platforms or TLS. It is a tool for a different job. Drone-based flight operations provide flexibility and the potential to take advantage of unexpected opportunities. Drones may enable access to locations where traditional flight operations are challenging, either because sites are remote or because permissions are difficult to secure. The diversity of lightweight laser scanners and drone platforms creates opportunities to collect focused measurements of relatively small areas at low cost and the ability to characterize large landscapes that rival the extent of traditional airborne laser scanning. Measurements from drones are qualitatively distinct from traditional airborne laser scanning, owing to the large point densities and wide scan angles that are made possible by low-altitude flight.

Critical to global validation of the AGBD data products being produced by current and forthcoming space missions is access to globally representative samples of airborne lidar that characterize the diversity of relationships between vertical forest structure and AGBD. Drone acquisitions can contribute to producing these calibration and validation data sets.

Acknowledgements This research was funded by Brown University and the National Aeronautics and Space Administration of the United States of America.

Author contributions JRK conceived the paper and the analysis and wrote the manuscript with input from all authors. JRK, MB, KCC, BI, and CZ collected and processed the data. JRK performed the analysis. CE, CF, MB, BI and CZ designed the aircraft and flight control system and integrated the sensors. BI was the pilot in command. CZ and MB were payload operators. KK and TV obtained permission from the Civil Aviation Authority of the Czech Republic for flight operations. MK, JT and KK performed analyses of individual tree segmentation.

References

Åkerblom M, Raumonen P, Kaasalainen M et al (2015) Analysis of geometric primitives in quantitative structure models of tree stems. Remote Sens 7:4581–4603. https://doi.org/10.3390/rs70404581

Anderson-Teixeira KJ, Davies SJ, Bennett AC et al (2015) CTFS-ForestGEO: a worldwide network monitoring forests in an era of global change. Glob Change Biol 21:528–549. https://doi.org/10.1111/gcb.12712

Aouf N, Ollero A, Sasiadek JZ (2009) Special issue on: airborne simultaneous localisation and map building (A-SLAM). J Intell Robot Syst 55:247–248. https://doi.org/10.1007/s10846-009-9322-1

Asner GP, Heidebrecht KB (2002) Spectral unmixing of vegetation, soil and dry carbon cover in arid regions: comparing multispectral and hyperspectral observations. Int J Remote Sens 23:3939–3958. https://doi.org/10.1080/01431160110115960

Asner GP, Knapp DE, Kennedy-Bowdoin T et al (2007) Carnegie airborne observatory: in-flight fusion of hyperspectral imaging and waveform light detection and ranging for three-dimensional studies of ecosystems. J Appl Remote Sens 1:013536. https://doi.org/10.1117/1.2794018

Brede B, Lau A, Bartholomeus HM, Kooistra L (2017) Comparing RIEGL RiCOPTER UAV LiDAR Derived Canopy Height and DBH with Terrestrial LiDAR. Sensors 17:2371. https://doi.org/10.3390/s17102371

Calders K, Newnham G, Burt A et al (2015) Nondestructive estimates of above-ground biomass using terrestrial laser scanning. Methods Ecol Evol 6:198–208. https://doi.org/10.1111/2041-210X.12301

Calders K, Disney MI, Armston J et al (2017) Evaluation of the range accuracy and the radiometric calibration of multiple terrestrial laser scanning instruments for data interoperability. IEEE Trans Geosci Remote Sens 55:2716–2724. https://doi.org/10.1109/TGRS.2017.2652721

CEOS Working Group on Calibration and Validation (2017) Land product validation subgroup. https://lpvs.gsfc.nasa.gov/about.html. Accessed 23 Feb 2017

CEOS Strategy for Carbon Observations from Space (2014) The Committee on Earth Observation Satellites (CEOS) Response to the Group on Earth Observations (GEO) Carbon Strategy

Chave J, Rejou-Mechain M, Burquez A et al (2014) Improved allometric models to estimate the aboveground biomass of tropical trees. Glob Change Biol 20:3177–3190. https://doi.org/10.1111/gcb.12629
Clark DA (2002) Are tropical forests and important carbon sink? Reanalysis of the long-term plot data. Ecol Appl 12:3–7
Clark DB, Kellner JR (2012) Tropical forest biomass estimation and the fallacy of misplaced concreteness. J Veg Sci 23:1191–1196. https://doi.org/10.1111/j.1654-1103.2012.01471.x
Disney MI, Vicari MB, Burt A et al (2018) Weighing trees with lasers: advances, challenges and opportunities. Interface Focus 8:20170048. https://doi.org/10.1098/rsfs.2017.0048
Dubayah R, Goetz SJ, Blair JB, et al (2014) The Global Ecosystem Dynamics Investigation. AGU Fall Meeting Abstracts 14
Duncanson L, Armston J, Disney M, et al (2019) The importance of consistent global forest aboveground biomass product validation. Surv Geophys (this issue)
Gonzalez de Tanago J, Lau A, Bartholomeus H et al (2018) Estimation of above-ground biomass of large tropical trees with terrestrial LiDAR. Methods Ecol Evol 9:223–234. https://doi.org/10.1111/2041-210X.12904
Hancock S, Armston J, Hofton M et al (2019) The GEDI simulator: a large-footprint waveform lidar simulator for calibration and validation of spaceborne missions. Earth Space Sci 6:294–310. https://doi.org/10.1029/2018EA000506
Held A, Phinn S, Soto-Berelov M, Jones S (2015) AusCover good practice guidelines: a technical handbook supporting calibration and validation activities of remotely sensed data products. 1.1. TERN AusCover
Janík D, Král K, Adam D et al (2016) Tree spatial patterns of Fagus sylvatica expansion over 37years. For Ecol Manag 375:134–145. https://doi.org/10.1016/j.foreco.2016.05.017
Jenkins JC, Chojnacky DC, Heath LS, Birdsey RA (2003) National-scale biomass estimators for United States tree species. For Sci 49:12–35
Kellner JR, Hubbell SP (2017) Adult mortality in a low-density tree population using high-resolution remote sensing. Ecology 98:1700–1709. https://doi.org/10.1002/ecy.1847
Kellner JR, Clark DB, Hofton MA (2009) Canopy height and ground elevation in a mixed-land-use lowland Neotropical rain forest landscape. Ecology 90:3274. https://doi.org/10.1890/09-0254.1
Lefsky MA, Cohen WB, Harding DJ et al (2002) Lidar remote sensing of above-ground biomass in three biomes. Glob Ecol Biogeogr 11:393–399
Lichti DD, Jamtsho S (2006) Angular resolution of terrestrial laser scanners. Photogram Rec 21:141–160. https://doi.org/10.1111/j.1477-9730.2006.00367.x
Lin Y, Hyyppä J, Kaartinen H, Kukko A (2013) Performance analysis of mobile laser scanning systems in target representation. Remote Sens 5(7):3140–3155
Lowe DG (1999) Object recognition from local scale-invariant features. In: Proceedings of the seventh IEEE international conference on computer vision, vol 2, pp 1150–1157
Metcalf CJE, Clark JS, Clark DA (2009) Tree growth inference and prediction when the point of measurement changes: modelling around buttresses in tropical forests. J Trop Ecol 25:1–12. https://doi.org/10.1017/S0266467408005646
Milenković M, Ressl C, Karel W et al (2018) Roughness Spectra derived from multi-scale LiDAR point clouds of a gravel surface: a comparison and sensitivity analysis. ISPRS Int J Geo-Inf 7:69. https://doi.org/10.3390/ijgi7020069
Morton DC, Nagol J, Carabajal CC et al (2014) Amazon forests maintain consistent canopy structure and greenness during the dry season. Nature 506:221–224. https://doi.org/10.1038/nature13006
Muukkonen P (2007) Generalized allometric volume and biomass equations for some tree species in Europe. Eur J For Res 126:157–166
NISAR Mission Science Handbook (2018) NASA-ISRO SAR (NISAR) Mission Science Users' Handbook version 1
Pan Y, Birdsey RA, Fang J et al (2011) A large and persistent carbon sink in the world's forests. Science 333:988–993. https://doi.org/10.1126/science.1201609
Patterson PL, Healey SP, Ståhl G, et al (2019) Statistical properties of hybrid estimators proposed for GEDI – NASA's global ecosystem dynamics investigation. Environmental Research Letters
Qiang F, Xin-sheng H (2013) A unified approach of observability analysis for airborne SLAM. J Cent South Univ 20:2432–2439. https://doi.org/10.1007/s11771-013-1753-y
Raumonen P, Kaasalainen M, Akerblom M et al (2013) Fast automatic precision tree models from terrestrial laser scanner data. Rem Sens 5:491–520. https://doi.org/10.3390/rs5020491
Réjou-Méchain M, Tymen B, Blanc L et al (2015) Using repeated small-footprint LiDAR acquisitions to infer spatial and temporal variations of a high-biomass Neotropical forest. Remote Sens Environ 169:93–101. https://doi.org/10.1016/j.rse.2015.08.001

Saleska SR, Didan K, Huete AR, da Rocha HR (2007) Amazon forests green-up during 2005 drought. Science 318:612. https://doi.org/10.1126/science.1146663
Samanta A, Ganguly S, Hashimoto H et al (2010) Amazon forests did not green-up during the 2005 drought. Geophys Res Lett 37:L05401. https://doi.org/10.1029/2009GL042154
Scipal K, Arcioni M, Chave J, et al (2010) The BIOMASS mission #x2014; An ESA Earth Explorer candidate to measure the BIOMASS of the earth's forests. In: 2010 IEEE international geoscience and remote sensing symposium, pp 52–55
Stavros EN, Schimel D, Pavlick R, et al (2017) ISS observations offer insights into plant function. In: Nature Ecology & Evolution. https://www.nature.com/articles/s41559-017-0194. Accessed 14 May 2018
Takoudjou SM, Ploton P, Sonké B et al (2018) Using terrestrial laser scanning data to estimate large tropical trees biomass and calibrate allometric models: a comparison with traditional destructive approach. Methods Ecol Evol 9:905–916. https://doi.org/10.1111/2041-210X.12933
Tang H, Dubayah R (2017) Light-driven growth in Amazon evergreen forests explained by seasonal variations of vertical canopy structure. PNAS 114:2640–2644. https://doi.org/10.1073/pnas.1616943114
Trochta J, Krůček M, Vrška T, Král K (2017) 3D Forest: an application for descriptions of three-dimensional forest structures using terrestrial LiDAR. PLoS ONE 12:e0176871. https://doi.org/10.1371/journal.pone.0176871
Wallace L, Lucieer A, Watson C, Turner D (2012) Development of a UAV-LiDAR system with application to forest inventory. Remote Sens 4:1519–1543. https://doi.org/10.3390/rs4061519

Affiliations

James R. Kellner[1,2] · John Armston[3] · Markus Birrer[4] · K. C. Cushman[1,2] · Laura Duncanson[3] · Christoph Eck[4] · Christoph Falleger[4] · Benedikt Imbach[4] · Kamil Král[5] · Martin Krůček[5] · Jan Trochta[5] · Tomáš Vrška[5] · Carlo Zgraggen[4]

[1] Institute at Brown for Environment and Society, Brown University, 85 Waterman Street, Providence, RI 02912, USA

[2] Department of Ecology and Evolutionary Biology, Brown University, 80 Waterman Street, Providence, RI 02912, USA

[3] Department of Geographical Sciences, University of Maryland College Park, 2181 LeFrak Hall, College Park, MD 20740, USA

[4] Aeroscout GmbH, Hengstrain 14, 6280 Hochdorf, Switzerland

[5] The Silva Tarouca Research Institute, Department of Forest Ecology, Lidicka 25/27, 602 00 Brno, Czechia

Surveys in Geophysics (2019) 40:979–999
https://doi.org/10.1007/s10712-019-09538-8

The Importance of Consistent Global Forest Aboveground Biomass Product Validation

L. Duncanson[1] · J. Armston[1] · M. Disney[2] · V. Avitabile[3] · N. Barbier[4] · K. Calders[5] · S. Carter[6] · J. Chave[7] · M. Herold[6] · T. W. Crowther[8] · M. Falkowski[9] · J. R. Kellner[10,11] · N. Labrière[7] · R. Lucas[12] · N. MacBean[13] · R. E. McRoberts[14] · V. Meyer[15] · E. Næsset[16] · J. E. Nickeson[17] · K. I. Paul[18] · O. L. Phillips[19] · M. Réjou-Méchain[4] · M. Román[25] · S. Roxburgh[18] · S. Saatchi[15] · D. Schepaschenko[20] · K. Scipal[21] · P. R. Siqueira[22] · A. Whitehurst[23] · M. Williams[24]

Received: 16 October 2018 / Accepted: 27 April 2019 / Published online: 30 May 2019

Abstract
Several upcoming satellite missions have core science requirements to produce data for accurate forest aboveground biomass mapping. Largely because of these mission datasets, the number of available biomass products is expected to greatly increase over the coming decade. Despite the recognized importance of biomass mapping for a wide range of science, policy and management applications, there remains no community accepted standard for satellite-based biomass map validation. The Committee on Earth Observing Satellites (CEOS) is developing a protocol to fill this need in advance of the next generation of biomass-relevant satellites, and this paper presents a review of biomass validation practices from a CEOS perspective. We outline the wide range of anticipated user requirements for product accuracy assessment and provide recommendations for the validation of biomass products. These recommendations include the collection of new, high-quality in situ data and the use of airborne lidar biomass maps as tools toward transparent multi-resolution validation. Adoption of community-vetted validation standards and practices will facilitate the uptake of the next generation of biomass products.

Keywords Map validation · Biomass mapping · Reference data · Committee on Earth Observing Satellites · Remote sensing · Lidar

1 Introduction

Forest biomass has been recognized as a Global Climate Observing System (GCOS) Essential Climate Variable (ECV), a critical input to the United Nations' Reducing Emissions from Deforestation and Degradation-plus (REDD+) program, and an important input to Earth system models (Herold et al. 2019). Spatially continuous maps of

✉ L. Duncanson
lduncans@umd.edu

Extended author information available on the last page of the article

forest biomass are therefore important inputs for decreasing uncertainties in the global carbon cycle, particularly for areas where insufficient ground or airborne lidar data are available. Accurate biomass products are of great importance for forest management and climate mitigation. However, due to a previous dearth of satellite data specifically designed for producing accurate estimates of forest structure (Goetz et al. 2009) few global-scale forest biomass products are currently available, and the assessment of their accuracy is challenged by a lack of appropriate reference data. To overcome this critical carbon accounting gap, several upcoming Earth Observation (EO) missions will collect satellite data sensitive to forest structure and aboveground biomass, defined as the dry-weight of the live or dead woody component of aboveground vegetation. We anticipate a multitude of new global forest biomass products in the coming decade, but foresee challenges in intercomparison and validation across biomass products. These challenges have already been highlighted by several studies comparing the few existing continental or global-scale biomass products (Avitabile et al. 2016; Avitabile and Camia 2018; Baccini et al. 2012; Huang et al. 2015; Mitchard et al. 2013; Saatchi et al. 2011; Santoro et al. 2015; Thurner et al. 2014) and may hinder the effective adoption of biomass products for various policy, management and science applications.

A specific example of the importance of independent biomass product validation comes from comparisons of two widely known pantropical biomass maps (Baccini et al. 2012; Saatchi et al. 2011). By independent, we mean using reference data that were not included in the generation of products and ideally conducted by a third party. Despite having been produced from the same core satellite datasets (the Geoscience Laser Altimeter System [GLAS] and the Moderate-resolution Imaging Spectroradiometer [MODIS]), these maps differ substantially in several tropical areas (Avitabile et al. 2016; Mitchard et al. 2013, 2014) potentially because they employed different empirical modeling approaches, calibration datasets and extrapolation techniques. However, determination of the exact causes of discrepancies between these products, or indeed a determination of the more accurate product for a given application, is impossible without common approaches to independent validation. Aboveground biomass product validation is challenging, primarily because of the paucity of high-quality, publicly available and globally representative Fiducial Reference Measurements (FRM, https://earth.esa.int/web/sppa/activities/frm) with well-characterized uncertainties and challenges related to the fact that these reference data are not direct measurements but rather estimates based on tree-level allometric model predictions (Clark and Kellner 2012). Indeed, in the pantropical case, the map producers themselves had limited available validation datasets, and Baccini et al. (2012) and Saatchi et al. (2011) performed cross-validation of their mapped products using a subset of GLAS data that were deliberately left out of their biomass model training, rather than validating with an independent dataset. While Saatchi et al. (2011) conducted an error propagation for the final estimated uncertainties associated with their pantropical product and Baccini et al. (2012) reported confidence intervals on their estimates per continent, determination of the degree of accuracy of these products in geographic areas outside the calibration range, or at the various resolutions needed for policy implementation (Herold et al. 2019), was not possible. These products have been compared to the Intergovernmental Panel on Climate Change (IPCC) Tier 1 biomass estimates, following the 2006 IPCC Good Practices Guidelines (GPG), and although a composite of the two pantropical products was suitable to replace IPCC Tier 1 estimates when national inventories were not available, it was recommended that national estimates be favored over these remote sensing-based estimates, given the large disparities between the products and national inventories (Langner et al. 2014).

The issue of product validation will become even more pressing as the number of spaceborne datasets specifically designed to map ecosystem structure increases (e.g., from NASA's GEDI, NASA/ISRO's NISAR, ESA's BIOMASS, JAXA's ALOS-4) and approaches to biomass prediction using these data diversify. Previous biomass products have varied in terms of the spatial and temporal resolution, modeling approach, geographic scope of calibration data, scaling, error propagation and uncertainty reporting (Goetz et al. 2009; Huang et al. 2015; Mitchard et al. 2013).To effectively meet the goals of scientists and decision makers, the global change community requires well-tested validation approaches that are transparent and flexible (with respect to geographic scope and spatial resolution).

The Committee on Earth Observing Satellites (CEOS) is an international body that works to coordinate Earth Observation programs and data collected by space agencies. For nearly two decades, the CEOS Working Group on Calibration and Validation (WGCV) has had a subgroup specifically focused on Land Product Validation (LPV). In close coordination with CEOS member agencies, the LPV subgroup has recognized the need for good practices and protocols to guide biomass product validation in advance of the expected suite of upcoming biomass products. This LPV subgroup launched the biomass focus area in 2017 to help gather community support in developing a validation protocol for the products that will be generated from the upcoming biomass-related missions. This paper presents the conceptual development of the CEOS biomass protocol, reviews the specifications of the biomass products expected from new mission datasets and outlines the importance of biomass product validation for various biomass product user communities (e.g., for climate and carbon cycle modeling, policy applications and ecologists). We present the CEOS biomass protocol structure, including sources of errors inherent in biomass products, how these errors can be propagated and reported and which fiducial reference measurements are required to estimate product uncertainty. Finally, we discuss challenges envisioned by the authors in the implementation of the CEOS LPV validation protocol and its key recommendations.

2 Background on Validation of Biomass Products

Biomass products have been generated across a range of spatial extents, from local and regional to global (Ahmed et al. 2013; Andersen et al. 2011; Avitabile et al. 2016; Baccini et al. 2008; Blackard et al. 2008; Boudreau et al. 2008; Duncanson et al. 2015a, b; Gregoire et al. 2011; Huang et al. 2015; Margolis et al. 2015; Neigh et al. 2013; Su et al. 2016). True biomass product validation requires the physical, destructive harvesting, drying, and weighing of trees. This is extremely difficult in practice, and typically undesirable or logistically impossible in many cases. As a result, validation of satellite-based AGB estimates relies on independent verification (i.e., inter-comparison of estimates made across scales with consistent and well-characterized uncertainties). For local biomass maps, validation is often conducted by the map producer, either through statistical cross-validation (i.e., using reference data not included in model calibration) or through comparison to independent data (e.g., from national forest inventories). These accuracy statistics are usually included in the publication of a biomass model, often expressed as a root mean squared error (RMSE), %RMSE and/or coefficient of determination (R^2). For biomass maps, the accuracies are often expressed as a standard error or coefficient of variation. In local examples, while some researchers are diligent about reporting the specific validation methods used to generate their maps, others do not include these details, leaving map users unsure of whether the reported accuracies represent over-fitting to the calibration dataset, whether models exhibit systematic error, or whether model prediction residuals were heteroscedastic. Increasingly, these local biomass maps are used as reference data for larger area maps, and the errors associated with these reference products propagate in the final map.

In national, continental or global product validation, the challenge becomes greater, with local maps only representing a small proportion of the globe. Currently, there is no globally representative set of ground plots or local biomass maps sufficient for validation at the appropriate resolution of spaceborne products. The few global or continental-scale biomass products that have been released have relied on spaceborne estimates of biomass from NASA's GLAS instrument for validation, often the same instrument used to generate the biomass product in the first place. In the case of Saatchi et al. (2011), the GLAS data used to validate the GLAS biomass model were not included when fitting the model, and the associated mapped products used a Monte Carlo error propagation to perform a product accuracy assessment. However, there are known issues with the GLAS biomass estimates in areas of high biomass or complicated terrain (Duncanson et al. 2010; Simard et al. 2011; Sun et al. 2008), and thus, any validation depending on GLAS data likely cannot capture the true errors associated with a GLAS-based product. Conversely, researchers have used forest inventory data to validate forest structure products (Simard et al. 2011), but these inventory datasets represent small sample of the landscape that may not be representative of the relatively coarse pixels (500 m–1 km) often associated with wide area biomass products.

We argue that it cannot be the sole responsibility of a dataset producer (be it a mission team or an independent academic) to conduct validation exercises over the range of scales necessary for all possible users; beyond the due diligence of transparently reporting the methods used to produce the map, and associated errors, an independent validation process is required. However, independent groups involved in validation exercises may vary in terms of the datasets available to them, the size of field inventory plots, availability of airborne lidar data, geographic scope of validation datasets, etc. Further, since some users may require different errors reporting, e.g., for IPCC compliance, national-scale estimates are desirable. Alternatively, ecosystem modelers will tend to require aggregated pixel-based errors. Additionally, users will inevitably be interested in different scales, be it local areas for forest management, national reporting efforts, continental carbon balance accounting, global ecosystem models, etc. To date, there is no clear guiding document focused on addressing the validation needs of many biomass product users and, for this reason, we present work toward a protocol that addresses this need.

3 Earth Observation Missions for Biomass Mapping

NASA's ICESat was launched in 2003, and its GLAS instrument collected global waveform lidar measurements over vegetation that were used to estimate forest height and structure until the last ICESat laser failed in 2009 (Abshire et al. 2005). GLAS data were not designed to study forest structure, but these data have nevertheless become popular for biomass mapping. These data are relatively sparse in spatial sampling, and each lidar footprint illuminated a nominally 65-m-diameter circle, which resulted in the mixing of reflected signal from ground and canopy surfaces, ultimately presenting challenges for estimating biomass in areas of high relief or structural complexity (Duncanson et al. 2010). Despite these challenges, many wide area biomass maps used GLAS data to map forest structure (e.g., Baccini et al. 2008; Margolis et al. 2015; Saatchi et al. 2011; Simard et al. 2011; Su et al. 2016).

Several upcoming missions (e.g., GEDI, BIOMASS, NISAR) should provide improved data for biomass mapping compared to those earlier sensors as they are designed with a primary science goal of mapping forest biomass. Official mission biomass products are expected from each of these missions, but because of the publicly available nature of these mission datasets we also expect a host of other new biomass products through data fusion and alternative algorithms, etc. We therefore anticipate the release of products with a range of spatial resolutions, geographic extents and temporal domains. Table 1 shows the expected resolution of core biomass products from upcoming spaceborne missions themselves, but fusion products will likely present both coarser and higher resolution maps.

Many of these upcoming missions have specific biomass product accuracy requirements as part of the criteria by which mission success is determined (Table 1). Independent validation of these products at their nominal resolution would help demonstrate that requirements have been met. This is particularly useful if validation of biomass products from each mission or estimation approach can be conducted at the same set of sites, allowing direct comparisons between accuracies of each product in different forest types, environments, disturbance histories, etc. Comparisons between official mission products and other new biomass maps will also allow the community to appreciate the accuracy impacts of algorithmic improvements, data fusion approaches, etc., on product accuracy and ultimately reduce confusion and latency in the adoption of new biomass mapping approaches.

4 The Importance of Validation for Biomass Product Users

While biomass product validation is important for product inter-comparison toward improved map products, and for demonstrating that missions have met their design requirements, it is also key for the uptake of biomass maps by many other communities. Here, we highlight several communities that will likely use the next generation of biomass products for a wide range of applications. Although this list is not exhaustive, it enables an exploration of many biomass product and validation needs. We briefly discuss considerations for biomass validation for policy applications, carbon accounting in 'non-forest' ecosystems (woodlands, savannas), belowground biomass estimation, and for modeling activities.

4.1 Policy Applications

Aboveground biomass is an important input to several current and future policy initiatives and user groups, and all of these groups require information on biomass product accuracy, albeit in different ways. Uncertainty estimates are critical for achieving goals and commitments related to forest management, climate change and sustainable development. For example, the Paris Agreement on Climate Change requires transparent reporting on national greenhouse gases (GHG) emissions, and measuring, reporting and verification of forest-related mitigation activities, for which improvements in the quality of forest biomass and carbon stocks information is essential. Biomass products are also relevant for Goal 15 of the United Nations Sustainable Development Goals (SDGs) that includes improved forest carbon management, reduced deforestation, afforestation and reforestation at a global scale. Additional policy initiatives which will use the next generation of biomass products include the IPCC assessment processes and various sustainable land use initiatives

Table 1 The expected upcoming biomass-relevant missions. Note that only NISAR, GEDI and BIOMASS are approved missions with a formal requirement for biomass mapping accuracy. Missions with no official biomass product are marked Not Applicable (NA), and missions that have yet to publish these requirements are listed at To Be Determined (TBD)

Mission	Funding agency	Launch Date (Expected or past)	Data type	Biomass product resolution	Geographic domain	Accuracy requirement
NISAR	NASA/ISRO	2021	L-band SAR	1 ha (< 100 Mg/ha)	Global	< 20% RMSE for < 100 Mg/ha
GEDI	NASA	December 5, 2018	1064 nm waveform lidar	1 km	ISS (± ~ 51.6°)	< 20% standard error for 80% of forested 1 km cells
BIOMASS	ESA	2022	P-band SAR	4 ha	Global except western Europe and North America	< 20% RMSE for AGB > 50 Mg/ha; 10 Mg/ha for AGB ≤ 50 Mg/ha
MOLI	JAXA	~ 2022	1064 nm waveform lidar	500 m	ISS (± ~ 51.6°)	TBD
SAO-COM1A	CONAE	October, 2018	L-band SAR	TBD	Global	TBD
ICESat-2	NASA	September 15, 2018	532 nm photon counting lidar	NA	Global	NA
ALOS-4	JAXA	2021	L-band SAR	TBD	Global	TBD
TanDEM-L	DLR	Planned ~ 2023	L-band SAR	1 ha	Global	20% accuracy or 20 Mg/ha

as well as a wide range of local/subnational stakeholders. IPCC good practices guidelines are focused on national scale validation and reporting and recommend that (1) nations neither over- nor underestimate biomass so far as can be judged and (2) that uncertainties are reduced as far as practicable (IPCC, 2006, Volume 1, Chapter 1, Section 1.2). These recommendations are primarily geared at countries for their national reporting and suggest that biomass maps would be suitable as long as their predictions exhibit no systematic error at a national-scale and have lower errors than current estimation strategies. There is a wide range of considerations with respect to the stakeholders' needs of biomass validation, including the spatial resolution, geographic extent and temporal domain of product validation, and the definition of forest or land type and biomass definition over which the validation is considered. Indeed, the use of biomass products by these policy communities is tightly linked to error reporting (Romijn et al. 2018). To suit this wide range of policy needs, independent product validation will need to be flexible enough to allow user-specified geographic scopes and spatial scales, including consistent and traceable error reporting at the national and/or regional levels, as well as by land cover class (Herold et al. 2019)

4.2 Non-forest Vegetation Carbon Accounting

While non-forest vegetation contributes less than 20% of global biomass, they represent one half of terrestrial productivity and cover some 70% of the Earth's land surface (Pan et al. 2013). These non-forest ecosystems include savannas, woodlands, chaparral and shrublands. They are ecologically distinct from forests both in terms of a) the climatic, edaphic and disturbance histories that prevent them from becoming intact forest and b) the plant form and function of trees growing outside of closed-canopy conditions. The functional and structural dissimilarities of these systems lead to special considerations for biomass product validation. For example, many of the allometric models developed for species typically found in forests do not apply to the same species growing in open canopy conditions. Therefore, error propagation through allometric models will not account for potential biases that may occur from applying the models outside of the range of environmental conditions for which they were calibrated. In some regions, e.g., northern Eurasia, issues of low canopy density have been accounted for by direct inclusion of tree density in biomass conversion factors (Schepaschenko et al. 2018). However, many regions lack the necessary data to perform such corrections. We therefore recommend more attention to the development of new allometric biomass models in non-forest regions.

Non-forest vegetation also presents challenges for validation when many biomass products will be designed specifically for ecosystems classified as forests. Global biomass map producers may even remove many non-forest classes from their mapping domains or may adopt different forest–non-forest masks at different resolutions which will complicate product inter-comparisons. Additionally, due to land use and land cover change dynamics, 'non-forest' vegetation at the boundaries of intact forest changes over time, where disturbance and recovery may lead to pixels being somewhat arbitrarily classified as forest or 'non-forest' through time. This would influence the magnitudes of forest biomass stocks and fluxes more than the actual carbon flux associated with disturbance agents. Regardless of the methods used to define boundaries between 'forest' and 'non-forest,' it is important to be able to track biomass dynamics across these boundaries consistently and without bias (not systematically over or underestimating forest area) as far as possible. In an independent validation process, a clear indication of which forest mask was applied should be clearly indicated, and biomass products that include 'non-forest' biomass should be validated specifically in these other ecosystems rather than merging 'non-forest' accuracies with non-ecosystem specific error reporting.

4.3 Belowground Biomass

The CEOS LPV biomass validation protocol is focused on aboveground biomass. However, we also recognize the importance of belowground biomass and therefore will provide a summary of both how soil scientists will use aboveground biomass products and also the state of the art of belowground biomass validation techniques. Global soil carbon has been estimated to be in excess of 2000 Pg in the top 1 m of soil, much of that outside of forests (Batjes 1996). These carbon pools include both root biomass, which is typically related to aboveground stocks, while Soil Organic Carbon (SOC) is often not correlated to aboveground structure and thus cannot be estimated from space.

There is a strong need to link belowground stock and flux measurements with the aboveground focus of EO missions and forest inventories. Current practices for estimating belowground biomass include localized field work (e.g., digging pits) or applying indirect (remote sensing) methods such as ground penetrating radar to detect a range of soil characteristics including soil depth (Wollschläger et al. 2010), carbon storage (Hruska et al. 1999) or root quantities and distributions (Comas et al. 2017). These field-based methods are not applicable to wide area mapping and can only serve to calibrate or validate spaceborne estimates of belowground biomass. When field data are not available, typically simple root-to-shoot ratios are applied to estimate belowground biomass, although these are known to be poorly constrained in many systems and require much more attention to develop globally representative models (Mokany et al. 2006). Spatially continuous approaches use spectral data to map forest type and function and assume relationships between vegetation reflectance and belowground biomass (e.g., Fisher et al. 2016; Hengl et al. 2014, 2017). Despite the considerable importance of belowground biomass to the global carbon cycle, and the link between aboveground and belowground carbon stocks, this field is relatively less mature in terms of product validation. Although the CEOS LPV biomass protocol will only touch on some of the considerations of belowground validation, we expect a more specific assessment regarding belowground biomass validation to follow as aboveground biomass good practices continue to mature.

4.4 Modeling Community Needs

Reliable estimates of biomass, particularly through time, present opportunities to reduce current knowledge gaps in carbon cycle modeling. For example, the allocation of carbon to different plant tissues, as well as the timescale of turnover of these tissues, remains one of the key uncertainties in vegetation modeling and projections of the terrestrial carbon sink. Many different types of models would benefit from biomass products and thus validation activities, but the nature of the models will influence their validation requirements. For example, carbon-pool models operating at a specific spatial resolution (e.g., 1 ha, or 1 degree postings) would use pixel-based predictions of aboveground biomass (and error) as inputs. Conversely, cohort models may require resolving a biomass map into classifications of land cover and consider error as an average for a given land cover class (e.g., different forest types). Similarly, forestry and economic land use models require administrative level error reporting (e.g., at national extents). Finally, individual-based models such as gap models operate within larger forest patches and thus will likely require biomass (and associated height) errors at finer resolution and smaller geographic extent than the broader-scale model types. To provide modeling groups with informative biomass product errors, error should be reported across multiple resolutions and spatial extents. In addition, producers of biomass maps should be encouraged to document and quantify errors associated with not only the biomass products, but also from the input data (e.g., height error from lidar) and empirical models as they will ultimately inform modeling results and improve the integration of biomass estimates into modeling activities. This recommendation is in addition to the requirement for validation over multiple spatial scales and geographic extents, which is also important to a range of other communities and applications.

5 CEOS LPV Validation Protocol Structure

There are clear overlaps between the needs of several of the communities highlighted above. Flexibility of validation in terms of spatial scale, geographic scope, error reporting, forest definition selection and considerations of both biomass stocks and fluxes are recurring themes when discussing user community requirements (e.g., policy, modeling, soil science, dryland ecology). Therefore, we aim to develop a biomass validation protocol with the flexibility to account for the breadth of user needs. The goal of the CEOS WGCV LPV biomass protocol is to provide a good practices guidance document with recommendations for how to conduct independent biomass product validation. Here, we present what we see as the primary considerations for data collection and validation design in this framework. We also highlight some challenges biomass map producers face with respect to enabling reliable, consistent and transparent product validation.

5.1 Reference Datasets

Any validation depends on the availability of high-quality reference datasets that are independent from the data used to generate an AGB product. Large area AGB datasets used for validation are usually from either forest inventories, forest monitoring plots or local airborne lidar biomass maps, all of which include error in their estimates of biomass. Forest inventory datasets typically consist of large numbers of small, geographically dispersed plots designed to sample across a defined geographic extent or region, often national or local, and are typically updated at some systematic time interval. Conversely, forest monitoring plots tend to be fewer in number but larger in area and collect a broader range of environmental datasets, with individual trees carefully identified and measured over several years to track biological and ecological trends as well as estimation of biomass. Although arguably more useful for pixel-level comparison to AGB maps, these plots typically do not represent probability samples. Finally, airborne lidar biomass maps provide wall-to-wall coverage that can be used to estimate biomass at multiple spatial resolutions. These lidar maps are usually calibrated with local field measurements, either from inventory plots, monitoring plots or field data often collected specifically as part of the lidar collection. These include (by definition) more error at the pixel-level than the field plots used to calibrate them.

None of these three types of reference data represent actual measurements of forest biomass, which is an important distinction from many of the other GCOS ECVs (e.g., Land Surface Temperature) where validation is performed using fiducial reference measurements. Fiducial reference measurements or FRM are a specific kind of reference data, traceable to International System of Units (SI) working, developed with national metrology institutes and produced together with documented uncertainties. As mentioned above, direct aboveground biomass measurements can only be collected through the destructive harvest and weighing of trees, which is of course inherently inappropriate for monitoring plots and logistically unfeasible for inventory plots (Clark and Kellner 2012). Instead of directly measuring biomass, field estimates rely on one of two methods—the measurements of tree and forest plot data and application of allometric models (e.g., Chave et al. 2014a, b; Jenkins et al. 2003), or the use of a Terrestrial Laser Scanning (TLS) instrument (Calders et al. 2015) combined with estimated wood density. In the former, individual tree biomass is predicted through the application of empirical models of relationships between measurable tree attributes (e.g., species identity, stem diameter, height) to aboveground biomass. These allometric size-to-mass models are estimated from destructively sampling trees and often from small samples (sometimes fewer than 30 trees) with a prevalence of small-size trees (Duncanson et al. 2015a, b). Although it is recognized that these models often exhibit systematic prediction errors (Ahmed et al. 2013; Chave et al. 2014b; Gonzalez de Tanago et al. 2018) and that these errors increase with tree size, they are currently the most practical approach for estimating field biomass in a cost-effective fashion. Indeed, statistical packages are now available to both estimate and propagate errors in field biomass estimates to forest plot levels (Réjou-Méchain et al. 2017).

The second (and more recent) approach for estimating field plot volume (and biomass where wood density information is available) is to use TLS instruments to scan forest plots with a ground-based lidar instrument and reconstruct detailed 3D scenes of a forest plot that include individual tree woody materials (Disney et al. 2018; Newnham et al. 2015). TLS datasets, when processed to predict plot-level tree volumes, may produce models whose predictions are more accurate relative to allometric biomass predictions, largely because they represent the full tree size distribution that may be underrepresented in allometric models (Calders et al. 2015; Gonzalez de Tanago et al. 2018). Although TLS data may be preferred when available, the relatively large cost and effort required to collect high-quality TLS data from remote forests, and the current challenges in data processing (including large computational demands and relatively immature software availability, (Newnham et al. 2015; Trochta et al. 2017), mean that these measurements are typically only available over a relatively small set of forest monitoring plots (Disney et al. 2018). It should also be noted that TLS data require wood density information from tree species identification and thus will never fully replace field surveys but add more complete structure measurements of forest plots over manual diameter and height measurements. Attention is required to ensure that TLS data do not overestimate the woody volume of hollow trees, and we recommend further research into TLS allometric model development and correction factors for varying wood density and hollow trees. Instead, TLS data may be most useful to check and re-calibrate allometric models with larger sample sizes and across the full range of tree sizes, thereby reducing underestimates of biomass, particularly for large trees. Considering that ~ 50% of landscape-scale forest biomass may be stored in a few large trees, particularly in the tropics, reducing errors and biases related to large tree biomass stocks is particularly important (Lutz et al. 2012, 2018).

5.2 Geolocation, Temporal and Spatial Scale

Once plot-level biomass has been predicted, spaceborne products must be spatially linked to these reference plots. This process adds error to validation efforts because spaceborne and field datasets have inherent geolocation errors associated with them. This is especially true for field measurements where large errors may occur due to important GNSS multipathing effects in forest ecosystems (Barton and Johnson 2004). Even with the use of high-grade GNSS devices, some unfavorable conditions due to the number and position of satellites or to local forest structure and topography may generate large geolocation errors (~ 10 m). Increasing the number of GNSS measurements in space and time (e.g., > 15 measurements per plot) would be one way to achieve an accurate location, as would the use of a pole-mounted GPS antenna. Another major problem is that field plots sizes and shapes usually differ from the sizes and/or shapes of map units characterizing spaceborne biomass products. This is particularly problematic when attempting to validate relatively coarse (500 m–1 km) biomass products with smaller forest inventory plots. For instance, Réjou-Méchain et al. (2014) found that large errors (typically 30%) can be introduced to the calibration/validation procedure from using small forest plots (less than 0.25 ha) in this manner. Importantly, they demonstrated that this error induces bias into the estimator of RS model parameters and, in turn, systematic error in RS model predictions (Réjou-Méchain et al. 2014).

Small plots are also challenging even when validating finer-resolution (e.g., 30 m) remote sensing products for several reasons. At such resolutions, a small spatial or configuration mismatch between the field and the EO product may generate important errors due to the large local heterogeneity of forest biomass. Further, in areas with large tree crowns, satellite data for a plot may correspond to a part of a crown for a tree whose stem may not fall within the plot (Mascaro et al. 2011). As plot size increases, the relative errors associated with so-called edge effects and geolocation decrease. Thus, higher correlations are typically found between field plots and remote sensing datasets when large (e.g., ≥ 1 ha) plots are used (Huang et al. 2013; Zolkos et al. 2013). One solution to the small plot problem is to implement a screening procedure to remove plots that are not representative of the larger pixels and to obtain a subset that also maintains the frequency distribution (i.e., histogram) of the original dataset (Avitabile and Camia 2018).

Where this screening procedure is not practicable, and in regions with relatively large errors associated with plot size, (e.g., large crowns, greater geolocation uncertainties), such as in the tropics, we recommend the use of relatively large field plots to calibrate models based on satellite datasets, where possible. Additionally, we recommend validating with plots that are the same spatial resolution as satellite-based products to avoid issues from dilution bias (Réjou-Méchain et al. 2014). However, relying on large plots may reduce the number of plots that can be measured for validating a biomass product. Thus, there is a trade-off between the use of large plots and the need for capturing environmental gradients that drive variability in forest structure, and/or errors in spaceborne data acquisitions (e.g., canopy density, topography, soil moisture). Although errors from co-registration, temporal differences and edge effects all decrease with increasing plot size, typically sampling errors will increase with plot size because the sample size (number of plots) will decrease, assuming constant sampling design budget constraints.

The temporal discrepancies between field plots, airborne datasets and spaceborne datasets are also critical for product validation. Different forest ecosystems will change at different rates, due to inherent differences in productivity, as well as stochastic disturbances, ecosystem recovery and natural demographic shifts. The rate of change of a given ecosystem therefore determines the importance of collecting validation data that is temporally coincident with satellite biomass products. In slow growing areas with little disturbance, validation data from up to ~ 5 years before or after satellite data collection may be suitable, but in more productive ecosystems and forests that are degraded or regenerating, or have undergone disturbance validation data will require more frequent updating. Assuming plots and satellite data can be precisely geo-referenced, one solution to determining whether validation data remains useful is to use a Landsat-based change detection dataset, e.g., Harris et al. (2012), Kennedy et al. (2010), to flag pixels that have undergone significant change in the time elapsed between field/airborne and satellite data acquisitions. Alternative approaches capable of resolving finer-scale disturbances (e.g., individual tree fall) may also emerge due to the increased availability of high spatial and temporal optical datasets, such as from Planet. Admittedly these approaches will likely not resolve disturbances in the understory, and thus, we recommend that where possible validation data should be as close in time as practicable to satellite data collection. Older data should only be considered when they represent geographic gaps, and local knowledge and spaceborne optical sensors suggest no significant change in forest biomass or cover.

5.3 Scaling from Ground Plots to Airborne Lidar Maps

A strategy to minimize the trade-off between plot size and representativeness can be the use of locally calibrated airborne lidar biomass maps to scale between field plots and satellite datasets. These maps are increasingly used as reference data for spaceborne products, primarily because they cover a much larger spatial extent than field plots and thus greatly increase the area over which in situ information is mapped if their estimates are adequately calibrated across environmental gradients with respect to forest structural and species variability. Further, provided data are acquired with acceptable sensor and survey configurations, lidar data can easily be aggregated to multiple resolutions and thus a single airborne lidar campaign can serve to validate multiple biomass products with different spatial resolutions.

It is generally recommended to generate the airborne lidar maps at large (e.g., 0.25 ha or larger) pixel sizes in the tropics because errors related to geolocation and edge effects are reduced as plot sizes increase. This is also true of many temperate systems (e.g., Huang et al. 2013), but small plots may be appropriate in many temperate and boreal forests, depending on the demographic structure of the forest and quality of plot geolocation. In the case of the tropics, for example, if one calibrates a 25-m lidar map with 25 m plots, and aggregates predictions to 1 ha, errors will likely be greater than if calibrating a 1-ha lidar map with 1 ha plots (Labriere et al. 2018). If errors are propagated from ground plots to local lidar maps, finer-resolution lidar models and maps are also appropriate, e.g., McRoberts et al. (2018).

5.4 Error Characterization and Propagation

A key aspect of robust product validation relates to error characterization and propagation. We recommend the transparent reporting of which errors are considered in a validation process, and what error propagation technique is used to produce error estimates on the reference datasets used. It is essential to consider that even the highest quality field data and airborne lidar maps contain errors. The CEOS LPV biomass protocol will include a thorough discussion outlining the types of errors that should be considered (many of which are discussed above) and present three basic methods of error propagation. These recommendations are in line with the IPCC Good Practice Guidelines, which are discussed in (McRoberts et al. 2019). We define 'error' as a deviation from field-estimated AGB values, and 'uncertainty' as the range of values within which the field value lies, with some reported level of confidence. The quantification and propagation of errors results in a reported product uncertainty, and there are several different approaches adopted at reporting product uncertainty. Validation with independent data will quantify error (deviation from reference data) with associated uncertainty in that error (from uncertainties in the reference datasets themselves). In the case of a product generated with careful attention to error, we would expect that the outcome of the validation would be an error with an associated uncertainty that is within the range of uncertainty reported by the map producer.

There are several different approaches to quantifying uncertainty in a product (both by the map producers and the map validators). Design-based uncertainty estimation is appropriate where probability-based validation datasets are available and collected specifically to sample the population of interest. These datasets are typically only available for forest inventory plots, which present challenges related to small plot sizes, as discussed above. Also, such inventories rarely exist in tropical forest environments, and the effort required to establish them is prohibitive. A second form of uncertainty estimation is in model-based inference, where the validation datasets are often not probability samples of a forest, but a network of opportunistically collected plots (e.g., collected to calibrate an airborne lidar campaign). In model-based inference, the uncertainties are based on the effects of sampling variability as reflected in the data used to construct the models and residual variability of sample observations around their model predictions. A third technique for calculating errors and associated uncertainties is from hybrid inference, which uses a combination of design-based and model-based inference. McRoberts et al. (2019) present a statistically rigorous test based of hybrid inference that can be used to assess the validation of a coarse resolution global biomass map using finer-resolution, higher-quality local biomass maps. Hybrid inference thus provides a possible statistical solution for validation using airborne lidar biomass maps.

5.5 Validation and Reporting

Once a reference dataset has been collected and scaled to match the resolution of the biomass product in question and the associated errors have been characterized, quantified and propagated, the actual product validation is conducted. The reported accuracies (defined as a lack of error) and uncertainties for the product in question depend on the user requirements for the biomass product, as discussed in Sect. 4. Ecosystem models running at a global scale may need the average coefficient of variation, or confidence interval, per 1-degree grid cell. However, models running at a local or regional scale might require pixel level maximum or minimum estimates of biomass (e.g., from a confidence interval). Conversely, REDD+activities will likely focus on national-level biomass reporting and thus focus on the estimated errors (e.g., biases from the truth) and associated uncertainties at an aggregated national scale. We expect most users will aggregate uncertainty assessments to the appropriate scale for their application. As biomass product accuracies and uncertainties are likely to vary as a function of forest structure and geography (e.g., differing between tropical and temperate systems, mature and recovering forests), we recommend reporting at a variety of scales. For example, if a reference dataset is a geographically dispersed set of linked field and airborne lidar datasets, uncertainties should be reported at global, continental and ecoregion scales. It is unlikely that such a dataset would be representative at a national scale, but according to IPCC good practices the standard error per ecoregion could be used by a country in tandem with national and local data or available ecoregion maps to use the biomass product toward improved REDD+reporting (https://reddcompass.org/). The CEOS LPV biomass protocol will include specific recommendations for reference datasets and error and uncertainty reporting, but it should be stressed that any independent validation of AGB maps should be sufficiently flexible enough to allow errors and uncertainties to be reported at a variety of spatial resolutions, geographical and thematic scopes. It is important to note that we do not include recommendations for product harmonization or inter-comparison and instead suggest comparing products to fiducial reference datasets in order to avoid confusion when discrepancies are found among biomass products.

6 Implementation Considerations

Independent validation requires the availability of independent reference datasets. Ideal reference datasets can be thought of as 'super sites' with highly accurate field measurements (Chave et al. 2019), TLS and airborne or drone lidar data (Kellner et al. 2019) (i.e., sites that have been surveyed in a manner allowing aggregation of lidar data at multiple resolutions and collected at the same time as the satellite data used to construct a given biomass product). To conduct a thorough product validation, these sites would have to be distributed across the geographic scope of interest and capture the important environmental gradients driving the relationships between biomass and error, such as disturbance history, climatic and topographic gradients, etc. However, these sites are relatively rare, particularly in the tropics, where much of the aboveground biomass is stored, and where they do exist they are costly to maintain and repeat survey. Including reference data with larger geographic and temporal coverage is, thus, important for the purpose of capturing more regional variation in biomass estimates and associated uncertainties. Such data can be provided by forest plot observation networks, research plots and forest inventories. Setting up and maintaining a permanent plot network of biomass supersites is key to conducting meaningful validation activities.

Multiple upcoming missions with biomass mapping requirements also have requirements for product and/or algorithm validation, and these activities include budgets for new data acquisitions. These mission teams (namely GEDI, NISAR, ESA BIOMASS and ICESat-2) are actively working together to streamline calibration and validation data collections and maximize the geographic coverage of validation data roughly representative of forest conditions in the year 2020 (Fig. 1). However, these mission activities will not sufficiently cover many geographic domains. Of particular note is the lack of sites in continental Asia and expanding collaborations with other CEOS member agencies who may be able to help expand data collection activities is highly desirable.

It is also important to note that enabling a transparent and reproducible map validation requires open data. Open data allow transparency to validation activities, and reproducibility of results which we feel is essential for the wide adoption of the CEOS LPV recommendations for biomass product validation. This is however a complicated issue, as balance must be met between credit for the cost, skills and labor required for field data collection and curation, and the importance of open data for the scientific process. The FOS system has struck compromises between open data and maintaining census data privacy by publicly releasing only plot estimates of biomass with consistency propagated error (see Chave et al. 2019), and in situ data collected through funds from NASA and ESA are open by definition; hopefully more space agencies and plot networks will follow suit. A more equitable solution to this problem can be found if adequate long-term funding is provided to plot networks to ensure not simply that tree-by-tree data are collected and processed, but that the (mostly tropical) people doing this are adequately rewarded, trained and respected

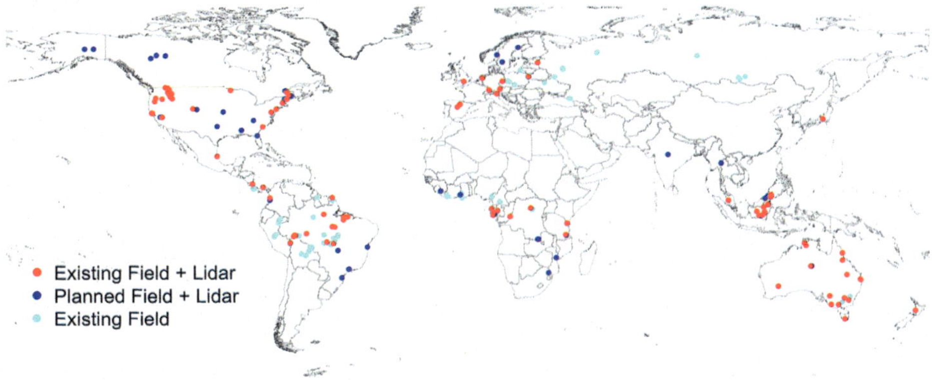

Fig. 1 Existing or planned field and airborne lidar datasets for use in spaceborne mission biomass model fitting or product validation. The red points have been compiled by the NASA GEDI team and are used in their biomass model development. The blue points, compiled by the NISAR team, are planned sites for both model development and product validation. The turquoise sites, compiled by the Forest Observation System (FOS, Chave et al. 2019), represent high-quality standardized field estimates of biomass

for their skills and efforts. In a more equal and secure environment, we argue that open publication of datasets will become a natural expectation.

Once global field and airborne lidar biomass reference datasets are compiled, a final challenge lies in actually conducting an independent validation and reporting the results. Ideally, fiducial reference data will be freely available so that any organization or researcher could conduct a transparent product validation. It is desirable that online tools become available that enable not only data access, but also error characterization and propagation in the generation of reference datasets such as airborne lidar maps. Existing tools such as Google Earth Engine provide unparalleled capabilities and potential for user-friendly product generation, but have yet to develop the necessary tools for validating products. Indeed, this requires a relatively specific set of tools and we recommend the development of an independent toolkit to perform biomass product validation following CEOS WGCV LPV biomass protocol recommendations. The BIOMASS package (Réjou-Méchain et al. 2017) is already widely used by many researchers and mission teams to propagate errors from individual tree estimates to field plots in the tropics, but comparable tools are not yet available to expand this to airborne lidar maps, particularly at multiple resolutions as is needed by the breadth of expected biomass product users. The new ESA/NASA Multi-Mission Algorithm and Analysis Platform (MAAP) is an activity focused on developing cloud-based tools for biomass product generation from active remote sensing datasets (primarily from GEDI, NISAR, and the ESA BIOMASS Mission). The MAAP is currently under development and is anticipated to be a useful platform for both housing and implementing biomass validation data. The Forest Observation System (FOS, https://forest-observation-system.net/) is also collating field validation datasets with preprocessed plot data that include propagated errors, and it is currently expanding to include airborne lidar-based biomass maps. The ESA Climate Change Initiative (CCI) Biomass Project (building on the ESA GlobBiomass project), similarly plans to support these activities by encouraging collaboration and cooperation in the application of the CEOS LPV protocol through the Global Plant Biomass Facility. The intention is to encourage and facilitate unified approaches to the collection of forest inventory data (including in near real time) and democratize access to in situ biomass estimates. All of these new activities illustrate the existing gap in biomass product tools and present exciting potential opportunities for implementing the CEOS LPV biomass protocol.

7 Summary and Next Steps

The CEOS WGCV LPV biomass protocol document (to be published in 2019) will cover a thorough range of validation considerations, including the collection of new reference datasets, and recommendations for using airborne lidar biomass maps as tools for scaling between field and satellite data. The protocol will also consider the characterization, propagation and ultimately reporting of errors throughout the validation process, as well as the existing limitations for the implementation of product validation. The protocol document is intended to guide both biomass map producers and users toward consistent interpretation of product errors, with the ultimate goals of reducing confusion about which biomass products are more accurate or appropriate for a given application and increasing the suitability of new data collections for product validation. Considering the time and cost-intensive nature of collecting high-quality in situ and airborne datasets, a document that can facilitate, to the degree possible, standardization of data collection and methods for product validation will hopefully help streamline new acquisition plans across different agencies, and allow for easier interpretation of mapped products and errors. Equally, establishing and maintaining permanent reference sites for biomass, including the ongoing acquisition of field and airborne data, and where in situ partners are fully involved, will be key for conducting meaningful and sustainable biomass product validation. We are entering an exciting era in Earth Observation, with a wealth of new datasets from forthcoming missions leading to the anticipation of many new biomass products. The societal and scientific needs for accurate biomass products are clear—and with the promotion of transparent independent product validation at the range of scales required by data users we expect the next generation of mission datasets to successfully fill what has been a limiting knowledge gap in the global carbon cycle.

Acknowledgements The authors gratefully acknowledge support from by NASA's Terrestrial Ecology (CEOS Land Product Validation Subgroup Activities) under NASA Carbon Monitoring System Grant #15-CMS15-0055. We also thank the International Space Science Institute for hosting the fruitful meeting that led to this paper, and the workshop organizers. Additional thanks to the NASA GEDI Science Definition team, and specifically Suzanne Marselis and David Minor and Steve Hancock for assistance in data compilation and processing.

References

Abshire JB, Sun X, Riris H, Sirota JM, McGarry JF, Palm S, Yi D, Liiva P (2005) Geoscience laser altimeter system (GLAS) on the ICESat mission: on-orbit measurement performance. Geophys Res Lett 32:15–20. https://doi.org/10.1029/2005GL024028

Ahmed R, Siqueira P, Hensley S, Bergen K (2013) Uncertainty of forest biomass estimates in north temperate forests due to allometry: implications for remote sensing. Remote Sens 5:3007–3036. https://doi.org/10.3390/rs5063007

Andersen H-E, Strunk J, Temesgen H (2011) Using airborne light detection and ranging as a sampling tool for estimating forest biomass resources in the Upper Tanana Valley of Interior Alaska. West J Appl For 26:157–164

Avitabile V, Camia A (2018) An assessment of forest biomass maps in Europe using harmonized national statistics and inventory plots. For Ecol Manag 409:489–498. https://doi.org/10.1016/j.foreco.2017.11.047

Avitabile V, Herold M, Heuvelink GBM, Lewis SL, Phillips OL, Asner GP, Armston J, Ashton PS, Banin L, Bayol N, Berry NJ, Boeckx P, de Jong BHJ, DeVries B, Girardin CAJ, Kearsley E, Lindsell JA, Lopez-Gonzalez G, Lucas R, Malhi Y, Morel A, Mitchard ETA, Nagy L, Qie L, Quinones MJ, Ryan CM, Ferry SJW, Sunderland T, Laurin GV, Gatti RC, Valentini R, Verbeeck H, Wijaya A, Willcock S (2016) An integrated pan-tropical biomass map using multiple reference datasets. Global Change Biol 22:1406–1420. https://doi.org/10.1111/gcb.13139

Baccini A, Laporte N, Goetz SJ, Sun M, Dong H (2008) A first map of tropical Africa's above-ground biomass derived from satellite imagery. Environ Res Lett 3:045011. https://doi.org/10.1088/1748-9326/3/4/045011

Baccini A, Goetz SJ, Walker WS, Laporte NT, Sun M, Sulla-Menashe D, Hackler J, Beck PSA, Dubayah R, Friedl MA, Samanta S, Houghton RA (2012) Estimated carbon dioxide emissions from tropical deforestation improved by carbon-density maps. Nat Clim Change 2:182–185. https://doi.org/10.1038/nclimate1354

Barton MD, Johnson DA (2004) Footprints of Fe-oxide (-Cu-Au) systems. University of Western Australia Special Publication 33, pp 112–116

Batjes NH (1996) Total carbon and nitrogen in the soils of the world. Eur J Soil Sci 47(2):151–163

Blackard J, Finco M, Helmer E, Holden G, Hoppus M, Jacobs D, Lister A, Moisen G, Nelson M, Riemann R (2008) Mapping U.S. forest biomass using nationwide forest inventory data and moderate resolution information. Remote Sens Environ 112:1658–1677. https://doi.org/10.1016/j.rse.2007.08.021

Boudreau J, Nelson R, Margolis H, Beaudoin A, Guindon L, Kimes D (2008) Regional aboveground forest biomass using airborne and spaceborne LiDAR in Québec. Remote Sens Environ 112:3876–3890. https://doi.org/10.1016/j.rse.2008.06.003

Calders K, Newnham G, Burt A, Murphy S, Raumonen P, Herold M, Culvenor D, Avitabile V, Disney M, Armston J, Kaasalainen M (2015) Nondestructive estimates of above-ground biomass using terrestrial laser scanning. Methods Ecol Evol 6:198–208. https://doi.org/10.1111/2041-210X.12301

Chave J, Réjou-Méchain M, Búrquez A, Chidumayo E, Colgan MS, Delitti WBC, Duque A, Eid T, Fearnside PM, Goodman RC, Henry M, Martínez-Yrízar A, Mugasha WA, Muller-Landau HC, Mencuccini M, Nelson BW, Ngomanda A, Nogueira EM, Ortiz-Malavassi E, Pélissier R, Ploton P, Ryan CM, Saldarriaga JG, Vieilledent G (2014a) Improved allometric models to estimate the aboveground biomass of tropical trees. Global Change Biol 20:3177–3190. https://doi.org/10.1111/gcb.12629

Chave J, Réjou-Méchain M, Búrquez A, Chidumayo E, Colgan MS, Delitti WBC, Duque A, Eid T, Fearnside PM, Goodman RC, Henry M, Martínez-Yrízar A, Mugasha WA, Muller-Landau HC, Mencuccini M, Nelson BW, Ngomanda A, Nogueira EM, Ortiz-Malavassi E, Pélissier R, Ploton P, Ryan CM, Saldarriaga JG, Vieilledent G (2014b) Improved allometric models to estimate the aboveground biomass of tropical trees. Global Change Biol 20:3177–3190. https://doi.org/10.1111/gcb.12629

Chave J, Davies SJ, Phillips OL, Lewis SL, Sist P, Schepaschenko D, Armston J, Baker TR, Coomes D, Disney M, Duncanson L, Hérault B, Labrière N, Meyer V, Réjou-Méchain M, Scipal K, Saatchi S (2019) Ground data are essential for biomass remote sensing missions. Surv. Geophys. https://doi.org/10.1007/s10712-019-09528-w

Clark DB, Kellner JR (2012) Tropical forest biomass estimation and the fallacy of misplaced concreteness. J Veg Sci 23:1191–1196. https://doi.org/10.1111/j.1654-1103.2012.01471.x

Comas X, Terry N, Hribljan JA, Lilleskov EA, Suarez E, Chimner RA, Kolka RK (2017) Estimating belowground carbon stocks in peatlands of the Ecuadorian paramo using groundpenetrating radar (GPR). J Geophys Res Biogeosci 122:370–386. https://doi.org/10.1002/2016JG003550

Disney MI, Boni Vicari M, Burt A, Calders K, Lewis SL, Raumonen P, Wilkes P (2018) Weighing trees with lasers: advances, challenges and opportunities. Interface Focus 8:20170048. https://doi.org/10.1098/rsfs.2017.0048

Duncanson LI, Niemann KO, Wulder MA (2010) Estimating forest canopy height and terrain relief from GLAS waveform metrics. Remote Sens Environ 114:138–154. https://doi.org/10.1016/j.rse.2009.08.018

Duncanson L, Rourke O, Dubayah R (2015a) Small sample sizes yield biased allometric equations in temperate forests. Sci Rep 5:17153. https://doi.org/10.1038/srep17153

Duncanson LI, Dubayah RO, Cook BD, Rosette J, Parker G (2015b) The importance of spatial detail: assessing the utility of individual crown information and scaling approaches for lidar-based biomass density estimation. Remote Sens Environ 168:102–112. https://doi.org/10.1016/j.rse.2015.06.021

Fisher JB, Sweeney S, Brzostek ER, Evans TP, Johnson DJ, Myers JA, Bourg NA, Wolf AT, Howe RW, Phillips RP (2016) Tree-mycorrhizal associations detected remotely from canopy spectral properties. Global Change Biol 22:2596–2607. https://doi.org/10.1111/gcb.13264

Goetz SJ, Baccini A, Laporte NT, Johns T, Walker W, Kellndorfer J, Houghton RA, Sun M (2009) Mapping and monitoring carbon stocks with satellite observations: a comparison of methods. Carbon Balance Manag 4:15–20. https://doi.org/10.1186/1750-0680-4-2

Gonzalez de Tanago J, Lau A, Bartholomeus H, Herold M, Avitabile V, Raumonen P, Martius C, Goodman RC, Disney M, Manuri S, Burt A, Calders K (2018) Estimation of above-ground biomass of large tropical trees with terrestrial LiDAR. Methods Ecol Evol 9:223–234. https://doi.org/10.1111/2041-210X.12904

Gregoire TG, Ståhl G, Næsset E, Gobakken T, Nelson R, Holm S (2011) Model-assisted estimation of biomass in a LiDAR sample survey in Hedmark County, Norway. Can J For Res 41:83–95. https://doi.org/10.1139/X10-195

Harris NL, Brown S, Hagen SC, Saatchi SS, Petrova S, Salas W, Hansen MC, Potapov PV, Lotsch A (2012) Baseline map of carbon emissions from deforestation in tropical regions. Science 336:1573–1576. https://doi.org/10.1126/science.1217962

Hengl T, de Jesus JM, MacMillan RA, Batjes NH, Heuvelink GB, Ribeiro E, Samuel-Rosa A, Kempen B, Leenaars JG, Walsh MG, Gonzalez MR (2014) SoilGrids1 km–global soil information based on automated mapping. PLoS ONE 9:e105992. https://doi.org/10.1371/journal.pone.0105992

Hengl T, De Jesus JM, Heuvelink GBM, Gonzalez MR, Kilibarda M, Blagotić A, Shangguan W, Wright MN, Geng X, Bauer-Marschallinger B, Guevara MA, Vargas R, MacMillan RA, Batjes NH, Leenaars JGB, Ribeiro E, Wheeler I, Mantel S, Kempen B (2017) SoilGrids250 m: global gridded soil information based on machine learning. PLoS ONE 12:1–40. https://doi.org/10.1371/journal.pone.0169748

Herold M, Carter S, Avitabile V, Espejo AB, Jonckheere I, Lucas R, McRoberts RE, Næsset E, Nightingale J, Petersen R, Reiche J, Romijn E, Rosenqvist A, Rozendaal DMA, Seifert FM, Sanz MJ, De Sy V (2019) The role and need for space-based forest biomass-related measurements in environmental management and policy. Surv. Geophys. https://doi.org/10.1007/s10712-019-09510-6

Hruska J, Cermák JAN, Sustek s (1999) Mapping tree root systems with ground-penetrating radar. Tree Physiol 19:125–130

Huang W, Sun G, Dubayah R, Cook B, Montesano P, Ni W, Zhang Z (2013) Mapping biomass change after forest disturbance: applying LiDAR footprint-derived models at key map scales. Remote Sens Environ 134:319–332. https://doi.org/10.1016/j.rse.2013.03.017

Huang W, Swatantran A, Johnson K, Duncanson L, Tang H, O'Neil Dunne J, Hurtt G, Dubayah R (2015) Local discrepancies in continental scale biomass maps: a case study over forested and non-forested landscapes in Maryland, USA. Carbon Balance Manag. https://doi.org/10.1186/s13021-015-0030-9

Jenkins JC, Chojnacky DC, Heath LS, Birdsey RA (2003) National-scale biomass estimators for united states tree species. For Sci 49:12–35

Kellner JR, Armston J, Birrer M, Cushman KC, Duncanson L, Eck C, Falleger C, Imbach B, Král K, Krůček M, Trochta J, Vrška T, Zgraggen C (2019) New opportunities for forest remote sensing through ultra-high-density drone lidar. Surv. Geophys. https://doi.org/10.1007/s10712-019-09529-9

Kennedy RE, Yang Z, Cohen WB (2010) Detecting trends in forest disturbance and recovery using yearly Landsat time series: 1. LandTrendr—temporal segmentation algorithms. Remote Sens Environ 114:2897–2910. https://doi.org/10.1016/j.rse.2010.07.008

Labriere N, Tao S, Chave J, Scipal K, Toan TL, Abernethy K, Alonso A, Barbier N, Bissiengou P, Casal T, Davies SJ, Ferraz A, Herault B, Jaouen G, Jeffery KJ, Kenfack D, Korte L, Lewis SL, Malhi Y, Memiaghe HR, Poulsen JR, Rejou-Mechain M, Villard L, Vincent G, White LJT, Saatchi S (2018) In situ reference datasets from the TropiSAR and AfriSAR campaigns in support of upcoming spaceborne biomass missions. IEEE J Sel Top Appl Earth Obs Remote Sens 10:1–11. https://doi.org/10.1109/JSTARS.2018.2851606

Langner A, Achard F, Grassi G (2014) Can recent pan tropical biomass maps be used to derive alternative Tier 1 values for reporting REDD + activities under UNFCCC? Environ Res Lett 9:124008. https://doi.org/10.1088/1748-9326/9/12/124008

Lutz JA, Larson AJ, Swanson ME, Freund JA (2012) Ecological importance of large-diameter trees in a temperate mixed-conifer forest. PLoS ONE 7:e36131. https://doi.org/10.1371/journal.pone.0036131

Lutz JA, Furniss TJ, Johnson DJ, Davies SJ, Allen D, Alonso A, Anderson-Teixeira KJ, Andrade A, Baltzer J, Becker KML, Blomdahl EM, Bourg NA, Bunyavejchewin S, Burslem DFRP, Cansler CA, Cao K, Cao M, Cárdenas D, Chang L-W, Chao K-J, Chao W-C, Chiang J-M, Chu C, Chuyong GB, Clay K, Condit R, Cordell S, Dattaraja HS, Duque A, Ewango CEN, Fischer GA, Fletcher C, Freund JA, Giardina C, Germain SJ, Gilbert GS, Hao Z, Hart T, Hau BCH, He F, Hector A, Howe RW, Hsieh C-F, Hu Y-H, Hubbell SP, Inman-Narahari FM, Itoh A, Janík D, Kassim AR, Kenfack D, Korte L, Král K, Larson AJ, Li Y, Lin Y, Liu S, Lum S, Ma K, Makana J-R, Malhi Y, McMahon SM, McShea WJ, Memiaghe HR, Mi X, Morecroft M, Musili PM, Myers JA, Novotny V, de Oliveira A, Ong P, Orwig DA, Ostertag R, Parker GG, Patankar R, Phillips RP, Reynolds G, Sack L, Song G-ZM, Su S-H, Sukumar R, Sun I-F, Suresh HS, Swanson ME, Tan S, Thomas DW, Thompson J, Uriarte M, Valencia R, Vicentini A, Vrška T, Wang X, Weiblen GD, Wolf A, Wu

S-H, Xu H, Yamakura T, Yap S, Zimmerman JK (2018) Global importance of large-diameter trees. Global Ecol Biogeogr 27:849–864. https://doi.org/10.1111/geb.12747

Margolis HA, Nelson RF, Montesano PM, Beaudoin A, Sun G, Andersen H-E, Wulder MA (2015) Combining satellite lidar, airborne lidar, and ground plots to estimate the amount and distribution of aboveground biomass in the boreal forest of North America. Can J For Res 45:838–855. https://doi.org/10.1139/cjfr-2015-0006

Mascaro J, Detto M, Asner GP, Muller-Landau HC (2011) Evaluating uncertainty in mapping forest carbon with airborne LiDAR. Remote Sens Environ 115:3770–3774. https://doi.org/10.1016/j.rse.2011.07.019

McRoberts RE, Stehman SV, Liknes GC, Næsset E, Sannier C, Walters BF (2018) The effects of imperfect reference data on remote sensing-assisted estimators of land cover class proportions. ISPRS J Photogramm Remote Sens 142:292–300. https://doi.org/10.1016/j.isprsjprs.2018.06.002

McRoberts RE, Næsset E, Liknes GC, Chen Q, Walters BF, Saatchi S, Herold M (2019) Using a finer resolution biomass map to assess the accuracy of a regional, map-based estimate of forest biomass. Surv Geophys 10:15–20. https://doi.org/10.1007/s10712-019-09507-1

Mitchard ET, Saatchi SS, Baccini A, Asner GP, Goetz SJ, Harris NL, Brown S (2013) Uncertainty in the spatial distribution of tropical forest biomass: a comparison of pan-tropical maps. Carbon Balance Manag 8:10. https://doi.org/10.1186/1750-0680-8-10

Mitchard ETA, Feldpausch TR, Brienen RJW, Lopez-Gonzalez G, Monteagudo A, Baker TR, Lewis SL, Lloyd J, Quesada CA, Gloor M, ter Steege H, Meir P, Alvarez E, Araujo-Murakami A, Aragão LEOC, Arroyo L, Aymard G, Banki O, Bonal D, Brown S, Brown FI, Cerón CE, Chama Moscoso V, Chave J, Comiskey JA, Cornejo F, Corrales Medina M, Da Costa L, Costa FRC, Di Fiore A, Domingues TF, Erwin TL, Frederickson T, Higuchi N, Honorio Coronado EN, Killeen TJ, Laurance WF, Levis C, Magnusson WE, Marimon BS, Marimon Junior BH, Mendoza Polo I, Mishra P, Nascimento MT, Neill D, Núñez Vargas MP, Palacios WA, Parada A, Pardo Molina G, Peña-Claros M, Pitman N, Peres CA, Poorter L, Prieto A, Ramirez-Angulo H, Restrepo Correa Z, Roopsind A, Roucoux KH, Rudas A, Salomão RP, Schietti J, Silveira M, de Souza PF, Steininger MK, Stropp J, Terborgh J, Thomas R, Toledo M, Torres-Lezama A, van Andel TR, van der Heijden GMF, Vieira ICG, Vieira S, Vilanova-Torre E, Vos VA, Wang O, Zartman CE, Malhi Y, Phillips OL (2014) Markedly divergent estimates of Amazon forest carbon density from ground plots and satellites: divergent forest carbon maps from plots & space. Global Ecol Biogeogr 23:935–946. https://doi.org/10.1111/geb.12168

Mokany K, Raison RJ, Prokushkin AS (2006) Critical analysis of root : shoot ratios in terrestrial biomes. Global Change Biol 12:84–96. https://doi.org/10.1111/j.1365-2486.2005.001043.x

Neigh CSR, Nelson RF, Ranson KJ, Margolis HA, Montesano PM, Sun G, Kharuk V, Næsset E, Wulder MA, Andersen H-E (2013) Taking stock of circumboreal forest carbon with ground measurements, airborne and spaceborne LiDAR. Remote Sens Environ 137:274–287. https://doi.org/10.1016/j.rse.2013.06.019

Newnham GJ, Armston JD, Calders K, Disney MI, Lovell JL, Schaaf CB, Strahler AH, Danson FM (2015) Terrestrial laser scanning for plot-scale forest measurement. Curr For Rep 1:239–251. https://doi.org/10.1007/s40725-015-0025-5

Pan Y, Birdsey RA, Phillips OL, Jackson RB (2013) The structure, distribution, and biomass of the world's forests. Annu Rev Ecol Evol Syst 44:593–622. https://doi.org/10.1146/annurev-ecolsys-110512-135914

Réjou-Méchain M, Muller-Landau HC, Detto M, Thomas SC, Le Toan T, Saatchi SS, Barreto-Silva JS, Bourg NA, Bunyavejchewin S, Butt N, Brockelman WY, Cao M, Cárdenas D, Chiang J-M, Chuyong GB, Clay K, Condit R, Dattaraja HS, Davies SJ, Duque A, Esufali S, Ewango C, Fernando RHS, Fletcher CD, Gunatilleke IAUN, Hao Z, Harms KE, Hart TB, Hérault B, Howe RW, Hubbell SP, Johnson DJ, Kenfack D, Larson AJ, Lin L, Lin Y, Lutz JA, Makana J-R, Malhi Y, Marthews TR, McEwan RW, McMahon SM, McShea WJ, Muscarella R, Nathalang A, Noor NSM, Nytch CJ, Oliveira AA, Phillips RP, Pongpattananurak N, Punchi-Manage R, Salim R, Schurman J, Sukumar R, Suresh HS, Suwanvecho U, Thomas DW, Thompson J, Uríarte M, Valencia R, Vicentini A, Wolf AT, Yap S, Yuan Z, Zartman CE, Zimmerman JK, Chave J (2014) Local spatial structure of forest biomass and its consequences for remote sensing of carbon stocks. Biogeosci Discuss 11:5711–5742. https://doi.org/10.5194/bgd-11-5711-2014

Réjou-Méchain M, Tanguy A, Piponiot C, Chave J, Hérault B (2017) biomass: an r package for estimating above-ground biomass and its uncertainty in tropical forests. Methods Ecol Evol 8:1163–1167. https://doi.org/10.1111/2041-210X.12753

Romijn E, De Sy V, Herold M, Böttcher H, Roman-Cuesta RM, Fritz S, Schepaschenko D, Avitabile V, Gaveau D, Verchot L, Martius C (2018) Independent data for transparent monitoring of greenhouse

gas emissions from the land use sector—what do stakeholders think and need? Environ Sci Policy 85:101–112. https://doi.org/10.1016/j.envsci.2018.03.016
Saatchi SS, Harris NL, Brown S, Lefsky M, Mitchard ET, Salas W, Zutta BR, Buermann W, Lewis SL, Hagen S et al (2011) Benchmark map of forest carbon stocks in tropical regions across three continents. Proc Natl Acad Sci 108:9899–9904
Santoro M, Beaudoin A, Beer C, Cartus O, Fransson JES, Hall RJ, Pathe C, Schmullius C, Schepaschenko D, Shvidenko A, Thurner M, Wegmüller U (2015) Forest growing stock volume of the northern hemisphere: spatially explicit estimates for 2010 derived from Envisat ASAR. Remote Sens Environ 168:316–334. https://doi.org/10.1016/j.rse.2015.07.005
Schepaschenko D, Moltchanova E, Shvidenko A, Blyshchyk V, Dmitriev E, Martynenko O, See L, Kraxner F (2018) Improved estimates of biomass expansion factors for russian forests. Forests 9:312. https://doi.org/10.3390/f9060312
Simard M, Pinto N, Fisher JB, Baccini A (2011) Mapping forest canopy height globally with spaceborne lidar. J Geophys Res 116:15–20. https://doi.org/10.1029/2011JG001708
Su Y, Guo Q, Xue B, Hu T, Alvarez O, Tao S, Fang J (2016) Spatial distribution of forest aboveground biomass in China: estimation through combination of spaceborne lidar, optical imagery, and forest inventory data. Remote Sens Environ 173:187–199. https://doi.org/10.1016/j.rse.2015.12.002
Sun G, Ranson KJ, Kimes DS, Blair JB, Kovacs K (2008) Forest vertical structure from GLAS: an evaluation using LVIS and SRTM data. Remote Sens Environ 112:107–117. https://doi.org/10.1016/j.rse.2006.09.036
Thurner M, Beer C, Santoro M, Carvalhais N, Wutzler T, Schepaschenko D, Shvidenko A, Kompter E, Ahrens B, Levick SR, Schmullius C (2014) Carbon stock and density of northern boreal and temperate forests: carbon stock and density of boreal and temperate forests. Global Ecol Biogeogr 23:297–310. https://doi.org/10.1111/geb.12125
Trochta J, Krůček M, Vrška T, Král K (2017) 3D Forest: an application for descriptions of three-dimensional forest structures using terrestrial LiDAR. PLoS ONE 12:e0176871. https://doi.org/10.1371/journal.pone.0176871
Wollschläger U, Gerhards H, Yu Q, Roth K (2010) Multi-channel ground-penetrating radar to explore spatial variations in thaw depth and moisture content in the active layer of a permafrost site. Cryosphere 4:269–283. https://doi.org/10.5194/tc-4-269-2010
Zolkos SG, Goetz SJ, Dubayah R (2013) A meta-analysis of terrestrial aboveground biomass estimation using lidar remote sensing. Remote Sens Environ 128:289–298. https://doi.org/10.1016/j.rse.2012.10.017

Affiliations

L. Duncanson[1] · J. Armston[1] · M. Disney[2] · V. Avitabile[3] · N. Barbier[4] · K. Calders[5] · S. Carter[6] · J. Chave[7] · M. Herold[6] · T. W. Crowther[8] · M. Falkowski[9] · J. R. Kellner[10,11] · N. Labrière[7] · R. Lucas[12] · N. MacBean[13] · R. E. McRoberts[14] · V. Meyer[15] · E. Næsset[16] · J. E. Nickeson[17] · K. I. Paul[18] · O. L. Phillips[19] · M. Réjou-Méchain[4] · M. Román[25] · S. Roxburgh[18] · S. Saatchi[15] · D. Schepaschenko[20] · K. Scipal[21] · P. R. Siqueira[22] · A. Whitehurst[23] · M. Williams[24]

1 Department of Geographical Sciences, University of Maryland, College Park, 2181 Lefrak Hall, College Park, MD 20742, USA

2 Department of Geography, University College London, Gower Street, London WC1E 6BT, UK

3 European Commission, Joint Research Centre (JRC), Via E. Fermi 2749, 21027 Ispra, Italy

4 AMAP, IRD, CIRAD,CNRS, INRA, Montpellier University, TA A51/PS2, 34398 Montpellier cedex 5, France

5 CAVElab – Computational and Applied Vegetation Ecology, Ghent University, Room A2.089, Coupure Links 653, 9000 Ghent, Belgium

[6] Laboratory of Geo-Information Science and Remote Sensing, Wageningen University and Research, Droevendaalsesteeg 3, 6708 PB Wageningen, The Netherlands

[7] Laboratoire Evolution et Diversit. Biologique, UMR 5174, CNRS, Universit. Toulouse Paul Sabatier, 118 route de Narbonne, 31062 Toulouse cedex 9, France

[8] Institute of Integrative Biology, ETH Zürich, Univeritätstrasse 16, 8006 Zurich, Switzerland

[9] Department of Ecosystem Science and Sustainability, Colorado State University, Fort Collins, CO 80523, USA

[10] Institute at Brown for Environment and Society, Brown University, Providence, RI 02912, USA

[11] Department of Ecology and Evolutionary Biology, Brown University, Providence, RI 02912, USA

[12] Earth Observation and Ecosystem Dynamics Research Group, Department of Geography and Earth Sciences (DGES), Aberystwyth University, Aberystwyth, Wales SY23 3DB, UK

[13] Department of Geography, Indiana University, 701 E. Kirkwood Ave., Bloomington, IN 47405, USA

[14] USDA Forest Service, Northern Research Station, Saint Paul, 1992 Folwell Ave, St Paul, MN 55108, USA

[15] Jet Propulsion Laboratory, California Institute of Technology, Pasadena, CA 91109, USA

[16] Faculty of Environmental Sciences and Natural Resource Management, Norwegian University of Life Sciences, NMBU, P.O. Box 5003, 1432 Ås, Norway

[17] NASA Goddard Space Flight Center/Science Systems and Applications Inc., 10210 Greenbelt Rd #600, Lanham, MD 20706, USA

[18] CSIRO Land and Water, GPO Box 1700, Canberra, ACT 2601, Australia

[19] School of Geography, University of Leeds, Leeds LS2 9JT, UK

[20] International Institute for Applied Systems Analysis, Schlossplatz 1, 2361 Laxenburg, Austria

[21] European Space Agency, ESTEC, Keplerlaan 1, 2201 AZ Noordwijk, The Netherlands

[22] Department of Electrical and Computer Engineering, 201 Marcus Hall, University of Massachusetts, 100 Natural Resources Road, Amherst, MA 01003, USA

[23] Arctic Slope Federal Technical Services, 7000 Muirkirk Meadows Dr #100, Laurel, MD 20707, USA

[24] School of GeoScience, University of Edinburgh, Drummond St, Edinburgh EH8 9XP, UK

[25] Earth from Space Institute, Universities Space Research Association, Columbia, MD, USA

Surveys in Geophysics (2019) 40:1001–1015
https://doi.org/10.1007/s10712-019-09507-1

Using a Finer Resolution Biomass Map to Assess the Accuracy of a Regional, Map-Based Estimate of Forest Biomass

Ronald E. McRoberts[1] · Erik Næsset[2] · Greg C. Liknes[1] · Qi Chen[3] · Brian F. Walters[1] · Sassan Saatchi[4] · Martin Herold[5]

Received: 21 July 2018 / Accepted: 8 January 2019 / Published online: 8 February 2019

Abstract

National greenhouse gas inventories often use variations of the *gain–loss* approach whereby emissions are estimated as the products of estimates of areas of land-use change characterized as *activity data* and estimates of emissions per unit area characterized as *emission factors*. Although the term *emissions* is often intuitively understood to mean release of greenhouse gases from terrestrial sources to the atmosphere, in fact, emission factors can also be negative, meaning removal of the gases from the atmosphere to terrestrial sinks. For remote and inaccessible forests for which ground sampling is difficult if not impossible, emission factors may be based on map-based estimates of biomass or biomass change obtained from regional maps. For the special case of complete deforestation, the emission factor for the aboveground biomass pool is simply mean aboveground, live-tree, biomass per unit area prior to the deforestation. If biomass maps are used for these purposes, estimates must still comply with the first IPCC good practice guideline regarding accuracy relative to the true value and the second guideline regarding uncertainty. Accuracy assessment for a map-based estimate entails comparison of the estimate to a second estimate obtained using independent reference data. Assuming ground sampling is not feasible, a map of greater quality than the regional map may be considered as a source of reference data where greater quality connotes attributes such as finer resolution and/or greater accuracy. For a local, sub-regional study area in Minnesota in the USA, the accuracy of an estimate of mean aboveground, live-tree biomass per unit area (AGB, Mg/ha) obtained from a coarser resolution, regional, MODIS-based biomass map was assessed using reference data sampled from a finer resolution, local, airborne laser scanning (ALS)-based biomass map. The rationale for a local assessment of a regional map is that, although assessment of a regional map would be difficult for the entire extent of the map, it can likely be assessed for multiple local sub-regions in which case expected local regional accuracy for the entire map can perhaps be inferred. For this study, the local assessment was in the form of a test of the hypothesis that the local sub-regional estimate from the regional map did not deviate from the local true value. A hybrid approach to inference was used whereby design-based inferential techniques were used to estimate uncertainty due to sampling from the finer resolution map, and model-based inferential techniques were used to estimate uncertainty resulting from using the finer resolution map unit values which were subject to prediction error as reference data. The test revealed no statistically significant difference between the MODIS-based and ALS-based map estimates, thereby indicating that for the local sub-region, the regional, MODIS-based estimate complied with the first IPCC good practice guideline for accuracy.

Extended author information available on the last page of the article

Keywords Hybrid inference · Design-based inference · Model-based inference · Greenhouse gas inventory · IPCC good practice guidelines

1 Introduction

National greenhouse gas (GHG) inventories assess the scale of emissions for multiple land uses including the Agriculture, Forestry and other land use sectors (IPCC 2006) and are typically implemented using either the *stock-change* or *gain–loss* approach (IPCC 2006, p. 4.11; GFOI 2016, p. 22). The stock-change approach, for which emissions are estimated as differences in stocks for two dates, is well suited for countries with established forest sampling programs such as national forest inventories (NFI). For countries with remote and inaccessible forests that are difficult to sample, the gain–loss approach is more often used. With the latter approach, emissions are estimated as products of land-use change area estimates, characterized as *activity data*, and estimates of emissions per unit area, characterized as *emission factors* or *removal factors*. Although the term *emissions* is often intuitively understood to mean release of greenhouse gases from terrestrial sources to the atmosphere, in fact, emission factors can also be negative, meaning removal of the gases from the atmosphere to terrestrial sinks.

For purposes of estimating activity data, methods that address ground sampling difficulties for remote regions have been widely reported (e.g., Olofsson et al. 2013, 2014; McRoberts et al. 2018b), but such is not case for estimating emission factors. One approach that circumvents ground sampling is to use global or regional emission factors (e.g., Pearson et al. 2017), but the adverse effects on bias and precision are mostly unknown (Pelletier et al. 2012). Another approach is to convert estimates obtained from regional biomass maps to carbon estimates where the term *regional* could refer to a global map, a global map within latitude limits, a national map or simply a large area map (Saatchi et al. 2011; Baccini et al. 2012). For the special case of complete deforestation as the land-use change class, the emission factor for the aboveground biomass pool is simply mean aboveground, live-tree biomass per unit area (AGB, Mg/ha) prior to the deforestation. Regardless of the approach, the Intergovernmental Panel on Climate Change (IPCC) specifies two good practice guidelines for GHG inventories: (1) "neither over- nor underestimates so far as can be judged," and (2) "uncertainties are reduced as far as is practicable" (GFOI 2016, p. 15). For the first guideline, the standard for assessing under- or over-estimation was assumed to be the true value; for the second guideline, the presupposition was that uncertainty must first be rigorously estimated before it can be reduced.

Multiple methods can be used to assess the degree to which an estimate from a regional map satisfies the first IPCC good practice guideline regarding accuracy relative to the true value. Generally, such methods entail comparing the regional map-based estimate to an estimate based on independent reference data of at least greater quality than the regional map data. For this study, an underlying assumption was that ground sampling was not a feasible source of reference data for reasons such as the remoteness and/or inaccessibility of the forests. For the latter situation, an alternative is a method analogous to that used for acquiring reference data for estimating activity data, i.e., reference data in the form of visual interpretations of aerial or satellite imagery, albeit subject to the constraint that the interpretations are of greater quality than the map data (Stehman 2009; GFOI 2016, p. 125; Olofsson et al. 2013; McRoberts et al. 2018b). Direct extension of this approach to visual assessment of biomass from interpreted imagery would likely fail to satisfy the criterion that the reference data must be of greater quality than the map data. A potentially more viable approach would be to use a finer resolution biomass map as a source of reference data subject to the criterion that the finer resolution map data are of greater accuracy than the regional map data. However, even if reference data from a finer resolution map are of greater quality than the regional map unit values, they are still subject to uncertainty.

The study objective was to illustrate a statistically rigorous method for testing the hypothesis that an estimate of mean AGB from a regional map for a sub-regional study area complies with the first IPCC good practice guideline, i.e., the estimate does not deviate from the true value. Although ground reference data were, in fact, available for the study area, for illustrative purposes the analyses assumed such data were not available such as could be the case for tropical forest countries lacking sufficiently extensive ground plot networks. Thus, the test took the form of a comparison of the regional map-based estimate for the sub-region to an estimate based on a sample from a finer resolution map.

2 Data

2.1 Study Area

The 7583-km^2 study area consisted of the entirety of Itasca County in north central Minnesota in the USA (Fig. 1). Land cover includes water, wetlands and approximately 80% forest consisting of uplands with deciduous mixtures of pines (*Pinus* spp.), spruce (*Picea* spp.) and balsam fir (*Abies balsamea* (L.) Mill.) and lowlands with spruce (*Picea* spp.), tamarack (*Larix laricina* (Du Roi) K. Koch), white cedar (*Thuja occidentalis* (L.)) and black ash (*Fraxinus nigra* Marsh.). Forest stands in the study area are typically naturally regenerated, uneven-aged, and mixed species.

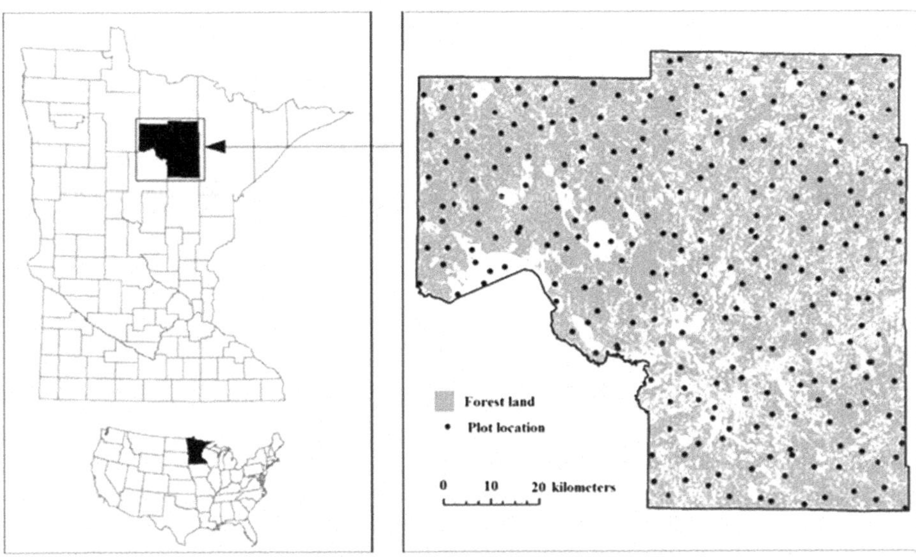

Fig. 1 Study area in Itasca County, Minnesota, USA

2.2 Maps

2.2.1 *Coarser Resolution (CR) Regional Map*

The 250-m × 250-m coarse resolution (CR) regional map used for this study was constructed by the Forest Inventory and Analysis (FIA) program of the US Forest Service which conducts the NFI of the USA (McRoberts et al. 2010). The map was based on regression trees, AGB for FIA plots measured between 1990 and 2003, and predictor variables from multiple sources including 2001 Moderate Resolution Imaging Spectrometer (MODIS) image products (Hansen et al. 2003; Huete et al. 2002; Vermote and Vermueulen 1999), the 1992 National Land Cover Dataset (Vogelmann et al. 2001), and topographic and climatic variables (Blackard et al. 2008). Validation techniques included pixel-level comparisons and comparisons of mean plot-level AGB and mean model AGB predictions for polygons of various sizes. In general, the map tended to over-predict for areas of small AGB and under-predict for areas of large AGB. Reported correlations between aggregations of FIA plot AGB values and map values ranged from 0.31 to 0.73, depending on the region of the country; for the region that included the study area the correlation was 0.46. Blackard et al. (2008) provide more details regarding the regional CR map.

2.2.2 *Finer Resolution (FR) Map*

A 13-m × 13-m, local, finer resolution (FR) AGB map was constructed using 2012 airborne laser scanning (ALS) data and FIA plot data obtained for an equal probability sample. Allometric models were used to predict AGB for the central, circular, 7.32-m (24-ft) radius subplots for 541 FIA plots measured between 2010 and 2014. For this study, uncertainty in the allometric model predictions was considered negligible relative to the effects of sampling variability (McRoberts et al. 2016). The forest inventory data are described in greater detail in McRoberts et al. (2018a). Wall-to-wall airborne laser scanning (ALS) data were acquired in April 2012 with a nominal pulse density of 0.67 pulses/m^2. Distributions of all first return heights were constructed for the 168.3-m^2 circular plots and the 169-m^2 square cells that tessellated the study area and served as FR map units. Standard ALS metrics included the mean, quadratic mean, standard deviation, skewness, kurtosis of the ALS height distributions and deciles of the height and canopy density distributions. The ALS data are described in greater detail in McRoberts et al. (2018a).

A model of the relationship between plot-level AGB as the dependent variable and the ALS metrics as independent variables was formulated as

$$\begin{aligned} \mathrm{AGB}_i &= f\left(\boldsymbol{x}_i; \boldsymbol{\beta}\right) + \varepsilon_i \\ &= \beta_0 \cdot x_{1i}^{\beta_1} \cdot \exp\left(\beta_2 \cdot x_{2i} + \cdots + \beta_{\mathrm{p}} \cdot x_{\mathrm{p}i}\right) + \varepsilon_i \end{aligned} \tag{1}$$

where i indexes plots, x_{ji} is an ALS metric, $\beta_j \in \boldsymbol{\beta}$ is a parameter to be estimated and ε_i is a residual assumed to be distributed $N(0,\boldsymbol{\Sigma})$. The model was fit to the data using weighted nonlinear least squares where $\widehat{\boldsymbol{\Sigma}}^{-1}$ was used to weight the observations. A forward selection procedure was used to select independent variables for inclusion in the model if they statistically significantly increased the quality of fit of the model to the data at a nominal $\alpha = 0.05$ level. The resulting model was used to predict AGB for all 13-m × 13-m cells in the study area, thereby producing the FR map. For illustrative purposes for this study, the FR map was henceforth considered to be the only available source of reference data for assessing the accuracy of the regional, CR map-based estimate of mean AGB. Although the local FR map was constructed using ground plot data acquired from within the study area, this is not a necessary condition (McRoberts et al. 2014).

Heteroscedastic (nonconstant) residual variance as required for the weight matrix, $\widehat{\boldsymbol{\Sigma}}^{-1}$, was estimated using a 4-step procedure (McRoberts et al. 2016): (1) for $\hat{\varepsilon}_i = y_i - \hat{y}_i$, the sample pairs $\left(\hat{y}_i, \hat{\varepsilon}_i\right)$ were ordered with respect to $\hat{y}_i$, (2) the pairs were aggregated into groups of size 15; (3) within each group, g, the mean of the predictions, $\bar{\hat{y}}_g$, and the standard deviation, $\hat{\sigma}_g$, of the estimated residuals, $\hat{\varepsilon}_i$, were calculated; and (4) the relationship between the group standard deviations and the group prediction means was represented using a model of the form

$$\hat{\sigma}_{\mathrm{g}} = \gamma \cdot \bar{\hat{y}}_{\mathrm{g}}, \tag{2}$$

where γ is a parameter to be estimated. To accommodate heteroscedasticity when fitting the model, the *i*th diagonal element of $\widehat{\boldsymbol{\Sigma}}$ was $\hat{\sigma}_i^2 = \left(\hat{\gamma} \cdot \hat{y}_i\right)^2$ from Eq. (2) where $\hat{y}_i$ is the model prediction for the *i*th plot. Because distances between plot locations mostly exceeded the range of spatial correlation, the off-diagonal elements of $\widehat{\boldsymbol{\Sigma}}$ were set to 0.

3 Methods

3.1 The Test

Three assumptions underlay the study: (1) the only information available for the local portion of the regional CR map was the map unit values, (2) ground sampling as a source of reference data for the study area was not feasible, and (3) the FR map included sufficient meta-data to estimate mean AGB and the uncertainty of the estimate.

The objective was to test the hypothesis that the regional CR map-based estimate of mean AGB for the local sub-region complies with the first IPCC good practice guideline. In particular, the test focused on the accuracy of the CR map-based estimate where *accuracy* is related to *bias* and refers to agreement between the true value and the average of repeated independent estimates (IPCC, 2006, p. 3.7). Because the true value is not known, the best alternative is an estimate obtained using reference data that are at least of greater quality than the CR map data (Stehman 2009; GFOI 2016, p. 125). Further, because typically only a single sample of model calibration data is available, the test result can only be expressed probabilistically, i.e., as an inference. An intuitive test statistic is

$$t = \frac{\hat{\mu}^{\mathrm{CR}} - \hat{\mu}^{\mathrm{FR}}}{\sqrt{\widehat{\mathrm{MSE}}\left(\hat{\mu}^{\mathrm{CR}}\right) + \widehat{\mathrm{MSE}}\left(\hat{\mu}^{\mathrm{FR}}\right)}}. \tag{3}$$

where the uncertainties associated with the CR and FR map-based estimates of the mean are expressed in terms of *mean square error* (*MSE*) rather than *variance* because the map-based estimators of the means are not necessarily unbiased as a result of possible systematic map error (Cochran 1977, p. 15). The exact forms of the two $\widehat{\mathrm{MSE}}$s are described in the sections that follow. Under the assumption that the data used to construct the FR map are independent of the data used to construct the CR map, no covariance term in the denominator of Eq. (3) is necessary.

A crucial issue is that reference data acquired from any map, regardless of map accuracy, cannot be assumed to be without error. Mowrer and Congalton (2000) characterize reference data subject to non-negligible error as *imperfect reference data*. In the event of imperfect reference data, compliance with the IPCC good practice guidelines requires incorporation of the effects of this source of uncertainty into $\widehat{\text{MSE}}$s. In particular, the $\widehat{\text{MSE}}$ must incorporate uncertainty due to sampling from the map used as the source of reference data and uncertainty due to the imperfect nature of the sample reference data, i.e., the FR map unit values are subject to error.

The term *hybrid inference* characterizes recently developed methods that combine design-based inferential techniques for assessing the effects of sampling variability and model-based inferential techniques for assessing the effects of imperfect reference data (Fattorini 2012; Corona et al. 2014; McRoberts et al. 2016; Ståhl et al. 2016). Hybrid inference has four key features: (1) A probability sample of population units for which only auxiliary information is available; (2) a prediction technique that uses the auxiliary information to predict the reference data values for the sample units; (3) a design-based estimator of the population parameter using the reference data predictions for the sample units; and (4) estimation of uncertainty using a design-based estimator to accommodate the effects of sampling variability and a model-based estimator to accommodate the effects of uncertainty in the reference data.

For this study, hybrid inference entailed sampling from the FR map, using design-based estimators with the FR sample map reference data to estimate the mean and the component of $\widehat{\text{MSE}}$ due to sampling variability (Sect. 3.2), and using model-based estimators to estimate the component of $\widehat{\text{MSE}}$ due to the effects of the imperfect FR map reference data (Sect. 3.3). The hybrid $\widehat{\text{MSE}}$ is then expressed additively as,

$$\widehat{\text{MSE}}^{\text{Hyb}}\left(\hat{\mu}^{\text{FR}}\right) = \widehat{\text{MSE}}^{\text{DB}}\left(\hat{\mu}^{\text{FR}}\right) + \widehat{\text{MSE}}^{\text{MB}}\left(\hat{\mu}^{\text{FR}}\right), \tag{4}$$

where the design-based (DB) component, $\widehat{\text{MSE}}^{\text{DB}}\left(\hat{\mu}^{\text{FR}}\right)$, estimates the effects of sampling from the FR map and the model-based (MB) component, $\widehat{\text{MSE}}^{\text{MB}}\left(\hat{\mu}^{\text{FR}}\right)$ estimates the effects of uncertainty in the imperfect FR map reference data. $\widehat{\text{MSE}}^{\text{Hyb}}\left(\hat{\mu}^{\text{FR}}\right)$ then replaces $\widehat{\text{MSE}}\left(\hat{\mu}^{\text{FR}}\right)$ in Eq. (3). The technical objectives in Sects. 3.2 and 3.3 were to address $\hat{\mu}^{\text{FR}}$, $\widehat{\text{MSE}}^{\text{DB}}\left(\hat{\mu}^{\text{FR}}\right)$ and $\widehat{\text{MSE}}^{\text{MB}}\left(\hat{\mu}^{\text{FR}}\right)$ with $\hat{\mu}^{\text{CR}}$ and $\widehat{\text{MSE}}\left(\hat{\mu}^{\text{CR}}\right)$ addressed in Sect. 3.4.

3.2 Design-Based Component of Hybrid Inference

Three primary assumptions underlie *design-based inference*, also characterized as *probability-based inference* (Hansen et al. 1983). First, the basis for validity is a probability sample that incorporates some form of randomization. Second, each population unit is assumed to have one and only one possible value, apart from negligible observation or measurement error. Third, the probability of selection for each population unit into the sample is positive and known. Much of the effort for design-based inference involves selecting an appropriate combination of sampling design and corresponding estimator. Familiar sampling designs include simple random, systematic, stratified, multi-phase and multi-stage sampling designs. Estimators corresponding to these designs are generally unbiased or at least approximately unbiased, and uncertainty estimation typically entails comparing observations to their corresponding means or model predictions. All design-based estimators assume that observations or measurements of the response variable have at most negligible uncertainty (Snedecor and Cochran 1967, p. 164).

A relatively simple sampling design used for the current study entailed selecting the central FR map unit within each CR map unit. Cochran (1997, Sect. 8.13) characterizes this design as a *systematic aligned* sampling design. The FR map-based estimate of the mean and the corresponding $\widehat{\text{MSE}}$ were calculated using the simple *expansion* (*Exp*) estimators (Royall and Herson 1973),

$$\hat{\mu}^{\text{FR}} = \frac{1}{N_{\text{CR}}} \sum_{i=1}^{N_{\text{CR}}} \hat{y}_i^{\text{FR}}, \tag{5a}$$

and

$$\widehat{\text{MSE}}^{\text{DB}}\left(\hat{\mu}^{\text{FR}}\right) = \frac{1}{N_{\text{CR}} \cdot \left(N_{\text{CR}} - 1\right)} \sum_{i=1}^{N_{\text{CR}}} \left(\hat{y}_i^{\text{FR}} - \hat{\mu}^{\text{FR}}\right)^2, \tag{5b}$$

where i indexes the CR map units, $\hat{y}_i^{\text{FR}}$ is the value for the central FR map unit within the ith CR map unit and N_{CR} is the FR sample size. Because the FR map units are considerably smaller than the CR map units, the total number of CR map units, N_{CR}, is much larger than the total number of FR map units; thus, N_{CR} and N_{FR} are not interchangeable. In particular, with the systematic aligned sampling design with one FR sample unit for each CR sample unit, the FR sample size is exactly equal to the total number of CR map units, N_{CR}. Although when used with systematic sample data, $\widehat{\text{MSE}}$ for the Exp estimator may be biased, it tends to be conservatively biased in the sense that $\widehat{\text{MSE}}$s may be slightly too large (Särndal et al. 1992, p. 83). Other sampling designs were considered, but preliminary analyses indicated that the resulting $\widehat{\text{MSE}}$s were larger than for the systematic design. In addition, post-stratified (McRoberts et al. 2013) and model-assisted difference estimators (Ståhl et al. 2016) were also investigated for use with the systematic sample data, but preliminary analyses again indicated there was little to be gained relative to the Exp estimators.

Overall, the advantages of the Exp estimators are that they are simple, intuitive and easy to implement. A disadvantage is that variances are frequently large, particularly for small sample sizes and/or populations with large variability among population unit values. Because $\hat{\mu}^{\text{FR}}$ is an independent estimate of the sub-regional population mean, it can serve as the independent estimate required for comparison to $\hat{\mu}^{\text{CR}}$ using Eq. (3).

3.3 Model-Based Component of Hybrid Inference

Assumptions underlying *model-based inference,* also characterized as *model-dependent inference*, differ considerably from the more familiar design-based inference. First, the basis for validity is correct specification of the model, not a probability sample. Second, model-based inference assumes an entire distribution of possible values for each population unit, not just a single value. Third, randomization occurs via realization of observations from the distributions characterizing the population units selected for the sample, not via the sampling design.

Despite being considered observations without error for the design-based component of hybrid inference (Sect. 3.2), the FR map unit values are model predictions (Sect. 3.2), and the uncertainty of these predictions must be incorporated into the final hybrid $\widehat{\text{MSE}}$. For regression models, the form of the model-based component of $\widehat{\text{MSE}}$ can be formulated as

$$\widehat{\mathrm{MSE}}^{\mathrm{MB}}\left(\hat{\mu}^{\mathrm{FR}}\right) = \frac{1}{N_{\mathrm{CR}}^2}\sum_{i_1=1}^{N_{\mathrm{CR}}}\sum_{i_2=1}^{N_{\mathrm{CR}}} z'_{i_1}\cdot\widehat{\mathbf{V}}_{\hat{\beta}}\cdot z_{i_2} + \frac{1}{N_{\mathrm{CR}}^2}\sum_{i=1}^{N_{\mathrm{CR}}}\hat{\sigma}_i^2 + \frac{1}{N_{\mathrm{CR}}^2}\sum_{i_1\neq}^{N_{\mathrm{CR}}}\sum_{i_2}^{N_{\mathrm{CR}}}\hat{\sigma}_{i_1}\cdot\hat{\sigma}_{i_2}\cdot\hat{\rho}_{i_1 i_2} \tag{6a}$$

where N_{CR} is again the number of CR map units,

$$z_{ij} = \frac{\partial f\left(\boldsymbol{x}_i;\hat{\boldsymbol{\beta}}\right)}{\partial \beta_j}, \tag{6b}$$

$$\widehat{\mathbf{V}}_{\hat{\mathbf{B}}} = \left(\mathbf{Z}'\cdot\hat{\boldsymbol{\Sigma}}^{-1}\cdot\mathbf{Z}\right)^{-1} \tag{6c}$$

is the first-order Taylor's series estimator of the covariance matrix for the model parameter estimates, $\widehat{\boldsymbol{\Sigma}}$ is the covariance matrix for the model prediction residuals, $\hat{\varepsilon}_i$, $\hat{\sigma}_i^2$ is the estimate of the heteroscedastic residual variance for the FR map from Sect. 2.2.2 and $\hat{\rho}_{i_1 i_2}$ is the estimate of spatial correlation for the FR map residuals.

The first term of Eq. (6a) is further denoted

$$\widehat{\mathrm{MSE}}_{\mathrm{pre}}^{\mathrm{MB}}\left(\hat{\mu}^{\mathrm{FR}}\right) = \frac{1}{N_{\mathrm{CR}}^2}\sum_{i_1=1}^{N_{\mathrm{CR}}}\sum_{i_2=1}^{N_{\mathrm{CR}}} z'_{i_1}\cdot\widehat{\boldsymbol{V}}_{\hat{\beta}}\cdot z_{i_2}, \tag{6d}$$

and estimates the effect on model predictions (*pre*) of variability in the sample data used to construct the model, i.e., each model calibration sample would produce different model parameter estimates and, therefore, different model predictions. The computational intensity associated with the double sums in Eq. (6d) can be reduced, using the alternate form,

$$\widehat{\mathrm{MSE}}_{\mathrm{pre}}^{\mathrm{MB}}\left(\hat{\mu}^{\mathrm{FR}}\right) = \sum_{j_1=1}^{J}\sum_{j_2=1}^{J}\bar{z}_{j_1}\cdot v_{j_1 j_2}\cdot\bar{z}_{j_2}, \tag{6e}$$

where J is the number of model parameters, j_1 and j_2 index the parameters, $v_{j_1 j_2}\in\widehat{\mathbf{V}}_{\hat{\beta}}$, $\bar{z}_j = \frac{1}{N_{\mathrm{CR}}}\sum_{i=1}^{N_{\mathrm{CR}}} z_{ij}$ and z_{ij} is defined by Eq. (6b) (Saarela et al. 2015; McRoberts et al. 2018c). For nonparametric prediction techniques such as Random Forests, Monte Carlo bootstrap procedures can be used to estimate $\widehat{\mathrm{MSE}}_{\mathrm{Pre}}^{\mathrm{MB}}\left(\hat{\mu}^{\mathrm{FR}}\right)$ (Efron and Tibshirani 1994, pp. 47–48, 113; McRoberts et al. 2018c).

The second term of Eq. (6a) is further denoted

$$\widehat{\mathrm{MSE}}_{\mathrm{pre}}^{\mathrm{MB}}\left(\hat{\mu}^{\mathrm{FR}}\right) = \frac{1}{N_{\mathrm{CR}}^2}\sum_{i=1}^{N_{\mathrm{CR}}}\hat{\sigma}_i^2, \tag{6f}$$

and is the result of residual (*res*) variability of observations around their model predictions where $\hat{\sigma}_i^2$ is estimated as described in Sect. 2.2 The third term of Eq. (6a) is denoted

$$\widehat{\mathrm{MSE}}_{\mathrm{spa}}^{\mathrm{MB}}\left(\hat{\mu}^{\mathrm{FR}}\right) = \frac{1}{N_{\mathrm{CR}}^2}\sum_{i_1\neq}^{N_{\mathrm{CR}}}\sum_{i_2}^{N_{\mathrm{CR}}}\hat{\sigma}_{i_1}\cdot\hat{\sigma}_{i_2}\cdot\hat{\rho}_{i_1 i_2} \tag{6g}$$

and is the result of spatial (*spa*) correlation among residuals. Spatial correlation, ρ, is often estimated via a correlogram using the model prediction residuals obtained when fitting the model. However, when plot separation distances exceed the range of spatial correlation which is usually the case for efficient sampling designs, construction of the correlogram for small distances and estimation of the range are not possible. For this study, an estimate of the range of spatial correlation was selected as 200 m, the maximum of values reported in the literature for similar studies (Breidenbach et al. 2008, 2016; McRoberts et al. 2007; Mauro et al. 2017). An exponential correlogram of the form $\rho = \exp(\lambda \cdot d)$ was assumed for which d is distance and $\lambda = \frac{\ln(0.05)}{\nu}$ where ν is the range of spatial correlation defined to be the distance for which $\rho = 0.05$.

3.4 Implementing the Statistical Test

In addition to $\hat{\mu}^{\mathrm{FR}}$ and $\widehat{\mathrm{MSE}}^{\mathrm{Hyb}}(\hat{\mu}^{\mathrm{FR}})$, the test statistic of Eq. (3) requires $\hat{\mu}^{\mathrm{CR}}$ and $\widehat{\mathrm{MSE}}(\hat{\mu}^{\mathrm{CR}})$. The CR map-based estimate, $\hat{\mu}^{\mathrm{CR}}$, is easily calculated using the *synthetic* estimator as

$$\hat{\mu}^{\mathrm{CR}} = \frac{1}{N_{\mathrm{CR}}} \sum_{i=1}^{N_{\mathrm{CR}}} \hat{y}_i^{\mathrm{CR}}, \tag{7a}$$

where i indexes CR map units, N_{CR} is the number of CR map units and $\hat{y}_i^{\mathrm{CR}}$ is the value for the ith CR map unit. $\widehat{\mathrm{MSE}}(\hat{\mu}^{\mathrm{CR}})$ is obtained using Eq. (6a), albeit with local meta-data provided by the regional map authors, and replaces $\widehat{\mathrm{MSE}}(\hat{\mu}^{\mathrm{CR}})$ in Eq. (3).

The test corresponding to the first IPCC guideline was that the CR map-based estimate, $\hat{\mu}^{\mathrm{CR}}$, was neither an over- nor under-estimate relative to the true value and took the form of a comparison of $\hat{\mu}^{\mathrm{CR}}$ to the independent FR map-based estimate, $\hat{\mu}^{\mathrm{FR}}$. The exact form of the test depends on how the CR map is considered. If the CR map values and their mean are considered constants without inherent uncertainty apart from being correct or incorrect, then $\widehat{\mathrm{MSE}}^{\mathrm{MB}}(\hat{\mu}^{\mathrm{CR}}) = 0$ in which case Eq. (3) reduces to

$$t = \frac{\hat{\mu}^{\mathrm{CR}} - \hat{\mu}^{\mathrm{FR}}}{\sqrt{\widehat{\mathrm{MSE}}^{\mathrm{Hyb}}(\hat{\mu}^{\mathrm{FR}})}}. \tag{7b}$$

If the CR map values are considered predictions with inherent uncertainty, then $\widehat{\mathrm{MSE}}^{\mathrm{MB}}(\hat{\mu}^{\mathrm{CR}}) \neq 0$ in which case the test statistic expressed by Eq. (3) must be used.

Because map authors have access to the original data used to construct the map, they can readily calculate $\widehat{\mathrm{MSE}}^{\mathrm{MB}}(\hat{\mu}^{\mathrm{CR}})$ for the entire regional map or any local sub-regional portion of it. However, for map users to calculate $\widehat{\mathrm{MSE}}^{\mathrm{MB}}(\hat{\mu}^{\mathrm{CR}})$ for only a local sub-region, the regional map authors must provide the meta-data specific to the sub-region of the CR map of interest. For a CR map consisting of N_{CR} units, the number of covariance values necessary to calculate $\widehat{\mathrm{MSE}}^{\mathrm{MB}}(\hat{\mu}^{\mathrm{CR}})$ is on the order of N_{CR}^2. Expecting map authors to provide this many values in an easily accessible manner is impractical. A solution would be a method for expressing the local meta-data required for calculating $\widehat{\mathrm{MSE}}^{\mathrm{MB}}(\hat{\mu}^{\mathrm{CR}})$ in a summarized form or as functions of the map values. Although such summaries or functions have been proposed for $\widehat{\mathrm{MSE}}_{\mathrm{res}}^{\mathrm{MB}}(\hat{\mu}^{\mathrm{CR}})$ and $\widehat{\mathrm{MSE}}_{\mathrm{spa}}^{\mathrm{MB}}(\hat{\mu}^{\mathrm{CR}})$ (McRoberts et al. 2018c), such is not the case for $\widehat{\mathrm{MSE}}_{\mathrm{pre}}^{\mathrm{MB}}\left(\hat{\mu}^{\mathrm{CR}}\right)$. Until such summaries or functions are developed, accurate calculation of $\widehat{\mathrm{MSE}}_{\mathrm{pre}}^{\mathrm{MB}}\left(\hat{\mu}^{\mathrm{CR}}\right)$ for a sub-region of a regional CR map is not possible.

In lieu of $\widehat{\mathrm{MSE}}^{\mathrm{MB}}\left(\hat{\mu}^{\mathrm{CR}}\right)$, two alternatives may be considered. First, the test statistic can be calculated using $\widehat{\mathrm{MSE}}^{\mathrm{MB}}\left(\hat{\mu}^{\mathrm{CR}}\right) = 0$ which yields the same test statistic as Eq. (7a). Second, if the FR map is of greater quality than the CR map, then it is reasonable to assume that $\widehat{\mathrm{MSE}}^{\mathrm{MB}}\left(\hat{\mu}^{\mathrm{FR}}\right) \le \widehat{\mathrm{MSE}}^{\mathrm{MB}}\left(\hat{\mu}^{\mathrm{CR}}\right)$. If so, then substituting $\widehat{\mathrm{MSE}}^{\mathrm{MB}}\left(\hat{\mu}^{\mathrm{FR}}\right)$ for $\widehat{\mathrm{MSE}}^{\mathrm{MB}}\left(\hat{\mu}^{\mathrm{CR}}\right)$ in Eq. (3) yields

$$t = \frac{\hat{\mu}^{\mathrm{CR}} - \hat{\mu}^{\mathrm{FR}}}{\sqrt{\widehat{\mathrm{MSE}}^{\mathrm{MB}}\left(\hat{\mu}^{\mathrm{FR}}\right) + \widehat{\mathrm{MSE}}^{\mathrm{Hyb}}\left(\hat{\mu}^{\mathrm{FR}}\right)}}. \tag{7c}$$

If $\hat{\mu}^{\mathrm{CR}}$ is not statistically significantly different from $\hat{\mu}^{\mathrm{FR}}$ using either Eqs. (7b) or (7c) to calculate the test statistic, then it will also likely not be statistically significantly different when using Eq. (3), because use of $\widehat{\mathrm{MSE}}^{\mathrm{MB}}\left(\hat{\mu}^{\mathrm{CR}}\right)$ rather than $\widehat{\mathrm{MSE}}^{\mathrm{MB}}\left(\hat{\mu}^{\mathrm{FR}}\right)$ in the denominator would presumably produce an even smaller and less statistically significant |t|. However, if $\hat{\mu}^{\mathrm{CR}}$ is statistically significantly different from $\hat{\mu}^{\mathrm{FR}}$ using either Eqs. (7b) or (7c), then the statistical significance of the difference between $\hat{\mu}^{\mathrm{CR}}$ and $\hat{\mu}^{\mathrm{FR}}$ is undetermined because the degree to which use of $\widehat{\mathrm{MSE}}^{\mathrm{MB}}\left(\hat{\mu}^{\mathrm{CR}}\right)$ rather than $\widehat{\mathrm{MSE}}^{\mathrm{MB}}\left(\hat{\mu}^{\mathrm{FR}}\right)$ alters $|t|$ and therefore alters the significance is unknown.

4 Results and Discussion

4.1 The Fine Resolution (FR) Map

The model on which the FR map was based was fit to the FIA plot AGB data using weighted nonlinear least squares techniques where the weighting matrix was $\widehat{\boldsymbol{\Sigma}}^{-1}$ (Sect. 2.2). Following selection of mean ALS height as the metric for the first power component of the model, no other metric statistically significantly improved the fit. Quality of fit of the model to the data was assessed using unweighted pseudo-$R^2 = 0.72$ and mean weighted residual of 0.004.

4.2 Estimates

The CR map consisted of 121,236 map units. The synthetic estimate of mean AGB for the CR map was $\hat{\mu}^{\mathrm{CR}} = 53.77$ Mg/ha. The hybrid estimate of the mean using reference data sampled from the FR map was $\hat{\mu}^{\mathrm{FR}} = 51.49$ Mg/ha with standard error, $\mathrm{SE}\left(\hat{\mu}^{\mathrm{FR}}\right) = \sqrt{\widehat{\mathrm{MSE}}^{\mathrm{Hyb}}\left(\hat{\mu}^{\mathrm{FR}}\right)} = 1.34$ Mg/ha. The individual $\widehat{\mathrm{MSE}}$ component contributions to $\mathrm{SE}\left(\hat{\mu}^{\mathrm{FR}}\right)$ expressed as square roots of the respective $\widehat{\mathrm{MSE}}$s were $\sqrt{\widehat{\mathrm{MSE}}^{\mathrm{DB}}\left(\hat{\mu}^{\mathrm{FR}}\right)} = 0.15$ Mg/ha, $\sqrt{\widehat{\mathrm{MSE}}_{\mathrm{pre}}^{\mathrm{MB}}\left(\hat{\mu}^{\mathrm{FR}}\right)} = 1.33$ Mg/ha, $\sqrt{\left(\widehat{\mathrm{MSE}}_{\mathrm{res}}^{\mathrm{MB}}\left(\hat{\mu}^{\mathrm{FR}}\right)\right)} = 0.09$ Mg/ha, $\sqrt{\widehat{\mathrm{MSE}}_{\mathrm{spa}}^{\mathrm{MB}}\left(\hat{\mu}^{\mathrm{FR}}\right)} < 0.01$ Mg/ha with $\sqrt{\widehat{MSE}^{MB}\left(\hat{\mu}^{FR}\right)} = \sqrt{\widehat{MSE}^{MB}_{pre}\left(\hat{\mu}^{FR}\right) + \widehat{MSE}^{MB}_{res}\left(\hat{\mu}^{FR}\right) + \widehat{MSE}^{MB}_{spa}\left(\hat{\mu}^{FR}\right)} = 1.33$ Mg/ha. Using the Eq. (7b) form of the test statistic, $t = 1.70$, and using the Eq. (7c) form, $t = 1.21$. Assuming a t-distribution for the test statistic and that $|t| < 2.0$ indicates a nonsignificant difference, the hypothesis that $\hat{\mu}^{CR}$ complies with the first IPCC good practice guideline for the sub-region cannot be rejected, regardless of which form of the test statistic was used and regardless of whether the CR map units were considered to have or not have inherent uncertainty.

Among all $\widehat{MSE}$ components, $\widehat{MSE}^{MB}_{pre}\left(\hat{\mu}^{FR}\right)$ was dominant. One consequence for this study was that although post-stratified and model-assisted regression estimators would tend to reduce the $\widehat{MSE}^{DB}\left(\hat{\mu}^{FR}\right)$ component, neither estimator would have had more than a negligible effect on $SE\left(\hat{\mu}^{FR}\right)$. A second consequence was that the uncertainty in the FR map reference data, expressed by $\widehat{MSE}^{MB}\left(\hat{\mu}^{FR}\right)$ and dominated by $\widehat{MSE}^{MB}_{pre}\left(\hat{\mu}^{FR}\right)$, was a greater contributor to the uncertainty of $\hat{\mu}^{FR}$ than was $\widehat{MSE}^{DB}\left(\hat{\mu}^{FR}\right)$. The small relative value of $\widehat{MSE}^{DB}\left(\hat{\mu}^{FR}\right)$ can be attributed to the very large FR map sample size of $N_{CR} = 121{,}236$. Small relative values of $\widehat{MSE}^{MB}_{res}\left(\hat{\mu}^{FR}\right)$ have been reported previously (Ståhl et al. 2016; McRoberts et al. 2018c), as have negligible values of $\widehat{MSE}^{MB}_{spa}\left(\hat{\mu}^{FR}\right)$ (McRoberts et al. 2018c). In particular, $\widehat{MSE}^{MB}_{spa}\left(\hat{\mu}^{FR}\right)$ will always be small when the proportion of pairs of sample units separated by distances less than the range of spatial correlation is small relative to the total number of pairs of sample units.

Caution should be exercised when extrapolating these results to situations for which the map accuracies and relative sizes of the CR and FR map units may be substantially different than for this study. Nevertheless, for study areas with very large numbers of CR map units, the systematic aligned sampling design should work well. In addition, $SE\left(\hat{\mu}^{FR}\right)$ will likely be dominated by $\widehat{MSE}^{MB}_{pre}\left(\hat{\mu}^{FR}\right)$.

4.3 Additional Considerations

Consideration should be given to the effects of different sources of remotely sensed auxiliary information used to construct the CR and FR maps. For example, for this study, the underlying ALS data used to construct the FR map were filtered to remove returns from man-made objects such as buildings, water towers and utility poles. In addition, because water absorbs ALS pulses, no ALS returns were available for lakes. For the Itasca study area, water area is substantial at approximately 9% of the total county area. Missing ALS metrics for water and man-made objects were set to 0.

As previously noted, reference data acquired from a map, regardless of map accuracy, are subject to error. The hybrid $\widehat{MSE}$ accommodates the effects of this reference data error via $\widehat{MSE}^{MB}\left(\hat{\mu}^{FR}\right)$. However, there is no accommodation for the effects of map error on $\hat{\mu}^{FR}$. Thus, regardless of quality of the FR map, design-based estimators such as Eq. (5a) cannot be assumed to be unbiased when used with reference data subject to error. The effects of this bias on significance levels for tests of hypothesis are generally unknown but likely small.

An assumption underlying the analyses is the availability of meta-data that can be used to calculate at least $\widehat{MSE}^{MB}_{pre}\left(\hat{\mu}^{FR}\right)$ as the dominant component of $\widehat{MSE}^{MB}\left(\hat{\mu}^{FR}\right)$ for the sub-regional portion of the regional CR map. For an FR map constructed using a regression model, $\widehat{\text{MSE}}_{\text{pre}}^{\text{MB}}\left(\hat{\mu}^{\text{FR}}\right)$ can be readily calculated using the covariance matrix for the model parameter estimates and the map unit values for the model predictor variables. For nonparametric prediction techniques such as Random Forests, Monte Carlo bootstrap techniques can be considered (Efron and Tibshirani 1994, pp. 47–48, 113; McRoberts et al. 2018c). However, the exact form of the resampling for these techniques must mimic the sampling design used to acquire the data used to construct the FR map.

As noted, accurate calculation of $\widehat{\text{MSE}}^{\text{MB}}\left(\hat{\mu}^{\text{CR}}\right)$ requires the meta-data necessary to calculate at least $\widehat{\text{MSE}}_{\text{pre}}^{\text{MB}}\left(\hat{\mu}^{\text{FR}}\right)$ for the local sub-region portion of the CR map. Although the map authors could provide the meta-data, the number of required data elements makes this an impractical option. Alternatively, the map authors could provide the meta-data in a summarized form, but methods for doing so in a simple and straightforward manner are currently not available.

Supplementary clarification on the IPCC good practice guidelines would be beneficial. For example, the first guideline refers to neither over- nor under-estimation, but what is the standard for comparison? For this study, the standard was assumed to be the true value, but true values are seldom known. As a second example, how should uncertainty as expressed in the second guideline be interpreted? Should it be in terms of the deviation between an estimate and the true value, or should it relate to a concept such as precision? As a third example, should the regional map be considered a stand-alone constant product whose map unit values have no inherent uncertainty apart from being correct or incorrect, or should the unit values be considered the mean of an entire distribution of possible values? Resolution of the three issues expressed by these examples is important, if not also crucial, but is beyond the scope of this study.

Finally, accuracy assessment for the entirety of a regional CR biomass map is likely not feasible, if for no other reason than that reference data that span the entire CR map are likely neither available nor feasible to acquire. The best that can be achieved is likely independent local accuracy assessments for each of a set of individual sub-regional sites for which reference data are available or can be acquired (Duncanson et al., in review). Ideally, these sites would be systematically distributed across the CR map such as via the global hexagon-based approach proposed by White et al. (1992). Regardless of how the sub-regional sites are distributed, methods for aggregating their individual accuracy assessments based on potentially quite different site-specific reference data into a comprehensive inference for the entire CR map require attention.

5 Conclusions

Four conclusions were drawn from the study, all based on the assumption that ground reference data cannot be acquired. First, when reference data for assessing the accuracy of a map-based estimate are imperfect, such as when they are obtained from a second map, even if it is of greater quality, hybrid inferential techniques must be used to accommodate both sampling variability and non-negligible errors in the reference data. Second, for this study, the spatially aligned systematic sample consisting of the central fine resolution map unit in each coarse resolution map unit worked well. The design was easy to implement, the expansion estimators were easy to apply, and the $\widehat{\text{MSE}}$

component was dominated by a single component, the model-based prediction component. Third, for this study, the regional, coarse resolution map-based estimate of mean biomass per unit area complied with the first IPCC good practice guideline for the local, Itasca County sub-region. Fourth, future research priorities include developing methods for estimating the model-based component of MSE for only a local portion of a regional map and developing methods for extending inferences from individual sub-regions to the entirety of the regional map.

The methods developed and illustrated for this study are important for tropical forest countries whose ground plot networks are so limited as to preclude plot-based, IPCC compliant estimation, particularly with respect to uncertainty. In addition, they are important for biomass map authors who wish to convey to users general information regarding accuracies of map-based estimates that can be expected for local areas of their regional maps.

References

Baccini A, Goetz SJ, Walker WS, Laporte NT, Sun M, Sulla-Menashe D, Hackler J, Beck PSA, Dubayah R, Freidl MA, Samanta S, Houghton RA (2012) Estimated carbon dioxide emissions from tropical deforestation improved by carbon-density maps. Nat Clim Change 2:182–185

Blackard JA, Finco MV, Helmer EH, Holden GR, Hoppus ML, Jacobs DM, Lister AJ, Moisen GG, Nelson MD, Riemann R, Ruefenacht B, Salajanu D, Weyermann DL, Winterberger KC, Brandeis TJ, Czaplewski R, McRoberts RE, Patterson PL, Tymcio RP (2008) Mapping U.S. forest biomass using national forest inventory data and moderate resolution information. Remote Sens Environ 112:1658–1677

Breidenbach J, Kublin E, McGaughey R, Andersen H-E, Reutebuch S (2008) Mixed-effects models for estimating stand volume by means of small footprint airborne laser scanner data. Photogramm J Finl 21(1):4–15

Breidenbach J, McRoberts RE, Astrup R (2016) Empirical coverage of model-based variance estimators for remote sensing assisted estimation of stand-level timber volume. Remote Sens Environ 173:274–281

Cochran WG (1977) Sampling techniques, 3rd edn. Wiley, New York, p 428

Corona P, Fattorini L, Franceschi S, Scrinz G, Torresan C (2014) Estimation of standing wood volume in forest compartments by exploiting airborne laser scanning information: model-based, design-based, and hybrid perspectives. Can J For Res 44:1303–1311

Duncanson L, Armston J, Disney M, Avitabile V, Barbier N, Calders K, Carte, S, Chave J, Herold M, Crowther T, Falkowski M, Kellner J, Labrière N, Lucas R, MacBean N, McRoberts RE, Meye, V Næsset E, Nickeson JE, Paul KI, Phillips O., Réjou- Méchain M, Román M, Roxburgh S, Saatchi S, Schepashenko D, Scipal K, Siqueira PR, Williams M, Whitehurst A. In review. The importance of global land product validation: towards a standardized protocol for aboveground biomass. Surveys in Geophysics. This issue

Efron B, Tibshirani R (1994) An introduction to the bootstrap. Chapman and Hall/CRC, Boca Raton

Fattorini L (2012) Design-based or model-based inference? The role of hybrid approaches in environmental surveys. In: Fattorini L (ed) Studies in Honor of Claudio Scala, Department of Economics and Statistics. University of Siena, Siena, Italy, pp 173–214

GFOI (2016) Integration of remote-sensing and ground-based observations for estimation of emissions and removals of greenhouse gases in forests: methods and guidance from the global forest observations initiative, 2nd edn. Food and agriculture organization, Rome 224 p.https://www.reddcompass.org/download-the-mgdAccessed July 2017

Hansen MH, Madow WG, Tepping BJ (1983) An evaluation of model-dependent and probability-sampling inferences in sample surveys. J Am Stat Assoc 78:776–793

Hansen M, DeFries R, Townshend JR, Carroll M, Dimiceli C, Sohlberg R (2003) 500 m MODIS vegetation continuous fields: tree cover. GLCF, University of Maryland, College Park

Huete A, Didan K, Miura T, Rodriguez E, Gao X, Ferreira L (2002) Overview of the radiometric and biophysical performance of the MODIS vegetation indices. Remote Sens Environ 83:195–213

IPCC (2006) 2006 IPCC guidelines for national greenhouse gas inventories, volume 4: agriculture, forestry and other land use. Eggleston HS, Buendia L, Miwa K, Ngara T, Tanabe K (eds). Published: Institute for Global Environmental Strategies, Japan. http://www.ipcc-nggip.iges.or.jp/public/2006gl/index.html. Accessed February 2018

Mauro F, Monleon VJ, Temesgen H, Ruiz LA (2017) Analysis of spatial correlation in predictive models of forest variables that use LiDAR auxiliary information. Can J For Res 47:788–799

McRoberts RE, Tomppo EO, Finley AO, Heikkinen J (2007) Estimating areal means and variances of forest attributes using the k-nearest neighbors technique and satellite imagery. Remote Sens Environ 111:466–480

McRoberts RE, Hansen MH, Smith WB (2010) United States of America. In: Tomppo E, Gschwantner T, Lawrence M, McRoberts RE (eds) National forest inventories, pathways for common reporting. Springer, Berlin 610 p

McRoberts RE, Næsset E, Gobakken T (2014) Estimation for inaccessible and non-sampled forest areas using model-based inference and remotely sensed auxiliary information. Remote Sens Environ 154:226–233

McRoberts RE, Chen Q, Domke GM, Ståhl G, Saarela S, Westfall JA (2016) Hybrid estimators for mean aboveground carbon per unit area. For Ecol Manag 378:44–56

McRoberts RE, Chen Q, Gormanson DD, Walters BF (2018a) The shelf-life of airborne laser scanning data for enhancing forest inventory inferences. Remote Sens Environ 206:254–259

McRoberts RE, Stehman SV, Liknes GC, Næsset E, Sannier C, Walters BF (2018b) The effects of imperfect reference data on remote sensing-assisted estimators of land cover class proportions. ISPRS J Photogramm Remote Sens 142:292–300

McRoberts RE, Næsset E, Gobakken T, Chirici G, Condés S, Hou Z, Saarela S, Chen Q, Ståhl G, Walters BF (2018c) Assessing components of the model-based mean square error estimator for remote sensing-assisted forest applications. Can J For Res 48:642–649

Mowrer HT, Congalton RG (eds) (2000) Quantifying spatial uncertainty in natural resources: theory and applications for GIS and remote sensing. Sleeping Bear Press, Ann Arbor

Olofsson P, Foody GM, Stehman SV, Woodcock CE (2013) Making better use of accuracy data in land change studies: estimating accuracy and area and quantifying uncertainty using stratified estimation. Remote Sens Environ 129:122–131

Olofsson P, Foody GM, Herold M, Stehman SV, Woodcock CE, Wulder MA (2014) Good practices for estimating area and assessing accuracy of land change. Remote Sens Environ 148:42–57

Pearson TRH, Brown S, Murra L, Sidman G (2017) Greenhouse gas emissions from tropical forest degradation: an underestimated source. Carb Balance Manag 12:3

Pelletier J, Kirby KR, Potvin C (2012) Significance of carbon stock uncertainties on emission reductions from deforestation and forest degradation in developing countries. For Policy Econ 24:3–11

Royall RM, Herson J (1973) Robust estimation in finite populations II. J Am Stat Assoc 68(344):890–893

Saarela S, Schnell S, Grafström A, Tutominen S, Nordkvist K, Hyppä J, Kangas A, Ståhl G (2015) Effects of sample size and model form on the accuracy of model-based estimators of growing stock volume. Can J For Res 45:1524–1534

Saatchi SS, Harris NL, Brown S, Lefsky M, Mitchard ETA, Salas W, Zutta BR, Buermann W, Lewis SL, Hagen S, Petrova S, White L, Silman M, Morel A (2011) Benchmark map of forest carbon stocks in tropical regions across three continents. Proc Natl Acad Sci 108:9899–9904

Särndal C-E, Swensson B, Wretman J (1992) Model assisted survey sampling. Springer, New York, p 694

Snedecor GW, Cochran WG (1967) Statistical methods, 6th edn. The Iowa State University Press, Lowa

Ståhl G, Saarela S, Schnell S, Holm S, Breidenbach J, Healey SP, Patterson PL, Magnussen S, Næsset E, McRoberts RE, Gregoire TG (2016) Use of models in large-area forest surveys: comparing model-assisted, model-based and hybrid estimation. For Ecosyst 3:5

Stehman SV (2009) Sampling designs for accuracy assessment of land cover. Int J Remote Sens 30(20):5243–5272

Vermote EF, Vermueulen A (1999) Atmospheric correction algorithm: spectral reflectances (MOD09). University of Maryland, College Park

Vogelmann JE, Howard S, Yang L, Larson C, Wylie B, Van Driel N (2001) Completion of the 1990s national land cover data set for conterminous United States from landsat thematic mapper data and ancillary data sources. Photogramm Eng Remote Sens 67:650–661

White D, Kimerling AJ, Overton WS (1992) Cartographic and geometric components of a global sampling design for environmental monitoring. Cartogr Geogr Inf Syst 19(1):5–22

Affiliations

Ronald E. McRoberts[1] · Erik Næsset[2] · Greg C. Liknes[1] · Qi Chen[3] · Brian F. Walters[1] · Sassan Saatchi[4] · Martin Herold[5]

✉ Ronald E. McRoberts
mcrob001@umn.edu

1 Northern Research Station, U.S. Forest Service, Saint Paul, MN, USA

2 Faculty of Environmental Sciences and Natural Resource Management, Norwegian University of Life Sciences, Ås, Norway

3 Department of Geography and Environment, University of Hawai'i at Mānoa, Honolulu, HI, USA

4 Joint Propulsion Laboratory, California Institute of Technology, Pasadena, CA, USA

5 Laboratory for Geoinformation Science and Remote Sensing, Wageningen University and Research, Wageningen, The Netherlands

Surveys in Geophysics (2019) 40:1017–1027
https://doi.org/10.1007/s10712-019-09541-z

A Joint ESA-NASA Multi-mission Algorithm and Analysis Platform (MAAP) for Biomass, NISAR, and GEDI

Clément Albinet[1] · Amanda S. Whitehurst[2] · Laura Alisic Jewell[3] · Kaylin Bugbee[4] · Henri Laur[1] · Kevin J. Murphy[5] · Bjorn Frommknecht[1] · Klaus Scipal[6] · Gabriella Costa[1] · Benhan Jai[3] · Rahul Ramachandran[7] · Marco Lavalle[3] · Laura Duncanson[8]

Received: 31 October 2018 / Accepted: 7 May 2019 / Published online: 1 June 2019

Abstract
In order to help scientists in the above ground biomass community and to support the science behind the upcoming BIOMASS, NISAR, and GEDI satellite missions, ESA and NASA are collaborating on the Multi-Mission Algorithm and Analysis Platform (MAAP). The MAAP is a jointly developed and implemented platform that will include not only data (satellite, airborne, in situ data, and products), but also computing capabilities and sets of tools and algorithms developed to support this specific field of research. To best ensure that users are able to collaborate across the platform and to access needed resources, the MAAP requires all data, algorithms, and software to conform to open access and open source policies. In addition to aiding researchers, the MAAP exercise is establishing a collaboration framework between ESA and NASA that focuses on sharing data, science algorithms, and computable resources in order to foster and accelerate scientific research.

Keywords Biosphere · Data dissemination · Data processing · Open access · Polarimetric synthetic aperture radar · Lidar

✉ Clément Albinet
clement.albinet@esa.int

[1] ESA-ESRIN, 00044 Frascati, Italy

[2] Arctic Slope Federal Technical Services, Washington, D.C 20546, USA

[3] Jet Propulsion Laboratory, California Institute of Technology, Pasadena 91109, USA

[4] The University of Alabama, Huntsville 35899, USA

[5] NASA-Headquarters, Washington, D.C 20546, USA

[6] ESA-ESTEC, 2201 Noordwijk, The Netherlands

[7] NASA-Marshall Space Flight Center, Huntsville 35808, USA

[8] University of Maryland, College Park 20742, USA

1 Introduction

1.1 Background

For over 40 years, Earth Observation (EO) satellites developed or operated by ESA and NASA have provided users access to a wealth of data. In the coming years, access to an unprecedented amount of data will continue to grow as new missions will further extend the routine monitoring of the Earth system at the global scale. Exponential data growth is a significant factor in the Earth sciences and carbon monitoring community with the launch of the several high data volume missions including the ESA BIOMASS mission (Le Toan et al. 2011), the NASA-ISRO SAR (NISAR) mission (Rosen et al. 2015), and the NASA Global Ecosystem Dynamics Investigation (GEDI) mission (Stavros et al. 2017). The data from these missions, which expand the operational capability of global monitoring from space, combined with data from long-term EO archives (e.g., Landsat, ERS, ENVISAT), in situ networks, and models, will provide users with unprecedented insight into how the oceans, atmosphere, land, and cryosphere function and interact as part of an interconnected Earth System.

While this growing volume of environmental data from space represents a unique opportunity for both research science and applications, it also poses a major challenge for these communities to exploit the full potential of these data. First, the emergence of large volumes of data (Petabytes era) raises new issues related to the discovery, access, exploitation, visualization, and cost of these data with profound implications on how users conduct "data-intensive" Earth Science (Hey 2009) research. For example, the NISAR mission will generate around 40 PB of data per year, forcing users to consider new ways of exploiting these data.

Second, the growing diversity and complexity of both data and users demands cooperation among various communities. Each community has different needs, methods, languages, and protocols in order to understand a wealth of heterogeneous data provided in different structures and formats. Although the mission of space-based data providers is to provide stand-alone information on bio-geophysical parameters, and there have been some advances in terms of data sharing (Duncanson et al. 2019), further collaborations are needed to improve those estimates. Providers/programme funding organizations should work collaboratively to provide complementary, relevant, transparent, and validated in situ biomass data and information (Herold et al. 2019). Uncertainties prediction in biomass maps propagate to larger area estimates and can lead to substantial uncertainties in national emissions estimation if not properly considered, particularly in relation to the effect of spatial autocorrelation (McRoberts et al. 2019). However, whether used alone or in combination, the calibration of the products is likely to rely on pre-stratification of forested lands, with enough reference data in each stratum, as it is unlikely that a single universal model can be transferable across forest types and regions without biases (Réjou-Méchain et al. 2019). The researchers could easily address these current limitations of the data and error issues with the help of a dedicated tool. In addition, bridging the gap between these various communities of practice will be essential in order to fully exploit these valuable both ground and remotely sensed data (Chave et al. 2019). The scientific community is eager to address this gap and has discussed this issue of international collaboration, not seeing each biomass mission as an individual exercise but highlighting the combination and integration in one way or another at multiple conferences and science meetings, including the recent workshop on Space-Based Measurement of Forest Properties held at the International Space Science Institute in Bern, Switzerland (Scipal et al. 2017).

Responding to both these technological and community challenges requires the development of new ways of working, capitalizing on Information and Communication Technology (ICT) developments to facilitate the exploitation, analysis, sharing, mining, and visualization of massive EO datasets and high-level products within Europe, the USA, and beyond. Evolution in information technology and the consequent shifts in user behaviour and expectations provide new opportunities for increasing exploitation of EO data. In particular, Infrastructure-as-a-Service (IaaS)—providing infrastructure as shared scientific collaboration platforms across large communities—enables data and resource sharing and allows for massively scalable ICT infrastructure under pay-per-use models; this facilitates access to these services by users who would otherwise not be able to afford them. Additionally, social networking and related tools make a new level of online collaboration among communities of practice not only possible but also mainstream.

1.2 Proposed Solution

In light of these challenges, ESA and NASA are collaborating together to lower obstacles related to increased data volumes and to encourage open data policies by collaboratively developing a Multi-Mission Algorithm and Analysis Platform (MAAP) to help improve research of global aboveground terrestrial carbon dynamics (Albinet et al. 2018). This initiative aims at capitalizing on Ground Segment (GS) capabilities and ICT technologies in order to maximize the exploitation of EO data from the BIOMASS, GEDI, and NISAR missions, the data which represent a highly complementary set of measurements to determine Earth's aboveground woody biomass with unprecedented resolution and accuracy. The capabilities envisioned for the MAAP will enable for the first time efficient and customizable inter-comparison, cross-calibration, and fusion of data from these missions by a broad community of scientists and other users.

The principal idea underpinning the MAAP is to bring the user to the collocated data and processing, rather than the data to the users, thereby enabling ultra-fast data access and processing (i.e., transferring a few megabytes of results rather than several tera/petabytes of data to the user). Its utility has been proven by ESA and NASA with, for example, the ESA Grid Processing on Demand (G-POD)[1] environment or by private players with some services like Google Earth Engine (GEE) (Gorelick et al. 2016) or Amazon Web Services (AWS).[2]

Leveraging collocated data and processing is a goal for many scientific users in ecosystems research. However, the supporting infrastructure required for such research is currently limited. In order to lower these limitations and to accelerate scientific research, the MAAP seeks to develop a collaboration framework between ESA and NASA in order to more easily share date, science algorithms and computer resources for both ESA and NASA scientists.

[1] A. Biscuso, A. Marin, A. Cuomo, J. Farres, P. G. Marchetti, S. Pinto, G. Rivolta and E. Mathot. "ESA Grid Processing on Demand G-POD user manual". http://wiki.services.eoportal.org/tiki-index.php?page=GPOD%20Wiki.

[2] Amazon Web Services (AWS)—Cloud Computing Services. https://aws.amazon.com/.

As stated in Albinet et al. (2018), the MAAP will meet this goal by:

1. Enabling scientists and other users to easily discover, process, visualize and analyse large volumes of biomass relevant data from both ESA and NASA;
2. Providing data in the same coordinate reference frame in order to facilitate comparison, analysis, data evaluation, and data generation;
3. Providing a version-controlled science algorithm development environment that leverages tools, co-located data, and the provided processing resources;
4. Addressing intellectual property and sharing concerns related to both collaboratively developing algorithms and the subsequent sharing of those algorithms and the supporting data.

2 Platform Overview

2.1 Objective

The goal of this joint ESA-NASA project is to create a virtual working environment that shall enable (Albinet et al. 2018):

- Rapid data access by avoiding the movement of large amounts of data on the network.
- Increased data usability by easing the exploitation process so that users do not spend time on ICT matters.
- Synergistic use of different EO data sources (i.e, in situ data for validation or airborne campaign data, ancillary data (Total Electron Content, land cover), Digital Terrain Models or spaceborne data from other sensors) in order to better assess and exploit the mission data.
- Community building by fostering a spirit of resource and knowledge sharing.
- Improved science by exposing the official processing algorithm to every user and making them fully transparent.
- Rapid benchmarking of processing algorithms (more representative validation based on a wider dataset without having to host the data and faster validation made possible by a scalable ICT).
- Collaborative processing algorithm repository enabling algorithm module management, sharing, and deployment for production. This forms the basis for a processing algorithm library for rapid and iterative development and deployment of algorithms in the MAAP.
- A fully automated data processing framework allowing generation of products.
- Managed services providing expert support (thematic and ICT/GS) for complex exploitation tasks.
- Replicability of results and traceability of workflow and processes.
- A cost-effective approach for scalable ICT resources capitalizing on economies of scale through infrastructure pooling (generally cheaper than an "in-house" user investment); development of new business models such as "data rental" and new pricing models such as pay-per-use.

2.2 ESA-NASA Multi-mission Algorithm and Analysis Platform Approach

The joint ESA-NASA Multi-Mission Algorithm and Analysis Platform approach is described in Fig. 1.

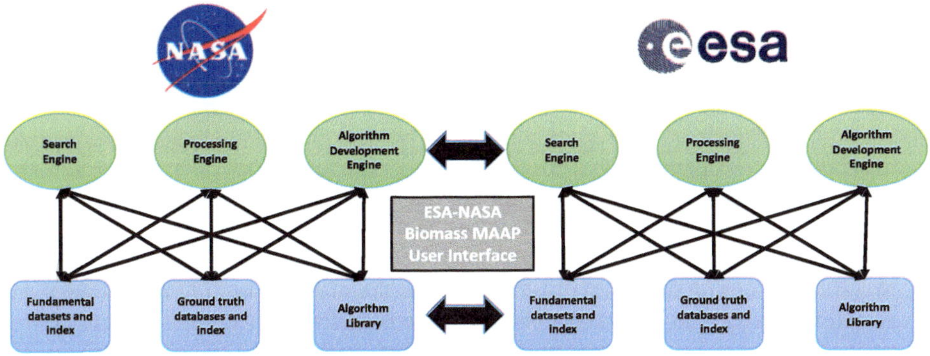

Fig. 1 Multi-mission algorithm and analysis platform functions and services

Table 1 MAAP user types and data access permissions. Data access permissions are accumulative

User types	Data access permissions
General users ("*EO data explorers*")	Discover science data products Access data using visualization/analysis tools Access all science data products related to the BIOMASS, NISAR, and GEDI missions
Science users ("*EO specialists*")	Generate new data products
Mission scientists/PIs ("*Algorithm developers*")	Discover science algorithms Create/edit algorithms
MAAP data curators	Access all metadata
MAAP operators	Access all MAAP data

The approach of the MAAP is to facilitate the exploitation of the commonalities described in Fig. 1 by providing to users a common virtual working environment (Albinet et al. 2018). This commonality does not imply a single system but rather a federation of two (or more) platforms which would allow both USA and European users transparent access to the content of both platforms. This project will allow users to access up-to-date data and algorithms for biomass estimation and to increase the cohesion of the transatlantic P/L-band Synthetic Aperture Radar (SAR) and biosphere communities. It will also further strengthen the collaboration between ESA and NASA around the Earth Observation communities, notably with respect to the ground segment and Calibration/Validation (Cal/Val) activities.

2.3 Typology of ESA-NASA MAAP Users

Five types of users have been defined for the MAAP. The first three types of users (i.e., EO Data Explorers, EO Specialists, and Algorithm Developers) are based on the generic typology approach of users of Exploitation Platforms, as described in footnote 3.[3] The other two user types (Operators and Other Systems) are focused on platform operations, administration, and resource access.

Whenever possible, the MAAP shall enforce a principle of non-discriminatory data access so that all users will be treated equally. For data products supplied from an international partner or other agency, ESA and NASA will restrict access only to the extent required by the appropriate Memorandum of Understanding. The MAAP shall implement the following user roles and data access permissions, which accumulate at each level (see Table 1).

Non-remote sensing experts will be able to use the platform as *"EO Data Explorers."* They will have access to basic functionalities, like data discovery and visualization. More advanced users, such as specialists in remote sensing, will want to switch to "EO Specialists" in order to access additional functionalities like data generation or download. Finally, the specialists in SAR or LIDAR will switch to "Algorithm developers" to be able to modify and generate algorithms and to validate the corresponding product.

2.4 MAAP Data and Information Policy

All data and algorithms made available in the MAAP shall conform to ESA's Revised Earth Observation Data Policy[4] and NASA's Data and Information policy.[5] This includes data and algorithms from ESA, NASA, and other data providers. The MAAP promotes the full and open sharing of all data with the research and applications communities, private industry, academia, and the general public. The greater the availability of the data, the more quickly and effectively the user communities can utilize the information to address basic Earth science questions and provide the basis for developing innovative practical applications to benefit the general public. For data access, the MAAP's data policy includes the following:

- The MAAP commits to the full and open sharing of Earth science data obtained from Earth observing satellites, sub-orbital platforms, and field campaigns with all users as soon as such data become available.
- There will be no period of exclusive access to MAAP Earth science data. Following a post-launch checkout period, all science data will be made available to the user community.
- The MAAP will make available all standard products along with the source code for algorithm software, coefficients, and ancillary data used to generate these products.
- The MAAP will enforce a principle of non-discriminatory data access so that all users will be treated equally.
- The MAAP will engage in ongoing partnerships with other institutions to increase the effectiveness of the MAAP. This interagency cooperation shall include: sharing of data from satellites and other sources, mutual validation and calibration data, and consolidation of duplicative capabilities and functions.

[3] Thematic Exploitation Platforms: Statement of Work, PGSI-GSEG-EOPG-SW-14-0034, 17 June 2014.

[4] ESA Data Policy for ERS, Envisat and Earth Explorer missions, October 2012. https://earth.esa.int/c/document_library/get_file?folderId=296006&name=DLFE-3602.pdf.

[5] NASA's Earth Science Data Systems (ESDS) Data and Information Policy. https://earthdata.nasa.gov/earth-science-data-systems-program/policies/data-information-policy.

- The MAAP will negotiate and implement arrangements with its international partners, with an emphasis on meeting the data acquisition, distribution, and archival needs of the MAAP.

3 Platform Implementation

ESA and NASA will jointly develop technical and data requirements for the MAAP. Based on these requirements, ESA and NASA will develop and manage independent systems that will be interoperable with each other and present users with a unified user interface. This joint approach aligns ground system development activities without impacting each agency's independent efforts and timelines (Albinet et al. 2018).

3.1 Phasing Approach

The development of the MAAP will consist of two phases: A Pilot Phase (phase 1) and a Full Phase (phase 2).

The Pilot Phase will focus on the use of AfriSAR campaign data (Wasik et al. 2018; Pourshamsi et al. 2018), other key campaign data, including INDREX-2 (Hajnsek et al. 2009), BioSAR (Tebaldini 2009), BioSAR-2 (Tebaldini and Rocca 2012), TropiSAR (Dubois-Fernandez et al. 2012), and BioSAR-3 (Sandberg et al. 2014), and relevant ancillary data. During this phase, pre-defined functionalities, including deployment of Level-2 (CEOS) algorithms for biomass based on the airborne campaign data, will be added sequentially to the MAAP. Additionally, user requirements will be defined during the Pilot Phase that will feed into the Full Phase of platform deployment. Initially, use of the Pilot platform will be open only to a limited community of science users.

During the Full Phase, additional, related airborne, field, and satellite data from both ESA and NASA will be incorporated into the MAAP. During development of the Full platform, data from the BIOMASS, GEDI, and NISAR missions will be deployed in the MAAP and access to the platform will be extended beyond the science users to a broader user community.

3.2 Guiding Principles

Collaborative development of the ESA and NASA MAAP system will be based on the following core principles (Albinet et al. 2018):

1. ESA and NASA will develop separate infrastructures of the MAAP system according to jointly developed requirements and interface specifications.
2. An Application and Programing Interface (API) will be used to provide interoperability.
3. The MAAP will provide an entry point for users that will enable access to the ESA and the NASA infrastructures.
4. All data, information, and results, including satellite, airborne, field, and Cal/Val data from Level 1A to Level 4, will be fully open and shared between the two implementations.
5. All source code will be developed as open source software. Official algorithms that are archived by NASA and/or ESA and included in the MAAP Algorithm Store (MAS) will be subject to the open source requirement. MAAP user-developed algorithms that are shared and published to the MAAP will also be subject to the open source requirement.

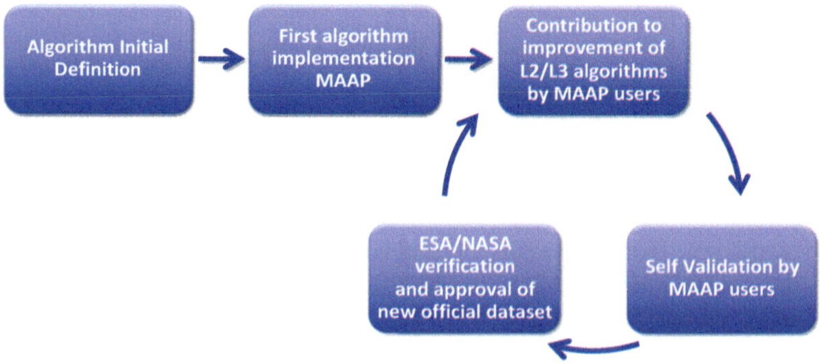

Fig. 2 The concept of product algorithm laboratory (PAL) to support algorithm evolution

6. Requirements and system architectures will be jointly developed and approved to ensure interoperability. This includes protocols and standards between different components of the system. Future modifications of requirements will be agreed to by both agencies.
7. Organizations, researchers, and developers will be given attribution for contributions.
8. Metrics on users, use, and performance will be captured and shared.

4 Benefits

4.1 Concept of Product Algorithm Laboratory

The concept of Product Algorithm Laboratory (PAL) is a new type of governance for the evolution of the EO mission algorithms. This concept will be allowed by the MAAP and be experimented, at least for the BIOMASS mission.

As described in Fig. 2, the Product Algorithm library begins with a traditional initial definition and implementation of a mission algorithm. Any user can modify this algorithm in its working environment, generate the corresponding dataset and validate it using the validation tools and in situ measurements provided within the MAAP. The user can then submit the new version of the algorithm for verification and approval to ESA/NASA who will then consider whether the algorithm will become the new baseline of the mission algorithm.

The processing algorithms evolution is easier within the MAAP since the development and implementation are made within the same environment. This rapid evolution should allow scientists to more quickly develop stable algorithms for R&D missions by leveraging this user cooperative approach. In addition, users outside the core science team will be able to contribute to the product improvement cycle. Finally, this approach aims at "breaking the wall between the world of the Science and the world of the operations."

4.2 Benefits to Users

The goal of this analysis platform addressing EO science missions is to better address scientific community needs in term of:

- Ease of data access.
  The users will have direct access to all the remote sensing data from BIOMASS, NISAR, and GEDI missions (and complementary/similar missions), together with relevant ground data from ESA/NASA campaigns (field data and Cal/Val).
- Ease of data sharing.
  The users will have the possibility of easily sharing large amounts of official data from ESA/NASA missions and from communities/projects that may have complementary data.
- Ease of data transport.
- Joint code/algorithm development (Concept of PAL), addressing intellectual property rights issues.
- Enable interoperability of data/code/algorithms.
- Supported transparency in research, development, and validation.

In addition, such a platform providing EO data, algorithms, processing resources, and documentation could be an incredible tool to support education through training materials, webinars, summer schools, etc.

4.3 Benefits to ESA and NASA

To ensure joint development, implementation, and platform interoperability, ESA and NASA are creating a framework for collaboration. This activity will enable both organizations to gain a better understanding of each other's programmatic processes as well as identify and develop best practices relating to interagency collaboration. Thus, it will help lay a foundation for joint projects in the future.

In addition to assisting the global scientific community, this project supports many of both NASA and ESA's strategic goals. The MAAP supports the NASA strategic goal to "advance knowledge of Earth as a system to meet the challenges of environmental change and to improve life on our planet," by providing an opportunity to advance new technology related to data sharing and analysis,[6] that focuses on improving and even more especially our understanding of global carbon dynamics. The MAAP also meets the goals of the ESA Space 4.0 concept and strategic goals of evolving the ground segment and data management of Earth Observation (EO) missions towards increased possibilities for users.[7] And through the coordination between both agencies for both development and management of the platform, it will facilitate the expansion of "partnerships with international, intergovernmental, academic, industrial, and entrepreneurial communities, recognizing them as important contributors of skill and creativity to our missions and for the propagation of our results."[8]

Current and future efforts of ESA's Directorate of Earth Observation Programmes (EOP) and NASA's Earth Science Data Systems Program (ESDS) will directly benefit from the work required to complete the development and management of the MAAP. Due to the diversity of EO missions under ESA management, EOP must adapt to be able to serve a wide variety of user types (research, operational, general public). Specifically, for

[6] NASA 2014 Science Plan. https://science.nasa.gov/about-us/science-strategy#2014.

[7] Space 4.0, ESA's Ministerial Council, 1 and 2 December 2016 in Lucerne, Switzerland. http://www.esa.int/About_Us/Ministerial_Council_2016/What_is_space_4.0.

[8] NASA 2014 Strategic Plan. https://www.nasa.gov/sites/default/files/files/2014_NASA_Strategic_Plan.pdf.

R&D EO missions which focus on innovative sensors (Earth Explorers), there is a need to further support the collaboration between remote sensing scientists. Although, the ESA/NASA MAAP focuses specifically on the Global Aboveground Terrestrial Carbon Dynamics, these efforts could help identify other geophysical areas and EO missions where ESA and NASA could collaborate along the same concept. This could include, but is not limited to possibly expanding to other institutions in the CEOS framework.

The NASA Earth Science Data Systems (ESDS) archives provide a long-term, continuously updated global environmental record, which currently contain over 24 PB of Earth observation data. As new missions are launched and instruments commissioned, especially over the next 5 years, the archive is anticipated to grow to over 150 PB of data, expanding at an annual rate of around 50 PB of data per year. This immense growth in data presents unique science opportunities while at the same time provides significant challenges for data management and access. To continue providing support for NASA researchers and data users worldwide, ESDS is conducting a variety of activities, including developing prototype infrastructure and analysis platforms. To help further these activities, the ESA-NASA MAAP will leverage and evaluate components and infrastructure through its development and implementation process (Albinet et al. 2018).

5 Conclusion

The MAAP is being developed in preparation for the BIOMASS, NISAR, and GEDI missions. Prior to the launch of these missions, scientists are collecting airborne and field data to support Cal/Val analyses and develop algorithms (e.g., the AfriSAR campaign). Initial MAAP development activities will be focused on supporting the science teams' preparatory efforts by developing capabilities to share and evaluate airborne, field and Cal/Val data. These early activities will help inform platform development and pave the way for an operating platform in the BIOMASS, NISAR, and GEDI mission timeframe.

This platform will give the opportunity, for the first time, to build from a community of user of this new Earth Observation mission around this innovative concept and explore new concepts of data and information access by scientists across agencies without the need to mirror data across institutions.

References

Albinet C, Whitehurst AS, Laur H, Murphy KJ, Frommknecht B, Scipal K, Mitchell AE, Jai B, Ramachandran R, Lavalle M (2018) ESA-NASA multi-mission analysis platform for improving global aboveground terrestrial carbon dynamics. In: 2018 IEEE international geoscience and remote sensing symposium, Valencia, Spain

Chave J, Davies SJ, Phillips OL, Lewis SL, Sist P, Schepaschenko D, Armston J, Baker TR, Coomes D, Disney M, Duncanson L, Hérault B, Labrière N, Meyer V, Réjou-Méchain M, Scipal K, Saatchi S (2019) Ground data are essential for biomass remote sensing missions. Surv Geophys. https://doi.org/10.1007/s10712-019-09528-w

Dubois-Fernandez PC et al (2012) The TropiSAR Airborne Campaign in French Guiana: objectives, description, and observed temporal behavior of the backscatter signal. IEEE Trans Geosci Remote Sens 50(8):3228–3241. https://doi.org/10.1109/TGRS.2011.2180728

Duncanson L, Armston J, Disney M, Nickeson J, Roman M (2019) The importance of global land product validation: towards a standardized protocol for aboveground biomass. Surv Geophys. https://doi.org/10.1007/s10712-019-09538-8

Gorelick N, Hancher M, Dixon M, Ilyushchenko S, Thau D, Moore R (2016) Google Earth engine: planetary-scale geospatial analysis for everyone. Remote Sens Environ. https://doi.org/10.1016/j.rse.2017.06.031

Hajnsek I, Kugler F, Lee S, Papathanassiou KP (2009) Tropical-forest-parameter estimation by means of Pol-InSAR: the INDREX-II Campaign. IEEE Trans Geosci Remote Sens 47(2):481–493. https://doi.org/10.1109/TGRS.2008.2009437

Herold M, Carter S, Avitabile V, Espejo AB, Jonckheere I, Lucas R, McRoberts RE, Næsset E, Nightingale J, Petersen R, Reiche J, Romijn E, Rosenqvist A, Rozendaal DMA, Seifert FM, Sanz MJ, De Sy V (2019) The role and need for space-based forest biomass-related measurements in environmental management and policy. Surv Geophys. https://doi.org/10.1007/s10712-019-09510-6

Hey T, Tansley S, Tolle K (2009) The fourth paradigm: data-intensive scientific discovery, Microsoft Res., Redmond, Wash, p 286. https://www.microsoft.com/en-us/research/publication/fourth-paradigm-data-intensive-scientific-discovery/

Le Toan T, Quegan S, Davidson M, Balzter H, Paillou P, Papathanassiou K, Plummer S, Rocca F, Saatchi S, Shugart H, Ulander L (2011) The BIOMASS mission: mapping global forest biomass to better understand the terrestrial carbon cycle. Remote Sens Environ 115(11):2850–2860

McRoberts R, Næsset E, Liknes GC, Chen Q, Walters BF, Saatchi S, Herold M (2019) Using a finer resolution biomass map to assess the accuracy of a regional map-based estimate of forest biomass. Surv Geophys. https://doi.org/10.1007/s10712-019-09507-1

Pourshamsi M, Garcia M, Lavalle M, Balzter H (2018) A machine-learning approach to PolInSAR and LiDAR data fusion for improved tropical forest canopy height estimation using NASA AfriSAR Campaign data. IEEE J Sel Top Appl Earth Observ Remote Sens 11(10):3453–3463. https://doi.org/10.1109/JSTARS.2018.2868119

Réjou-Méchain M, Barbier N, Couteron P, Ploton P, Vincent G, Herold M, Mermoz S, Saatchi S, Chave J, de Boissieu F, Féret JB, Takoudjou SM, Pélissier R (2019) Upscaling forest biomass from field to satellite measurements: sources of errors and ways to reduce them. Surv Geophys. https://doi.org/10.1007/s10712-019-09532-0

Rosen PA, Hensley S, Shaffer S, Veilleux L, Chakraborty M, Misra T, Bhan R, Raju Sagi V, Satish R (2015) The NASA-ISRO SAR mission: an international space partnership for science and societal benefit. IEEE Radar Conf (Radar Conf), 1610–1613

Sandberg G, Ulander LMH, Wallerman J, Fransson JES (2014) Measurements of forest biomass change using P-Band synthetic aperture radar backscatter. IEEE Trans Geosci Remote Sens 52(10):6047–6061

Scipal K (2017) The ISSI forest structure workshop: summary and recommendations. In: ESA document, Ref. ESA-EOPSM-BIOM-3212

Stavros EN, Schimel D, Pavlick R, Serbin S, Swann A, Duncanson L, Fisher JB, Fassnacht F, Ustin S, Dubayah R, Schweiger A (2017) ISS observations offer insights into plant function. Nat Ecol Evolut. https://doi.org/10.1038/s41559-017-0194

Tebaldini S (2009) Algebraic synthesis of forest scenarios from multibaseline PolInSAR data. IEEE Trans Geosci Remote Sens 47(12):4132–4142. https://doi.org/10.1109/TGRS.2009.2023785

Tebaldini S, Rocca F (2012) Multibaseline Polarimetric SAR Tomography of a Boreal Forest at P- and L-Bands. IEEE Trans Geosci Remote Sens 50(1):232–246. https://doi.org/10.1109/TGRS.2011.2159614

Wasik V, Dubois-Fernandez PC, Taillandier C, Saatchi SS (2018) The AfriSAR Campaign: tomographic analysis With phase-screen correction for P-band acquisitions. IEEE J Sel Top Appl Earth Observ Remote Sens 11(10):3492–3504. https://doi.org/10.1109/JSTARS.2018.2831441